Functions of the
Natural Immune System

Functions of the
Natural Immune System

Edited by
Craig W. Reynolds
and
Robert H. Wiltrout
Frederick Cancer Research Facility
National Cancer Institute
Frederick, Maryland

Springer Science+Business Media, LLC

Library of Congress Cataloging in Publication Data

Functions of the natural immune system / edited by Craig W. Reynolds and Robert H. Wiltrout.

   p.   cm.

Includes bibliographies and index.

ISBN 978-1-4612-8046-0   ISBN 978-1-4613-0715-0 (eBook)

DOI 10.1007/978-1-4613-0715-0

1. Natural immunity. 2. Immune system — Physiology. I. Reynolds, Craig W. II. Wiltrout, Robert H.

QR185.2.F86   1988   88-22488

616.07′9 — dc19   CIP

© 1989 Springer Science+Business Media New York
Originally published by Plenum Press New York in 1989
Softcover reprint of the hardcover 1st edition 1989

Contributors

JOSEPH F. ALBRIGHT, Department of Microbiology, George Washington University School of Medicine, Washington, DC 20037

JULIA W. ALBRIGHT, Department of Microbiology, George Washington University School of Medicine, Washington, DC 20037

IGNACIO ANEGON, Wistar Institute of Anatomy and Biology, Philadelphia, Pennsylvania 19104

CHRISTINE A. BIRON, Department of Pathology, University of Massachusetts Medical School, Worcester, Massachusetts 01655

JACK F. BUKOWSKI, Department of Pathology, University of Massachusetts Medical School, Worcester, Massachusetts 01655

WING C. CHAN, Department of Pathology and Laboratory Medicine, Emory University School of Medicine, Atlanta, Georgia 30322

KEN LEE CHOI, Departments of Medicine and Microbiology/Immunology, University of Colorado School of Medicine, Denver, Colorado 80262

HENRY N. CLAMAN, Departments of Medicine and Microbiology/Immunology, University of Colorado School of Medicine, Denver, Colorado 80262

JOHN CLANCY JR., Department of Anatomy, Loyola University Medical Center, Maywood, Illinois 60153

LORAN T. CLEMENT, Department of Pediatrics, University of California at Los Angeles School of Medicine, Los Angeles, California 90024

MARIA CRISTINA CUTURI, Wistar Institute of Anatomy and Biology, Philadelphia, Pennsylvania 19104

BERNADETTE FERRY, Transplantation Laboratory, University of Helsinki, Helsinki, Finland

PATRICIA FITZGERALD-BOCARSLY, Department of Pathology, School of Medicine and Dentistry, New Jersey Medical School, Newark, New Jersey 07103

ELIESER GORELIK, Pittsburgh Cancer Institute, Pittsburgh, Pennsylvania 15213, and Department of Pathology, University of Pittsburgh School of Medicine, Pittsburgh, Pennsylvania 15213

ARNOLD H. GREENBERG, Manitoba Institute of Cell Biology, University of Manitoba, Winnipeg, Manitoba, Canada R3E 0V9

PEKKA HÄYRY, Transplantation Laboratory, University of Helsinki, Helsinki, Finland

RONALD B. HERBERMAN, Pittsburgh Cancer Institute, Pittsburgh, Pennsylvania 15213, and Departments of Medicine and Pathology, University of Pittsburgh School of Medicine, Pittsburgh, Pennsylvania 15213

JAMES H. HOLDA, Departments of Medicine and Microbiology/Immunology, University of Colorado School of Medicine, Denver, Colorado 80262

MARITA JAAKKOLA, Transplantation Laboratory, University of Helsinki, Helsinki, Finland

KATSUO KUMAGAI, Department of Microbiology, Tohoku University School of Dentistry, Sendai, Japan

ALAN T. LEFOR, Surgery Branch, National Cancer Institute, National Institutes of Health, Bethesda, Maryland 20892

CARLOS LOPEZ, Viral Exanthems and Herpesvirus Branch, Division of Viral Diseases, Center for Infectious Diseases, Centers for Disease Control, Atlanta, Georgia 30333

TOM MAIER, Departments of Medicine and Microbiology/Immunology, University of Colorado School of Medicine, Denver, Colorado 80262

KIM W. MCINTYRE, Department of Pathology, University of Massachusetts Medical School, Worcester, Massachusetts 01655

JEAN E. MERRILL, Department of Neurology, UCLA School of Medicine, Los Angeles, California 90024

JAMES J. MULÉ, Surgery Branch, National Cancer Institute, National Institutes of Health, Bethesda, Maryland 20892

JUNEANN W. MURPHY, Department of Botany and Microbiology, University of Oklahoma, Norman, Oklahoma 73019

MARIANNE MURPHY, Wistar Institute of Anatomy and Biology, Philadelphia, Pennsylvania 19104

ICHIRO NAKAMURA, Department of Pathology, School of Medicine and Biomedical Sciences, State University of New York at Buffalo, Buffalo, New York 14214

ROBERT J. NATUK, Department of Pathology, University of Massachusetts Medical School, Worcester, Massachusetts 01655

ARTO NEMLANDER, Transplantation Laboratory, University of Helsinki, Helsinki, Finland

YRJÄNÄ NIETOSVAARA, Transplantation Laboratory, University of Helsinki, Helsinki, Finland

JOHN R. ORTALDO, Laboratory of Experimental Immunology, Biological Response Modifiers Program, Division of Cancer Treatment, National Cancer Institute–FCRF, Frederick, Maryland 21701-1013

BICE PERUSSIA, Wistar Institute of Anatomy and Biology, Philadelphia, Pennsylvania 19104

STEVEN A. ROSENBERG, Surgery Branch, National Cancer Institute, National Institutes of Health, Bethesda, Maryland 20892

FERGUS SHANAHAN, Department of Medicine, UCLA School of Medicine, Los Angeles, California 90024

JOAN STEIN-STREILEIN, Departments of Medicine and Microbiology/Immunology, University of Miami School of Medicine, Miami, Florida 33136

RYUJI SUZUKI, Department of Microbiology, Tohoku University School of Dentistry, Sendai, Japan

SATSUKI SUZUKI, Department of Microbiology, Tohoku University School of Dentistry, Sendai, Japan

STEPHAN TARGAN, Department of Medicine, UCLA School of Medicine, Los Angeles, California 90024

JUSSI TARKKANEN, Transplantation Laboratory, University of Helsinki, Helsinki, Finland

ARABELLA B. TILDEN, Department of Medicine, Division of Hematology and Oncology, University of Alabama at Birmingham, and Veterans Administration Medical Center, Birmingham, Alabama 35294

GIORGIO TRINCHIERI, Wistar Institute of Anatomy and Biology, Philadelphia, Pennsylvania 19104

JARKKO USTINOV, Transplantation Laboratory, University of Helsinki, Helsinki, Finland

RAYMOND M. WELSH, Department of Pathology, University of Massachusetts Medical School, Worcester, Massachusetts 01655

ELLIOT F. WINTON, Division of Hematology and Oncology, Department of Medicine, Emory University School of Medicine, Atlanta, Georgia 30322

BRUCE A. WODA, Department of Pathology, University of Massachusetts Medical Center, Worcester, Massachusetts 01605

HYEKYUNG YANG, Department of Pathology, University of Massachusetts Medical School, Worcester, Massachusetts 01655

Preface

The study of the reactivity of nonimmune leukocytes to foreign antigens extends at least as far back as Metchnikoff's work in the late 1800s. These pioneering experiments suggested that the ingestion and destruction of microbes by polymorphonuclear leukocytes (PML) and macrophages may lead to a resistance within the host to infectious agents. Since those initial observations, it has become clear that additional leukocyte populations may also contribute to this overall natural immune response. In addition, it has become obvious that the reactivity of these natural effector cells is not limited to microbes only and is not simply mediated by phagocytosis alone.

The purpose of the present volume is to examine a variety of functions reportedly mediated by cells within the natural immune system. A number of previous volumes have extensively reviewed the functional activity of granulocytes and macrophages. However, no single volume has ever attempted to review the diverse functional activities of the lymphoid cells within the natural immune system. Therefore, the major emphasis of the present volume is on the reactivity of natural effector cells specifically within the lymphoid populations. Such a selection is by no means intended to downplay the importance of granulocytes and macrophages as natural effector cells. Instead, we simply wish to avoid duplicating the previous efforts of other editors and thereby provide a novel review of the natural immune system.

Even within the lymphoid population, a variety of different cell types can be termed *natural effector cells*. Terms used to describe the lymphocyte subpopulations include *natural killer (NK) cells, naturally cytotoxic (NC) cells, natural suppressor (NS) cells,* and *large granular lymphocytes (LGL)*. In spite of the variety of names given to these natural effector cells, one feature is common to all—the ability to mediate a variety of immunological responses in the absence of prior sensitization. It is this functional capability that is the unifying characteristic of these natural effector cells. For the above reason, the present volume is divided into chapters based on the specific functional activity of these cells. Each author will review a particular function of the natural immune system, including discussions of the various cell types mediating this function, biological situations where this function is best observed, and any data pertaining to the specific relevance of this function to the overall immune response.

It will become obvious when reading these chapters that the study of lymphocytes within the natural immune system is a relatively new science. A great deal of "suggestive" data are consistent with the hypothesis that these cells contribute substantially to the host's overall immune defenses. However, in most cases the present state of this science is such that few "definitive" answers can be provided as to how, when, and where the natural immune system is most active. One of the major purposes of this volume is to provide not only comprehensive reviews of previous research on the functional activity of natural effector cells but also an objective summary of where we stand at present and where we need to go in the future. By providing this information, we hope that the scientific community will appreciate the variety and number of functional activities attributable to natural effector cells. Ultimately, an appreciation of these functions should lead to a clearer understanding of how modulation of the natural immune system might be used clinically to prevent or fight a variety of neoplastic, microbial, and pathological conditions.

Craig W. Reynolds
Robert H. Wiltrout

Frederick, Maryland

Contents

PART II. INVOLVEMENT OF THE NIS IN THE CONTROL OF MICROBIAL INFECTIONS

CHAPTER 3

Natural Effector Cells in Influenza Virus Infection 67

JOAN STEIN-STREILEIN

CHAPTER 4

Natural Host Defense Systems Active against Herpes Simplex
Virus Infections 85

CARLOS LOPEZ AND PATRICIA FITZGERALD-BOCARSLY

CHAPTER 5

RAYMOND M. WELSH, ROBERT J. NATUK, KIM W. MCINTYRE,
HYEKYUNG YANG, CHRISTINE A. BIRON, AND JACK F. BUKOWSKI

CHAPTER 6

JULIA W. ALBRIGHT AND JOSEPH F. ALBRIGHT

CHAPTER 7

JUNEANN W. MURPHY

CHAPTER 8

Role of Natural Killer Cells in Inflammation and Antibacterial
Activity   185

ARNOLD H. GREENBERG

PART III.   IMMUNOREGULATORY PROPERTIES OF THE NIS

CHAPTER 9

Role of the Natural Immune System in the Antibody Response:
Regulatory Effect of NK Cells   213

KATSUO KUMAGAI, SATSUKI SUZUKI, AND RYUJI SUZUKI

CHAPTER 13

JOHN R. ORTALDO

PART IV. INVOLVEMENT OF THE NIS IN BONE MARROW
AND ALLOGRAFT REJECTION

CHAPTER 14

ICHIRO NAKAMURA

CHAPTER 15

PEKKA HÄYRY, ARTO NEMLANDER, JUSSI TARKKANEN, BERNADETTE
FERRY, MARITA JAAKKOLA, YRJÄNÄ NIETOSVAARA, AND
JARKKO USTINOV

PART V. INVOLVEMENT OF THE NIS IN VARIOUS PATHOLOGICAL CONDITIONS

CHAPTER 16

Involvement of Natural Effector Cells in Graft versus Host Disease 361

JOHN CLANCY JR.

CHAPTER 17

The Natural Immune System and Hematocytopenias 381

WING C. CHAN AND ELLIOT F. WINTON

CHAPTER 18

The Natural Immune System in Autoimmune and Neurological
Disease 411

JEAN E. MERRILL

CHAPTER 19

Natural Killer Cells and Diabetes Mellitus 433

BRUCE A. WODA

CHAPTER 20

Role of Natural Effector Cells in Human Gastrointestinal
Disease 455

FERGUS SHANAHAN AND STEPHAN TARGAN

PART I
ROLE OF THE NATURAL IMMUNE SYSTEM IN THE CONTROL OF TUMOR CELL GROWTH

1
Role of the Natural Immune System in Control of Primary Tumors and Metastasis

RONALD B. HERBERMAN and ELIESER GORELIK

1. Introduction

Intensive studies of antitumor resistance during the past 25 years have revealed that some tumors in experimental animals can evoke high levels of immune responses against tumor-associated antigens. This immunological response was mostly mediated by T lymphocytes and was found to be extremely efficient in the specific destruction of the tumor target cells *in vitro* and in inhibition of tumor growth and metastatic spread.[1-4]

In parallel with the extending search for effective antitumor immunity, two major conclusions were made: (1) Various spontaneous and induced tumors in animals were nonimmunogenic, and any attempts to immunize animals with tumor cells failed to induce specific antitumor cytotoxic T lymphocytes[5,6]; (2) Some lymphoid cells distinguishable from T lymphocytes were cytotoxic against a variety of tumor cells.

The main characteristics of these tumoricidal non-T effector cells were that (1) they had cytotoxic activity without apparent immunization, and (2) their cytotoxic effects were not specific for a particular tumor but were observed with nonimmunogenic as well as immunogenic tumor cells, without restriction by the major histocompatibility complex.[7,8]

RONALD B. HERBERMAN • Pittsburgh Cancer Institute, Pittsburgh, Pennsylvania 15213, and Departments of Medicine and Pathology, University of Pittsburgh School of Medicine, Pittsburgh, Pennsylvania 15213. ELIESER GORELIK • Pittsburgh Cancer Institute, Pittsburgh, Pennsylvania 15213, and Department of Pathology, University of Pittsburgh School of Medicine, Pittsburgh, Pennsylvania 15213.

At the beginning of the 1970s, it was found that macrophages, in addition to their phagocytic activity and other functions, could be activated by various factors and express high levels of cytotoxic activity against a variety of malignant cells.[9–11] The role of macrophages in the control of tumor and metastatic growth was demonstrated by numerous investigations.[2,12]

From the occasional early observations of tumoricidal activity by normal nonsensitized lymphocytes, and mostly based on the intensive studies of this phenomenon in the past 10 years, investigators have documented the existence of lymphocytes with a preexistent ability to destroy tumor cells without a need for immunization against tumor cells. These lymphocytes have been termed natural killer (NK) cells.[13,14] NK cells have been detected in humans as well as in experimental animals and, in fact, have been found in almost all vertebrate species examined.[7,8]

Within a quite short period of time, studies on NK cells have expanded into a broad and multifaceted research area, ranging from a series of problems in rather fundamental immunobiology to practical issues related to host resistance against tumors and infectious diseases, immune surveillance, and immunotherapy.[7,8,15] Intensive studies have been performed to identify and characterize NK cells and to investigate the mechanisms of their cytotoxic activity.

2. Role of NK Cells in Host Resistance to Primary-Tumor Growth

The most important practical issue to be settled is the role of NK cells *in vivo*. Most of the attention has been directed toward the role of NK cells in antitumor defenses. Using experimental tumors, it was found that NK cells were able to destroy tumor cells *in vivo* and inhibit tumor growth. This inhibition of tumor growth depends on the level of susceptibility of the tumor cells to NK cell-mediated cytotoxicity and the level of NK reactivity of mice.[15–21] Using NK-susceptible and NK-resistant tumor cell variants of B16 melanoma, it was found that tumor growth was inhibited when NK-susceptible B16 melanoma cells were transplanted into normal C57BL/6 mice.[22] Inhibition of NK-sensitive tumor cells was more profound in nude or thymectomized mice, which have higher NK cell activity than euthymic mice.[15,18,20] Using F_1 hybrids between different strains of mice and A mice, a positive correlation was found between the levels of NK activity in the F_1 mice and their resistance to the growth of YAC, a lymphoma of A strain mice.[17,20] Similar results were obtained when bone marrow cells from high or low NK strains were transplanted to lethally irradiated mice. Recipients of cells from high NK donors developed high NK activity and had increased resistance to growth of YAC lymphoma cells.[23]

In beige mice with low NK reactivity, the growth of NK-susceptible, but not NK-resistant, tumor lines was accelerated.[16,22] Various syngeneic, allogeneic, and human tumors induced tumors with increased frequency and had a more rapid growth rate in nude mice whose NK cell activity was suppressed by pretreatment with anti–asialo GM_1 serum.[24]

Using an assay of local transplantation of tumor cells labeled with [^{125}I]dUrd, it was found that inhibition of tumor growth was due to the destruction of tumor cells following their transplantation. Elimination of NK-sensitive tumor cells occurred very quickly (1–2 days), and the rate of tumor cell clearance increased in mice with augmented NK reactivity.[25] NK cells transplanted together with tumor cells were able to destroy *in vivo* NK-sensitive tumor cells but had no effect on NK-resistant tumor cells.[25] Transplantation of NK-sensitive lymphoma cells admixed with normal spleen cells or NK cell-enriched spleen subpopulations resulted in suppression of the local tumor growth.[26,27]

Thus, these data confirm the potential importance of NK cells in antitumor defenses. However, the results also demonstrate some limitations of the effectiveness of NK cells in *in vivo* tumor cell destruction.

Although inhibition of local tumor growth by NK cells has been detected mostly against tumor cells with apparent susceptibility to the cytotoxic action of NK cells, accumulating experimental data indicate that NK cell antitumor activity is highly expressed in the bloodstream against a variety of tumor cells, including some that appear to be NK-resistant. These data shed new light on the importance of NK cells in the control of metastatic spread and growth of tumors.

3. NK Cells and Metastatic Growth

3.1. Elimination of Metastatic Cells and Prevention of Metastasis Formation

Numerous experimental data demonstrated that 90–99% of intravenously (IV) inoculated tumor cells were eliminated during the first 24 h. The small fraction of surviving tumor cells (often < 0.1%) were then able to form tumor metastases.[28–30] Recent investigations have provided some insight into the mechanisms of intravascular death of tumor cells. It was found that NK cells play an important role in the intravascular elimination of tumor cells.[31–35] This conclusion was based on the following findings:

1. Elimination of the radiolabeled tumor cells after intravenous inoculation positively correlated with the level of NK reactivity in the recipients. Strains of mice with high levels of NK cell activity were more efficient for tumor cell elimination than mice of strains

with lower NK reactivity.[31-35] Beige mice or very young mice (3 weeks old), which have low levels of splenic NK reactivity, showed relatively high survival of IV-inoculated tumor cells and more metastatic foci developed in the lungs in these mice.[31-33,36] Furthermore, athymic nude mice, which had higher levels of NK cell activity than euthymic mice, were found to be more efficient in the elimination of intravascular tumor cells and in resisting the development of experimental pulmonary metastases.[31,32,37]

2. Treatments that inhibited NK cell activity in parallel inhibited elimination of the tumor cells from the blood. NK cell function has been depressed by pretreatment of mice with various substances like cyclophosphamide, β-estradiol, corticosteroids, or urethane.[31,32,38,39] Treatment of mice with these agents before IV inoculation of tumor cells resulted in depression of tumor cell elimination and increased formation of tumor foci in the lungs.[28,32,40-42] Pretreatment of mice with anti-asialo GM_1 serum inhibited NK cell activity and also increased the survival of IV-inoculated tumor cells and the yield of experimental metastases.[31] Similar effects were observed when NK activity of mice was depressed by treatment with NK1.1 or NK1.2 antiserum.[43] It is of interest that in mice with depressed NK reactivity (i.e., beige mice or mice after treatment with cyclophosphamide or anti-asialo GM_1 serum), the dramatic increases in the number of the pulmonary B16 melanoma metastases were accompanied by increased formation of extrapulmonary, especially liver, metastases which rarely occur, if at all, in mice with normal NK reactivity.[31,32,44] These data indicate that B16 melanoma cells have the inherent ability to settle and proliferate in the liver, but ordinarily their growth is efficiently prevented by NK cells.

3. Augmentation of NK cell activity could influence the rate of tumor cell elimination and metastasis formation. Cytotoxic activity of NK cells can be stimulated by pretreatment of mice with *C. parvum*, BCG, and poly I:C, probably via the induction of interferon production.[45-48] Treatment of mice with these agents was found to be associated with an increase in the clearance of radiolabeled tumor cells from the pulmonary vasculature and a decrease in the number of the detectable experimental metastases.[28,32,44] NK cell activity has also been stimulated by exposure to various pathogenic and nonpathogenic viruses and other microorganisms.[47,48] Thus, NK reactivity of mice depends not only on their age and genotype but also on housing conditions. Mice maintained in isolators in barrier facilities display lower levels of NK reactivity than age- and strain-matched mice raised in a conventional facility. These differences in environmental factors may be responsible for the differences in the formation of experimental metastases in mice maintained under various conditions.[33,36,37]

4. More direct evidence for the involvement of NK cells in intravascular tumor cell destruction and inhibition of metastasis formation has come from experiments with adoptively transferred NK-enriched or depleted lymphoid cell populations. NK reactivity of mice depressed by Cy treatment could be reconstituted by adoptive transfer of lymphoid cells.[35] In parallel, adoptively transferred lymphoid cells were able to abrogate the metastasis-augmenting effects of Cy treatment.[49] The cells responsible for reconstitution of antimetastatic activity of Cy-treated mice were shown to be distinct from T- or B-lymphocytes or macrophages, and exerted NK activity *in vitro*. The NK activity of transfused spleen cells was depleted by pretreatment of donors with Cy or by direct incubation of spleen cells with anti-NK 1.2 serum and complement. Spleen cells with depleted NK cell activity failed to restore antimetastatic defense in Cy-treated recipients.[49]

Similarly, the augmentation of metastasis formation in recipients treated with anti-asialo GM_1 serum was abrogated by transfusion of normal spleen cells. This effect was mediated by asialo GM_1-positive spleen cells. Spleen cells of donors treated with anti-asialo GM_1 lost their NK activity and were unable to affect metastasis formation in recipients with depressed NK reactivity.[31]

Treatment of rats with anti-asialo GM_1 has been associated with elimination of large granular lymphocytes (LGL) as well as NK activity.[50] The NK activity of PBL and spleen cells of rats treated with anti-asialo GM_1 could be reconstituted by IV transfusion of purified populations of LGL. Furthermore, rats reconstituted with LGL recovered their ability to eliminate IV-inoculated radiolabeled tumor cells and inhibited the formation of experimental tumor metastases.[50]

These experimental data strongly support the concept that NK cells can be extremely efficient in the intravascular elimination of tumor cells and in prevention of metastasis formation, although it is impossible to exclude the involvement of other cells like neutrophils or monocytes or of humoral factors such as natural antibodies or complement in these processes.[51–53]

Detailed analysis of the elimination of the IV-inoculated radiolabeled tumor cells has demonstrated that alterations in NK reactivity did not affect the initial arrest of inoculated tumor cells but substantially modified the rate of their subsequent elimination. Tumor cell elimination is characterized by a biphasic curve, with an initial rapid loss of radioactivity during the first 24 h followed by a more gradual decrease over subsequent days.[28,29,54] The rapid phase of tumor cell elimination is probably associated with the intravascular destruction of tumor cells. Since extravasation of tumor cells has been shown to start at about 4 h after tumor cell inoculation and most surviving tumor cells leave the vascular lumen during the first 1–2 days, the slope of the decrease in radioactivity after

1–2 days appears to reflect the survival of tumor cells in the lung parenchyma.[28,29,54] Differences in the rate of tumor cell elimination, before and following their extravasation, might indicate that NK cells are mostly efficient in the blood and are less effective in the lung parenchyma. This suggestion can be supported by the following observations:

1. The treatment of mice with Cy resulted in the depression of NK cell function and decreased tumor cell elimination. Conversely, C. parvum treatment stimulated NK cell activity and increased tumor cell elimination. Thus, during the first day after tumor cell inoculation, the number of surviving tumor cells in these mice differed by a factor of 10^4. However, at later times, the slopes of the tumor cell elimination became parallel in all mice regardless of their treatment. Such observations could indicate that at this time the rate of tumor cell elimination was similar in normal, Cy, and C. parvum-treated mice.[28] These results are consistent with the possibility that NK cells are unable to influence the survival of tumor cells once they have migrated into the perivascular area.

2. Increased tumor cell elimination and inhibition of metastasis formation by C. parvum can be achieved only if the treatment is applied 1–14 days before tumor cell inoculation. C. parvum treatment applied 1–7 days after tumor cell inoculation had no antimetastatic effect.[28,55,58] Similarly, inhibition of NK reactivity in mice by Cy irradiation was associated with increased tumor cell survival and increased pulmonary metastases only when these treatments were given before IV inoculation of tumor cells. When Cy was given after tumor cell inoculation, there was no detectable effect on the number of tumor metastases in the lungs.[28,42] Treatment of mice with poly I:C or anti-asialo GM_1 serum 1 day before IV inoculation of tumor cells inhibited or augmented, respectively, the formation of the lung tumor colonies. However, postponing the treatment of mice with poly I:C or anti-asialo GM_1 to just 4–24 h after IV inoculation of B16 melanoma cells abolished the effect of these treatments on the incidence and number of experimental metastases in the lungs.[59]

3. Although C. parvum or Cy treatment decreased or increased, respectively, the number of metastatic foci that developed in the lungs, the comparison of the individual metastatic foci in these mice revealed no differences in their size or volume.[28,42] Thus, suppression or augmentation of NK cell activity did not appear to reflect the degree of proliferation of metastatic cells in the lung parenchyma. In contrast, in mice specifically immunized, the number and size of the experimental pulmonary metastatic foci produced by immunogenic tumors are significantly lower than in control unimmunized mice.[28] These results can be attributed to T lympho-

cytes, which are able to penetrate into the lung parenchyma and inhibit the proliferation of the metastatic cells in the immune mice.

These data indicate that NK cells express their antitumor effect preferentially in the blood. This may be due to the relatively higher concentration and/or higher activity of the NK cells in the blood than in the perivascular tissues. Indeed, in comparison to other lymphoid organs, peripheral blood lymphocytes (PBLs) express the highest levels of NK cytotoxicity,[60] whereas the NK activity of spleen cells decreases in mice or rats older than 3 months. PBL maintained high levels of NK cytotoxicity, and the clearance of IV-inoculated tumor cells was not impaired in animals at 6–8 months of age.[25,60]

It should be noted, however, that the presence of NK cells in the parenchyma of various organs such as lungs, liver, and intestines has been demonstrated by several investigations.[61–64] The cytotoxic activity of NK cells in these organs can be dramatically augmented by treatment with various biological response modifiers. It is of interest that stimulation of NK cell activity in the lungs and liver of mice treated with maleic anhydride divinyl ether (MVE-2) was much higher than in the spleen and blood.[65] It seems likely that prolonged maintenance of high levels of NK cell activity in the various organs and tissues, after treatment with such biological response modifiers (BRMs), might influence the growth of the extravasated metastatic tumor cells.

3.2 NK Cells and Formation of Spontaneous Tumor Metastases

In order to adequately evaluate the role of NK cells in antimetastatic defenses, it is necessary to also assess their involvement in the formation of spontaneous metastases in tumor-bearing hosts. Involvement of NK cells in the control of metastatic spread in tumor-bearing animals can be assessed by comparing metastatic growth in animals with high or low NK cell activity.

Treatment with C. parvum of mice bearing 3LL tumors protected against the development of postoperative pulmonary metastases. However, the antimetastatic effect of C. parvum was observed when treatment was performed at least 3–4 days before, but not after, local tumor excision.[66] Inoculation of poly I:C into mice bearing B16 melanoma, at 4–5 days before tumor removal, also inhibited the formation of postoperative pulmonary metastases. In contrast, no effect was found when poly I:C treatment was started 1 day after surgery (Gorelik, unpublished observation).

These data suggest that the antimetastatic effect of C. parvum or poly I:C was mediated by the increase in tumor cell elimination rather than by inhibition of their proliferation and delay in the appearance of the visible metastatic foci. Indeed, the differences in numbers of postoperative metastases in the control and C. parvum-treated mice did not diminish when metastases were observed at various periods after surgery.[66]

When the formation of spontaneous metastases was investigated in mice with low or depressed NK reactivity, an increase in the number of visible metastases was found.[22,31,37] In beige mice bearing B16 melanoma cells, the number of pulmonary metastases was significantly higher than in the control mice.[22,31] Low levels of NK cell activity could be maintained by multiple (every 4–5 days) injections of anti-asialo GM_1 serum into C57BL/6 mice with transplanted B16 melanoma or 3LL carcinoma cells. After surgical removal of the established local B16 melanoma or 3LL carcinoma, there was a substantial increase in the number of postoperative pulmonary metastases in these mice.[31] Similar effects on the formation of the spontaneous pulmonary metastases in mice were observed when NK reactivity of tumor-bearing mice was depressed by implantation of pellets containing 17β-estradiol.[41] Thus, these data demonstrate that NK cells can play an important role in the prevention of the metastatic spread and formation of spontaneous tumor metastases in experimental animals.

3.3. NK Cells and the Selection of Metastatic Cell Variants

Since NK cells could participate in the elimination of tumor cells from the blood, resistance of tumor cells to the cytotoxic action of NK cells would be expected to be beneficial for their survival and subsequent development of metastatic foci. Also, one might predict that metastatic tumor cells would have higher levels of resistance to lysis by NK cells than tumor cells derived from local sites of tumor growth.

Indeed, when the cytotoxic activity of normal spleen cells was tested against 3LL tumor cells derived from the local tumor or from spontaneous pulmonary metastases, higher resistance was found among metastatic cells.[67] Furthermore, serial subcutaneous transfer of 3LL tumor cells mixed with normal spleen cells in a ratio of 1:100 resulted in the selection of tumor cells that displayed higher resistance to the cytotoxic action of NK cells in vivo and in vitro and higher metastatic ability.[57,68]

In analogous studies of rat spontaneous mammary adenocarcinomas and their metastases,[58] tumor cells derived from the pulmonary metastases were also found to be more resistant to NK activity. In addition, tumor cells derived from lymph node metastases or from pleural effusions of tumor-bearing rats were also highly resistant to NK activity, whereas tumor cells from pericardial metastases showed high levels of susceptibility to natural cytotoxicity. The resistance of metastatic cells to NK cells was observed when the tumor cells were isolated directly from the metastatic nodules, but this resistance disappeared after *in vitro* culturing.[58]

Using metastatic cells in a cold target competition of NK activity assay, it was found that they had an average of only 59% of the competitive activity of the primary mammary adenocarcinoma cells. However, since five other NK-resistant metastatic lines had normal competitive activity, NK resistance of the metastatic cells may not be solely attributable to the loss of target structures.[58]

Metastatic cells derived from two lung foci and tumor cells from the primary hamster histiocytic tumor were compared for their susceptibility to NK activity in the experiments performed by Teale *et al.*[69] The parental tumor cells (HSV-2-333-2-26) were weakly metastatic and highly susceptible to the cytotoxic action of PBL or spleen cells. In contrast, tumor cells derived from metastatic nodules in the lungs (Met A and Met B) were highly metastatic and exhibited high levels of resistance to *in vitro* cytotoxicity mediated by PBL or spleen cells.[69]

In all of the above-mentioned experiments, the resistance to lysis by NK cells was demonstrated when tumor cells were derived from spontaneous metastases. In contrast, when the susceptibility or resistance of tumor cells to NK activity was compared with their ability to develop artificial metastatic nodules after IV inoculation, rather divergent results were obtained. B16 melanoma sublines (B16F1, B16F10, and B16F10^{Lr}) vary in their lung colonizing potential but have similar levels of susceptibility to NK cell-mediated cytotoxicity.[70] Clones selected from K-1735 melanoma or UV-2237 fibrosarcoma had a high or low ability to develop tumor nodules in the lungs following IV inoculation into syngeneic recipients. No correlation was found between the number of pulmonary experimental metastases and the susceptibility of the investigated clones to cytotoxic activity of NK cells.[70] However, the importance of NK resistance for metastatic potential was demonstrated when the metastatic properties of tumor cells were compared in nude mice, which are characterized by relatively high levels of NK reactivity.[70] Repeated incubation of UV-2237 tumor cells with nylon-wool-nonadherent spleen cells resulted in the selection of tumor cell variants that expressed high levels of resistance to NK-mediated cytotoxicity and, in parallel, had high metastatic properties in nude mice. Whereas parental UV-2237 tumor cells failed to develop experimental metastases in 8-week-old nude mice, IV inoculation of NK-resistant tumor cells resulted in the development of 92 metastatic foci per lung. Thus, the metastatic advantage of NK-resistant tumor cells was observed only in mice with relatively high NK reactivity. In contrast, the differences in the metastatic potential between sensitive and resistant tumor cell lines was not observed when these lines were inoculated into 3-week-old nude mice with relatively low NK reactivity.[70]

By cloning of the 3LL tumor, NK-sensitive and NK-resistant clones were obtained. NK-resistant clones, after IV inoculation, developed more metastatic foci in the lungs than NK-sensitive clones. However, no differences in the number of metastases were found when NK-resistant and NK-sensitive 3LL clones were inoculated into beige mice.[71] Poupon *et al.*[72] investigated the metastatic properties and the NK susceptibility of a nickel-induced rat rhabdomyosarcoma. Although tumor cells derived from the spontaneous lung metastases and from the locally growing rhabdomyosarcoma were similar in their susceptibility to NK cells, tumor cells derived from the spontaneous lumbar aortic metastases displayed higher levels of NK resistance than the cells from the primary tumor. Studies of the experimental metastases that developed after IV inoculation of the various clones of the rat rhabdomyosarcoma revealed that all NK-resistant tumor clones were highly metastatic. However, NK-susceptible clones could be either highly or weakly metastatic.[72]

In general, although some experimental data indicate that NK resistance should be an important property of the metastatic phenotype, NK resistance cannot completely account for the ability of tumor cells to metastasize. Metastatic cells have to possess a complex series of properties which allow them to fulfill the entire sequence of steps involved in the metastatic cascade. However, some apparent contradictions in the studies of the relationship between NK sensitivity and metastatic capability may be attributable to several methodological issues:

1. The NK resistance of metastatic cells would only be found to be important in conditions where NK cells are actively involved in protection from development of metastases. In conditions in which the levels of NK activity of the host is below a certain threshold level, metastases would be able to develop regardless of the susceptibility of the metastatic cells to NK cell-mediated cytotoxicity. Indeed, NK-resistant tumor variants have been shown to have some advantage for metastasis formation in nude mice with high NK activity but not in beige mice with low NK activity.[70,71] In parallel with progression of the local tumor, a substantial decline in the NK reactivity of tumor-bearing mice has been found.[73,74] Thus, even NK-susceptible tumor cells that metastasized during the later stages of tumor growth were able to develop metastases in the tumor-bearing host. Surgical excision of the local tumor could prevent the decline of NK reactivity that occurs in animals bearing large tumors, and therefore postoperative tumor metastases might be selected for higher levels of NK resistance than tumor cells from the local tumor site.[67,69]

2. *In vitro* cultivation of tumor cells results in an increase in their susceptibility to NK activity.[15] Therefore, using tumor cells obtained directly from *in vivo* growing local or metastatic tumors as targets in the cytotoxic assay should give more precise information about their NK susceptibility than using long-term cultivated pri-

mary and metastatic tumor cells. Indeed, after culturing the NK resistance of metastasis-derived tumor cells was lost, and the differences in NK susceptibility between tumor cells from local or metastatic tumors completely disappeared.[58]

3. The results of studies of the relationship between NK susceptibility and metastatic activity of tumor cells in some way depend on the criteria used for distinguishing between NK-susceptible and NK-resistant tumor cells or for classifying tumors as having high or low metastatic ability. In the study performed by Poupon *et al.*,[72] clones of the rhabdomyosarcoma were considered to be NK-resistant if the percent of lysis by spleen cells at an effector to target ratio of 200:1 was <10%. When the percent of lysis was 15–20%, clones were considered as NK-susceptible. The ability of the investigated clones to develop spontaneous or experimental metastases varied between four and 35 metastatic foci. This small difference in the metastatic potential made it quite difficult to separate clones into two groups with high and low metastatic ability and, in general, to assess the relationship between NK susceptibility of tumor clones and their metastatic characteristics.

4. Elimination of metastatic cells can be mediated by various immune mechanisms. Some cloned and uncloned tumor cell populations could display high levels of NK resistance and still have low metastatic capabilities as a result of their immunogenic properties. For immunogenic tumors, T cell-mediated immunity appears to be of most importance for the control of metastatic spread and growth. Suppression of NK reactivity of mice by anti-asialo GM$_1$ serum or by the use of 3-week-old mice with low NK activity was not associated with an increase in metastasis formation by immunogenic tumor cells.[37,75]

5. Since some NK-susceptible tumor cells are able to establish distant metastases, there must be some *in vivo* mechanisms that allow tumor cells to escape NK cell-mediated destruction. This might be due to mechanisms that hamper the intravascular recognition and NK-mediated destruction of tumor cells or mechanisms that help tumor cells to extravasate and thereby escape elimination by NK cells in the blood.

4. Role of NK Cells in the Antimetastatic Effects of Anticoagulant Drugs

Metastatic cells in the blood have been observed to be associated with platelets or fibrin attached to the vascular wall.[76–78] Platelet-aggregating and procoagulant activity of tumor cells has been demonstrated in numerous investigations.[79–81]

The interaction between tumor cells and platelets or other components of the hemostatic system has been considered to be an important step in the metastatic process facilitating attachment of tumor cells to the vascular endothelium and subsequent extravasation. This conclusion has been supported by observations that anticoagulant drugs have substantial antimetastatic effects.[80,82–85]

However, an alternative or additional possibility is that platelet aggregation and/or fibrin deposition on the surface of tumor cells might prevent their adequate recognition and lytic interaction with NK or other cytotoxic cells. From this hypothesis, one would predict that the antimetastatic activity of NK cells would be increased when anticoagulant drugs prevent the coating of tumor cells with platelets or fibrin and that the antimetastatic effects of anticoagulant drugs would be augmented in animals with high NK cell activity and diminished or undetectable in animals with low NK reactivity. To evaluate these predictions, antiplatelet (prostacycline or PGI_2) and anticoagulant drugs (heparin and warfarin) were used.[44,59]

In a study of the antimetastatic effect of PGI_2, heparin, and warfarin in mice with stimulated or depressed NK reactivity,[44,59] treatment of mice with anticoagulant drugs had a significant antimetastatic effect. This effect required the presence of active NK cells, since in mice with suppressed NK reactivity after anti-asialo GM_1 treatment, the antimetastatic effect of PGI_2, warfarin, or heparin was abrogated. The importance of NK cells in the antimetastatic effect of anticoagulant drugs was further supported by the observation that the antimetastatic effect of the anticoagulant drugs was potentiated in mice with augmented NK cell activity following poly I:C stimulation. This combination of treatments resulted in a decrease in the number of metastatic foci and an increase in the percent of mice completely free from visible metastases.

The antimetastatic effect of heparin was also abrogated in mice in which NK reactivity was suppressed by Cy treatment (200 mg/kg).[59] In some experiments the treatment of mice with anti-asialo GM_1 serum or Cy just partially diminished the antimetastatic effects of heparin or warfarin. Thus, the number of the pulmonary metastases in these mice decreased to the level observed in the control, nontreated group of mice. The partial reduction of the antimetastatic activity of the anticoagulants could be explained by partial depletion of NK cells and the activity of the residual NK cells whose efficacy was potentiated by prevention of coagulation.[59] It is also of note that the NK reactivity of beige or young C57BL/ 6 could be stimulated by poly I:C. In this condition the antimetastatic effect of heparin was substantially augmented.[59]

The antimetastatic effect of heparin and poly I:C was time-dependent and was not observed if the treatment was given 24 h after tumor cell inoculation.[59] At this time, surviving B16 melanoma cells have almost completed their extravasation and settled in the lung parenchyma.[2] These results indicate that the main effects on the treatments occurred in the blood, before extravasation of tumor cells. The antimetastatic effects of the anticoagulants were probably mediated by the acceleration of tumor cell elimination from the lung vasculature.

The data indicate that (1) NK cells are crucial for the antimetastatic effects of anticoagulant drugs, (2) platelet aggregation and fibrin coagulation on the tumor cell membrane surface may be one of the mechanisms responsible for the protection of tumor cells from destruction by NK cells, and (3) anticoagulant drugs seem to make tumor cells more vulnerable to the cytotoxic action of NK cells and increase the rate of tumor cell elimination from the blood, resulting in a decrease in metastasis formation. A combination of anticoagulant drugs and NK augmenting agents has been found to be particularly effective for the prevention of development of tumor metastases.[86] Ability of fibrin coating to protect tumor cells from destruction by NK or IL2-stimulated spleen cells was confirmed by testing their cytotoxic activity in the presence of plasma. The cytotoxic activity of NK or LAK cells substantially decreased in the presence of plasma diluted 1:20–1:80. The level of inhibition of the cytotoxicity was in parallel with the level of coagulation—i.e., the higher the level of coagulation induced by tumor cells, the higher the level of tumor cell protection observed.[87] When plasma coagulation was presented by adding 2 μg/ml of heparin, the cytotoxic activity of the effector cells was fully exerted.[87]

5. Major Histocompatibility Complex (MHC) Antigen Expression and NK Sensitivity of Tumor Cells

The functional importance of MHC antigens on the cell surface is mostly associated with "self" identification of somatic cells in the multicellular organism by controlling the recognition of cell membrane components of normal or malignant cells by T cell-mediated immunity.

MHC antigens can guide the specific immune lymphocyte for recognition of foreign antigens and elimination of "nonself" cells. Therefore, MHC loss variants of tumor cells are not recognizable by the T-cell immune system and thus could escape immune destruction. Similarly, virus-infected cells missing MHC-restricted elements are less likely to be eliminated by the host. Cells that partially or completely lose their MHC antigens may be termed "no-self" cells.[88]

These cells could be eliminated by mechanisms independent of the T-cell system. It is considered that NK cells and macrophages can destroy malignant or virus-infected cells in an MHC-nonrestricted manner.[89] Numerous experimental data indicate that NK cells could represent the effector mechanism that can restrict the survival and propagation of normal or malignant cells lacking all or some elements of the MHC system.

Many experiments have been performed in which F_1 hybrid mice were inoculated with parental cells that express only one of the H-2 haplotypes expressed by F_1 hybrid host. Semisyngeneic F_1 mice have higher resistance against parental tumor or bone marrow grafts than the syngeneic host.[90–94] The hybrid resistance to normal bone marrow cells was shown to be due to recognition of Hh antigens controlled by H-2D-linked recessive genes.[91]

The rejection of semisyngeneic normal or malignant cells by F_1 recipients was found to be mediated not by T-cell immunity but mostly by NK cells.[92–95] Elimination of semisyngeneic grafts in F_1 hosts was substantially reduced in beige mice and in mice in which NK reactivity was depressed by pretreatment with anti-asialo GM_1 serum.[95]

Based on this finding Kärre et al.[88] hypothesized that the hybrid resistance phenomenon reflected an in vivo mechanism in which NK cells destroyed normal cells that were unable to present in full the MHC antigens of the host.

Similarly, H-2-negative tumor cells could also be recognized as the "absence of self" and thereby be a prime subject of attack by NK cells, whereas T lymphocytes would be unable to destroy such tumor cells.[88] This hypothesis includes an additional assumption that MHC products can diminish or switch off susceptibility to the cytotoxic effects of NK cells.[88] Based on this suggestion, H-2-negative normal or malignant cells were postulated to have higher levels of susceptibility to the cytotoxic action of NK cells than H-2-positive cells. Several experimental data support this hypothesis. Cortical thymocytes that have low H-2 antigen expression are more sensitive than H-2 high medullary thymocytes to the cytotoxic activity of NK cells.[96] H-2-negative teratocarcinoma cells can be destroyed by NK cells, but their susceptibility is reduced with the appearance of H-2 antigens.[97]

To confirm these observations, Ljunggren and Kärre[98] selected H-2-negative variants of EL4 and RBL-5 lymphomas after mutagenesis with ethane methane sulfonate (EMS) and exposure to anti-H-2 antibodies and complement. These variants were rather stable under in vitro or in vivo conditions. Although H-2-positive and H-2-negative variants of EL4 or RBL-5 lymphomas had similar growth characteristics in vitro, H-2-negative variants, in contrast to H-2-positive variants, failed to grow in the syngeneic mice after inoculation of low numbers of tumor cells. It seems that non-T-cell-mediated immunity is responsible for lysis of H-2-negative cells, in contrast to H-2-positive cells, which were able to induce specific immune resistance. Rejection of H-2-negative cells of low numbers of EL4 or RBL-5 lymphoma cells was also observed in nude mice, but these H-2-negative variants grew nude mice whose NK reactivity was depressed by pretreatment with anti-asialo GM_1 serum.

An inverse correlation between H-2 antigen expression and NK sensitivity of tumor cells was also found when these parameters were investigated with YAC-1 lymphoma cells. YAC-1 cells maintain high levels of sensitivity to NK cells in parallel with low levels of H-2^a antigen expression during *in vitro* cultivation. After a single *in vivo* passage, YAC-1 cells increased the level of H-2^a antigen expression and became resistant to NK-mediated cytotoxicity. Explantation of YAC-1 ascites into tissue culture was associated with a decrease in H-2^a antigen expression and an increase in their NK sensitivity.[99]

Using mutagenic EMS treatment and selection with anti-H-2^a antibodies and complement, two H-2-negative variants of YAC lymphoma cells were obtained.[100] These variants and low H-2^a parental YAC-1 cells were similar in their sensitivity to NK cells. However, parental YAC-1 cells, after interferon treatment or *in vivo* passage, increased the H-2^a antigen expression and became NK-resistant, whereas H-2^a-negative variants under interferon treatment or *in vivo* passage did not increase either H-2^a expression or resistance to NK-mediated killing.[100]

Similar observations were made when B16 melanoma cells were investigated. B16 melanoma cells derived from *in vivo* growing tumor expressed relatively high levels of H-2 antigens and resistance to NK activity. During *in vitro* culture, the level of H-2 antigen expression declined, and there was a parallel increase in sensitivity to NK cells.[101] These changes were also associated with the alteration of their metastatic ability.

Similar results were obtained when human Epstein-Barr virus–transformed B-cell lines were investigated. The original tumor line that expressed HLA antigens was resistant to NK cell-mediated cytotoxicity. Cell variants that lack class I HLA molecules were highly NK-sensitive.[102] Based on these data, it was suggested that H-2 gene products could participate in the regulation of tumor cell sensitivity to NK cell cytotoxicity. Since the presence of H-2 class I molecules on some tumor cells correlated with resistance to NK lysis, it was assumed that class I MHC gene products could switch off the cytolytic machinery of NK cells.[88,98]

The inverse correlation between the H-2 antigen expression and NK sensitivity of some tumor cells could shed light on some previously unexplainable associations between MHC antigen expression and metastatic ability of tumor cells. In contrast to the expectation that H-2-negative variants have a better chance to escape immune destruction, some metastatic cells showed a higher level of H-2 expression than the parental tumor cells.[103] This observation could be explained by the assumption that NK cells can participate in the formation of the population of metastatic cells by elimination of relatively sensitive tumor cells. Preferential negative selection of NK-sensitive H-2-negative tumor cells would then be expected to result in the appearance of metastases that expressed higher levels of MHC gene products than primary tumor cells did. Indeed, locally growing T-10 tumor of (H-2^b × H-2^k) F$_1$ origin expressed only H-2^b and lacked H-2^k antigens, whereas metastases from this tumor had both H-2^b/H-2^k antigens.[103,104]

Comparison of the metastatic ability and H-2 antigen expression among the individual clones from T-10 tumor again demonstrated that cells with absence of one of the parental haplotypes were unable to develop pulmonary metastases. Although all T-10 clones were resistant to *in vitro* NK activity, the results of *in vivo* rapid clearance of tumor cells substantiate the prediction of the relationship between H-2 antigen expression, NK sensitivity, and metastatic capability.[103,104] Involvement of NK cells in the control of metastasis formation by T-10 clones was supported by the finding that stimulation of NK reactivity by poly I:C inhibited the metastatic growth, whereas suppression of NK cell function by cyclophosphamide caused a substantial increase in the formation of pulmonary metastases by metastatic and nonmetastatic T-10 clones.[104]

Although higher NK sensitivity of MHC-negative tumor cell variants was demonstrated in several experimental tumor models, some data indicate the existance of an opposite, i.e, positive, correlation between NK sensitivity and MHC antigen expression.

It was found that E1A gene products of human adenovirus type 12, but not type 5, were able to inhibit the expression of MHC antigens on the transformed tumor cells. Lack of class I MHC antigens on the surface of tumor cells was associated with resistance to NK cell killing. In contrast, tumor cells bearing E1A gene from adenovirus type 5 had a high level of MHC antigen expression and showed high sensitivity to NK cells.[105,106]

In previous experiments we have found that highly invasive and metastatic BL6 melanoma cells did not express class I H-2^b antigens. After treatment of these cells with N-methyl-N-nitro-nitrosoguanidine (MNNG), the expression of class I H-2^b antigen and immunogenecity dramatically increased.[107] Poly I:C-stimulated spleen cells of nude mice were highly cytotoxic for BL6T2, whereas H-2-negative BL6 cells were less sensitive to NK activity in a 16-h ^{51}CR-release assay. Similar results were obtained after 4-h incubation of radiolabeled tumor cells with IL-2-activated effector cells. In contrast, the two lines were equally sensitive to lysis by purified granules derived from rat LGL or by macrophages. Using various clones selected from BL6 or BL6T2 cells, it was found that BL6 or BL6T2 clones with low H-2^b antigen expression were less sensitive to lysis by NK cells than H-2^b-positive clones.

After interferon treatment of either BL6 or BL6T2, the target cells became more resistant to lysis by either NK cells or by purified LGL granules. Interferon-treated BL6 cells had substantially increased expression of H-2^b antigens and in this respect became similar to untreated BL6T2. However, interferon-treated Bl6 cells were more resistant than BL6T2 cells to lysis by NK cells and LGL granules, suggesting that augmentation of H-2 antigen expression and NK resistance could be two independent interferon-induced effects. Using a cold target inhibition assay, it was found that BL6T2 and YAC-1 were highly competitive and inhibited the cytotoxic activity of NK and LAK cells against radiolabeled

YAC-1 and BL6T2, whereas BL6 melanoma cells showed poor competitive ability.

Individual BL6 clones with low levels of H-2^b expression and resistance to NK lysis also failed to inhibit the cytotoxicity of spleen cells against YAC-1 cells, but BL6T2 clones with high H-2^b antigen expression were rather efficient in cold target inhibition assay (Gorelik *et al.*, unpublished observations).

Thus, these data demonstrate that NK resistance of H-2-negative BL6 cells could be due to a paucity of NK recognizable determinants. MNNG treatment of BL6 melanoma cells was associated with an increase in class I H-2^b antigen expression and NK sensitivity, suggesting the involvement of class I MHC antigens in sensitivity of tumor cells to NK cell-mediated cytotoxicity.

The results from our studies with the BL6 tumor lines and those with adenovirus-induced tumors are quite divergent from results of the earlier studies cited above. The mechanisms responsible for this phenomenon are mostly unknown. It seems that the effects of MHC gene products on NK sensitivity differ with the target cells and might influence the level of target cell recognition or postbinding events. It is unlikely that NK cells are able to recognize the MHC molecules but rather recognize non-MHC structures that are associated with class I MHC molecules. Such an association was found for the insulin receptor[108] and the epidermal growth factor receptor.[109] Certainly the observed positive or negative correlation between the level of MHC antigen expression and NK sensitivity of tumor cells does not necessarily mean that these two events are causally related. On the other hand, the fact that this association was observed among various rodent or human tumor cell lines indicates that MHC molecules might play some role in the regulation of sensitivity of tumor cells to NK cell cytotoxicity. To obtain more direct evidence of involvement of MHC gene products in determination of NK sensitivity of tumor cells, it will be necessary to develop more specific methods for up- and down-regulation of MHC antigen expression and thereby investigate directly the effect of these alterations on sensitivity of tumor cells to NK cell mediated lysis.

6.   Role in Immune Surveillance against Tumors

Of paramount interest is whether NK cells may be involved in immune surveillance against the initial development of spontaneous or carcinogen-induced tumors. There are several pieces of circumstantial evidence that are consistent with, or suggestive of, a role of NK cells.

1. Patients with the genetically determined Chediak-Higashi syndrome have a high risk of development of lymphoproliferative diseases.[110] In recent detailed studies on several patients with this

disease,[111,112] all were found to have profound deficits in NK and K cell activities, whereas a variety of other immune functions, including cytotoxicity against tumor cells by T cells, monocytes, and granulocytes, was essentially normal.

2. Similarly, beige mice, which have an analogous genetic defect, also have a substantial,[113,114] but incomplete,[115] selective deficiency in NK activity. Aged beige mice have recently been reported to have a high incidence of lymphomas, and heterozygous mice without the beige phenotype had a lower incidence.[117]

3. Another human genetic abnormality, X-linked lymphoproliferative disease,[118] has been associated with a defect in the ability to control proliferation of B cells infected with Epstein–Barr virus (EBV). Low NK activity has recently been found in such individuals, and this deficit has been suggested to be involved in the pathogenesis of the disease.[119,120] In support of this possibility, cells with the characteristics of NK cells have been found to inhibit the *in vitro* proliferation of autologous EBV-infected B cells.[121]

4. Patients on immunosuppressive therapy after allotransplants have a high risk of developing tumors, both lymphoproliferative disease and also a variety of carcinomas.[122] Patients on such treatment regimens have recently been found to have very low NK activity, and this has been suggested as a contributing factor to the subsequent development of tumors.[123]

5. Patients with paroxysmal nocturnal hemoglobinuria also have a high risk of developing leukemia and have recently been reported to have deficient NK activity.[124]

6. Normal individuals from families with a high incidence of malignant melanoma [125] were found to have low NK activity. Similarly, it has recently been reported that normal adults with a family history of various types of carcinomas had significantly lower NK activity than normal donors without such a history.[126] Each of these lines of evidence fits one of the major predictions of the immune surveillance theory—that tumor development would be associated with, and in fact preceded by, depressed immunity.

A related prediction of the immune surveillance theory is that carcinogenic agents would cause depressed immune function, thereby impairing the ability of the host to reject the transformed cells. This postulate has been examined by many investigators in regard to the possible role of mature T cells and humoral immunity, and conflicting results have been obtained.[127] In contrast, the initial and still fragmentary data on this point in relation to NK cells suggest a contribution of these effector cells to resistance against the development of at least certain types of tumor:

1. Urethane, which produces lung tumors in only some strains of mice, caused transient and marked depression of NK activity in a

susceptible strain but not in resistant strains.[128,129] Administration of normal bone marrow cells, which as discussed earlier can reconstitute NK activity, to urethane-treated mice reduced the subsequent development of lung tumors.

2. Carcinogenic doses of dimethylbenzanthracene were also found to produce depression of NK activity during the latent period.

3. Sublethal irradiation of mice has been found to cause considerable depression of NK activity.[133] Of particular interest, the schedule of multiple, low doses of irradiation of C57BL mice, which has been highly effective in inducing thymic lymphomas in this strain, was found to produce a substantial deficit in NK activity.[133,134] The depressed NK activity could be restored by transfer of normal bone marrow cells,[133,135] a procedure that has been reported to interfere with radiation-induced leukemogenesis.[136] Transfer of cloned lymphoid cells with NK-like activity, beginning immediately after the last dose of irradiation, also strongly inhibited the development of lymphomas.[136a]

4. Similarly, AKR mice, which develop a high incidence of spontaneous leukemias and have low NK activity, could be protected against tumor development by adoptive transfer of bone marrow from high-NK mice but not by marrow from low-NK mice.[137]

5. Low doses of methylcholanthrene were shown to produce a significantly higher incidence of sarcomas in mice treated shortly after birth with diethylstilbestrol, which was accompanied by persistently depressed NK activity.[138]

All of these observations support the possibility that one of the requisites for tumor induction by carcinogenic agents may be interference with host defenses, including those mediated by NK cells. However, there have been several studies in which the incidence of spontaneous or carcinogen-induced tumors in mice did not show an inverse correlation with levels of NK activity.[128,139] Some of these apparent discrepancies may be attributable to the inherent resistance of the tissues of some strains of mice to transformation by a particular carcinogen. For example, the resistance of beige mice to urethane-induced lung tumors may be due to the known resistance of the lung tissues of some strains to transformation by urethane.[140] Thus, if a particular strain is resistant to the induction of neoplastic cells, we would not expect to see the development of tumors even in the absence of any surveillance mechanism. Another point to consider is that exposure to some carcinogens may induce profound depression of natural immunity in all recipients, and in such a situation one could not expect to observe a correlation between tumor incidence and levels of NK activity in normal, untreated animals.

7. Stimulation of Natural Cell-Mediated Immunity and Treatment of Tumor Metastases

7.1. Therapy with Immunomodifiers

Progression of the local tumor growth is paralleled by an increase in tumor cell release into the circulation and a decline in NK cell activity.[74] Both of these processes could increase the probability of development of distant tumor metastases. Thus, stimulation of NK reactivity by various immunomodifiers might be helpful for the control of metastatic spread and growth. However, the available experimental data indicate that stimulation of NK cell function is most efficient for the prevention rather than the eradication of established tumor metastases.[28,31,26] In the clinical situation, metastases can be detected at the time of diagnosis, or metastatic cells have already settled at different anatomical locations and become detectable after surgical excision of the primary tumor. This raises the question whether NK cells may be considered a potential effector mechanism for the treatment of metastases in cancer. A further question is whether tumor cells from most patients are sufficiently susceptible to NK activity for this mechanism to play a role in resistance to metastatic spread.

In this regard, it has become apparent that stimulation of NK cells with interferon or interferon inducers not only can increase their cytotoxic ability against NK-sensitive tumor target cells but also can make them able to destroy tumor cells that demonstrate resistance to the cytotoxic action of spontaneous NK cells.[7,8]

In general, determination of the level of resistance or susceptibility of tumor cells to NK-mediated cytotoxicity is mostly based on the results of in vitro testing using a ^{51}Cr-release assay. However, in vivo assays of IV inoculation of radiolabeled tumor cells indicate that in vitro resistant tumor cells display substantial susceptibility to NK cell-mediated destruction in the blood. Depression of NK reactivity of mice with anti-asialo GM_1 serum or Cy resulted in the increase of the survival of the "resistant" tumor cells. On the contrary, stimulation of NK cell function by treatment of mice with poly I:C increased the elimination of the "resistant" tumor cells and inhibition of metastasis formation.[44,86] It may indicate that the tumoricidal effect of NK cells in the blood could be more efficient than is demonstrated by the in vitro assay, although it is not possible to exclude the in vivo contribution of other cells or humoral factors.

Although there are no data to demonstrate that stimulation of NK cells by immunomodifiers can inhibit or eradicate established metastatic foci, experiments with adoptively transferred IL2-activated NK cells indicate that they have the capability to do this (see below, Section 7.2). All of the experiments described above were performed with a single injection of the immunomodifier. It is well established that stimulation of NK reactivity with interferon inducers had short transient effects.[37,48] Seven days after C. parvum treatment, NK cell activity of mice returned to normal levels.[37] There are substantial difficulties for immunomodifiers to maintain elevated levels of NK cell activity. Furthermore, mice became unresponsive to the second boosting of NK cells by *C. parvum*, which parallels the failure of the second *C. parvum* injection to inhibit experimental metastasis formation.[37]

Similarly, in cancer patients, the first injections of interferon stimulated NK cell activity, but they became unresponsive to subsequent treatment with interferon. Continued, frequent interferon treatment often caused a decline in the NK cell activity.[141] Refractoriness of NK cells to the multiple boosting by interferon inducers in mice may be due, at least in part, to the appearance of suppressor cells. Indeed, suppressor cells for NK cell function were found in mice treated with various immunomodifiers. These suppressors were characterized as macrophages[37,48,142] and/or nonmacrophage cells.[143] Although these data were obtained using *in vitro* experimental systems, it seems possible that macrophages might exert similar *in vivo* suppressive effects on NK cell activity. Peritoneal thioglycollate-elicited macrophages inoculated IV into mice inhibited spleen NK cell function, increased the survival of tumor cells in the blood, and caused a dramatic augmentation of the experimental pulmonary metastasis formation.[144–146] Furthermore, in mice inoculated IV with thioglycollate-elicited macrophages, poly I:C treatment was unable to stimulate NK cell activity and exert antimetastatic effects.[145,146] Augmentation of metastasis formation was observed in mice inoculated IV with nontumoricidal and even tumoricidal thioglycollate macrophages. These macrophages induced intravascular reactions with blood cells (preferentially neutrophils) and switched on a cascade of inflammatory reactions with a possible increase in vascular permeability. All these processes might help tumor cells to escape from the blood and to increase tumor cell survival and metastasis formation.[144]

Understanding the mechanisms of the regulation of NK cells might provide the basis for developing protocols for stimulating and maintaining high levels of NK reactivity in cancer patients. The importance of this issue is supported by the observation that in order for activated macrophages to be able to eradicate established pulmonary metastases in mice, it was essential to maintain high levels of tumoricidal activity of the alveolar macrophages, by IV inoculation of liposomes containing muramyl dipeptides, for a substantial time after tumor excision.[12]

It is of note that the effect of NK cells on established metastatic foci has been investigated mainly with pulmonary metastasis as the experimental model. The effect of NK cells on metastases in other anatomical locations remains unclear. Spontaneous NK activity in the lymph nodes is relatively low. However, the cytotoxic activity of NK cells in lymph nodes can be stimulated. This stimulation depends on the type, dose, and route of administration of the immunomodifiers. Stimulation of lymph node NK activity resulted in the destruction of B16 melanoma cells metastasizing into lymph nodes.[37]

Some evidence indicates that NK cells could play an important role in the control of extrapulmonary metastasis formation. B16 melanoma cells inoculated IV develop tumor foci mainly in the lungs but in some cases also in the liver. Although suppression of NK cell activity by anti-asialo GM_1 serum increased the number of B16 melanoma pulmonary metastases, a more dramatic increase in liver metastases was found.[31] The presence of hepatic NK cells with the morphological and phenotypic characteristics of the LGL was recently demonstrated.[64] The number and activity of the hepatic NK cells substantially increased after stimulation of mice with *C. parvum* or MVE-2. This stimulation of NK cell activity was more profound in the lungs and liver parenchyma than in spleen and blood. Under these conditions, the formation of the B16 melanoma metastasis in the liver was completely prevented, and a dramatic decrease in the lung metastases was found.[65].

There is considerable room for optimism that stimulation and maintenance of high levels of activity of natural cell-mediated immunity with appropriate BRMs may contribute to the treatment of the cancer patients. This approach is especially attractive for the prevention or eradication of postoperative tumor metastases in cancer patients. Our data demonstrate that potentiation of NK reactivity and treatment of mice with anticoagulant drugs had synergistic antimetastatic effects. Several attempts have been made to assess the effect of anticoagulant drugs on the treatment of cancer patients, and some positive results have been reported.[147] Our experimental data suggest the value of treatment with anticoagulant drugs in combination with a BRM that is able to maintain high levels of NK cell activity.

Another possible therapeutic approach using natural cell-mediated immunity is related to the involvement of NK cells in antibody-dependent cell cytotoxicity (ADCC). ADCC against most tumor cells depends on the presence of functional NK cells and is eliminated by depletion of NK cells.[7,8,148] Although there is no unequivocal *in vivo* evidence for the antitumor or antimetastatic effects of ADCC mechanisms, the therapeutic effects of some monoclonal antibodies for the treatment of postoperative tumor metastases might depend on ADCC mechanisms.

Based on the experimental data, the importance of NK cells in the control of the metastatic process is apparent. In addition, the involvement of macrophages in metastatic defenses and their potential therapeutic efficiency have been demonstrated by numerous investigations.[12, 25]

7.2. Treatment of Tumor Metastases by Adoptive Transfer of Natural Effector Cells

An alternative approach to the augmentation of natural cell-mediated immunity is the adoptive transfer of effector cells. The recent report by

Rosenberg et al.[149] on the preliminary results of a clinical trial at the National Cancer Institute, exploring the therapeutic efficacy in patients with advanced cancer of lymphokine-activated killer (LAK) cells, plus high doses of interleukin 2 (IL-2), has led to very widespread interest by the public at large[150,151] as well as by oncologists and tumor immunologists. In their initial studies, Grimm, Rosenberg, and their colleagues[152,153] thought LAK cells were a unique effector cell population, clearly distinct from NK cells. However, more recent studies, by a wide variety of investigators, have indicated that most human, mouse, and rat LAK cells generated from peripheral blood or splenic lymphocytes represent IL-2-activated NK cells.[154]

For a long time, investigators have entertained the concept of obtaining effector cells in vitro, stimulating and/or expanding them, and then utilizing them for systemic or local therapy of tumors.[155] The concept of employing this strategy with nonimmune effector cells, rather than specifically immune CTL, is considerably more recent.[156,157]

The initial studies with normal lymphocytes cultured in the presence of IL-2 indicated that such cells could have significant antitumor effects.[158] Normal murine lymphocytes cultured for 3 days in the presence of a high concentration of human recombinant IL-2 were then shown to have significant therapeutic activity against metastases from various syngeneic sarcomas, when injected intravenously at 3 and 6 days after tumor challenge, along with repeated high doses of human recombinant IL-2 (25,000 units intraperitoneally 3 times a day from day 3 through day 8). This treatment by high doses of LAK cells (10^8 cells) significantly reduced the number of metastases detectable in the lungs and liver but did not result in complete elimination of metastases or cures of the mice. Rosenberg and his colleagues then began a series of focused efforts to optimize the parameters for effective therapy of syngeneic murine sarcomas by this approach. Mule et al.[159] examined several of the parameters of the treatment protocol, to gain some insight into the requirements for effective therapy. Two injections of LAK cells, 3×10^7 or 10^8, were found to be more effective than one dose of cells. The cells cultured for 3 days in the presence of IL-2 were shown to have optimal therapeutic effects, and it is of interest that this length of culture also resulted in peak levels of cytotoxic reactivity. Therapeutic effects were seen in recipients pretreated with 500R X-irradiation, suggesting that the therapeutic effects were not due to a major host component. In contrast, X-irradiation of the LAK cells with 3000R resulted in a loss of efficacy, suggesting that the transferred cells had to not only remain viable but also be able to proliferate. However, since the transfer of allogeneic LAK cells appeared to be as effective as syngeneic cells, long-term survival and proliferation in the recipients were probably not required.

The administration of recombinant IL-2 along with LAK cells appeared to be required, with the presumption that the lymphokine treatment was needed for in vivo stimulation of proliferation of the donor cells. The use of 6000 units of recombinant IL-2 per dose together with LAK cells was highly effective, but 30,000 units per dose appeared to give better results. In contrast, the use of up to 34,000 units of recombinant IL-2 per dose, without administration of LAK cells, had no detectable effect on tumor metastases when the treatment was initiated on day 3 after tumor challenge. However, 20,000 units of IL-2 per dose was effective by itself in significantly reducing established pulmonary metastases when treatment was initiated at day 10 after tumor challenge. At both days 3 and 10, 100,000 units per dose of IL-2 were appreciably more effective. To understand the possible basis for the therapeutic effects of high doses of IL-2 by itself and also to assess the hypothesis regarding the need for administration of IL-2 to support *in vivo* proliferation of transferred LAK cells, studies have been performed on the ability of various doses of recombinant IL-2 to stimulate proliferation of lymphocytes *in vivo*. As a parallel to the dose of IL-2 shown to be required for significant therapeutic effects of LAK cells, 6000 units of IL-2 administered intraperitoneally three times per day was found to significantly increase the uptake of radiolabeled iododeoxyuridine, as a measure of proliferating lymphocytes, in a variety of organs, including the lungs, liver, kidneys, and mesenteric lymph nodes.

Substantially higher levels of proliferation in these organs were observed after administration of 100,000 units of IL-2 per dose. Although the cellular composition of the proliferating lymphocytes was not studied in detail, it is of note that augmented levels of cytotoxic reactivity against NK-resistant target cells were observed after the *in vivo* treatment with IL-2. As a further study along these lines, uptake of radiolabeled iododeoxyuridine was examined in mice treated with 6000 units of IL-2 per dose, alone or in combination with transferred LAK cells.[160] The combination of the two treatments gave higher levels of proliferation in the lungs and liver, but in other organs, such as spleen, kidneys, and lymph nodes, maximal uptake of radiolabel was observed with IL-2 alone. As another correlation with the therapy studies, expansion of lymphocytes was also seen in recipients pretreated with 500R X-irradiation.

In the above series of studies, although adoptive immunotherapy could induce an impressive reduction in established metastases of murine sarcomas, the therapeutic effects were usually transient, and few if any complete cures were achieved. To explore this approach to therapy further and to explore the treatment conditions that might be required for curative results, a model of transplantable murine adenocarcinoma of the kidney (Renca) was developed.[161] By inoculation of Renca cells under the kidney capsule, a course of tumor progression was initiated that closely mimicked the progression of human renal cell carcinoma, with spontaneous metastases to regional lymph nodes in the peritoneal cavity, liver, and lungs. This model has become of particular interest because of recent clinical studies indicating that adoptive immunotherapy with LAK cells plus recombinant IL-2 may be particularly effective for metastatic renal cell carcinoma (see discussion below). In the initial therapy experiments, treatment was initiated at 7 days after tumor challenge, when only occult metastases were present. Administration of cytotoxic lymphocytes, after 24 h of incubation with human recombinant IL-2, plus recombinant IL-2, resulted in a significant increase in survival but only a low percentage of cures. Similarly, treatment at this time with chemotherapy, either doxorubicin or cyclophosphamide, gave a low percentage of cures. In contrast, combination chemoimmunotherapy with doxorubicin and IL-2-stimulated cytotoxic lymphocytes plus IL-2 resulted in a cure of two-thirds of the tumor-bearing mice.

These results are quite interesting from several standpoints. First, they demonstrate that treatment with a combination of modalities may be considerably more effective than immunotherapy alone, with cytoreduction or other effects of chemotherapy leading to synergistic interactions with adoptive immunotherapy. Secondly, the protocol for immunotherapy that was required for these impressive results was considerably less intensive than those utilized by Rosenberg and his colleagues. The lymphocytes utilized for transfer were cultured with only 200 units of IL-2 for a shorter period, only 24 h, since these conditions were sufficient to induce strong cytotoxicity in vitro against Renca. The immunotherapy itself consisted of three daily inoculations of 3.5×10^7 cultured lymphocytes intravenously plus three daily intravenous inoculations of 10,000 units of IL-2. These doses were lower than those utilized in the above-described experiments, indicating that under some circumstances, effective therapy can be achieved with relatively modest doses of IL-2 and cytotoxic lymphocytes.

Similar results were achieved when a combination of doxorubicin and IL-2-stimulated lymphocytes plus IL-2 was utilized for treatment of intraperitoneal Renca tumor.[162] Administration of chemoimmunotherapy into the peritoneal cavity resulted in cures of 90% of the tumor-bearing mice. Based on such encouraging results, attempts were made to treat this transplantable tumor at a more advanced stage of disease, beginning at 3 weeks after tumor inoculation under the kidney capsule, when microscopically detectable peritoneal metastases were already present. It was possible to achieve cures in up to 80% of tumor-bearing mice by chemoimmunotherapy but only when the treatment was administered by both the intraperitoneal and intravenous routes. In addition, removal of the tumor-bearing kidney was required for effective therapy. Such impressive results with advanced disease are quite encouraging and suggest that maximal reduction of tumor burden, by surgery and chemotherapy as well as administration of immunotherapy into the regions of metastases, may be required for complete elimination of advanced metastatic tumors.

In most adoptive immunotherapy studies with immune T cells, the lymphocytes have been obtained from strongly immunized mice. However, Shu et al.[163] have found that spleen or lymph node lymphocytes from tumor-bearing mice could also be stimulated to develop therapeutic efficacy. Lymph node cells from the region draining the site of primary tumor growth were particularly effective, with therapeutically active cells obtained from mice with a broad range of tumor burdens. It is also of note that in this model of specific adoptive immunotherapy, curative effects were achieved with substantially lower doses of both lymphocytes and recombinant IL-2 than those required for partially effective therapy by LAK cells plus IL-2. It is also of interest that preceding chemotherapy or immunosuppression was not required for effective therapy of visceral metastases by this murine sarcoma,[164] in contrast to the requirement for cyclophosphamide pretreatment, to observe therapeutic effects with a murine lymphoma.[165–169] These differences may be attributable, at least in part, to the site of tumor growth, since effective therapy of intradermally inoculated MCA105 sarcoma was dependent on preceding sublethal X-irradiation.[164]

The extensive preclinical data and the preliminary clinical observations indicate that adoptive immunotherapy with IL-2-stimulated lymphocytes plus IL-2 can induce substantial antitumor effects, even when treatment is initiated after metastases are established. The number of effector cells actually required for therapeutic effects is not clear. The adoptive therapy studies reported to date have utilized unfractionated mononuclear cells. It may be possible to achieve a substantially more favorable situation if effector cells are first isolated and purified. This should substantially decrease the requirement for the total number of cells and may also help to eliminate suppressor cells as well as irrelevant T cells. This might also substantially alter the requirement for IL-2, since the dose needed to stimulate and expand a relatively small number of cells might be considerably less than that required for effects on a much larger number of total responding lymphocytes.

Recent studies in our laboratory have supported these suggestions. It has been possible to highly purify IL-2-activated NK cells by adherence to plastic[170] and therapy studies with such purified rat or mouse LAK cells have been initiated (N. Vujanovic, J. Hiserodt, E. Gorelik, and R. B. Herberman, unpublished observations). These effector cells, when administered to rats or mice with experimentally induced metastases, together with moderate doses of IL-2, had considerably more potent therapeutic effects than unfractionated populations of IL-2-stimulated spleen cells. Of particular interest has been the observation that even when equivalent total lytic activity was transferred, the purified, adherent LAK cells were more effective than standard LAK cells.

Since most LAK activity in IL-2-stimulated peripheral blood or splenic lymphocyte populations is attributable to highly activated NK cells, the major therapeutic effects of cells with LAK activity may be viewed as an important extension of the role of natural effector cells in host resistance against tumors, and a direct demonstration that NK cells have the potential not only to prevent but to treat metastatic deposits of tumor cells.

References

1. Baldwin, R. W., 1976, Role of immunosurveillance against chemically induced rat tumors, *Transplant Rev.* **28:**62.
2. Fidler, I. J., Gerstein, D. M., and Hart, I. R., 1978, The biology of cancer invasion and metastasis, *Adv. Cancer Res.* **28:**149.
3. Klein, G., and Klein, E., 1977, Rejectability of virus induced tumors and non-rejectability of spontaneous tumors-A lesson in contrasts, *Transplant. Proc.* **9:**1095.
4. Kripke, M., 1981, Immunologic mechanisms in UV radiation carcinogenesis, *Adv. Cancer Res.* **34:**69.
5. Hewitt, H. B., Blake, E. R., and Walder, A. S., 1976, A critique of the evidence for active host defense against cancer, based on personal studies of 27 murine tumours of spontaneous origin, *Br. J. Cancer* **33:**241.
6. Weiss, D., 1977, The questionable immunogenicity of certain neoplasms, *Cancer Immunol. Immunother.* **2:**11.
7. Herberman, R. B. (ed.), 1980, *Natural Cell-Mediated Immunity Against Tumors*, Academic Press, New York.
8. Herberman, R. B. (ed.), 1982, *NK Cells and Other Natural Effector Cells*, Academic Press, New York.
9. Alexander, P., 1976, The functions of the macrophage in malignant disease, *Annu. Rev. Med.* **27:**207.
10. Fidler, I. J., 1974, Inhibition of pulmonary metastases of intravenous injection of specifically activated macrophages, *Cancer Res.* **34:**1074.
11. Hibbs, J. B. Jr., 1974, Discrimination between neoplastic and nonneoplastic cells *in vitro* activated macrophages, *JNCI* **53:**1487.
12. Fidler, I., and Poste, G., 1982, Macrophage-mediated destruction of malignant tumor cells and new strategies for the therapy of metastatic disease, *Springer Semin. Immunopathol.* **5:**161.
13. Herberman, R. B., Nunn, M. E., and Lavrin, D. H., 1975, Natural cytotoxic reactivity of mouse lymphoid cells against syngeneic and allogeneic tumors. I. Distribution of reactivity and specificity, *Int. J. Cancer* **16:**216.
14. Kiessling, R., Klein, E., and Wigzell, H., 1975, Natural killer cells in the mouse. I. Cytotoxic cells with specificity for mouse Moloney leukemia cells: Specificity and distribution according to genotype, *Eur. J. Immunol* **5:**112.
15. Herberman, R. B., and Holden, H. T., 1978, Natural cell-mediated immunity, *Adv. Cancer* **27:**305.
16. Kärre, K., Klein, G. O., Kiessling, R., Klein, G., and Roder, J. C., 1980, Low natural *in vivo* resistance to syngeneic leukaemias in natural killer–deficient mice, *Nature* **284:**624.
17. Kiessling, R., Petranyi, G., Klein, G., and Wigzell, H., 1975, Genetic variation of *in vitro* cytolytic activity in *in vivo* rejection potential of nonimmunized semisyngeneic mice against a mouse lymphoma line, *Int. J. Cancer* **15:**933.
18. Kiessling, R., Petranyi, G., Klein, G., and Wigzell, H., 1976, Non-T-cell resistance against a mouse Moloney lymphoma, *Int. J. Cancer* **17:**275.
19. Kiessling, R., and Wigzell, H., 1979, An analysis of the murine NK cells as to structure, function, and biological relevance, *Immunol. Rev.* **44:**165.
20. Petranyi, G., Kiessling, R., Povey, S., Klein, G., Herzenberg, E., and Wigzell, H., 1976, The genetic control of natural killer cell activity and its association with *in vivo* resistance against a Moloney lymphoma isograft, *Immunogenetics* **3:**15.
21. Riesenfeld, I., Orn, A., Gidlund, M., Axberg, I., Alm, G. V. and Wigzell, H., 1980, Positive correlation between *in vitro* NK activity and *in vivo* resistance toward AKR lymphoma cells, *Int. J. Cancer* **25:**399.
22. Talmadge, J. E., Meyers, K. M., Prieur, D. J., and Starkey, J. R., 1980, Role of NK cells in tumor growth and metastasis in beige mice, *Nature* **284:**622.
23. Haller, O., Kiessling, R., Orn, A., Kärre, K., Nilsson, K., and Wigzell, H., 1977, Natural cytotoxicity to human leukemia mediated by mouse non-T cells, *Int. J. Cancer* **20:**93.

24. Habu, S., Fukui, H., Shimamura, K., Kasai, M., Nagai, Y., Okumura, K., and Ta-maoka, N., 1981, *In vivo* effects of anti–asialo GM₁. I. Reduction of NK activity and enhancement of transplanted tumor growth in nude mice, *J. Immunol.* **127**:34.
25. Gorelik, E., and Herberman, R. B., 1981, Radioisotope assay for evaluation of *in vivo* natural cell-mediated resistance of mice to local transplantation of tumor cells, *Int. J. Cancer* **27**:709.
26. Kasai, M., LeClerc, J. C., McVay-Boudreau, L., Shen, F. W., and Cantor, H., 1979, Direct evidence that natural killer cells in nonimmune spleen cell populations prevent tumor growth *in vivo*, *J. Exp. Med.* **149**:1260.
27. Tam, M. R., Emmons, S. L., and Pollack, S. B., 1980, FACS analysis and enrichment of NK effector cells, in: *Natural Cell-Mediated Immunity Against Tumors* R. B. Herber-man, (ed.), Academic Press, New York, pp. 265–276.
28. Brown, J., and Parker, E., 1979, Host treatments affecting artificial pulmonary metas-tases: Interpretation of loss of radioactivity labelled cells from lungs, *Br. J. Cancer* **40**:677.
29. Fidler, I., 1970, Metastasis: Quantitative analysis of distribution and fate of tumor em-boli labeled with ¹²⁵I-5-iodo-2'deoxyuridine, *JNCI* **45**:773.
30. Hofer, K., Prensky, W., and Hughes, W., 1969, Death and metastatic distribution of tumor cells in mice monitored with ¹²⁵I-iododeoxyuridine, *JNCI* **45**:763.
31. Gorelik, E., Wiltrout, R., Okumura, K., Habu, S., and Herberman, R., 1982, Role of NK cells in the control of metastatic spread and growth of tumor cells in mice, *Int. J. Cancer* **30**:107.
32. Hanna, N., and Fidler, I. J., 1980, The role of natural killer cells in the destruction of tumor emboli, *JNCI* **65**:801.
33. Hanna, N., and Fidler, I. J., 1981, Expression of metastatic potential of allogeneic circulating and xenogenic neoplasms in young nude mice, *Cancer Res.* **41**:438.
34. Riccardi, D., Puccetti, P., Santoni, A., and Herberman, R. B., 1979, Rapid *in vivo* assay of mouse NK cell activity, *JNCI* **63**:1041.
35. Riccardi, C., Barlozzari, T., Santoni, A., Herberman, R., and Cesarini, C., 1981, Trans-fer to cyclophosphamide-treated mice of natural killer (NK) cells and *in vivo* natural reactivity against tumors, *J. Immunol.* **126**:1284.
36. Hanna, N., 1980, Expression of metastatic potential of tumor cells in young nude mice correlated with low levels of natural killer cell–mediated cytotoxicity, *Int. J. Cancer* **26**:675.
37. Hanna, N., 1982, Role of natural killer cells in control of cancer metastasis, *Cancer Metastasis Rev.* **1**:45.
38. Gorelik, E., and Herberman, R. B., 1981, Inhibition of the activity of mouse NK cells by urethane, *JNCI* **66**:543.
39. Seaman, W., Blackman, M., Gindhart, T., Roubinia, J., Loeb, J., and Talal, N., 1978, β-Estradiol reduced natural killer cells in mice, *J. Immunol.* **12**:2193.
40. DeBrabancer, M., Aerts, F., and Borgers, M., 1974, The influence of a glucocorticoid on the lodgement and development in the lungs of intravenously injected tumour cells, *Eur. J. Cancer* **10**:755.
41. Hanna, N., and Schneider, M., 1983, Enhancement of tumor metastasis and suppres-sion of natural killer cell activity by β-estradiol treatment, *J. Immunol.* **130**:974.
42. Steele, G., and Adams, K., 1977, Enhancement by cytotoxic agents of artificial pulmo-nary metastasis, *Br. J. Cancer* **36**:653,
43. Pollack, S., 1982, Direct evidence for anti-tumor activity by NK cells *in vivo:* Growth of B16 melanoma in anti-NK 1.1 treated mice, in: *NK Cells and Other Natural Effector Cells* (R. B. Herberman, ed.), Academic Press, New York, pp. 1347–1352.
44. Gorelik, E., Bere, E., and Herberman, R., 1984, Role of NK cells in the antimetastatic effect of anticoagulant drugs, *Int. J. Cancer* **33**:87.
45. Djeu, J. Y., Heinbaugh, J. A., Holden, H. T., and Herberman, R. B., 1979, Role of macrophages in the augmentation of mouse natural killer cell activity by poly I:C and interferon, *J. Immunol.* **122**:182.

46. Herberman, R. B., Djeu, J. Y., Kay, H. D., Ortaldo, J. R., Riccardi, C., Bonnard, G. D., Holden, H. T., Fagnani, R., Santoni, A., and Puccetti, P., 1979, Natural killer cells: Characteristics and regulation of activity, *Immunol. Rev.* **44**:43.
47. Herberman, R. B., Ortaldo, J. R., Djeu, J. Y., Holden, H. T., Jett, J., Lang, N. P., and Pestka, S., 1980, Role of interferon in regulation of cytotoxicity of natural killer cells and macrophages, *Ann. N.Y. Acad. Sci.* **250**:63.
48. Herberman, R. B., Brunda, M. J., Djeu, J. Y., Domzig, W., Goldfarb, R. H., Holden, H., Ortaldo, J. R., Reynolds, C. W., Riccardi, C., Santoni, A., Stadler, B. M., and Timonen, T., 1982, Immunoregulation and natural killer cells, in: *Natural Killer Cells. Volume 4: Human Cancer Immunology* (B. Serrou, C. Rosenfeld, and R. B. Herberman, eds.), Elsevier/North-Holland, Amsterdam, pp. 37–52.
49. Hanna, N., and Burton, R., 1981, Definitive evidence that natural killer (NK) cells in experimental tumor metastasis *in vivo, J. Immunol.* **127**:1754.
50. Barlozzari, T., Reynolds, C., and Herberman, R., 1983, *In vivo* role of natural killer cells: Involvement of large granular lymphocytes in the clearance of tumor cells in anti–asialo GM$_1$–treated rats, *J. Immunol.* **131**:1024.
51. Chow, D., Brown, G., and Greenberg, A., 1982, NK cell and NAB antitumor activity *in vivo*, in: *NK Cells and Other Natural Effector Cells* (R. Herberman, ed.), Academic Press, New York, pp. 1379–1386.
52. Gale, R. P., and Zighelboim, J., 1975, Polymorphonuclear leukocytes in antibody-dependent cellular cytotoxicity, *J. Immuol* **114**:1047.
53. Korec, S., 1980, The role of granulocytes in host defense against tumors, in: *Natural Cell-Mediated Immunity Against Tumors* (R. B. Herberman, ed.), Academic Press, New York, pp. 1301–1307.
54. Liotta, L., and DeLisi, C., 1977, Method for quantitating tumor cell removal and tumor cell invasive capacity in experimental metastases, *Cancer Res.* **37**:4003.
55. Bomford, R., and Olivotto, M., 1974, The mechanism of inhibition by *Corynebacterium parvum* of the growth of lung nodules from intravenously injected tumor cells, *Int. J. Cancer* **14**:226.
56. Breard, J., Reinherz, E. L., O'Brien, C., and Schlossman, S. F., 1981, Delineation of an effector population responsible for natural killing and antibody-dependent cellular cytotoxicity in man, *Clin. Immunol. Immunopathol.* **18**:145.
57. Brodt, P., Feldman, M., and Segal, S., 1983, Differences in the metastatic potential of two sublines of tumor 3LL selected for resistance to natural NK-like effector cells, *Cancer Immunol. Immunother.* **16**:109.
58. Brooks, C., Flannery, G., Willmott, N., Austin, E., Kenwrick, S., and Baldwin, R., 1981, Tumour cells in metastatic deposits with altered sensitivity to natural killer cells, *Int. J. Cancer* **28**:191.
59. Gorelik, E., Bere, E., and Herberman, R., 1987, The mechanism for the antimetastatic effect of the anticoagulant drugs: Dependence on NK cell activity, in: *Hemostasis and Cancer* (L. Muszbek, eds.), CRC Press, Boca Raton, FL, pp. 37–46.
60. Lanza, E., and Djeu, J., 1982, Age-independent natural killer cell activity in murine peripheral blood, in: *NK Cells and Other Natural Effector Cells* (R. B. Herberman, ed.), Academic Press, New York, pp. 335–340.
61. Puccetti, P., Santoni, A., Riccardi, C., and Herberman, R., 1980, Cytotoxic effector cells with the characteristics of natural killer cells in the lungs of mice, *Int. J. Cancer* **25**:153.
62. Stein-Streilein, J., Bennett, M., Mann, D., and Kumar, V., 1983, Natural killer cells in mouse lungs: Surface phenotype, target preference, and response to local influenza virus infection, *J. Immunol.* **131**:2699.
63. Tagliabue, A., Befus, A., Clark, D., and Bienenstock, J., 1981, Characteristics of natural killer cells in the murine intestinal epithelium and lamina propria, *J. Exp. Med.* **155**:1785.
64. Wiltrout, R., Mathieson, B., Talmadge, J., Reynolds, C., Zhang, S., Herberman, R. B., and Ortaldo, J., 1984, Augmentation of organ-associated NK activity by biological re-

sponse modifiers: Isolation and characterization of large granular lymphocytes from the liver, *J. Exp. Med.* **160**:1431.

65. Wiltrout, R., Herberman, R., Zhang, S., Chirigos, M., Ortaldo, J., Green, K., and Talmadge, J., 1985, Role of organ-associated NK cells in decreased formation of experimental metastasis in lung and liver, *J. Immunol.* **134**:4267.

66. Sadler, T., and Castro, J., 1976, The effects of *Corynebacterium parvum* and surgery on the Lewis lung carcinoma and its metastases, *Br. J. Surg.* **63**:292.

67. Gorelik, E., Fogel, M., Feldman, M., and Segal, S., 1979, Differences in resistance of metastatic tumor cells and cells from local tumor growth to cytotoxicity of natural killer cells, *JNCI* **63**:1397.

68. Gorelik, E., Feldman, M., and Segal, S., 1982, Selection of 3LL tumor subline resistant to natural effector cells concomitantly selected for increased metastatic potency, *Cancer Immunol. Immunother.* **12**:105.

69. Teale, D., Rees, R., Clark, A., Walker, J., and Potter, C., 1983, Reduced susceptibility to natural killer cell lysis of hamster tumours exhibiting high levels of spontaneous metastasis, *Cancer Lett.* **19**:221.

70. Hanna, N., and Fidler, I. J., 1981, Relationship between metastatic potential and resistant natural killer cell–mediated cytotoxicity in three murine tumor system, *JNCI* **66**:1183.

71. Segal, S., Kingsmore, S., Gorelik, E., and Feldman, M., 1982, Control by NK cells of the generation of lung metastases by the Lewis lung carcinoma, in: *Current Concepts in Human Immunology and Cancer Immunomodulation* (B. Serrou *et al.*, eds.), Elsevier/North-Holland, Amsterdam, pp. 227–234.

72. Poupon, M., Judde, J., Pot-Deprun, J., Sweeney, F., and Lespinats, G., 1983, Variable susceptibility to NK activity of cloned cell lines derived from a primary rat rhabdomyosarcoma: Relationship to metastatic potential, *Br. J. Cancer* **48**:75.

73. Gerson, J. M., Varesio, L., and Herberman, R. B., 1981, Systemic and *in situ* natural killer and suppressor cell activities in mice bearing progressively growing murine sarcoma virus-induced tumors, *Int. J. Cancer* **27**:243.

74. Gorelik, E., 1983, Resistance of tumor-bearing mice to a second tumor challenge, *Cancer Res.* **43**:138.

75. Gorelik, E., 1985, H-2 antigen expression, immunogenicity and metastatic properties of BL6 melanoma cells treated with MNNG, in: *Treatment of Metastasis: Problems and Prospects* (K. Hellman, ed.), Taylor and Francis, London, pp. 355–359.

76. Jones, D., Wallace, A., and Fraser, E., 1971, Sequence of events in experimental metastases of Walker-256 tumor: Light, immunofluorescent and electron microscopic observations, JNCI **46**:493.

77. Sindelar, W., Tralka, T., and Ketcham, A., 1975, Electron microscopic observation on formation of pulmonary metastases, *J. Surg. res.* **18**:137.

78. Wood, J., 1958, Pathogenesis of metastasis formation observed *in vivo* in the rabbit ear chamber, *Arch. Pathol.* **66**:550.

79. Gasic, G., 1984, Role of plasma, platelets, and endothelial cells in tumor metastasis cancer, *Cancer Metastasis Rev.* **3**:99.

80. Karpatkin, S., and Pearlstein, E., 1981, Role of platelets in tumor-cell metastases, *Ann. Intern. Med.* **95**:636.

81. Rickles, R., and Edwards, R., 1983, Activation of blood coagulation in cancer: Trousseau's syndrome revisited, *Blood* **62**:14.

82. Donati, M., and Poggi, A., 1980, Malignancy and haemostasis, *Br. J. Haematol.* **44**:173.

83. Gasic, G., Gasic, T., Galanti, N., Johnson, T., and Murphy, S., 1973, Platelet-tumour-cell interactions in mice: The role of platelets in the spread of malignant disease, *Int. J. Cancer* **11**:704.

84. Hilgard, P., 1982, Blood platelets and tumor dissemination, *Prog. Clin. Biol.* **89**:143.

85. Honn, K., Cicone, B., and Skoff, A., 1981, Prostacyclin: A potent antimetastatic agent, *Science* **212**:1270.

86. Gorelik, E., 1987, Augmentation of the antimetastatic effects of anticoagulants by immunostimulations in mice, *Cancer Res.* **47**:809.

87. Gunji, Y., Lewis, J., Herberman, R. B., and Gorelik, E., 1987, Mechanisms of the antimetastatic action of anticoagulants. II. Fibrin coagulation protects tumor cells from destruction by NK or LAK cells. *Proc. Am. Assoc. Cancer Res.* **28:**72.
88. Kärre, K., Ljunggren, M., Piontek, G., Kiessling, R., Klein, G., Taniguchi, K., and Grönberg, A., 1984, Activation of cell-mediated immunity by absence or deleted expression of normal cellular gene products, i.e, by "no self" rather than "non self," *Immunobiology* **163:**43.
89. Herberman, R. B. (ed.), 1982, *NK Cells and Other Natural Effector Cells*, Academic Press, New York.
90. Snell, G., and Stevens, L., 1961, Histocompatibility genes of mice. III. H-1 and H-4, two histocompatibility loci in the first linkage group, *Immunology* **4:**366.
91. Cudkowicz, G., and Bennett, M., 1971, Peculiar immunobiology of bone marrow allografts. II. Rejection of parental grafts by resistant F1 hybrid mice, *J. Exp. Med.* **134:**1513.
92. Carlson, G., and Wegmann, T., 1977, Rapid *in vivo* destruction of semisyngeneic and allogeneic cells by nonimmunized mice as a consequence of nonidentity at H-2, *J. Immunol.* **118:**2130.
93. Carlson, G., Taylor, B., Marshall, S., and Greenberg, A., 1984, A genetic analysis of natural resistance to nonsyngeneic cells: The role of H-2, *Immunogenetics* **20:**287.
94. Klein, G. O., Klein, G., Kiessling, R., and Kärre, K., 1978, H-2 associated control of natural cytotoxicity and hybrid resistance against RBL-5, *Immunogenetics* **6:**561.
95. Kärre, K., Klein, G. O., Kiessling, R., Argov, S., and Klein, G., 1982, The beige model in studies of natural resistance against semisyngeneic, syngeneic and autochtonous tumors in NK cells and other natural effector cells, in: *NK Cells and Other Natural Effector Cells* (R. Herberman, eds.), Academic Press, New York, p. 1369.
96. Hansson, M., Kärre, K., Kiessling, R., Roder, J., Andersson, B., and Hayry, P., 1979, Natural NK cell targets in the mouse thymus: Characteristics of the sensitivity cell population, *J. Immunol.* **123:**765.
97. Stern, P., Gidlund, M., Orn, A., and Wigzell, H., 1980, Natural killer cells mediate lysis of embryonal carcinoma cells lacking MHC, *Nature* **285:**341.
98. Ljunggren, H., and Kärre, K., 1985, Host resistance directed selectively against H-2 loss lymphoma variants. Analysis of the mechanisms, *J. Exp. Med.* **162:**1745.
99. Becker, S., Kiessling, R., Lee, N., and Klein, G., 1978, Modulation of sensitivity to natural killer cell lysis: *In vitro* explanation of a mouse lymphoma, *JNCI* **1:**1495.
100. Piontek, G., Taniguchi, K., Ljunggren, H., Gronberg, A., Kiessling, R., Klein, G., and Kärre, K., 1985, YAC-1 MHC class I variants reveal an association between decreased NK sensitivity and increased H-2 expression following interferon treatment or *in vivo* passage, *J. Immunol.* **135:**4281.
101. Taniguchi, K., Kärre, K., and Klein, G., 1985, Lung colonization and metastasis by disseminated B16 melanoma cells: H-2 associated control at the level of the host and the tumor cell, *Int. J. Cancer* **36:**503.
102. Harel Bellan, A., Quillet, A., Marchiol, C., DeMars, R., Tursz, T., and Fradelizi, D., 1986, Natural killer susceptibility of human cells may be regulated by genes in the HLA region on chromosome 6, *Proc. Natl. Acad. Sci. USA* **83:**5688.
103. DeBaetselier, P., Katzav, S., Gorelik, E., Feldman, M., and Segal, S., 1980, Differential expression of H-2 gene products in tumour cells is associated with their metastogenic properties, *Nature* **288:**179.
104. Katzav, S., Segal, S., and Feldman, M., 1985, Metastatic capacity of cloned T-10 sarcoma cells that differ in H-2 expression: Inverse relationship to their immunogenic potency, *NJCI* **74:**307.
105. Sowada, Y., Fohring, B., Shenk, T., and Raska, K., 1985, Tumorigenicity of adenovirus-transformed cells: Region E1A of adenovirus 12 confers resistance to natural killer cells, *Virology* **147:**413.
106. Cook, J., Walker, T., Lewis, A., Fuley, H., Graham, R., and Pidler, S., 1986, Expression of the adenovirus E1A oncogene during cell transformation is sufficient to induce susceptibility to lysis by host inflammatory cells, *Proc. Natl. Acad. Sci. USA* **83:**6965.

107. Gorelik, E., Peppoloni, S., Overton, R., and Herberman, R., 1985, Increase in H-2 antigen expression and immunogenicity of BL6 melanoma cells treated with N-methyl-N-nitro-nitrosoguanidine, *Cancer Res.* **45**:5341.
108. Chratchko, Y., Van Obbengham, E., Kiger, N., and Feldman, M., 1983, Immunoprecipitation of insulin receptor by antibodies against class I antigens of the murine H-2 major histocompatibility complex, *FEBS Lett.* **163**:207.
109. Schreiber, A., Schlesenger, A., and Edidin, M., 1984, Interaction between major histocompatibility complex antigen and epidermal growth factor receptor on human cells, *J. Mol. Biol.* **98**:725.
110. Dent, P. B., Fish, L. A., White, J. F., and Good, R. A., 1966, Chediak-Higashi syndrome. Observations on the nature of the associated malignancy, *Lab Invest.* **15**:1634.
111. Roder, J. C., Haliotis, T., Klein, M., Korec, S., Jett, J. R., Ortaldo, J., Herberman, R. B., Katz, P., and Fauci, A. S., 1980, A new immunodeficiency disorder in humans involving NK cells, *Nature* **284**:553.
112. Roder, J. C., Laing, L., Haliotis, T., and Kozbor, D., 1982, Genetic control of human NK function, in: *Natural Killer Cells. Human Cancer Immunology*, Volume 4 (B. Serrou, C. Rosenfeld, and R. B. Herberman, eds.) Elsevier/North-Holland, Amsterdam, pp. 169–186.
113. Roder, J., and Duwe, A., 1979, The beige mutation in the mouse selectively impairs natural killer cell function, *Nature* **278**:451.
114. Roder, J. C., Lohmann-Matthes, M.-L., Domzig, W., and Wigzell, H., 1979, The beige mutation in the mouse. II. Selectivity of the natural killer (NK) cell defect, *J. Immunol* **123**:2174.
115. Brunda, M. J., Holden, H. T., and Herberman, R. B., 1980, Augmentation of natural killer cell activity of beige mice by interferon and interferon inducers, in: *Natural Cell-Mediated Immunity Against Tumors* (R. B. Herberman, ed.), Academic Press, New York.
116. Loutit, J. F., Townsend, K. M. S., and Knowles, J. F., 1980, Tumour surveillance in beige mice, *Nature* **285**:66.
117. Haliotis, T., Roder, J., and Dexter, D., 1982, Evidence for *in vivo* NK reactivity against primary tumors, in: *NK Cells and Other Natural Effector Cells* (R. B. Herberman, ed.), Academic Press, New York, pp. 1399–1404.
118. Purtilo, D. T., DeFlorio, D., Hutt, M. Bhawan, J., Yang, J. P. S., Otto, R., and Edward, W., 1977, Variable phenotypic expression of an X-linked recessive lymphoproliferative syndrome, *N. Engl. J. Med.* **297**:1077.
119. Sullivan, J. L., Byron, K. S., Brewster, F. E., and Purtilo, D. T., 1980. Deficient natural killer cell activity in X-linked lymphoproliferative syndrome, *Science* **210**:543.
120. Seeley, J. K., Bechtold, T., Purtilo, D. T., and Lindsten, T., 1982, Deficiency in X-linked lymphoproliferative syndrome, In: *NK Cells and Other Natural Effector Cells* (R. B. Herberman, ed.), Academic Press, New York, pp. 1211–1218.
121. Masucci, M. G., Bejarano, M. T., Vanky, F., and Klein, E., 1982, Cytotoxic and cytostatic activity of human large granular lymphocytes against allogeneic tumor biopsy cells and autologous EBV infected B lymphocytes, in: *NK Cells and Other Natural Effector Cells* (R. B. Herberman, ed.), Academic Press, New York, pp. 1047–1054.
122. Penn, I., and Starzl, T. E., 1972, A summary of the status of *do novo* cancer in transplant recipients, *Transplant. Proc.* **47**:719.
123. Lipinski, M., Tursz, T., Kreis, H., Finale, Y., and Amiel, J. L., 1980, Dissociation of natural killer cell activity and antibody-dependent cell-mediated cytotoxicity in kidney allograft recipients receiving high-dose immunosuppressive therapy, *Transplantation* **29**:214.
124. Yoda, Y., Abe, T., Mitamura, K., Saito, K., Kawada, K., Onozawa, Y., Adachi, Y., and Nomura, T., 1982, Deficient natural killer (NK) cell activity in paroxysmal nocturnal haemoglobinuria (PNH), *Br. J. Haematol.* **52**:559.
125. Hersey, P., Edwards, A., Honeyman, M., and McCarthy, W. H., 1979, Low natural killer cell activity in familial melanoma patients and their relatives, *Br. J. Cancer* **40**:113.
126. Strayer, D. R., Carter, W. A., Mayberry, S. D., Pequignot, E., and Brodsky, I., 1984,

Low natural cytotoxicity of peripheral blood mononuclear cells in individuals with high familial incidences of cancer, *Cancer Res.* **44:**370.

127. Stutman, O., 1975, Immunodepression and malignancy, in: *Advances in Cancer Research*, Volume 22 (G. Klein, S. Weinhouse, and B. Haddam, eds.), Academic Press, New York, pp. 261–422.

128. Gorelik, E., and Herberman, R. B., 1981, Inhibition of the activity of mouse NK cells by urethane, *JNCI* **66:**543.

129. Gorelik, E., and Herberman, R. B., 1981, Susceptibility of various strains of mice to urethane-induced lung tumors and depressed natural killer cell activity, *JNCI* **67:**1317.

130. Gorelik, E., and Herberman, R. B., 1981, Role of natural-cell-mediated immunity in urethane-induced lung carcinogenesis, in: *NK Cells and Other Natural Effector Cells* (R. B. Herberman, ed.), Academic Press, New York, pp. 1415–1421.

131. Ehrlich, R., Efrati, M., and Witz, I. P., 1980, Cytotoxicity and cytostasis medicated by splenocytes of mice subjected to chemical carcinogens and of mice bearing primary tumors, in: *Natural Cell-Mediated Immunity Against Tumors* (R. B. Herberman, ed.), Academic Press, New York, pp. 997–1010.

132. Hochman, P. S., Cudkowicz, G., and Dausset, J., 1978, Decline of natural killer cell activity in sublethally irradiated mice, *JNCI* **61:**265.

133. Gorelik, E., and Herberman, R. B., 1982, Depression of natural antitumor resistance of C57BL/6 mice by leukemogenic doses of radiation and restoration of resistance by transfer of bone marrow of spleen cells from normal, but not beige, syngeneic mice, *JNCI* **69:**89.

134. Parkinson, D. R., Brightman, R. P., and Waksal, S. D., 1981, Altered natural killer cell biology in C57BL/6 mice after leukemogenic split-dose irradiation, *J. Immunol.* **126:**1460.

135. Datta, S. K., Priest, E. L., and Trentin, J. J., 1982, Possible role of NK cells in radiation leukemogenesis: Adoptive repair of NK deficit of fractionally irradiated mice by marrow transfusion, in: *NK Cells and Other Natural Effector Cells* (R. B. Herberman, ed.), Academic Press, New York, pp. 1411–1414.

136. Kaplan, H. S., Brown, M. B., and Paull, J., 1966, Influence of bone marrow injections on involution and neoplasia of mouse thymus after systemic irradiation, *JNCI* **4:**303.

136a. Warner, J. F., and Dennert, G., 1982, *In vivo* function of a cloned cell line with NK activity: Effects on bone marrow transplants, tumor development and metastasis, *Nature* **300:**3.

137. Trentin, J. J., Datta, S. K., Priest, E. L., Gallagher, M. T., and Nasrallah, A. G., 1982, Transfer of NK activity and lymphoma resistance to AKR mice by marrow from high NK, lymphoma-resistant C57 × AKR F₁ mice, in: *NK Cells and Other Natural Effector Cells* (R. B. Herberman, ed.), Academic Press, New York, pp. 1405–1409.

138. Kalland, T., 1982, Effect of depression of NK activity by neonatal exposure to diethylstilbestrol in susceptibility to transplanted and primary carcinogen-induced tumors, in: *NK Cells and Other Natural Effector Cells* (R. B. Herberman, ed.), Academic Press, New York, pp. 1437–1444.

139. Stutman, O., 1983, The immunological surveillance hypothesis, in: *Basic and Clinical Tumor Immunology* (R. B. Herberman, ed.), Nijhoff, Boston, pp. 155–194.

140. Heston, W., and Dunn, T., 1951, Tumor development is susceptible strain A and resistant lung transplants in LAF1 host, *JNCI* **11:**1057.

141. Maluish, A., Ortaldo, J., Conlon, J., Sherwin, S., Leavitt, R., Strong, D., Weirnik, P., Oldham, R., and Herberman, R. B., 1983, Depression of natural killer cytotoxicity after *in vivo* administration of recombinant leukocyte interferon, *J. Immunol.* **131:**503.

142. Santoni, A., Riccardi, C., Barlozzari, T., and Herberman, R., 1980, Suppression of activity of mouse natural killer (NK) cells by activated macrophages from mice treated with pyran copolymer, *Int. J. Cancer* **26:**837.

143. Savary, C., and Lotzova, E., 1978, Suppression of natural killer cell cytotoxicity by splenocytes from *Corynebacterium parvum*–injected bone marrow tolerant and infant mice, *J. Immunol.* **120:**239.

144. Gorelik, E., Wiltrout, R., Brunda, M., Holden, H., and Herberman, R., 1982, Aug-

mentation of metastasis formation by thioglycollate-elicited macrophages, *Int. J. Cancer* **29**:575.

145. Gorelik, E., Wiltrout, R., Copeland, D., and Herberman, R., 1985, Modulation of formation of tumor metastases by peritoneal macrophages elicited by various agents, *Cancer Immunol. Immunother.* **19**:35.

146. Gorelik, E., Wiltrout, R., Brunda, M., Bere, W., and Herberman, R., 1985, Influence of adoptively transferred thioglycollate-elicited peritoneal macrophages on metastasis formation in mice with depressed or stimulated NK activity, *Clin. Exp. Metastasis* **3**:111.

147. Zacharski, L., 1981, Anticoagulation in the treatment of cancer in man, in: *Malignancy and the Hemostatic System* (M. Donati, J., Davidson, and S. Garattini, eds.), Raven Press, New York, pp. 113–127.

148. Landazuri, M. O., Silva, A., Alvarez, J., and Herberman, R. B., 1979, Evidence that natural cytotoxicity and antibody dependent cellular cytotoxicity are mediated in humans by the same effector cell populations, *J. Immunol.* **123**:252.

149. Rosenberg, S. A., Lotze, M. T., Muul, L. M., Leitman, S., Chang, A. E., Ettinghausen, S. E., Matory, Y. L., Skibber, J. M., Shiloni, E., Vetto, J. J., Seipp, C. A., Simpson, C., and Reichert, C. M., 1985, Observations on the systemic administration of autologous lymphokine-activated killer cells and recombinant interleukin-2 to patients with metastatic cancer, *N. Engl. J. Med.* **313**:1485.

150. Clark, M., Hager, M., Gosnell, M., Stadtman, N., and Shapiro, D., 1985, Search for a cure, *Newsweek* Dec. 16, p. 60.

151. Bylinsky, G., 1985, Science scores a cancer breakthrough, *Fortune* Nov. 25, p. 16.

152. Grimm, E. A., Mazumder, A., Zhang, H. Z., and Rosenberg, S. A., 1982, Lymphokine activated killer cell phenomenon, lysis of natural killer resistant fresh solid tumor cells by interleukin 2–activated autologous human peripheral blood lymphocytes, *J. Exp. Med.* **155**:823.

153. Rosenstein, M., Yron, I., Kaufmann, T., and Rosenberg, S. A., 1984, Lymphokine activated killer cells: Lysis of fresh syngeneic NK-resistant murine tumor cells by lymphocytes cultured in interleukin 2, *Cancer Res,* **44**:1946.

154. Herberman, R. B., Balch, C., Bolhuis, R., Golub, S., Hiserodt, J. C., Lanier, L., Lotzova, E., Phillips, J., Riccardi, C., Ritz, J., Santoni, A., Schmidt, R., Uchida, A., and Vujanovic, N., 1987, Lymphokine-activated killer cell activity: Characteristics of effector cells and their progenitors in blood and spleen, *Immunol. today* **8**:178.

155. Rosenberg, S. A., and Terry, W., 1977, Passive immunotherapy of cancer in animals and man, *Adv. Cancer Res.* **25**:323.

156. Herberman, R. B., 1982, Natural killer cells, *Hosp. Pract.* **17**:99.

157. Kedar, E., and Weiss, D. W., 1983, the *in vitro* generation of effector lymphocytes and their employment in tumor immunotherapy, *Adv. Cancer Res.* **38**:171.

158. Kedar, E., Ikejiri, B. L., Gorelik, E., and Herberman, R. B., 1982, Natural cell-mediated cytotoxicity *in vitro* and inhibition of tumor growth *in vivo* by murine lymphoid cells cultured with T cell growth factor (TCGF), *Clin. Immunol. Immunother.* **13**:14.

159. Mule, J. J., Shu, S., and Rosenberg, S. A., 1985, The anti-tumor efficacy of lymphokine-activated killer cells and recombinant interleukin 2 *in vivo, J. Immunol.* **135**:646.

160. Ettinghausen, S. E., Lipford, E. N., Mule, J. J., and Rosenberg, S. A., 1985, Recombinant interleukin 2 stimulates *in vivo* proliferation of adoptively transferred lymphokine-activated (LAK) cells, *J. Immunol.* **135**:3623.

161. Salup, R. R., and Wiltrout, R. H., 1986, Adjuvant immunotherapy of established murine renal cancer by interleukin 2–stimulated cytotoxic lymphocytes, *Cancer Res.* **46**:3358.

162. Salup, R. R., and Wiltrout, R. H., 1986, Treatment of adenocarcinoma in the peritoneum of mice: Chemoimmunotherapy with IL-2-stimulated cytotoxic lymphocytes as a model for treatment of minimal residual disease, *Cancer Immunol. Immunother.* **22**:31.

163. Shu, S., Chou, T., and Rosenberg, S. A., 1987, Generation from tumor-bearing mice with *in vivo* therapeutic efficacy, *J. Immunol.* **139**:295.

164. Chang, A. E., Shu, S., Chou, T., Lafreniere, R., and Rosenberg, S. A., 1986, Differ-

ences in the effects of host suppression on the adoptive immunotherapy of subcutaneous and visceral tumors, *Cancer Res.* **46:**3426.

165. Cheever, M. A., Greenberg, P. D., and Fefer, A., 1981, Specific adoptive therapy of established leukemia with syngeneic lymphocytes sequentially immunized *in vivo* and *in vitro* and nonspecifically expanded by culture with interleukin 2, *J. Immunol.* **126:**1318.

166. Cheever, M. A., Greenberg, P. D., Fefer, A., and Gillis, S., 1982, Augmentation of the anti-tumor therapeutic efficacy of long-term cultured T lymphocytes by *in vivo* administration of purified interleukin 2, *J. Exp. Med.* **155:**968.

167. Eberlein, T. J., Rosenstein, M., and Rosenberg, A., 1982, Regression of a disseminated solid tumor by systemic transfer of lymphoid cells expanded in interleukin 2, *J. Exp. Med.* **156:**385.

168. Eberlein, T. J., Rosenstein, M., Spiess, P., Wesley, R., and Rosenberg, S. A., 1982, Adoptive chemoimmunotherapy of a syngeneic murine lymphoma with long-term lymphoid cell lines expanded in T cell growth factor, *Cancer Immunol. Immunother.* **13:**5.

169. Mule, J. J., Rosenstein, M., Shu, S., and Rosenberg, S. A., 1985, Eradication of a disseminated syngeneic mouse lymphoma by systemic adoptive transfer of immune lymphocytes and its dependence upon a host component(s), *Cancer Res.* **45:**526.

170. Vujanovic, N. L., Herberman R. B., Al Maghazarchi, A., and Hiserodt, J. C., 1988, Lymphokine-activated killer cells in rats: III. A simple method for the purification of large granular lymphocytes and their rapid expansion and conversion into lymphokine activated killer cells, *J. Exp. Med.* **167:**15.

2

Lymphokine-Activated Killer Cells

Biology and Therapeutic Efficacy

ALAN T. LEFOR, JAMES J. MULÉ, and
STEVEN A. ROSENBERG

1. Introduction

Adoptive immunotherapy is defined as the transfer to the tumor-bearing host of immune cells with antitumor reactivity.[1] However, the generation of sufficient quantities of cells with specific antitumor reactivity has been a major obstacle to developing clinically useful adoptive immunotherapy regimens for the treatment of cancer in humans.

We have developed an approach to the adoptive immunotherapy of cancer utilizing cells with broad antitumor reactivity generated by the incubation of lymphocytes in the lymphokine interleukin 2 (IL-2). We have termed these cells lymphokine-activated killer (LAK) cells,[2] which are primarily defined by their capacity to lyse a variety of fresh tumor cells, but not normal cells, in short-term ^{51}CR-release assays.[3,4]

In this review, we will summarize our results, in both murine and human systems, discussing the *in vitro* characterization of LAK cells, as well as their biological effects *in vivo*. The LAK cell phenomenon has been extensively characterized,[2,5–7] and our own recent studies have shown that the adoptive transfer of LAK cells plus IL-2 has therapeutic benefit in the treatment of metastatic cancer in both the mouse[8,9] and humans.[10]

The role of endogenous LAK cells in natural immune function in normal animals remains unclear. We have hypothesized that LAK cells

ALAN T. LEFOR, JAMES J. MULÉ, and STEVEN A. ROSENBERG • Surgery Branch, National Cancer Institute, National Institutes of Health, Bethesda, Maryland 20892.

are part of the immunosurveillance mechanism to destroy cells that have undergone malignant transformation.

2. Characterization and *in Vitro* Activity of Murine Lymphokine-Activated Killer Cells

The incubation of murine lymphocytes in the lymphokine IL-2 for 3–4 days results in the generation of LAK cells capable of lysing fresh, natural killer (NK)-resistant tumor cells in short-term ^{51}Cr-release assays. Prior to the availability of recombinant IL-2, the lymphokine was extracted from the supernatants of stimulated cell cultures.

The generation of the LAK effector cell begins after 1–2 days of *in vitro* incubation with IL-2, with development of maximal lytic activity at 3–6 days.[5] Although the incubation of normal murine splenocytes in culture medium alone for 5 days does not result in cells capable of lysing fresh tumor targets, the addition of IL-2 in appropriate concentrations yields LAK cells.

In addition to functional activity of LAK cells, the cell surface markers of the LAK cell precursor and effector have been studied by selective depletion of cell subsets with specific antibodies and complement or with fluorescence-activated cell sorting. Nearly normal levels of LAK activity can be generated after the incubation of Thy 1.2-negative splenocytes in IL-2.[6] In the same study, only small and inconsistent LAK activity was generated from Thy 1.2-positive splenocytes. Ia-positive and surface immunoglobulin-positive splenocytes also had no LAK precursor activity. However, treatment of splenocytes with anti-asialo GM_1 sera and complement before IL-2 activation significantly diminished subsequent LAK activity *in vitro*. Furthermore, it was also shown that the *in vivo* antitumor efficacy of LAK cells was eliminated by selective removal of asialo GM_1-bearing cells at the precursor level.[11] Phenotypic analysis of the LAK effector cell has shown it to be Thy 1.2-positive and Ia-negative.[6] Separation of the Ia-depleted cells into subpopulations bearing or not bearing the Fc receptor demonstrated that the majority of cytotoxic activity is in the FcR-positive subpopulation. Thus, whereas the LAK cell precursors have neither T nor B cell surface markers, the LAK effector cell attains the Thy 1.2 surface marker during *in vitro* incubation with IL-2.

Whereas the spectrum of lysis of NK cells is limited to mainly cultured cell lines, with little ability to lyse fresh tumor cells, LAK cells are broadly lytic and able to mediate the lysis of a variety of syngeneic, allogeneic, and xenogeneic NK-resistant fresh tumor targets. In studies using congenitally immunodeficient mice,[12,13] it was found that splenocytes from certain strains failed to generate LAK activity after incubation with IL-2 while possessing significant NK activity. Conversely, other strains generated LAK cells but had little NK activity, suggesting that LAK cells and NK cells represent distinct populations in the spleen. The effect of a sublethal dose of cyclophosphamide, given *in vivo*, on the generation of cytotoxic T lymphocytes (CTL), NK cells, and LAK cells was recently reported.[14] In this study, the temporal recovery of each group of effector cells was determined after cyclophosphamide administration, with each group having a different time of reappearance. From 9 to 21 days, there was demonstrable NK activity, but not until after day 21 was LAK activity present. From these data it was concluded that LAK and NK represent distinct cell populations. Furthermore, LAK and NK cells have been shown to have a differential sensitivity to irradiation.[15] From these studies, LAK and NK cells appear to represent separate populations of effector cells, although considerable controversy exists in this area.

The existence of at least two distinct LAK precursor cells was recently reported.[16] In this study the lysis of trinitrophenyl (TNP)- modified syngeneic lipopolysaccharide (LPS)-stimulated splenocyte blasts was mediated by LAK cells generated from IL-2-activated, Thy 1-positive precursors. Depletion of Thy 1-bearing precursor cells before IL-2 activation completely abrogated the lysis of TNP-modified syngeneic LPS blasts by LAK cells, with no inhibition of the lysis of fresh tumor target cells. Recent work in our laboratory[17] has confirmed these observations. Furthermore, we observed that allogeneic LPS- or Con A-stimulated splenocyte blasts are also lysed by LAK cells with Thy 1-positive precursors. In cold-target inhibition experiments, we determined that the lysis of splenocyte blasts from mice of a given MHC haplotype can be inhibited only by blasts of a similar haplotype, suggesting that there are separate populations of LAK cells with Thy 1-positive precursors, each responsible for the lysis of blasts from a given strain of mouse. This is in contrast to the lysis of tumor cells by LAK cells with Thy 1-negative precursors, where we have observed[18] that the lysis of tumor cells can be inhibited by other types of tumor cells and is not MHC-restricted, suggesting the participation of a common effector cell possibly recognizing a common cell surface determinant.

A recent report from our laboratory[19] has demonstrated that LAK cells can also mediate antibody-dependent cellular cytotoxicity (ADCC). By adding specific antisera directed against the H-2 haplotype of the target cell, the lysis of target cells was enhanced up to 100-fold. In addition, normal tissues, not usually lysable by LAK cells, were lysed in the presence of appropriate antibody. Studies of antibody-depleted precursor populations gave strong evidence that both direct and indirect tumor cell lysis is mediated by the same effector cell.

The *in vitro* lysis of a wide variety of tumor targets by LAK cells has been extensively studied.[5,7,20] We have recently looked at the lysis of single-cell suspensions made from a variety of whole fresh normal murine tissues by LAK cells.[18] Although there was no lysis of murine kidney, intestinal mucosa, or peripheral blood mononuclear cells, we did observe a low but consistent lysis of lung, fetus, bone marrow, and liver. In cold-target inhibition studies, there was no inhibition of the lysis of tumor by lung, kidney, or bone marrow. However, these normal tissues were lysable with the addition of specific anti-H-2 antibody by an ADCC mechanism. In further studying the low level of lysis observed in some tissues, we separated the whole-organ single-cell suspension by plastic adherence. Although representing only about 5% of total cells, the adherent cell population was sensitive to lysis by LAK cells, whereas there was no detectable lysis of the remaining nonadherent cells. Cytologic examination showed enrichment of macrophages in the adherent population, suggesting that this tissue cell type may be a subset sensitive to LAK-mediated lysis.

3. Adoptive Immunotherapy in Murine Tumor Models

The availability of large amounts of purified recombinant IL-2 has made the *in vivo* study of LAK cells in tumor immunotherapy feasible, since the production of LAK cells requires incubation of cells with large amounts of IL-2. We have studied a variety of murine tumor models in our laboratory, including both pulmonary and hepatic metastasis models. The tumors used include several 3-methylcholanthrene-induced sarcomas, the B16 melanoma, and the MCA-38 adenocarcinoma, all syngeneic to the C57BL/6 mouse.

Early studies were performed using the pulmonary metastasis model, induced by the injection of fresh tumor cells into the tail vein of the mouse. In a study of LAK cells in the adoptive immunotherapy of pulmonary metastases in mice,[21] we concluded that the combination of LAK cells plus IL-2 is necessary for the successful reduction of established (3-day) pulmonary metastases. Although the administration of LAK cells alone decreased the number of tumor nodules,[22] the addition of IL-2 greatly enhanced this effect. The injection of normal splenocytes, cultured without Il-2, had no effect on the number of tumor nodules. In this study, the combination of LAK cells and IL-2 was found effective against both immunogenic and nonimmunogenic murine sarcomas.

The dependence of therapy on the dose of IL-2 administered has been evaluated.[23] The treatment of tumor-bearing mice with IL-2 alone in doses from 1200 to 30,000 units every 8 h showed little antitumor effect. However, the addition of LAK cells on days 3 and 6 after tumor injection showed significant reduction in the number of tumor nodules, but only at doses > 6000 units. The requirement for high doses of IL-2 is thought to be due to the relatively short half-life of IL-2 *in vivo*. The serum half-life of intraperitoneally injected IL-2 in the mouse is about 2 min.

The efficacy of LAK and IL-2 therapy has also been shown in immunosuppressed mice, having received 500 cGy total body irradiation prior to tumor cell injection.[23] Results showed effective reduction in the number of pulmonary metastases in both irradiated and normal mice. However, irradiation of LAK cells prior to injection with 3000 cGy significantly reduced or eliminated their ability to mediate antitumor effects *in vivo*, but they maintained activity *in vitro* in short-term ^{51}Cr-release assays against fresh tumor targets[23] performed immediately after irradiation. This suggests that expansion of the LAK cells is required for *in vivo* effects, which is prevented by cell irradiation, but cells may maintain *in vitro* cytotoxicity, which does not require cellular division.

More recent experiments have demonstrated the efficacy of high-dose IL-2 alone in the therapy of both micro- and macropulmonary metastases.[24] The spleens of mice receiving high-dose IL-2 were found to contain lymphocytes capable of lysing fresh tumor targets in short-term ^{51}Cr-release assays indicating the generation of LAK cells *in vivo* by high-dose IL-2 alone. The administration of 100,000 units IP three times daily to mice bearing 10-day established pulmonary metastases had a 79.5% reduction in the number of metastases, whereas mice receiving 20,000–50,000 units had a 22% reduction. This effect was highly reproducible. Preirradiated mice, however, had no reduction in the number of pulmonary metastases, suggesting that the IL-2 is inducing a radiosensitive host component rather than directly mediating antitumor effects. Interestingly, macrometastases were more sensitive than micrometastases to treatment with IL-2 alone. It is hypothesized that this effect is due to the increased relative number of infiltrating lymphoid cells and that it possibly reflects the state of activation of the host immune system.

The treatment of nonimmunogenic tumors with LAK cells and IL-2 was reviewed by Papa and co-workers.[20] Most of the previous studies described above involved the weakly immunogenic MCA-105 and MCA-106 sarcomas and the B16 melanoma, all in the C57BL/6 strain of mouse. In this study, nonimmunogenic tumors including the MCA-101 sarcoma and the MCA-38 adenocarcinoma in the C57BL/6 mouse and the M-3 melanoma in the C3H mouse were studied. The poor immunogenicity of human tumors was part of the underlying motivation for this study. In this study, adoptive immunotherapy with LAK cells and IL-2 or with high-dose IL-2 alone was found to be effective against pulmonary metastases from the nonimmunogenic MCA-101 sarcoma as well as from the MCA-38 adenocarcinoma. These effects were also seen against the M-3 melanoma in the C3H mouse, indicating the lack of a strain specificity in the observed therapeutic effect. An attempt was also made to predict *in vivo* efficacy of therapy based on the results of *in vitro* testing in short-term ^{51}Cr-release assays. However, no correlation was seen.

The ability of therapy with LAK cells and IL-2 to prolong survival was demonstrated in mice bearing both 3-day and 10-day pulmonary MCA-105 sarcoma metastases.[25] In mice bearing 3-day metastases, the average survival was approximately 20 days without therapy, but it increased to

32–35 days (in 2 experiments) when combined treatment with LAK cells and IL-2 was given. At autopsy a few tumor nodules were found in mice with prolonged survival. The persistence of a few nodules despite the eradication of a majority of tumor suggested a differential sensitivity to therapy, perhaps due to antigenic heterogeneity within the tumor inoculum. In this study, both *in vitro* and *in vivo* testing of cells from posttherapy nodules was performed. These nodules were excised and were found to be equally susceptible to *in vitro* lysis by LAK cells as fresh tumor cells. Reinjection of cells from nodules in animals that persisted after a course of therapy into new animals demonstrated continued sensitivity of the tumor cells to the therapy. There was no evidence that residual nodules escaped because of *in vivo* resistance to therapy. It was hypothesized that the inability to eradicate all tumor foci may therefore be due to altered traffic of LAK cells *in vivo*, sequestration of metastases in locations not accessed by LAK cells, inhibition of administered IL-2, production of murine antibodies to the human IL-2, and suboptimal therapeutic regimens.

The effect of various cell subsets on the *in vivo* therapy of metastatic tumor with LAK cells and IL-2 was recently reported from our laboratory.[26] Therapy was found to be effective in adult T cell-deficient mice (thymectomized mice that underwent lethal total-body irradiation followed by reconstitution with T cell-depleted bone marrow). This showed that no requirement exists for additional T lymphocytes of host origin for successful therapy with adoptively transferred LAK cells. It was previously observed[6] that treatment of splenocytes with anti-Thy 1.2 and complement prior to incubation with IL-2 had no effect on their ability to lyse tumor cells *in vitro*. In this study, it was similarly shown that depletion of Thy 1.2-positive cells prior to incubation with IL-2 had no effect on the reduction of the number of pulmonary metastases. However, whereas treatment of LAK effector cells with anti-Thy 1.2 and complement completely abrogated their ability to lyse tumor cells *in vitro*, the administration of IL-2 to these cells showed no diminution in their ability to reduce the number of pulmonary metastases in mice. Further *in vitro* incubation of these cells depleted by anti-Thy 1.2 and complement with IL-2 restored cytolytic activity along with reappearance of the Thy 1 antigenic marker. These studies demonstrate that a critical factor in successful immunotherapy is the *in vivo* maturation of LAK cells in the presence of exogenous IL-2.

Although all of the above studies of adoptive immunotherapy in murine models have concentrated on pulmonary metastasis models, other sites of metastatic disease are clearly of clinical significance. A reproducible model for the selective generation of liver metastases in mice was developed in our laboratory.[27] Metastases are induced by the intrasplenic injection of a tumor cell suspension, just under the splenic capsule, followed by immediate splenectomy. At the end of the study period, metastases are enumerated by intravenous administration of India ink, followed by bleaching of the whole liver in Fekete's solution. The tumor nodules on the surface were found to represent more than 90% of total nodules in the liver.

In a study of the adoptive immunotherapy of hepatic metastases from the nonimmunogenic MCA-102 sarcoma, the weakly immunogenic MCA-105 sarcoma, and the MCA-38 adenocarcinoma, treatment results similar to those for pulmonary metastases were observed.[9] Specifically, while low doses (5000–25,000 units 3 times daily) of IL-2 alone had little effect on reducing the number of hepatic metastases, the addition of LAK cells gave significant reduction in mice with both tumors studied. High doses of IL-2 alone (100,000 units) gave reduction in the number of nodules, but this reduction was markedly enhanced with the addition of LAK cells. Similarly, survival was increased in animals with hepatic metastases that were treated with IL-2 and LAK cells. In the same study, intraportal administration of LAK cells was compared to intravenous injection and found to be more effective, suggesting that local administration of LAK cells may improve their efficacy. This study demonstrated the efficacy of IL-2 and LAK therapy in a murine model of hepatic metastases with a variety of tumor types. These results and patterns of response are similar to those observed in pulmonary metastasis models. This is of additional interest, because we were concerned that the success of therapy of pulmonary nodules may have been due to the fact that the first capillary bed traversed by LAK cells is the lung. This study demonstrated that LAK cells are effective in other metastatic sites.

All of these studies in murine models have demonstrated the efficacy of adoptive immunotherapy in the treatment of experimental metastases in mice. Although high-dose IL-2 alone can reduce the number of metastatic foci, the administration of LAK cells markedly enhances the therapeutic benefit. The next section reviews studies that were performed to delineate mechanism of action of this therapy and to explore some of the difficulties associated with administration of LAK cells and IL-2.

4. Mechanisms of Adoptive Immunotherapy and Toxicity in the Murine Model

Because of the difficulty in obtaining sufficient numbers of autologous lymphocytes in human trials of LAK cells, the efficacy of allogeneic LAK cells was studied in our laboratory[28] in the murine model. The *in vitro* lysis of fresh tumor targets is not a major histocompatibility complex-restricted phenomenon.[28] In this study Shiloni demonstrated that LAK cells generated from DBA/2, BALB/c, and C3H mice can successfully reduce the number of hepatic and pulmonary metastases from the MCA-102 tumor in C57BL/6 mice. However, it was noted that in all three strains, more cells were required to reduce the number of metastases than were needed when LAK cells from the C57BL/6 mouse were used. Direct intraportal injection of allogeneic LAK cells was more effective than IV injection in mediating the regression of hepatic metastases. Prior immunization of the recipient mouse to histocompatibility antigens on the donor allogeneic cells used in tumor therapy led to a complete elimination of the antitumor activity of the allogeneic LAK cells against both hepatic and pulmonary metastases. The decreased effectiveness of allogeneic compared to syngeneic LAK cells and the abrogation of efficacy in immunized hosts appears to limit the success of this approach in humans.

A recent study from our laboratory[29] has demonstrated the efficacy of tumor-specific monoclonal antibody in the in vivo treatment of established B16 melanoma hepatic metastases. The antibody employed is of the IgG2b isotype and was developed against the B16 melanoma by Takami.[29] Whereas the use of antibody alone resulted in a reduction in the number of metastases, a significant synergism was observed when IL-2 was concurrently administered. It is postulated that this effect may be due to the generation of LAK cells which then kill tumor cells by an ADCC response similar to that observed in vitro. The combination of specific antibody treatment with LAK cells may be advantageous.

One of the significant problems associated with the administration of LAK cells and IL-2 is the development of a vascular leak syndrome (VLS), observed in both mice and humans. The extent of the VLS was studied by Rosenstein in our laboratory[30] using a murine model based on the extravasation of radiolabeled bovine serum albumin. Increased vascular permeability was most notable in thymus, spleen, lungs, liver, and kidneys. There was a strong dependence of the amount of increased vascular permeability on the number of days of treatment and on the dose of IL-2 administered. IL-2 produced a significant increase in the water weight of the lungs in treated animals. Immunosuppression by pretreatment irradiation, administration of cyclophosphamide or cortisone acetate, or the use of nude mice reduced or eliminated the development of the VLS.

Having observed the decreased VLS in mice receiving cortisone acetate and a decrease in clinically observed toxicity in patients receiving concurrent steroids for coexisting neurologic complications of their tumors along with IL-2, a study was made of the effect of steroids on the antitumor activity of LAK cells and IL-2 in mice.[31] Studies were performed in animals receiving high-dose IL-2 alone and in animals receiving both IL-2 and LAK cells. Administration of high-dose IL-2 to normal animals results in eventual death. In normal, non-tumor-bearing animals receiving IL-2 three times daily, survival was significantly increased by the concurrent administration of cortisone acetate. However, when combined with treatment in tumor-bearing mice, the antitumor effect of IL-2 alone was abrogated, as was the effect of IL-2 and LAK cells, although to a lesser degree. The generation of LAK cells was not affected, although there were fewer LAK precursor cells. Since the doses used, the decreased yield of lymphocytopheresis, and the abrogation of antitumor effect would be difficult to use safely in humans, the use of corticosteroids to decrease toxicity has not been applied clinically.

The mechanism of the in vivo action of IL-2 has been extensively investigated. The proliferation and migration of lymphoid cells was examined using ^{125}I-labeled deoxyuridine (IUdR), a thymidine analog, which labels the DNA of dividing cells.[32] Tissues were analyzed in a gamma counter, and a proliferation index was calculated by comparing treated mice to control animals. Although the administration of LAK cells alone had little effect, the concomitant administration of IL-2 markedly increased proliferation in lungs and liver. By studying preirradiated mice with reduced endogenous lymphoid proliferation, it was demonstrated that IL-2 also induced proliferation of transferred LAK cells. This effect was also demonstrated histologically. In the same study, it was found that lymphocytes recovered from the lungs of irradiated mice after LAK cell therapy maintained their in vitro activity in short-term ^{51}Cr-release assays against fresh tumor targets. The concurrent administration of IL-2 yielded LAK cells with a 32-fold greater activity in vitro than control animals receiving saline.

The same lymphoid proliferation model using radiolabeled IUdR was also used to optimize dosage schedules for the administration of IL-2.[33] The greatest lymphoid proliferation was observed when IL-2 was given at 50,000 units three times daily. Despite the same total daily dose, three-times-daily dosing gave higher proliferation than once-daily schedules, suggesting that prolonged exposure to lower levels of IL-2 is more effective than brief high peak levels in maximizing lymphoid proliferation. This same schedule which yielded greatest lymphoid proliferation also had maximal therapeutic effect in the treatment of lung metastases. Two infusions of LAK cells gave higher levels of IUdR uptake and a greater reduction of lung metastases than did a single LAK cell infusion.

5. Characterization of Human Lymphokine-Activated Killer Cells

The generation of LAK cells from human peripheral blood lymphocytes is similar to the methods used for generating murine LAK cells from splenocytes. Lymphocytes are incubated with IL-2 in vitro for 3–6 days, at which time they are capable of lysing fresh tumor targets in short-term ^{51}Cr-release assays.[7] There has been considerable study of the phenotype of the precursor cell responsible for LAK generation from human lymphocytes and its relationship to NK cells.

The precursor has been identified as distinct from T lymphocytes and monocytes; it is nonadherent, is T3–, and does not form rosettes with sheep erythrocytes (E rosettes).[7] Itoh and co-workers[34] showed that a majority of the LAK activity is present in Leu 11+ precursors after incubation with IL-2. Further fractionation showed the highest activity in the Leu 7− Leu 11+ fraction. These cells also had the highest NK activity. Leu 7+ cells had a high level of NK activity but lacked both a proliferative response to IL-2 and LAK activity. Both LAK and NK activity were absent in Leu 4+ and Leu-3a cell fractions.

Recent studies by Roberts and co-workers[35] in our laboratory focused on isolation and study of the human LAK precursor cell. Precursors were enriched from normal human peripheral blood mononuclear cells by sequential depletion of irrelevant cells. After plastic adherence to remove monocytes, a two- to threefold enrichment of LAK activity was observed. After rosetting with sheep erythrocytes, a majority of the LAK activity was observed in the E− fraction, whereas a minimal amount of LAK activity was found in the E+ fraction, which represented 81–92% of the cells. Contaminant cells were removed by antibody depletion, leaving the null cell fraction with enhanced LAK activity. This fraction also mediated higher levels of NK activity than the PBL they were separated from. The level of LAK activity was proportional to the amount of NK activity before incubation with IL-2. Most of the cells in the null fraction were Leu 11+. Only 2.2% were Leu 3+, similar to that reported for large granular lymphocytes, consistent with the observation that these preparations contain few mature T cells. The cells retained a Leu 11+, Leu 4− phenotype. There was no apparent requirement for accessory cells, with high levels of cytotoxicity generated by enriched populations. Furthermore, after incubation with IL-2, there were few null cells that expressed the IL-2 receptor, the transferrin receptor, or HLA-DR, which are usually found on activated T cells. Studies of conjugation showed that although required for cytolysis, cell–cell conjugation is not sufficient for target cell lysis.

Further definition of the phenotype of the LAK cell precursor and effector was recently reported by Skibber et al.[36] Separating precursor cells showed that those cells with the surface markers Leu 11 and Leu 15 had enriched LAK activity. At the effector level, a majority of the LAK activity is present in the Leu 4− Leu 5+ fraction. These markers have also been described on populations with NK activity. After sorting from NK cell-enriched fractions, incubation of Leu 11+ Leu 15+ in IL-2 gives enhanced LAK activity, whereas incubation of cells without those surface markers shows no enhancement of LAK activity. Although Grimm et al.[2] reported that the LAK effector was Leu 4+, recent results suggest that this may be due to a lectin-dependent cytotoxicity[37] observed in PHA-containing media. It has also been observed that Leu 11+ Leu15+ and Leu 4− Leu 5+ cells disappear from the peripheral circulation of patients after a single IV bolus of IL-2.[38]

In a recent study, both the LAK precursor and effector in human peripheral blood lymphocytes were shown to be Leu 4− Leu 19+, the phenotype of typical NK cells.[39] IL-2 was a sufficient stimulus for the lysis of fresh tumor targets by NK cells. There was greater proliferation of Leu 19+ T cells in response to IL-2 than NK cells, thus diminishing the number of NK cells proportional to the time in culture. They conclude that the LAK effector in peripheral blood is clearly an NK cell, differentiating this from animal studies where splenocytes are the predominant source of T cells used in experiments.

Another study of the phenotype of human LAK precursors[40] demonstrated that most of the LAK activity is present in the Leu 4− Leu 15+ NK-H1+ population of large granular lymphocytes (LGL). A small amount of cytotoxicity was noted in the Leu 4+ fraction. These workers conclude that several cell types generate LAK activity. Most LAK activity on a per-cell basis is felt to be mediated by NK cells, T cell receptor-negative LGLs.

One of the critical features of LAK cells is their apparent ability to mediate the lysis of tumor cells, while sparing normal cells. While Grimm and Rosenberg[7] observed some lysis of placenta and fetal tissues, there was no lysis of autologous lung, liver, pancreas, bowel, or colon in early experiments on human tissues. More recently, however, Sondel et al.[41] have reported the lysis of autologous human lymphocytes by LAK cells. The observed cytotoxicity was at a very low level and may be due to the lysis of a particularly sensitive subpopulation of cells. Sondel et al. suggest that the lysis of normal tissues by LAK cells may contribute to observed clinical toxicities.

6. Clinical Trials of Lymphokine-Activated Killer Cells and IL-2 in the Therapy of Metastatic Cancer

The goal of these studies has been the development of an adoptive immunotherapy regimen capable of causing the regression of malignancies in humans. After the demonstration that IL-2 and LAK cells are active in murine models, phase I trials in humans began in 1984 at the National Cancer Institute. Initial trials were directed at assessing the toxicity and maximum tolerated dose of LAK cells and IL-2 administered independently to patients with cancer.

In the first trial, 12 patients with advanced malignancies that had failed established forms of therapy were treated with Jurkat-derived purified IL-2.[42] Toxicities noted included fever, chills, malaise, and reversible hepatic dysfunction and were dose-related. IL-2 was given by bolus injection and continuous infusion. The serum half-life was about 5–7 min, with sustained levels attainable by continuous infusion. A decreased ability to generate LAK cells from the PBLs of treated patients was noted rapidly after infusion and did not return to normal levels until 48 h after infusion. There was no antitumor effect noted in these patients.

A limiting factor in these studies was the supply of IL-2 prior to the availability of recombinant material. Subsequent studies therefore employed recombinant IL-2. Twenty patients with a variety of primary malignancies were treated with IL-2.[43] Toxicities included those previously observed as well as fluid retention and consequent weight increases. This was observed only in patients receiving high doses of IL-2, more than 10^5 U/kg total cumulative dose. A rapid disappearance of LAK precursors was again noted. Interferon-γ levels were increased in patients treated with IL-2. In addition, a two- to 16-fold expansion of total lymphoid cells in the peripheral blood was observed. In this study, no patient had demonstrable antitumor effects.

Prior to the availability of large quantities of recombinant IL-2, initial clinical trials of adoptive transfer of cells were done using phytohemagglutinin-activated killer (PAK) cells, which demonstrated in vitro lysis of fresh tumor targets similar to that observed with LAK cells.[44] From January 1981 through June 1983, a total of 21 patients were treated with PAK cells alone. Toxicities observed included fever and chills. Some of these patients were also treated with intravenous cyclophosphamide, and some with concurrent administration of activated macrophages. Studies with peripheral lymphocytes obtained by leukophoresis and activated in vitro with IL-2 were begun in early 1984 with the availability of recombinant IL-2. Cell infusions with as many as 9×10^{10} cells were given with no apparent antitumor effects.[45]

A major technical problem in the treatment of patients with LAK cells has been the ability to culture lymphocytes obtained through leukaphoreses in IL-2 and prepare them for administration to the patient. Based on murine models, it was estimated that up to 1×10^{11} LAK cells would be needed for infusion. Optimal techniques for the generation of LAK cells for use in clinical trials have been developed.[46] Lymphocytes are separated on Ficoll-Hypaque gradients and incubated in roller bottles with 1000–1500 units IL-2/ml. Continued development of new techniques is under way including the use of automated equipment to generate LAK cells almost entirely in a closed system.

Having established the safety of administration of IL-2 and LAK cells independently to patients, our group began testing the combined administration of IL-2 and LAK cells to patients with advanced malignancies who failed conventional therapy at the end of 1984. The results of treatment in the first 25 patients were reported in 1985,[47] and those of the first 108 patients were reported in 1987.[10] All patients had evaluable disease either by routine imaging studies or by physical examination. These patients had a variety of primary lesions including malignant melanoma, colorectal cancer, soft-tissue sarcoma, renal cell cancer, lung cancer, and esophageal cancer. Patients underwent daily leukapheresis, and harvested cells were cultured with IL-2 for 3–4 days. Each patient received three doses of LAK cells. At the beginning of the study, IL-2 was given starting at the time of the first dose of LAK cells. As the study proceeded, the treatment protocol was modified, with patients receiving IL-2 prior to the first leukapheresis to increase the yield of LAK cells harvested.

The treatment of a single patient with an unresectable hepatoma with autologous LAK cells generated from the patient's splenocytes was recently reported.[48] This patient was treated with LAK cells alone via hepatic artery catheter. Chills and fever were reported as the only toxicities. The patient had a transient decrease in serum alpha-fetoprotein levels and decreased ascites. Studies in our laboratory have not shown that LAK cells alone can mediate antitumor effects. The addition of IL-2 to the treatment of patients with hepatoma may improve the result observed in this single case report.

The results in seven patients treated with intraperitoneal IL-2 were reported separately.[49] Total dose of IL-2 ranged from 800 to 3800×10^3 U/kg. Side effects noted included fever, chills, nausea, vomiting, diarrhea, and weight gain presumed secondary to a capillary leak syndrome. Serum IL-2 levels were maintained at 10–35 U/ml for up to 8 h following the administration of IL-2. There was marked decrease in pulmonary and hepatic metastases in a single patient treated with intraperitoneal IL-2. There were significant increases in the number and *in vitro* lytic ability of intraperitoneal cells following IL-2 administration. However, the use of intraperitoneal IL-2 has been discontinued owing to difficulties associated with Tenckhoff catheter management, the management of the massive ascites resulting from IL-2 administration, and the difficulty of infusing therapeutic agents to patients with intraabdominal adhesions and tumor bulk.

We recently reported our cumulative experience with the treatment of 157 patients with advanced cancer using LAK cells and IL-2.[10] A total of 108 patients were treated with the combination of LAK cells and IL-2, and 49 received high-dose IL-2 alone. A majority of patients had melanoma, colorectal cancer, or renal cell cancer, since these tumor types were most responsive in earlier trials. The treatment regimen consisted of IL-2 during the first week, with a plan for 14 total doses, although the total number of doses was limited by toxicity in many cases, requiring cessation of therapy. Thus, each patient was treated to his own tolerance limit. A majority of patients were treated with three-times-daily infusions of IL-2 at 100,000 U/kg, although some patients received 10,000 or 30,000 U/kg. Those patients receiving LAK cells and IL-2 underwent 5 days of leukapheresis during the second week. Those patients receiving IL-2 alone had no treatment during this time. At the end of the second week, LAK cells produced by incubating peripheral blood lymphocytes in IL-2 for 3–4 days were reinfused. At the same time, IL-2 was restarted and continued to the patient's own tolerance limit as determined by evidence of toxicity. The median treatment course in all patients was 16 days.

Of the 106 patients evaluable after treatment with LAK and IL-2, there were eight complete responses, 15 partial responses, and 10 minor responses. A partial response was defined as >50% reduction in the sum of the products of the greatest perpendicular diameters of all lesions. The responses lasted a median of 10 months in patients with complete response and 6 months in patients with partial response. Regression of tumor was seen at multiple anatomic sites, including bone, lung, subcutaneous tissue, lymph nodes, and bone marrow. Of the 46 patients evaluable after treatment with IL-2 alone, there were one complete response, five partial responses, and one minor response.

There were a number of toxic side effects, attributable to the administration of IL-2. Hypotension resulting from decreased systemic vascular resistance and a capillary leak was treated with pressors and cautious use of colloid. Respiratory distress was noted in 34 of 180 treatment courses, requiring intubation in 16 patients. Four patients suffered myocardial infarctions, and there were four treatment-related deaths.

To simplify the treatment regimen, a new prospective study is under way to compare treatment with LAK cells and IL-2 with IL-2 alone, although patient accrual to date has been insufficient to draw conclusions about this study. This report shows continued promise in this new form of therapy for advanced malignancies.

Another group has recently reported their findings in the treatment of 48 patients with advanced malignancies using LAK cells and continuous IV infusion of IL-2.[50] They used continuous-infusion IL-2 in an attempt to decrease the toxicity associated with drug administration, particularly the retention of fluid and associated problems. Of 40 evaluable patients, they report 13 partial responses and two minor responses. A variety of tumor types were treated including melanoma, renal cell cancer, Hodgkin's and non-Hodgkin's lymphomas, lung cancer, ovarian cancer, and parotid cancer.

Their treatment regimen lasted 15 days, with IL-2 given on days 1 through 5. In their first group, low doses of IL-2 were given. There were no responders in this group, and they concluded that at least 3×10^6 U/ m^2 body surface area is necessary for adequate lymphocytosis. Leukapheresis proceeded from days 7 through 10, with infusion of *in vitro*–cultured cells from days 13 through 15 with concurrent administration of IL-2. There were various forms of toxicity including skin rash, nausea, diarrhea, and stomatitis. Only five of 40 patients gained >10% of their body weight. Most patients did not require treatment in the intensive care unit. The authors attribute the relatively mild levels of observed toxicities to the continuous dosing of IL-2 in contrast to the bolus administration reported from our group.[49]

The sensitivity of renal cell cancer and melanoma to this form of treatment was noted. In addition, good response in a patient with adenocarcinoma of the lung was reported. Early responses to treatment were seen, with no late or evolving responses more than 1 month after therapy ceased. They found correlation of response to treatment with pretreatment performance status and to the baseline lymphocyte count as well as to the amount of lymphocytosis generated by the priming dose of IL-2. This study further supports the current efforts toward developing successful therapy programs with adoptive immunotherapy.

The results of clinical trials with IL-2 and LAK cells in the treatment of advanced malignancies are encouraging, but this modality is not without toxicity as currently administered. We feel that these results also demonstrate great promise for the future. With continued laboratory efforts, the spectrum of disease amenable to adoptive immunotherapy and the rates of response may increase.

7. Future Prospects and Conclusions

The efficacy of adoptive immunotherapy in the treatment of advanced malignancies in humans has been established. However, it remains an experimental form of therapy. Research efforts are continuing, to improve the effectiveness of adoptive immunotherapy with LAK cells and IL-2. In the laboratory, we are investigating the mechanism of the vascular leak syndrome induced by IL-2 to reduce the toxicity associated with this form of therapy. We are also investigating the use of other lymphokines, alone and in conjunction with IL-2, in improving the efficacy of therapy.

With the recent availability of monoclonal antibodies against determinants found on some human tumors, studies are being performed to evaluate the combined use of IL-2 and antibodies. This work is based on the in vitro[19] and in vivo[29] studies demonstrating the ADCC phenomenon mediated by LAK cells and the ability of antibodies to synergize with IL-2 in the treatment of murine hepatic metastases.

Clinical trials are now in progress to determine whether adoptive immunotherapy with LAK cells and IL-2 can improve survival and disease free interval when used in an adjuvant setting. This is being applied in patients with stage II melanoma and hepatic metastases from colorectal cancer rendered disease free after surgery. Patients are randomized to receive either surgery alone or surgery plus adoptive immunotherapy.

We anticipate that with further research in this new field, further improvement in adoptive immunotherapy will occur. Elucidation of the mechanisms of action of IL-2 will provide information about the role of IL-2 in the natural immune function of humans and lead to a better understanding of its uses in the treatment of disease. Ultimately, LAK cells and IL-2 may be used with monoclonal antibodies, as well as other biologic response modifiers or chemotherapeutic agents to treat patients with cancer.

References

1. Rosenberg, S. A., and Terry, W., 1977, Passive immunotherapy of cancer in animals and man, *Adv. Cancer Res.* **25**:323–388.
2. Grimm, E. A., Mazumder, A., Zhang, H., and Rosenberg, S. A., 1982, The lymphokine activated killer cell phenomenon: Lysis of NK resistant fresh solid tumor cells by IL-2 activated autologous human peripheral blood lymphocytes, *J. Exp. Med.* **155**:1823–1841.
3. Yron, I., Wood, T., Spiess, P., and Rosenberg, S. A., 1980, *In vitro* growth of murine T cells. V. The isolation and growth of lymphoid cells infiltrating syngeneic solid tumors, *J. Immunol.* **125**:238–245.
4. Lotze, M. T., Grimm, E., Mazumder, A., Strausser, J., and Rosenberg, S. A. 1981, *In vitro* growth of cytotoxic human lymphocytes. IV. Lysis of fresh and cultured autologous tumor by lymphocytes cultured in T cell growth factor (TCGF), *Cancer Res.* **41**:4420–4425.
5. Rosenstein, M., Yron, I., Kaufmann, Y., and Rosenberg, S. A., 1984, Lymphokine activated killer cells: Lysis of fresh syngeneic NK resistant murine tumor cells by lymphocytes cultured in interleukin-2, *Cancer Res.* **44**:1946–1953.
6. Yang, J., Mulé, J., and Rosenberg, S. A., 1986, Murine lymphokine activated killer (LAK) cells: Phenotypic characterization of the precursor and effector cells, *J. Immunol.* **137**:715–722.
7. Grimm, E. A., and Rosenberg, S. A., 1983, The human lymphokine activated killer cell phenomenon, in: *Lymphokines*, Volume 9 (E. Pick and M. Candy, eds.), Academic Press, New York, pp. 279–309.
8. Mulé, J., Shu, S., and Rosenberg, S. A., 1985, The anti-tumor efficacy of lymphokine-activated killer cells and recombinant interleukin 2 *in vivo*, *J. Immunol.* **135**:646–652.
9. Lafreniere, R., and Rosenberg, S. A., 1985, Adoptive immunotherapy of murine hepatic metastases with lymphokine activated killer (LAK) cells and recombinant interleukin-2 (RIL 2) can mediate the regression of both immunogenic and nonimmunogenic sarcomas and an adenocarcinoma, *J. Immunol.* **135**:4273–4280.
10. Rosenberg, S. A., Lotze, M., Muul, L., Chang, A., Avis, F., Leitman, S., Linehan, W. M., Robertson, C., Lee, R., Rubin, J., Seipp, C., Simpson, C., and White, D., 1987, A progress report on the treatment of 157 patients with advanced cancer using lymphokine activated killer cells and interleukin-2 or high dose interleukin-2 alone, *N. Engl. J. Med.* **316**:889–897.
11. Yang, J., Mulé, J., and Rosenberg, S. A., Requirement for asialo GM1 bearing cells in the generation of murine lymphokine-activated killer cells with therapeutic efficacy, *Cancer Res.* (in press).
12. Andriole, G., Mulé, J., Hansen, C., Linehan, W. M., and Rosenberg, S. A., 1985, Evidence that lymphokine-activated killer cells and natural killer cells are distinct based on an analysis of congenitally immunodeficient mice, *J. Immunol.* **135**:2911–2913.
13. Merluzzi, V., Smith, M., and Last-Barney, K., 1986, Similarities and distinctions between murine natural killer cells and lymphokine-activated killer cells, *Cell. Immunol.* **100**:563–569.
14. Ballas, Z., 1986, Lymphokine-activated killer (LAK) cells. I. Differential recovery of a LAK natural killer cells, and cytotoxic T lymphocytes after a sublethal dose of cyclophosphamide, *J. Immunol.* **137**:2380–2384.
15. Merluzzi, V., 1985, Comparison of murine lymphokine-activated killer cells, natural killer cells, and cytotoxic T lymphocytes, *Cell. Immunol.* **95**:95–104.
16. Ballas, Z., Rasmussen, W., and Van Otegham, J., 1987, Lymphokine-activated killer cells. II. Delineation of distinct murine LAK-precursor subpopulations, *J. Immunol.* **138**:1647–1652.
17. Lefor, A., Eisenthal, A., and Rosenberg, S. A., 1988, Heterogeneity of lymphokine ac-

tivated killer cells induced by interleukin-2: Separate lymphoid subpopulations lyse tumor, allogeneic blasts, and modified syngeneic blasts. *J. Immunol.* **140**:4062–4069.

18. Lefor, A., and Rosenberg, S. A., 1988, The specificity of lymphokine activated killer (LAK) cells *in vitro:* Fresh normal murine tissues are resistant to LAK mediated lysis. (Submitted.)

19. Shiloni, E., Eisenthal, A., Sachs, D., and Rosenberg, S. A., 1987, Antibody-dependent cellular cytotoxicity mediated by murine lymphocytes activated in recombinant interleukin 2, *J. Immunol.* **138**:1992–1998.

20. Papa, M., Mulé, J., and Rosenberg, S. A., 1986, Antitumor efficacy of lymphokine-activated killer cells and recombinant interleukin 2 *in vivo:* Successful immunotherapy of established pulmonary metastases from weakly immunogenic and nonimmunogenic murine tumors of three distinct histological types, *Cancer Res.* **46**:4973–4978.

21. Mulé, J., Shu, S., Schwarz, S., and Rosenberg, S. A., 1984, Adoptive immunotherapy of established pulmonary metastases with LAK cells and recombinant interleukin-2, *Science* **225**:1487–1489.

22. Mazumder, A., and Rosenberg, S. A., 1984, Successful immunotherapy of natural killer-resistant established pulmonary melanoma metastases by the intravenous adoptive transfer of syngeneic lymphocytes activated in vitro by interleukin 2, *J. Exp. Med.* **159**:495–507.

23. Mulé, J., Shu, S., and Rosenberg, S. A., 1985, The antitumor efficacy of lymphokine-activated killer cells and recombinant interleukin 2 *in vivo, J. Immunol.* **135**:646–652.

24. Rosenberg, S. A., Mulé, J., Spiess, P., Reichert, C., and Schwarz, S., 1985, Regression of established pulmonary metastases and subcutaneous tumor mediated by the systemic administration of high-dose recombinant interleukin 2, *J. Exp. Med.* **161**:1169–1188.

25. Mulé, J., Ettinghausen, S., Spiess, P., Shu, S., and Rosenberg, S. A., 1986, Antitumor efficacy of lymphokine-activated killer cells and recombinant interleukin-2 *in vivo:* survival benefit and mechanisms of tumor escape in mice undergoing immunotherapy, *Cancer Res.* **46**:676–683.

26. Mulé, J., Yang, J., Shu, S., and Rosenberg, S. A., 1986, The anti-tumor efficacy of lymphokine-activated killer cells and recombinant interleukin 2 *in vivo:* Direct correlation between reduction of established metastases and cytolytic activity of lymphokine-activated killer cells, *J. Immunol.* **136**:3899–3909.

27. Lafreniere, R., and Rosenberg, S. A., 1986, A novel approach to the generation and identification of experimental hepatic metastases in a murine model, *JNCI* **76**:309–322.

28. Shiloni, E., Lafreniere, R., Mulé, J., Schwarz, S., and Rosenberg, S. A., 1986, Effect of immunotherapy with allogeneic lymphokine-activated killer cells and recombinant interleukin 2 on established pulmonary and hepatic metastases in mice, *Cancer Res.* **46**:5633–5640.

29. Eisenthal, A., Lafreniere, R., Lefor, A., and Rosenberg, S. A., 1987, The effect of anti B16 melanoma monoclonal antibody on established murine B16 melanoma liver metastases, *Cancer Res.* **47**:2771–2776.

30. Rosenstein, M., Ettinghausen, S., and Rosenberg, S. A., 1986, Extravasation of intravascular fluid mediated by the systemic administration of recombinant interleukin 2, *J. Immunol.* **137**:1735–1742.

31. Papa, M., Vetto, J., Ettinghausen, S., Mulé, J., and Rosenberg, S. A., 1986, Effect of corticosteroid on the antitumor activity of lymphokine-activated killer cells and interleukin 2 in mice, *Cancer Res.* **46**:5618–5623.

32. Ettinghausen, S., Lipford, E., Mulé, J., and Rosenberg, S. A., 1985, Recombinant interleukin 2 stimulates *in vivo* proliferation of adoptively transferred lymphokine-activated killer (LAK) cells, *J. Immunol.* **135**:3623–3635.

33. Ettinghausen, S., and Rosenberg, S. A., 1986, Immunotherapy of murine sarcomas using lymphokine activated killer cells: Optimization of the schedule and route of administration of recombinant interleukin-2, *Cancer Res.* **46**:2784–2792.

34. Itoh, K., Tilden, A., Kumagai, K., and Balch, C., 1985, Leu-11 + lymphocytes with natural killer (NK) activity are precursors of recombinant interleukin 2 (rIL 2) induced activated killer (AK) cells, *J. Immunol.* **134**:802–807.
35. Roberts, K., Lotze, M., and Rosenberg, S. A., 1987, Separation and functional studies of the human lymphokine activated killer cell. *Cancer Res.* **47**:4366–4371.
36. Skibber, J., Lotze, M., Muul, L., Uppenkamp, I., Ross, W., and Rosenberg, S. A., 1987, Human lymphokine activated killer cells: Isolation and characterization of the precursor and effector cell, *Nat. Immun. Cell Growth Reg.* **6**:291–305.
37. Burns, G., Triglia, T., and Werkmeister, J., 1984, *In vitro* generation of human activated killer cells: Separate precursors and modes of generation of NK-like cells and "anomolous" killer cells, *J. Immunol.* **133**:1656–1665.
38. Lotze, M., Custer, M., and Rosenberg, S. A., 1988, Interleukin 2 (IL-2) administration to human results in rapid emigration of a specific lymphocyte subset (CD2 +, 3 –, 11 +, 16 +) from the peripheral blood. (Submitted.)
39. Philips, J., and Lanier, L., 1986, Dissection of the lymphokine activated killer phenomenon: Relative contribution of peripheral blood natural killer cells and T lymphocytes to cytolysis, *J. Exp. Med.* **164**:814–825.
40. Ortaldo, J., Mason, A., and Overton, R., 1986, Lymphokine activated killer cells: Analysis of progenitors and effectors, *J. Exp. Med.* **164**:1193–1205.
41. Sondel, P., Hank, J., Kohler, P., Chen, B., Minkoff, D., and Molenda, J., 1986, Destruction of autologous human lymphocytes by interleukin 2 activated cytotoxic cells, *J. Immunol.* **137**:502–511.
42. Lotze, M., Frana, L., Sharrow, S., Robb, R., and Rosenberg, S. A., 1985, *In vivo* administration of purified human interleukin 2. I. Half life and immunologic effects of the Jurkat cell line derived interleukin 2, *J. Immunol.* **134**:157–166.
43. Lotze, M., Matory, Y., Ettinghausen, S., Rayner, A., Sharrow, S., Seipp, C., Custer, M., and Rosenberg, S. A., 1985, *In vivo* administration of purified human interleuken 2. II. Half life, immunologic effects, and expansion of peripheral lymphoid cells *in vivo* with recombinant IL 2, *J. Immunol.* **135**:2865–2875.
44. Mazumder, A., Eberlein, T., Grimm, E., Lotze, M., and Rosenberg, S. A., 1984, Phase I study of the adoptive immunotherapy of human cancer with lectin activated autologous mononuclear cells, *Cancer* **53**:896–905.
45. Rosenberg, S. A., 1984, Immunotherapy of cancer by systemic administration of lymphoid cells plus interleukin-2, *J. Biologic Response Modifiers* **3**:501–511.
46. Muul, L., Director, E., Hyatt, C., and Rosenberg, S. A., 1986, Large scale production of human lymphokine activated killer cells for use in adoptive immunotherapy, *J. Immunol. Methods* **88**:265–275.
47. Rosenberg, S. A., Lotze, M., Muul, L., Leitman, S., Chang, A., Ettinghausen, S., Matory, Y., Skibber, J., Shiloni, E., Vetto, J., Seipp, C., Simpson, C., and Reichert, C., 1985, Observations on the systemic administration of autologous lymphokine activated killer cells and recombinant interleukin-2 to patients with metastatic cancer, *N. Engl. J. Med.* **313**:1485–1492.
48. Okuno, K., Takagi, T., Nakamura, N., Nakamura, Y., Iwasa, Z., and Yasutomi, M., 1986, Treatment for unresectable hepatoma via selective hepatic arterial infusion of lymphokine activated killer cells generated from autologous spleen cells, *Cancer* **58**:1001–1006.
49. Lotze, M., Custer, M., and Rosenberg, S. A., 1986, Intraperitoneal administration of interleukin-2 in patients with cancer, *Arch. Surg.* **121**:1373–1379.
50. West, W., Tauer, K., Yannelli, J., Marshall, G., Orr, D., Thurman, G., and Oldham, R., 1987, Constant infusion recombinant interleukin 2 in adoptive immunotherapy of advanced cancer, *N. Engl. J. Med.* **316**:898–905.

Summary of Part I

From the data presented in this section, it is apparent that lymphocytes derived from normal rodents or humans can mediate a variety of antitumor effects without prior stimulation. Since these antitumor effects can be observed without exposure to tumor antigens, such activity falls largely within the designation of natural immunity. It is important to affirm at this point that natural immunity as discussed in this book is focused largely on lymphocyte-mediated effects. Much has been written regarding the antitumor roles of other broadly cytotoxic cell types, chiefly macrophages. Since a number of recent reviews have focused on the antitumor effects of macrophages,[1] we have chosen not to reiterate that literature. This in no way minimizes the potential role of macrophages or for that matter, other cell types such as polymorphonuclear leukocytes (PMNs). However, it does allow us to provide more in-depth analysis of the lymphocyte-mediated events as defined by natural killer (NK) and lymphokine-activated killer (LAK) cells.

There are several important considerations in a discussion of the role of the natural immune system (NIS) in antitumor responses. First, there is a significant amount of very good evidence for a role of the NIS in several aspects of tumor control, including immune surveillance, prevention of metastasis formation, and effectiveness in the regression of existing tumor. Second, considerable evidence suggests that appropriate stimulation of the NIS by biological response modifiers (BRMs) is required for optimal antitumor activity. Third, there remains, however, some controversy regarding the characterization of the natural immune effector cell(s) that mediate various facets of antitumor activity and the precursor cells from which they derive. In this summary, we will review the major antitumor contributions of the NIS, discuss several areas of controversy, and consider future directions for the study of the role of the NIS in antitumor responses.

The most convincing evidence for a role of the NIS in immune surveillance is with single tumor cells or small aggregates of tumor cells in the peripheral blood. There is substantial evidence that NK cells do function in an immune surveillance role by inhibiting the formation of blood-borne metastases. The evidence for this role of NK cells has been well summarized by Herberman and Gorelik (Chapter 1) and suggests that NK cells recognize and kill tumor cells during the blood-borne phase of the metastatic cascade. Although the level of NK cell infiltration into tumors is generally low,[2] there is evidence that NK cells, by virtue of their ability to infiltrate into tissues and proliferate during inflammation (see Chapter 8), can also mediate antimetastatic responses during the extravasation/postextravasation phase of the metastatic process.[3]

In contrast, it is also clear that tumor cells are heterogeneous, and this is also undoubtedly related to metastatic capability at the single cell level.[4] Certainly, some tumor cells appear to have an advantage in establishing a metastasis. This may relate to increased resistance of some cells within a population to NK-mediated lysis directly or, rather, to a selective advantage in evading NK cells indirectly such as through enhanced ability to form multicell emboli with other tumor cells or host clotting components. Also, certain tumors may evade NK cells by simply metastasizing via the lymphatics instead of the bloodstream. Since NK cells are absent or at least very rare in lymph, such a route of tumor metastasis would be largely independent of NK-mediated effects.

As noted by Herberman and Gorelik, there is less evidence that cells with NK activity can function in primary immune surveillance. The evidence to support this conclusion rests largely on the well-known ability of these cells to recognize and lyse a variety of neoplastic cells *in vitro* without prior sensitization. Additionally, cells with NK activity can accumulate in sites of inflammation and to a lesser degree in sites of tumor growth. There is also evidence of an increased tumor incidence in mice and humans with depressed NK activity. In addition, most carcinogens depress NK activity early in the tumor induction process. In our view, this evidence is entirely consistent with the hypothesis that NK cells do contribute significantly to immune surveillance against primary tumors. However, in situations where primary neoplasms do arise, there is no direct evidence for the related hypothesis that this occurs primarily because of a failure in NK-mediated surveillance. This hypothesis is of course difficult if not impossible to adequately test experimentally. In fact, the formation of a primary tumor is likely the result of the completion of a complex sequence of events, of which avoidance of NK cells is only one. Thus, while avoiding destruction by NK cells may be critical, it seems clear that the successful initiation of a primary tumor is a result of an appropriate combination of circumstances of which NK cell participation is only one.

To obtain more definitive evidence for the role NK cells may play in the control of spontaneous, carcinogen-induced, or viral-induced neoplasms, more specific techniques for the augmentation and suppression of NK activity need to be developed. To date, there is no completely satisfactory model for the complete, specific, and well-maintained suppression of NK activity in experimental animals. Several approaches can be envisioned. Koo and colleagues[5] have reported that the repeated injection of anti-NK1.1 serum beginning at birth eliminated mature cytotoxic NK cells. Subsequently, Seaman et al.[6] demonstrated that such NK1.1(−) mice had impaired antitumor responses in spite of retaining normal cellular and humoral immune responses. Such a model could be utilized to test more extensively the role of NK cells in the induction of primary tumors by comparing the incidence of spontaneous, carcinogen-induced, or viral-induced tumors in normal versus NK1.1(−) mice. Similarly, Luster et al.[7] have reported that the administration of ochratoxin A, a naturally occurring mycotoxin, can suppress baseline NK activity in mice. T lymphocyte-mediated cytotoxicity and macrophage-mediated tumoricidal activity were unimpaired by this treatment. Therefore, such a model might prove useful by reconstituting ochratoxin A−treated mice with NK cells or by augmenting the reduced levels of NK activity with interferon or IL-2. Such approaches would be greatly aided by the development of agents that selectively augment NK activity. In addition, recent studies demonstrating the importance of compartmentalized effector cells within the NIS suggest that much more needs to be known regarding the regulation of NK activity and NK-mediated immune regulation in the actual sites where primary tumors develop. Therefore, studies on the relationship of carcinogen-induced suppression of NK activity should focus on changes in NK cell number and function in the actual organs where tumors are induced. Alternatively, transgenic mice on high and low NK backgrounds that express various oncogenes might be useful models to study the role of NK cells in primary-tumor formation. It seems likely that given the complexity of the process of carcinogenesis, the degree of involvement of NK cells in that process will be variable, based on the carcinogen used, the type of tumors induced, and the organ site of tumor formation.

In the near future a number of issues need to be addressed regarding the mechanism(s) involved in the recruitment and infiltration of NK cells into tissue sites. For example, what signals trigger NK cells to leave the circulation and enter tissue? Recent evidence for chemotactic responsiveness by LGL suggests that NK cells, or their progenitors, may be attracted to inflammatory sites by various mediators. However, at present we do not know either the nature or the source of such mediators. Alternatively, is it possible that BRMs induce changes in vascular endothelium in different anatomical compartments that could cause NK cells to arrest and extravasate in those sites? In addition, much remains to be known about how various BRM effect the production, distribution, and half-life of NK cells. The study of such issues should prove useful in determining whether NK cells can be induced by BRMs to play a broader role against established tumors.

The review by Herberman and Gorelik also suggests several other rewarding areas for future investigation. First, studies that focus on NK cell trafficking *in vivo* and the factors that regulate or alter the distribution of these cells may make it possible to determine whether enhanced accumulation of these cells in sites of tumor growth is beneficial. Second, further studies on the role of BRMs in preventing the formation of metastases may also be important, particularly since the multiple administration of several BRMs causes a depression of NK activity in the blood.[8] This observation suggests that metastases might be better prevented by administration of BRMs that do not result in a hyporesponsiveness of NK activity. Third, more needs to be known regarding the role of the clotting system in allowing tumor cells to evade blood-borne NK cells. Fourth, by understanding the regulation of NK cell localization and the optimization of NK activity by BRM, protocols can be designed to determine whether NK cells can actually be of therapeutic value for established metastases.

The evidence summarized above suggests that the NIS, under stimulation induced solely by the development and progression of a tumor, would not be expected to mediate regression of large tumors. However, it suggests that an optimally augmented and maintained NIS response may be effective in mediating regression of established tumor. Support for this hypothesis has been provided by the evidence that adoptive transfer of nonsensitized lymphocytes, following culture in IL-2, is at least partially efficacious for a variety of rodent and human cancers (see Lefor *et al.*, Chapter 2). These LAK cells have been shown to mediate regression of metastases in lungs and liver, as well as primary subcutaneous or intradermal tumors. In fact, recent studies have shown that the administration of IL-2, a potent augmentor of NK activity, in the absence of adoptive immunotherapy, can mediate regression of existing tumor.

It has become apparent that at least some of the antitumor activity characterized as LAK would fall under the auspices of activated natural effector cells. This conclusion is based on the fact that LAK activity is non-MHC-restricted and can be generated from nonsensitized normal lymphocytes by simple exposure to IL-2. Thus, such cytotoxic activity is distinct from that routinely ascribed to antigen-specific T lymphocytes. There now is considerable agreement, at least in the human and rat, that non-T lymphocytes with the characteristics of classical LGL mediate most of the LAK activity when peripheral blood leukocytes are stimulated by IL-2 *in vitro*. However, most authors agree that some LAK activity can also be generated from CD3+ T lymphocytes, suggesting that under the appropriate conditions, T lymphocytes can be induced to mediate a broader-spectrum tumor cell lysis.

Results in the mouse are more controversial but not completely divergent. Several groups have argued for and against the existence of a unique non-T, non-NK precursor for LAK activity. As discussed by Lefor *et al.*, these data have been obtained from a combination of experiments that suggest that the two activities can be discriminated from each other under a variety of experimental conditions. Since such a conclusion is based solely on functional criteria, future experiments need to study the possible differentiation events that could be required for generation of LAK activity from NK cells.[9]

It thus remains possible that the actual progenitor cells that give rise to LAK activity in the mouse have the same characteristics of LGL but are themselves unique based on other differentiation markers or functional properties. However, most phenotypic characterizations and molecular studies performed to date have failed to detect such a difference. In this regard, the need clearly exists to identify and clone the receptor(s) by which NK cells recognize antigen and to determine whether this is the same recognition structure utilized during the expression of LAK activity. Further attempts to better define non-T, non-B lymphocyte subsets in rodents by the generation of new cell surface–specific monoclonal antibodies should also be helpful.

From a practical standpoint, additional strategies need to be devised whereby the antitumor effects of NK or LAK cells can be enhanced *in vivo*. There are three general approaches by which the antitumor effects of such cells can be enhanced. One approach would be to perform combination treatments in which chemotherapeutic drugs[10] or cytokines are administered in concert with adoptive immunotherapy (AIT). Such an approach might enhance the effectiveness of AIT by reducing the tumor burden, increasing the localization of AIT to the tumor site(s), enhancing the susceptibility of the tumor to lysis by effector cells, modulating potential immunosuppressive effects, or a variety of other possibilities. It can also be envisioned that chemotherapeutic drugs or noncytokine BRMs might be useful when used in conjunction with IL-2. In fact, preliminary results have shown that the investigational agent flavone-8-acetic acid (FAA) synergizes with IL-2 for the treatment of murine renal cancer.[12] This observation with FAA is particularly interesting since, in addition to its direct tumoricidal activity, FAA is also a potent augmenter of systemic NK activity.[12–14]

Further studies are in progress to determine whether there is a relationship between the immunomodulatory effects of FAA and its therapeutic synergy with IL-2. A second approach to increasing the effectiveness of LAK cells would be to enhance the therapeutic effects mediated by IL-2 alone. It is relatively difficult to generate potent LAK activity *in vivo* unless repeated administration of large amounts of IL-2 is used. It is not clear whether this requirement for massive amounts of IL-2 results solely from pharmacokinetic considerations or rather relates to a need to overcome some form of negative immunoregulation. Alternatively, since cytokines often mediate immunological effects as part of a cascade, it is quite plausible that the approach of using other cytokines in conjunction with IL-2 will provide enhanced therapeutic effects. Some data have already been presented that rIFNα[11] can synergize with IL-2 for treatment of murine cancers. Other candidates for such an approach include the CSFs, TNF α or β, and IL-1.

At present, the mechanism(s) by which natural effector cells mediate their antitumor activity are not clear. As discussed by Ortaldo (Chapter 13), the direct cytotoxicity mediated by NK or LAK cells is but one parameter by which they can mediate their antitumor effects. It seems likely that the ability of these cells to produce immunoregulatory cytokines is more than casually related to the antitumor roles of such cells as T lymphocytes and macrophages. Future studies on further characterizing and optimizing the relationships between the NIS and other leukocytes should prove beneficial in the development of new clinical protocols that more effectively exploit the antitumor potential of these cells.

References

1. Herberman, R. B., Wiltrout, R. H., and Gorelik, E. (eds.), 1987, *Immune Responses to Metastases*, CRC Press, Boca Raton, FL.
2. Mantovani, A., Bottazzi, B., Allavena, P., and Balotta, C., 1987, Tumor associated leukocytes in metastasizing tumors, in: *Immune Responses to Metastases* (R. B. Herberman, R. H. Wiltrout, and E. Gorelik, eds.), CRC Press, Boca Raton, FL, pp. 106–118.
3. Wiltrout, R. H., Herberman, R. B., Zhang, S.-R., Chirigos, M. A., Ortaldo, J. R., Green, L. M., Jr., and Talmadge, J. E., 1985, Role of organ-associated NK cells in decreased formation of experimental metastases in lung and liver, *J. Immunol.* **134:**4267–4275.
4. Nicolson, G. L., 1987, Tumor cell instability, diversification, and progression to the metastatic phenotype: From oncogene to oncofetal expression, *Cancer Res.* **47:**1473–1487.
5. Koo, G. C., Dumont, F., Tutt, M., Hackett, J. Jr., and Kumar, V., 1986, The NK1.1(−) mouse: A model to study the differentiation of murine NK cells, *J. Immunol.* **138:**3742–3747.
6. Seaman, W. E., Sleisenger, M., Eriksson, E., and Koo, G. C., 1987, Depletion of natural killer cells in mice by monoclonal antibody to NK1.1. Reduction in host defense against malignancy without loss of cellular or humoral immunity, *J. Immunol.* **138:**4539–4544.
7. Luster, M. I., Germolec, D. R., Burleson, G. R., Jameson, C. W., Ackerman, M. F., Lamm, K. R., and Hayes, H. T., 1987, Selective immunosuppression in mice of natural killer cell activity by ochratoxin A, *Cancer Res.* **47:**2259–2263.
8. Talmadge, J. E., Herberman, R. B., Chirigos, M. A., Schneider, M. A., Adams, J. S., Phillips, H., Thurman, G. B., Varesio, L., Long, C. A., Oldham, R. K., and Wiltrout, R. H., 1985, Augmentation or induction of a hyporesponsiveness of murine NK activity by various classes of immunomodulators including recombinant interferons and interleukin 2, *J. Immunol.* **135:**2483–2491.
9. Salup, R. R., Mathieson, B. J., and Wiltrout, R. H., 1987, Precursor phenotype of lymphokine-activated killer cells in the mouse, *J. Immunol.* **138:**3635–3639.
10. Salup, R. R., Back, T. J., and Wiltrout, R. H., 1987, Successful treatment of advanced murine renal cell cancer by bicompartmental adoptive chemoimmunotherapy, *J. Immunol.* **138:**641–647.
11. Brunda, M. J., Bellantoni, D., and Sulich, V., 1987, *In vivo* antitumor activity of combinations of interferon alpha and interleukin 2 in a murine model. Correlation of efficacy with the induction of cytotoxic cells resembling natural killer cells, *Int. J. Cancer* **40:**365–371.
12. Wiltrout, R. H., Boyd, M. R., Back, T. T., Salup, R. R., and Hornung, R. L., 1988,

Flavone-8-acetic acid augments systemic natural killer cell activity and synergizes with interleukin 2 for treatment of murine renal cancer, *J. Immunol.* **140**:3261–3265.

13. Ching, L., and Baguley, B. C., 1987, Induction of natural killer cell activity by the antitumor compound flavone acetic acid (NSC 347512), *Eur. J. Cancer Clin. Oncol.* **23**:1047–1050.

14. Wiltrout, R. H. and Hornung, R. L., 1988, Natural products as antitumor agents: Direct versus indirect mechanisms of activity of flavonoids, *J. Natl. Cancer Inst.* **80**:21–23.

PART II
INVOLVEMENT OF THE NIS IN THE CONTROL OF MICROBIAL INFECTIONS

3
Natural Effector Cells in Influenza Virus Infection

JOAN STEIN-STREILEIN

It is known that specific IgA antibody is protective in preventing influenza infection.[1,2] Once infected, however, cellular mechanisms of the host are important and contribute toward limiting the infection.[3] The nature of the cellular immune effectors that develop in mice in response to influenza virus infection has been well described. Cytotoxic T lymphocytes (CTL) play a crucial role in preventing virus spread and dissemination of virus in mice,[4,5] and virus-specific delayed-type hypersensitivity (DTH) T lymphocytes contribute to the destructive intrapulmonary lesion that characterizes influenzal pneumonia disease.[6,7]

The role of the CTL in influenza virus infections was not as easily documented in the hamster as it was in mice. Investigators studying cytotoxic mechanisms in virus-infected hamsters detected increases in natural killer, activated macrophages, and antibody-dependent cell-mediated cytotoxic activity but failed to identify specific CTL activity.[8–12] Prior to the development of monoclonal antibodies to hamster T lymphocyte subpopulations, circumstantial evidence was published that supported the hypothesis that CTLs were produced in lungs of hamsters in response to parainfluenza virus infection.[13,14]

More recent studies have shown that hamsters respond to an influenza infection with a specific CTL response in the lung.[15] Nonspecific natural killer (NK) cytotoxicity is augmented early (3 days) after inoculation of PR/8/34 into the lung but is undetected when a specific cytotoxic T cell response appears at 6 days after inoculation. Depletion studies using rabbit anti-asialo GM_1 (to remove NK cells) and newly developed mouse

JOAN STEIN-STREILEIN • Departments of Medicine and Microbiology/Immunology, University of Miami School of Medicine, Miami, Florida 33136.

monoclonals Wi20 and Wi38 (to remove all T cells and CTLs, respectively) showed that both NK and T lymphocytes were important for protection against influenza virus.[15]

1. Historical Perspective of NK Cells and Influenza Infection

As early as 1978, human NK cells were shown to be an effector cell responsible for cytotoxicity against influenza virus-infected target cells.[16,17] In particular, Santoli and co-workers[17] concluded that the effector cell in the influenza system could not be distinguished from human NK cells and that interferon produced by the lymphocytes was responsible for the induction and enhancement of the NK activity. Earlier reports had shown that interferon could enhance specialized cellular functions,[18,19] but these workers showed that the interferon contained in supernatants of mixed cultures was responsible for increasing the cytotoxic efficiency of natural killer cells.[16,17,20,21] The observations fit with emerging information that interferon could directly interfere with virus replication and indirectly interfere with virus proliferation by enhancing cell-mediated effector mechanisms.

Because of studies associated with human patients, Ennis and co-workers stated that "despite the apparent need for immunologically specific killer T cells to recover from influenza, early control of virus infection may be achieved by interferon induction which may protect host cells against infection and may increase the activity of natural killer cell."[22] However, using a murine model for infecting with influenza virus by the intranasal route, these same workers were unable to show a change in virus titer in the lung by depleting NK activity with in vivo inoculations of a rabbit anti-asialo GM_1 serum ($RAGM_1$). Asialo GM_1 is a neutral glycosphingolipid found in high density on NK cells.[23] They concluded that primary murine influenza virus infection was not cleared by NK cells despite the enhanced activity noted early during infection.[24]

More recently Lewis and colleagues confirmed early work by Ennis and co-workers that alterations in peripheral blood lymphocytes occurred in a group of college students during an outbreak of influenza A/Philippines 2/82 (H3N2) virus infection.[25] Control subjects were college students with an acute febrile noninfluenzal respiratory illness that occurred during the same outbreak. It was reported that T cell responses to mitogens were reduced but NK activity was increased, whereas lymphocyte functions were virtually unchanged in the control group. Although the studies were not performed with lymphocytes from the lung, the assumption was made based on animal studies with influenza infection (infra vide) that the response in the periphery was representative of the local pulmonary response. Although this assumption is generally incorrect, one could hypothesize that during an acute influenzal infection, titers of factors such as interferon that augment pulmonary NK cells could also increase in peripheral fluids and augment NK activity in the peripheral blood.

2. Interferon and Influenza Infection

In 1983, Hoshino and co-workers reported that mice infected with an aerosol of influenza virus type A and subsequently treated with intranasal instillations of antiinterferon antiserum died in 7 days postinfection, whereas the mice that were infected but untreated survived.[26] Virus titers decreased after 3 days in the control group; virus titers in animals deficient in interferon continued to increase. These data support an important role for interferon in the early stages of influenza infection but do not differentiate whether the interferon is interfering with influenza virus replication directly or indirectly through augmentation of NK activity. It would be important to determine if antiinterferon treatment interferes with NK activity.

3. Role of NK Cells during Influenza Virus Infection

Using nude (nu^+/nu^+) and beige (bg^+/bg^+) mouse immunodeficiency models, Leung and Ada[27] reported that influenza virus infection increased the NK activity in the lung in both mutants. It is known that nude mice lack T lymphocytes but have normal to increased levels of NK activity.[28] Beige mice have defective granules and show low NK activity.[29] After intranasal inoculation of influenza virus, both beige and heterozygote littermates contained similar levels of infectious virus in their lungs. The authors stated that their results did not eliminate the possibility that NK cells may help to limit influenza replication. However, many workers in the field quoted the work with beige mice as evidence for there being no role for NK cells in protection against influenza infection in spite of the evidence that interferon production augmented the NK activity in beige mice.[27,29]

Because of our particular interest in the lung, we thought it reasonable to study local cell-mediated responses in the lung during infection and chose a virus that had a tropism for that tissue. Our first aim was to induce a local augmentation of NK activity.[30,31] When 0.125 hemagglutinating units (HAU) of influenza virus PR8/34 (HINI) was inoculated into mouse lungs via the trachea, we observed that there was a local (lung) augmentation of NK activity in the absence of a concurrent increase in the NK activity in the autologous spleen. This increase in NK activity was associated with a rise in the local interferon titers that could be measured in lung wash fluid and whole cell extract. We reasoned, like others before us, that this rise in interferon augmented the NK activity that was needed to control the influenza virus replication until the specific cytotoxic T lymphocytes arrived to curtail the infection.

3.1. Selective Depletion of NK Activity

The most direct approach used to study a physiological role for NK cells during virus infection has been to evaluate viral disease in animals selectively deficient in NK activity. As stated above, the beige (bg) mutation in mice in its homozygous state causes an important defect in NK activity.[32] The beige mouse is restricted in its potential as an NK-depleted model for the study of the role of NK cells in viral disease, because interferons induced by viruses have been shown to augment NK activity in the mutant mice.[27,29]

Other approaches that have been used to deplete NK cells have been to treat mice with strontium-89[33] or estradiol.[34] Both reagents destroy the bone marrow and the NK stem cells. Strontium-89 treatment is selective for NK cells but requires special animal handling; estradiol treatment is not selective for NK cells, and it affects T lymphocyte function as well as NK activity.

The identification of asialo GM_1 as a cell surface marker for murine NK cells permitted the development of an antibody reagent[23] that could eliminate NK activity *in vivo*.[35] Initially, it was demonstrated that $RAGM_1$ serum and complement can eliminate murine NK cell activity *in vitro* without affecting cytotoxic T cell activity. Kawase and co-workers[36] showed that a single IV injection of anti–asialo GM_1 serum caused a marked inhibition of splenic NK activity. Mice treated with the antiserum showed normal development of activated tumoricidal macrophages in response to poly I:C and normal development of cytotoxic T cells in response to allogeneic cells.[36] These results were somewhat surprising in view of the presence of asialo GM_1 on some T cells[37] including cytotoxic T cell precursors[38] but could be explained by a difference in distribution of the glycolipid on the plasma membranes of the cells susceptible and the cells resistant to antibody-mediated clearance.

Therefore, we concluded that treating animals *in vivo* with $RAGM_1$ would selectively deplete NK cells. Others have used this model to study the role of NK cells in both tumor[36] and virus disease states.[39] These workers reported that an inoculation of 10 μl whole $RAGM_1$ rendered the mice deficient in NK activity in the spleen as early as 24 h after inoculation. We were able to demonstrate that the IV inoculation of this same dose of $RAGM_1$ effectively removed the NK activity from the lung tissue as well as the spleen.[40] When the $RAGM_1$ was given intratracheally (IT), the NK activity in the lung was selectively depleted, leaving the NK activity in the spleen intact (Fig. 1). NK activity was depleted as early as 12 h

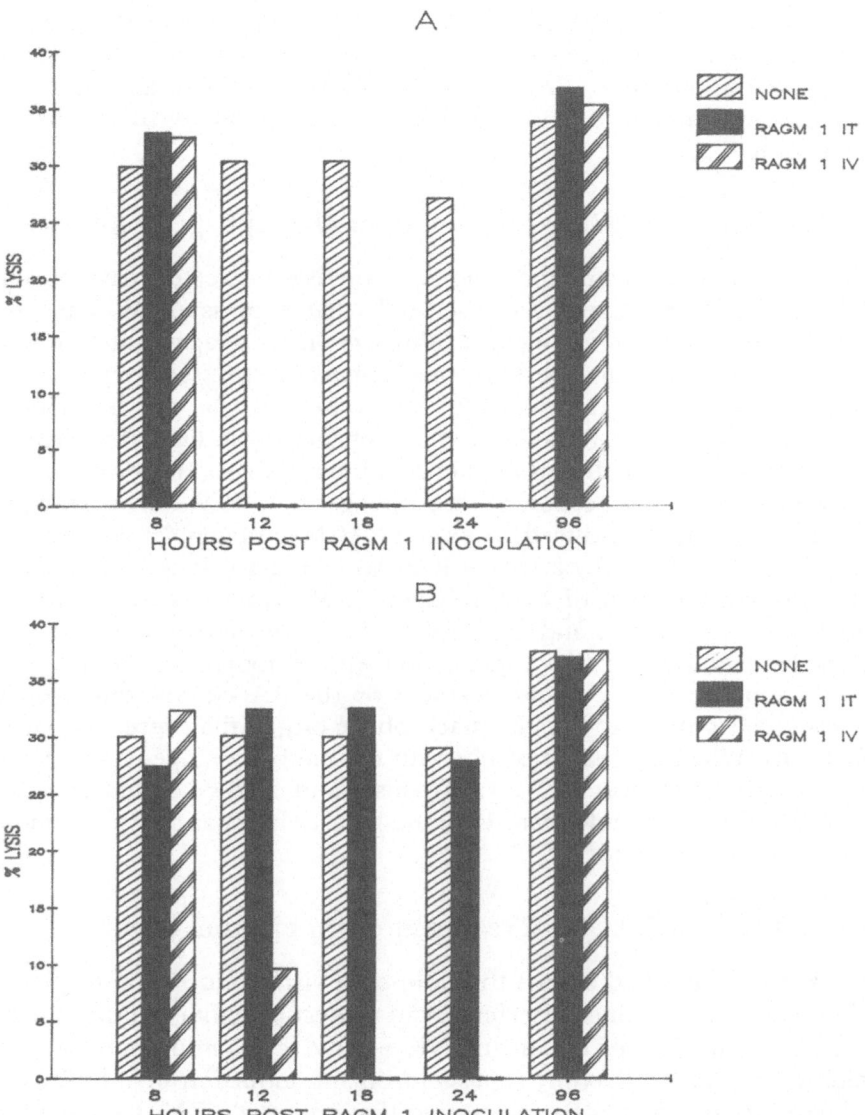

FIGURE 1. Effect of route of RAGM₁ on NK activity in lung and spleen. Mononuclear cells from lungs or spleen of mice treated with RAGM₁ by either intratracheal (IT) or intravenous (IV) route of administration tested for ability to lyse Cr-Yac-1 targets. Effector target ratios 50:1, 25:1, 12.5:1. E:T used in graph is 50:1. (A) NK activity in lung. (B) NK activity in spleen.

in the target organs by either route and returned by 96 h. Therefore, by inoculating $RAGM_1$ every 4 days, the animals remained deficient in NK activity. Mice and hamsters (MHA) were used in these studies, because they represented the two extremes of susceptibility to the influenza virus strain used in our laboratory. When mice ($B_6D_2F_1$) were given 0.125 HAU, 50% survived beyond 8 days. Hamsters became ill, but all survived a 256-HAU dose.

3.1.1. Effect of $RAGM_1$ Treatment on Macrophage Cytotoxicity

It was important for the validity of the NK-depleted model to study the effect of the reagent on other cells that express AGM_1 and could participate in the defense against influenza virus. Akagawa and co-workers[41] reported that mouse lung macrophages expressed asialo GM_1 on their cell surface, so the possibility was raised that *in vivo* treatment of animals with $RAGM_1$ removed a macrophage in the lung that normally contributed to early defense against influenza virus. In addition, these workers showed that peritoneal macrophages that don't normally express the antigen can be induced to express AGM_1 antigen.[42] However, we were unable to show any affect of $RAGM_1$ inoculated *in vivo* or *in vitro* on the cytotoxic function of macrophages.[40] Macrophages were harvested from influenza-infected and uninfected mice (or hamsters) and were tested immediately or after 24 h of incubation with endotoxin. In none of these experiments was there an impairment by the $RAGM_1$ treatment on the macrophage's ability to lyse the macrophage targets that were ^{51}Cr-labeled EL4 cells. Whether treated or not with the antibody reagent, the percent lysis of the target was 32.3 ± 1.6 (mouse) and 28.8 ± 1.3 (hamster), and when activated with endotoxin (LPS) the percent lysis was 53.9 ± 1.2 (mouse) and 48.7 ± 1.4 (hamster).

3.1.2. Effect of $RAGM_1$ Treatment on Interferon Titers

Early studies had shown that influenza virus induces the production of interferon.[16,17] Since interferon can protect by either limiting the replication of the virus or enhancing NK activity, it was important to investigate the effects of $RAGM_1$ treatment on the induction of interferon by influenza virus. Others have reported that anti-asialo GM_1 treatment had no effect on the interferon titers of virus-infected animals.[39] In collaboration with Gerald Sonnenfeld (University of Louisville, KY), we were unable to detect any difference in interferon titers in virus-infected mice that were or were not treated with $RAGM_1$ *in vivo*. In addition, we were unable to detect "early" interferon (released by 6 h) when mice were infected with influenza virus. The possibility that antibody treatment removed a preformed pool of interferon that was needed for early protection was ruled out. Although early interferon may be important in early defenses of some viruses, these observations do not support a role for early interferon during influenza virus infection in the lung.

3.1.3. *In Vivo* RAGM$_1$ Treatment and CTL Response

Previous workers reported that AGM$_1$ was on CTL precursors[38] and thymus cells[37] but that injections of antiserum *in vivo* had no effect on CTL activity. Recently, Stitz and co-workers showed that *in vivo* inoculation (IV) of RAGM$_1$ 6 days after infection with lymphocytic choriomeningitis virus (LCMV) or vaccinia virus reduced the specific T lymphocyte response.[43] These investigators also corroborated our report that inoculation of RAGM$_1$ *prior* to virus infection did not change the lytic activity of CTL generated during a primary infection. These observations suggest the possibility that asialo GM$_1$ is expressed on CTL. An alternate explanation of the data is that NK cells are needed to help CTLs. More recently the NK-1.1 monoclonal antibody has been used to selectively remove NK cells[44,45] *in vivo*. In light of the results reported by Stitz and co-workers, animals depleted of NK activity by inoculations of NK-1.1 antibody would be the model of choice to determine if NK cells are needed to help CTL cells. Seaman and co-workers recently reported that *in vivo* inoculation of monoclonal to NK-1.1 had no effect on the development of T lymphocyte cytotoxicity to alloantigen.[46]

3.2. Effect of NK Depletion on Survival following Influenza A Infection

Both mice and hamsters were treated with RAGM$_1$ and infected with influenza virus. In contrast to 50% of mice with normal NK activity dying by day 8 after virus infection, 100% of mice depleted of NK activity by intratracheal or intravenous inoculation of RAGM$_1$ succumbed to the infection by day 4 (Fig. 2). Hamsters with intact NK activity normally survive influenza infection, but hamsters depleted of NK activity became sick and 50% died by day 6 (Fig. 3). Because the intratracheal inoculation selectively removed the NK cells in the lung, we concluded not only that NK cells were important in the early defenses against influenza virus infection but that NK cells protecting the animals against the virus were in the lung. In studies done in collaboration with Drs. Bennett and Kumar at the University of Texas Health Science Center in Dallas, we demonstrated that the lung is a reservoir for NK cells as well as NK activity.[30] Other studies reported that virus infection and interferon could selectively augment pulmonary NK activity[31] but did not rule out the possibility that the activity was augmented by recruiting and immobilizing an NK population from the pulmonary vasculature. The observation that when RAGM$_1$ was inoculated intratracheally the splenic NK activity remained

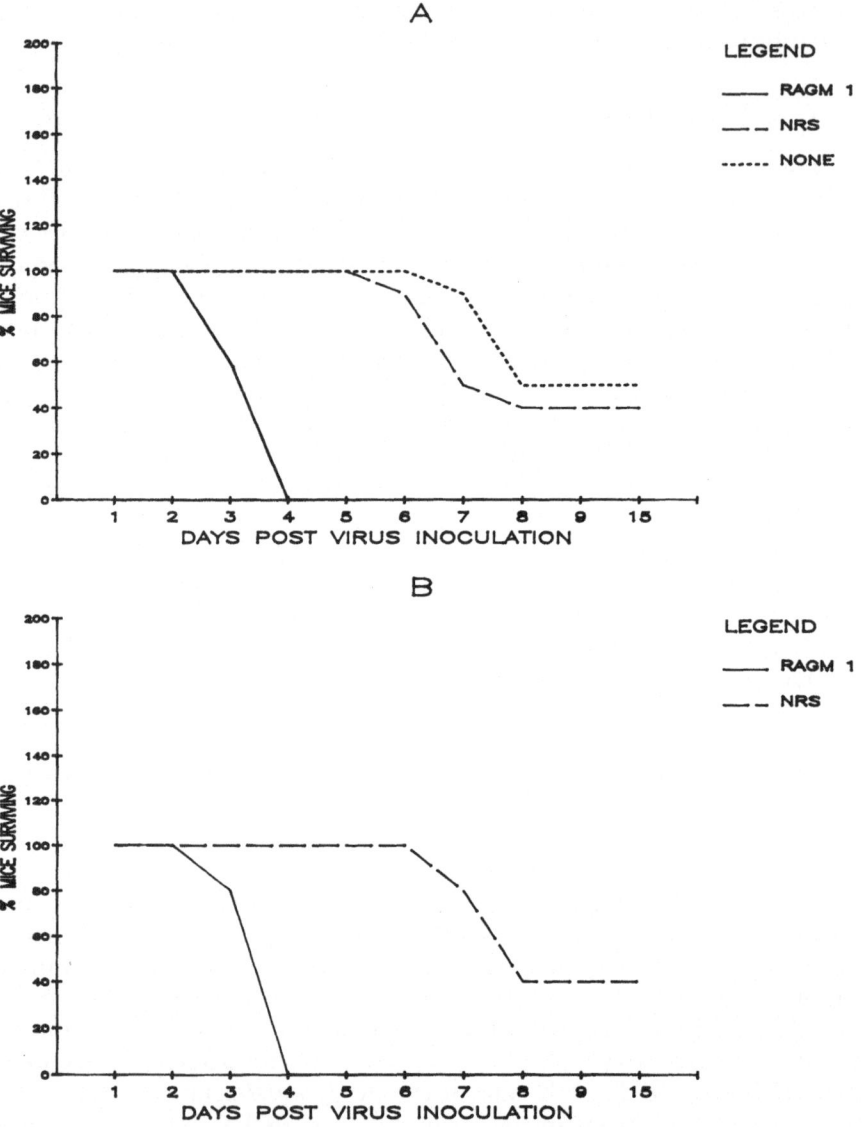

FIGURE 2. Effect of NK depletion on influenza virus infection in mice. (A) $B_6D_2F_1$ mice were depleted of NK cell activity in their lungs and spleens by IV inoculation of 50 μl of 1:5 dilution of rabbit antiserum to asialo GM_1. Control animals received 50 μl of 1:5 dilution of normal rabbit serum, IV, or no treatment (none). (B) $B_6D_2F_1$ mice were selectively depleted of NK cell activity in their lungs by intratracheal (IT) inoculation of 50 μl of 1:5 dilution of rabbit antiserum to asialo GM_1. Control animals received 50 μl of 1:5 dilution of normal rabbit serum IT. Antiserum was inoculated 1 day prior to and 3 days after virus (0.125 HAU of PR8/34) inoculation. There were 10 animals in each panel except the group that received the antiserum to asialo GM_1, $n = 20$. (Published with permission of *J. Immunol.* **136**:1435–1441.)

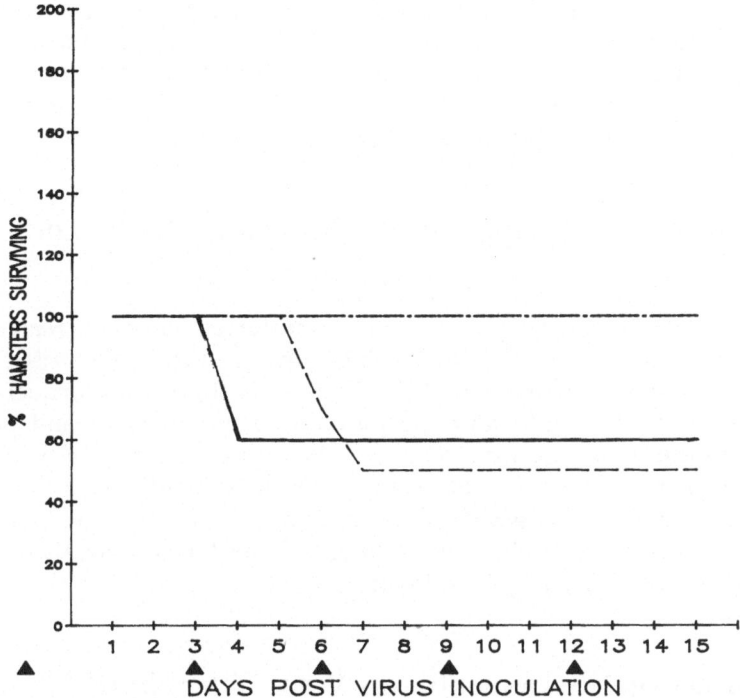

FIGURE 3. Effect of NK depletion on influenza virus infection in hamsters. MHA hamsters were inoculated with 50 μl of a 1:5 dilution of RAGM$_1$ or NRS 1 day prior to 256 HAU PR8/34 influenza virus given intratracheally (IT). RAGM$_1$ and NRS were repeated every third day through 15 days. Arrows indicate days animals were inoculated with serum.

intact is strong evidence for support of the hypothesis that NK cells are resident in the lung. These data clearly demonstrate that the NK cells in the lung are important in the early resistance to influenza virus infection.

However, the question remained, "Why does NK depletion alter survival of mice to IT inoculation of virus and not the survival of mice to IN inoculation?" We reasoned that the IV route of RAGM$_1$ administration used by previous investigators[24] may not deplete the NK cells in the upper respiratory tract. In a recent manuscript,[47] we report that the survival of the mice to an LD$_{50}$ dose of PR8/34 given IN was altered from 50% survival at eight days to 50% survival at four days if the RAGM$_1$ were given both IV and IN but not if only one route were used. We concluded that there is a local population of NK cells in the upper respiratory tract that can be depleted by RAGM$_1$ when given IN. However, since the area is essentially a passageway, the RAGM$_1$ may be flushed away and the NK activity can be replenished by NK cells arriving from the blood if the reagent is not given IV. Furthermore, a daily IN treatment of RAGM$_1$ from one day prior virus inoculation for a total of four days was able to change a survival of 50% at eight days to 0% survival at four days. These observations support the hypothesis that both locally and systemically derived NK cells participate in the early defense against influenza in the upper respiratory tract.

3.3. Effect of NK Depletion on Pulmonary Histologic Changes

When the lungs of infected and noninfected animals treated with RAGM$_1$ serum were examined histologically, lesions characteristic of influenza infection were observed[40,48] (Figs. 4–7). Erosion of bronchial epithelium, bronchial epithelial hyperplasia, and intraalveolar and perivascular infiltration of mononuclear and inflammatory cells was typically observed for the virus infection. Lungs of influenza-infected animals with intact NK activity consistently showed fewer lesions at time points observed. No lesions were observed in lungs of uninfected animals that were depleted of NK activity by RAGM$_1$ treatment.

3.4. Effect of NK Depletion on Virus Titers in the Lung

To further test the hypothesis that the viral pneumonia observed by histologic examination of the lungs was associated with increasing titers of influenza virus, the virus within the lungs was measured by standard procedures with hemagglutination of chicken red blood cells (CRBCs). Infected lungs from animals with intact NK activity consistently demonstrated a lower titer than infected lungs from animals depleted of NK activity (Table I). Increased virus titers correlated with altered survival curves recorded for euthymic ($B_6D_2F_1$) or athymic (nu/nu) mice given RAGM$_1$ IN/IV prior to virus or RAGM$_1$, IN, daily. The hemagglutination of the CRBC was specific for influenza virus, because prior treatment of lung extracts with antibodies to PR8/34 influenza virus (and not antibodies specific for Sendai virus) removed the hemagglutinating activity.

4. Pathological Consequences of NK Activity

In addition to controlling virus replication during influenza infection, it is known that lymphocytes can contribute to the pathology of the disease.[49–51] Wyde et al.[49] have shown that animals that lack DTH lymphocytes have less pneumonia than animals with DTH lymphoctyes. Furthermore, adoptive transfer of DTH restores the pneumonia.[51] However, few studies have demonstrated such a role for the NK cell in the development of pathology. Wabuke-Bunoti and co-workers reported that de-

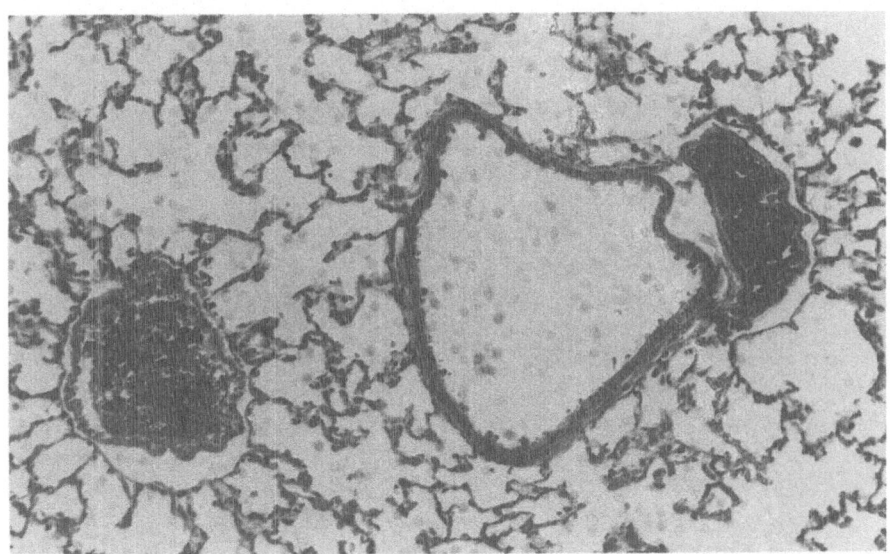

FIGURE 4. Lung of hamster 4 days after PR8/34 infection, 450×. The bronchiole is lined by a simple cuboidal epithelium, which is ciliated. Alveolar septae are thin and delicate. (Published with permission of Stein-Streilein and Guffee, *J. Immunol.* **136:**1435–1441.)

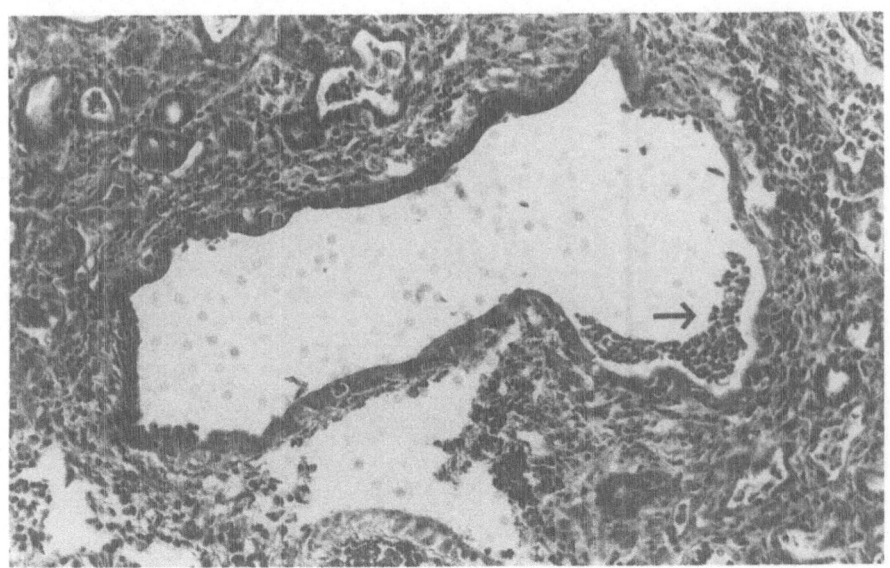

FIGURE 5. Lung of hamster treated with RAGM₁ 4 days after PR8/34 infection, 450×. The bronchiolar epithelium is ulcerated focally and overlaid with purulent exudate (arrow). The remaining epithelium is hyperplastic and lacks cilia. There is peribronchiolar infiltration of neutrophils and mononuclear cells that involves adjacent lung tissue. Alveolar lining cells have undergone adenomatous hyperplasia (arrow). (Published with permission of Stein-Streilein and Guffee, *J. Immunol.* **136:**1435–1441.)

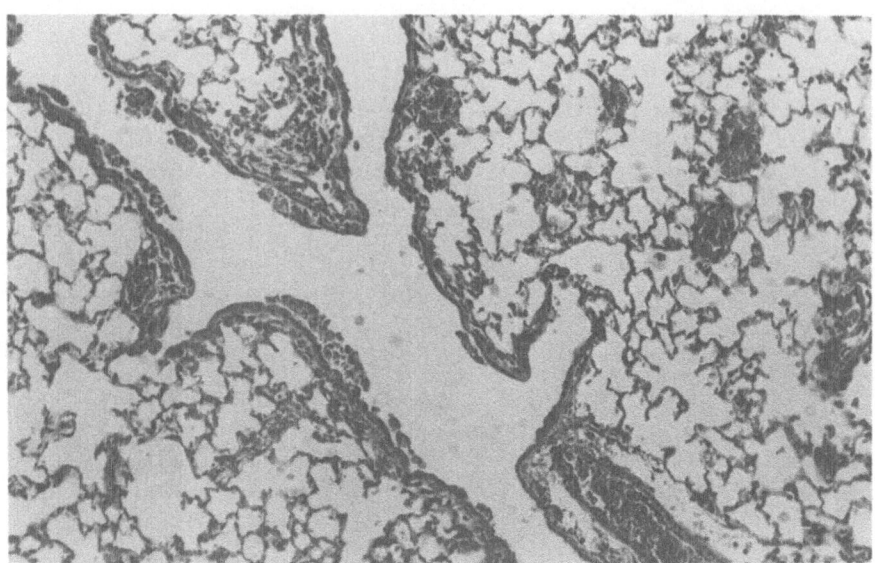

FIGURE 6. Lung of mouse 4 days after PR8/34 infection, 450×. The bronchiolar epithelium is normal in appearance, with occasional evidence of hyperplasia. A slight infiltration of inflammatory cells is in the peribronchiolar tissue. (Published with permission of Stein-Streilein and Guffee, *J. Immunol.* **136**:1435–1441.

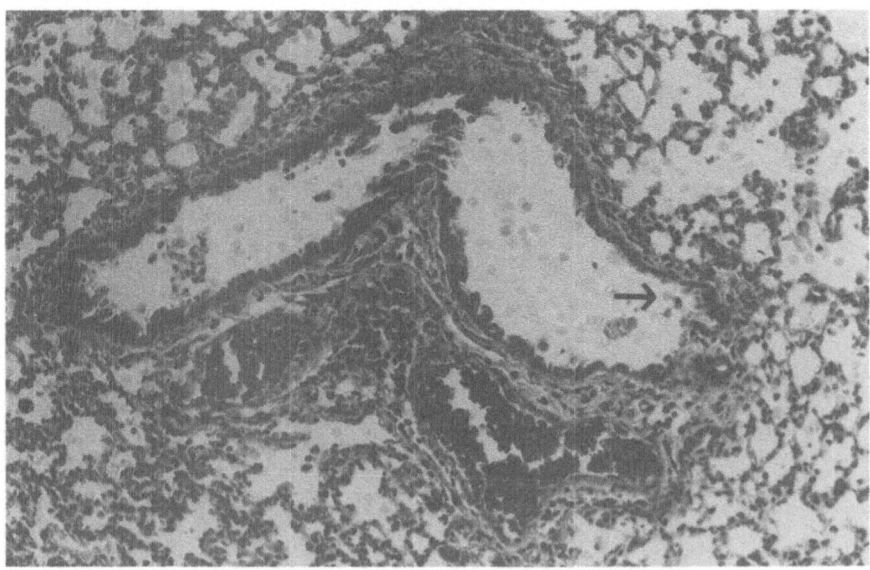

FIGURE 7. Lung of mouse treated with RAGM₁ 4 days after PR8/34 infection IT, 450×. The bronchiolar epithelium is focally ulcerated (arrow), and the lumen contains neutrophils and sloughed epithelial cells. There are hyperplasia and disorganization of the epithelium. Both the peribronchiolar tissue and adjacent alveolar septae are infiltrated by mixture of mononuclear cells and neutrophils. (Published with permission of Stein-Streilein and Guffee, *J. Immunol* **136**:1435–1441.)

TABLE I
Influenza Virus Hemagglutination of CRBC

Species	Days post-influenza	Treatment[a]	Virus titer hemagglutination CRBC[b]	
			$B_6D_2F_1$	Nude
		Experiment I		
Hamster	6	RAGM$_1$, only	0	—
		PR8/34, only	80	—
		RAGM$_1$, PR8/34	640	—
Mouse	4	RAGM$_1$, only	0	—
		PR8/34, only	80	—
		RAGM$_1$ and PR8/34	10,240	—
		Experiment II		
Mouse	4	PR8/34, IN	4	65
		IN/IV RAGM$_1$, PR8/34 (IT)	8,192	9,216
		IN/IV RAGM$_1$, PR8/34 (IN)	400	4,608
		IN daily, RAGM$_1$; PR8/34 (IN)	9,216	—

[a] Experiment I, animals were inoculated with RAGM$_1$ 24 h prior to IT inoculation with PR8/34 influenza virus. In experiment II the RAGM$_1$ was given IN/IV, prior to virus or RAGM$_1$, IN daily.
[b] A freeze-thaw extract lung tissue was tested for ability to agglutinate CRBC. Positive titers were inhibitable by antibodies specific for influenza and not Sendai virus. Virus titer is an average for 10 $B_6D_2F_1$ mice and 5 nu/nu mice.

pletion of NK cells by RAGM$_1$ ameliorated influenza virus-induced disease, reduced mortality, and effected changes in the relative proportion of inflammatory cell populations infiltrating the cerebrospinal fluid (CSF).[52] In their report the RAGM$_1$ was inoculated IV and the influenza virus, PR8/34, was inoculated intracranially (IC). NK activity was increased in the CSF 24–48 h later and the RAGM$_1$ treatment abolished the NK activity in both spleen and CSF. Interferon production was detected as early as 6 h postinfection, with peak activity observed at 12 h.

In general, CSF showed higher levels of IFN activity per milliliter than did serum. Antiinterferon treatment abolished NK activity in the spleen and reduced NK activity of CSF NK cells three- to fourfold. Coincident with the peak in NK activity, the investigators recorded an increase in large granular lymphocytes (LGLs) (12%, 12 h, to 37%, 24 h). Treatment of infected mice with RAGM$_1$ or anti-IFN reduced the numbers of LGLs seen at 24 h. As reported for other viruses, many cells infiltrating the CSF are either actively or passively involved in the inflammatory process, since there is normally no resident lymphoid tissue in the brain.

The treatment with RAGM$_1$ resulted in total abrogation of lytic activity in both the spleen and CSF exudate and in amelioration of the disease, with improved survival rate in lethally infected mice. The decline in NK activity was correlated with a decrease in LGLs and mature lymphocytes but had no effect on the levels of IFN detected in the fluids.

It is interesting when PR/8/34 is inoculated intracranially, it produces an abortive infection, but when large doses of virus are used, fatal lesions appear to be inflicted in the early phase of infection. Work reported by Wabuki-Bunoti clearly implicates both the virus and the host immune response in the pathogenesis of the influenza viral encephalitis and supports the possibility that other pathogenic effects of viral encephalitis may be augmented by NK cells.

5. Conclusions

A decade ago, reports indicated that NK cells participated in the defense against influenza virus infection. More recently, the development of antibody reagents specific for NK cells permitted studies that defined the physiological role of NK cells during influenza infection. These studies defined a role for NK cells during the early defenses against influenza virus infection in the lung. The results clearly show that the NK cells that protect the lung reside in the lung. Furthermore if NK cells are removed from both the upper respiratory tract and the blood, the survival and morbidity of mice that have been infected with influenza via the intranasal route are altered. It is interesting that, similarly to other cytotoxic cells, NK cells can participate in destruction of tissue when their activity is in excess. Just such a scenario is present when mice are inoculated intracranially with a superdose of virus.

The pulmonary NK cells are important in limiting influenza virus replication in the lung prior to the induction of the specific immune response. The initial defense by the NK cell in the lung is under the regulation of local mediators such as IFN gamma and IL-2. Immunosuppressed patients who may have little or no NK activity, perhaps because of a lack of NK growth factors produced by T lymphocytes, might benefit from exogenous biological mediators delivered locally to boost natural cytotoxicity against influenza and other susceptible virus infections.

References

1. Rossen, R. D., Kasel, J. A., and Couch, R. B., 1971, The secretory immune system: Its relation to respiratory viral infection, *Prog. Med. Virol.* **13:**194–238.
2. Waldman, R. H., Mann, J. S., and Small, P. A., 1969, Immunization against influenza: Prevention of illness in man by aerosolized inactivated vaccine, *JAMA* **207:**520–524.
3. Ada, G. L., Leung, K. N., and Ertl, H., 1981, An analysis of effector T cell generation and function in mice exposed to influenza A or Sendai virus, *Immunol. Rev.* **58:**524–531.
4. Yap, K. L., Ada, G. L., and McKenzie, I. F. C., 1978, Transfer of specific cytotoxic T lymphocytes protects mice inoculated with influenza virus, *Nature* **273:**238–239.
5. Yap, K. L., and Ada, G. L., 1978, The recovery of mice from influenza virus infection:

Adoptive transfer of immunity with immune T lymphocytes, *Scand. J. Immunol.* **7:**389–397.

6. Wyde, P. R., Couch, R. B., Mackler, B. F., Cate, R. R., and Levy, B. M., 1977, Effects of low- and high-passage influenza virus infection in normal and nude mice, *Infect. Immun.* **15:**221–229.

7. Wyde, P. R., and Cate, T. R., 1978, Cellular changes in lungs of mice infected with influenza virus: Characterization of the cytotoxic responses, *Infect. Immun.* **22:**423–429.

8. Nellis, M. J., and Streilein, J. W., 1980, Hamster T cells participate in MHC alloimmune reactions but do not affect virus induced cytotoxic activity. *Immunogenetics* **11:**75–86.

9. Nellis, M. J., Duncan, W. R., and Streilein, J. W., 1981, Immune response to acute virus infection in the syrian hamster. II. Studies on the identity of virus-induced cytotoxic effector cells, *J. Immunol.* **126:**214–218.

10. Yang, H., Cain, C., and Tompkins, W. A. F., 1983, Induction of thy 1.2$^+$ and thy 1.2$^-$ nonspecific cytotoxic lymphocytes in the hamster, *J. Immunol.* **131:**622–633.

11. Yang, H., and Tompkins, W. A. F., 1984, Nonspecific cytotoxicity of vaccinia-induced peritoneal exudates in hamsters is mediated by thy-1.2 homologue positive cells distinct from NK cells and macrophages, *J. Immunol.* **131:**2545–2550.

12. Chapes, S. K., and Tompkins, W. A. F., 1979, Cytotoxic macrophages induced in hamsters by vacinnia virus: Selective cytotoxicity for virus infected targets by macrophages collected late after immunization, *J. Immunol.* **123:**303–310.

13. Henderson, F. W., 1979, Pulmonary cell mediated cytotoxicity in hamsters with parainfluenza virus type 3 pneumonia, *Am. Rev. Respir. Dis.* **120:**41–47.

14. Kimmel, K., Wyde, P. R., and Glezen, W. P., 1982, Evidence of a T cell mediated cytotoxic response to parainfluenza virus type 3 pneumonia in hamsters, *J. Reticuloendothel. Soc.* **31:**71–83.

15. Stein-Streilein, J., Witte, P. L., Streilein, J. W., and Guffee, J., 1985, Local cellular defenses in influenza-infected lungs, *Cell. Immunol.* **95:**234–246.

16. Santoli, D., Trinchieri, G., and Lief, F. S., 1978, Cell-mediated cytotoxicity against virus infected target cells in humans. I. Characterization of the effector lymphocyte, *J. Immunol.* **121:**526–531.

17. Santoli, D., Trinchieri, G., and Koprowski, H., 1978, Cell-mediated cytotoxicity against virus infected target cells in humans. II. Interferon induction and activation of natural killer cells, *J. Immunol.* **121:**532–538.

18. Lindahl, P., Leary, P., and Gresser, I., 1972, Enhancement by interferon of the specific cytotoxicity of sensitized lymphocytes, *Proc. Natl. Acad. Sci. USA* **69:**721–725.

19. Heron, I., Berg, K., and Cantell, K., 1976, Regulatory effect of interferon on T cells *in vitro, J. Immunol.* **117:**1370–1373.

20. Trinchieri, G., and Santoli, D., 1978, Antiviral activity induced by culturing lymphocytes with tumor-derived and virus-transformed cells. Enhancement of human natural killer cell activity by interferon and antagonistic inhibition of susceptiblity of target cells to lysis, *J. Exp. Med.* **147:**1314–1333.

21. Trinchieri, G., Santoli, D., and Koprowski, H., 1978, Spontaneous cell mediated cytotoxicity in humans. Role of interferon and immunoglobulins, *J. Immunol.* **120:**1849–1855.

22. Ennis, F. A., Beare, A. S., Riley D., Schild, G. C., Meager, A., Yi-Hua, Q., Schwarz, G., and Rook, A. H., 1981, Interferon induction and increased natural killer cell activity in influenza infections in man, *Lancet* **246:**891–893.

23. Kasai, M., Iwamori, M., Nagi, Y., Okumura, K., and Tada, T., 1980, A glycolipid on the surface of mouse natural killer cells, *Eur. J. Immunol.* **10:**175–180.

24. Wells, M. A., Daniel, S., Kiley, S. C., Burlington, D. B., and Ennis, F. A., 1985, Absence of natural killer cell effects in murine influenza virus pneumonia, *J. Leukocyte Biol.* **38:**124 (abstract).

25. Lewis, D. E., Gilbert, B. E., and Knight, V., 1986, Influenza virus infection induces functional alterations in peripheral blood lymphocytes, *J. Immunol.* **137:**3777–3781.

26. Hoshino, A., Takenaka, H., Mizukoshi, O., Imanishi, J., Kishida, T., and Tovey, M. G., 1983, Effect of anti-interferon serum on influenza virus infection in mice, *Antiviral Res.* **3**:59–65.
27. Leung, K. N., and Ada, G. L., 1981, Induction of natural killer cells during murine influenza virus infection, *Immunobiology* **160**:352–366.
28. Clark, E. A., Schultz, L. D., and Pollack, S. B., 1981, Mutations in mice that influence natural killer (NK) cell activity, *Immunogenetics* **12**:601–613.
29. Vassalli, J. D., Granelli-Piperno, A., Grascelli, C., and Reich, E., 1978, Specific protease deficiency in polymorphonuclear leukocytes of Chediak–Higashi syndrome and beige mice, *J. Exp. Med.* **147**:1285–1290.
30. Stein-Streilein, J., Bennett, M., Mann, D., and Kumar, V., 1983, Natural killer cells in mouse lung: Surface phenotype, target preference and response to local influenza virus infection. *J. Immunol.* **131**:2699–2704.
31. Mann, D. W., Sonnenfeld, G., and Stein-Streilein, J., 1985, Pulmonary compartmentalization of interferon and natural killer cell activity, *Proc. Biol. Exp. Med.* **180**:224–230.
32. Roder, J., and Duive, A., 1979, The beige mutation in the mouse selectively impairs natural killer cell function, *Nature* **278**:451–453.
33. Kumar, V., Ben-Ezra, J., Bennett, M., and Sonnenfeld, G., 1979, Natural killer cells in mice treated with [89]Strontium: Normal target binding cell numbers but inability to kill even after interferon administration, *J. Immunol.* **123**:1832–1838.
34. Seaman, W. E., Gindhart, T. D., Greenspan, J. S., Blackman, M. A., and Talal, N., 1979, Natural killer cells, bone, and bone marrow: Studies in estrogen treated mice and in congenitally osteopetrotic (mi/mi) mice, *J. Immunol.* **122**:2541–2547.
35. Young, W. W., Jr., Hakomori, S. I., Durdrik, J. M., and Henney, C. S., 1980, Identification of ganglio-N-tetraosylceramide as a new cell surface marker for murine natural killer (NK) cells, *J. Immunol.* **124**:199–201.
36. Kawase, I., Urdal, D. L., Brooks, C. G., and Henney, C. S., 1982, Selective depletion of NK cell activity *in vivo* and its effect on the growth of NK-sensitive and NK-resistant tumor cell variants, *Int. J. Cancer* **29**:567–574.
37. Stein, K. E., Schwarting, G. A., and Marcus, D. M., 1978, Glycolipid markers of murine lymphocyte subpopulations, *J. Immunol.* **120**:676–679.
38. Beck, B. N., Gillis, S., and Henney, C. S., 1981, The display of the neutral glycolipid ganglioN-tetraosylceramide (asialo GM₁) on cells of the NK and T lineages, *Transplantation* **33**:118–122.
39. Bukowski, J. F., Woda, B. A., Habu, S., Okumura, K., and Welsh, R. M., 1983, Natural killer cell depletion enhances virus synthesis and virus-induced hepatitis *in vivo, J. Immunol.* **131**:1531–1538.
40. Stein-Streilein, J., and Guffee, J., 1986, *In vivo* treatment of mice and hamsters with antibodies to asialo GM₁ increases morbidity and mortality to pulmonary influenza infection, *J. Immunol.* **136**:1435–1441.
41. Akagawa, K. S., Maruyama, Y., Takano, M., Kasai, M., and Tokunaga, T., 1981, A cell surface antigen expressed on mouse lung macrophages, *Microbiol. Immunol.* **25**:1215–1220.
42. Akagawa, K. S., and Tokunaga, T., 1982, Appearance of a cell surface antigen associated with the activation of peritoneal macrophages in mice, *Microbiol. Immunol.* **26**:831–842.
43. Stitz, L., Baenziger, J., Pircher, H., Hengartner, H., and Zinkernagel, R. M., 1986, Effects of anti-asialo GM₁ treatment *in vivo* or with anti-asialo GM₁ plus complement *in vitro* on cytotoxic T cell activities, *J. Immunol.* **136**:4674–4680.
44. Koo, G. C., and Peppard, J. R., 1984, Establishment of monoclonal anti NK1.1 antibody, *Hybridoma* **3**:301–303.
45. Koo, G. C., Dumont, F. J., Tutt, M., Hackett, J. Jr., and Kumar, V., 1986, The NK-1.1(−) mouse: A model to study differentiation of murine NK cells, *J. Immunol.* **137**:3742–3747.

46. Seaman, W. E., Sleisenger, M., Eriksson, E., and Koo, G., 1987, Depletion of natural killer cells in mice by monoclonal antibody to NK-1.1, *J. Immunol.* **138:**4539–4544.
47. Stein-Streilein, J., Guffe, J., and Fan, W., 1988, Locally and systemically derived natural killer cells participate in defense against intranasally inoculated influenza virus, (in press).
48. Stein-Streilein, J., and Lipscomb, M. F., 1981, Immune response to influenza virus in guinea pigs, mice and hamsters, in: *Genetic Variation Among Influenza Virus* (D. Nyak, ed.), Academic Press, New York, pp. 567–576.
49. Cate, T. R., and Mold, N. G., 1975, Increased influenza pneumonia mortality of mice adoptively immunized with node and spleen cells sensitized by inactivated but not live virus, *Infect. Immun.* **11:**908–914.
50. Wyde, P. R., Couch, R. B., Mackler, B. F., Cate, T. R., and Levy, B. M., 1977, Effects of low- and high-passage influenza virus infection in normal and nude mice, *Infect. Immun.* **15:**221–229.
51. Yap, K. L., Ada, G. L., and McKenzie, I. F. C., 1978, Transfer of specific cytotoxic T lymphocytes protects mice inoculated with influenza virus, *Nature* **273:**238–239.
52. Wabuke-Bunoti, M. A. N., Bennink, J. R., and Plotkin, S. A., 1986, Influenza virus-induced encephalopathy in mice: Interferon production and natural killer cell activity during acute infection, *J. Virol.* **60:**1062–1067.

4

Natural Host Defense Systems Active against Herpes Simplex Virus Infections

Carlos Lopez and Patricia Fitzgerald-Bocarsly

1. Introduction

The strategies vary by which viruses invade a host and cause disease.[1] In return, the host defense systems that are called upon to intercept and sequester those pathogenic agents also differ.[2] The natural defense system of the infected animal constitutes the first barrier of active defense. These mechanisms act relatively nonspecifically and require no prior exposure to the invading microorganism in order to be active. Macrophages, interferon, and natural killer (NK) cells are three natural defense systems that are thought to be important in the control of herpes simplex virus (HSV) infections and will be the focus of this chapter.

Herpes simplex virus types 1 and 2 (HSV-1 and HSV-2) are closely related viruses that share about 50% DNA homology and that have many serologically cross-reactive antigens.[3,4] Two-thirds or more of primary HSV-1 and HSV-2 infections are asymptomatic.[5] When symptomatic, HSV-1 infections are usually associated with orolabial lesions, whereas HSV-2 usually causes genital lesions.[6] As with the other herpesviruses, HSV-1 and HSV-2 almost always establish a latent infection in sacral or trigeminal ganglia and can be reactivated at some later time.[4,7,8] Reactivation of virus can result in asymptomatic shedding or in recrudescent illness.[5]

CARLOS LOPEZ • Viral Exanthems and Herpesvirus Branch, Division of Viral Diseases, Center for Infectious Diseases, Centers for Disease Control, Atlanta, Georgia 30333. PATRICIA FITZGERALD-BOCARSLY • Department of Pathology, School of Medicine and Dentistry, New Jersey Medical School, Newark, New Jersey 07103

Since natural defense systems are immediately available for response against an invading microorganism, they are thought to play an especially important role in controlling primary infections. That most primary HSV-1 or HSV-2 infections in man are asymptomatic suggests that natural defense systems have controlled those infections prior to the development of lesions or symptomatic disease. Elements of natural defense should, therefore, play a pivotal role in the control of HSV infections prior to the development of antigen-specific immune responses such as neutralizing antibody, cytotoxic T cells, and delayed-type hypersensitivity.[9] Natural defense systems may play a direct role in controlling recrudescent illness as well. This could be either by an inhibitory effect on the reactivated herpesvirus infection or, indirectly, by reducing the number of latently infected ganglia at the time of primary infection.[10]

The basis for resistance against primary HSV infection has received considerable attention in the past few years. Animal models, especially those using inbred mice, have been used to explore different aspects of natural and adaptive immunity.[11] These model systems give the investigator the opportunity to do a variety of manipulations that cannot be done with outbred mice or with humans. There are, however, certain possible disadvantages to the use of these animal models which must be recognized. Since HSV-1 and HSV-2 are not indigenous to the mouse, the diseases caused by these viruses and the host defense mechanisms that control the infections may not be the same as those found in man. Thus, studies using mouse models must be corroborated by studies in man. In the following discussion, we shall describe mouse studies that suggest important roles for natural defense systems in the control of HSV infections and follow these with studies in man which attempt to demonstrate a similar role for those mechanisms in prevention of human disease.

2. Mouse Studies

2.1. Genetic Resistance to HSV-1

Using a recently isolated, highly virulent strain of HSV-1, Lopez[12] showed that inbred strains of mice differed in their ability to resist lethal infection. Three- to four-month-old adult mice were found to be resistant, moderately susceptible, or very susceptible to intraperitoneal (IP) challenge with HSV-1. The susceptible mice demonstrated hindleg paralysis 5–10 days after inoculation and died shortly thereafter. Eight different strains of HSV-1 have been tested in inbred mice.[12] Although the viruses tested varied greatly in their virulence, each killed the susceptible strain of mice at a lower concentration than the moderately susceptible animals, and none of the virus strains killed resistant mice. These observations have been confirmed by a series of studies by Kirchner and his associates[13–15] and by other groups.[16,17]

Studies have been undertaken to determine the genetics of resistance.[18] Reciprocal F_1 crosses were made between resistant and very susceptible strains and between resistant and moderately susceptible mice, and the animals were challenged with HSV-1. Both male and female progeny were challenged with virus to evaluate the possibility of a sex-linked resistance gene. If an X-linked gene is required for resistance, then male offspring of susceptible female by resistant male mice should be found to be susceptible. (See ref. 19 for how sex linkage is determined.) All of the male F_1 mice were found to be resistant to IP challenge with HSV-1. Female F_1 mice were also resistant to HSV-1 but not to the same high level as the male mice. These data demonstrated that genetic resistance to HSV-1 is a dominant trait and that it is not sex-linked[18] (C. Lopez, in preparation).

Studies from several laboratories indicate that genetic resistance to HSV-1 is immunologically mediated.[20–24] These studies used relatively nonspecific immunosuppression to show that immune mechanisms were required for controlling virus infections and limiting morbidity and mortality. Since immunosuppressed resistant mice and genetically susceptible mice died 5–8 days after challenge with HSV-1, the host defense mechanisms required for resistance were most likely aspects of the natural defense mechanisms rather than adaptive immune mechanisms, which are not usually operative until after this period of time.

Because studies had shown that genetic resistance to HSV-1 is immunologic in nature,[20] studies were undertaken to determine whether genes within the major histocompatibility region of the mouse influence resistance to HSV-1. Congenic mice with H-2 regions of various susceptible strains of mice on the resistant C57Bl/6 or C57Bl/10 background were challenged with HSV-1.[18] Since the congenic mice were found to be resistant, it was concluded that the H-2 of susceptible strains failed to diminish resistance in these mice. Similarly, experiments with congenic mice on the A/J (susceptible) background indicated that resistance was not transferred with H-2 genes of resistant mice. These and other studies clearly showed that genetic resistance to HSV-1 was not linked to the major histocompatibility genes of the mouse.

Genetic resistance has also been demonstrated for lethal infections with HSV-2.[16,25,26] In all of these studies, the C57Bl/6 or 10 series mice were found to be resistant to lethal infection, whereas most other strains of mice were found to be at least moderately susceptible. As with resistance to HSV-1, resistance to HSV-2 was shown to be a dominant genetic trait and to be abrogated by immunosuppression.[24]

Most studies of genetic resistance to HSV-1 have been carried out using IP inoculation of the mice. However, studies have shown that intravenously (IV), intraocularly, and intravaginally inoculated mice demonstrate results similar to those obtained by IP challenge.[26–28] Thus, C57BL/ 6 strain mice were more resistant than other strains of mice, and resistance could be diminished by X-irradiation, cyclophosphamide, or ^{89}Sr treatment.

Early in the study of genetic resistance to HSV-1, we noted that it bore a striking similarity to genetic resistance to bone marrow allographs.[29,30] The studies of Bennett[31] demonstrated that allogeneic resistance could be abrogated by treating mice with [89]Sr. Although ^{89}Sr treatment of mice confers susceptibility to allogeneic marrow, such treatment is not broadly immunosuppressive, since mice can still normally reject skin grafts and have almost normal T cell and B cell responses.[32] In our study,[33] [89]Sr treatment of resistant mice caused them to become as susceptible to IP challenge with HSV-1 as the genetically most susceptible strains.

Several observations suggest that genetic resistance to HSV-1 is mediated by natural defense mechanisms rather than adaptive immunity. First, susceptible mice challenged with HSV-1 usually died 5–8 days after challenge with virus.[12,33] In fact, by 3 days after IP inoculation, genetically susceptible animals have higher concentrations of virus in visceral organs than do genetically resistant mice.[33] Five days after inoculation, HSV-1 can be found in the brain and spinal cord of susceptible mice, whereas virus rarely if ever reached the central nervous system of genetically resistant animals. Thus, the differences between genetically susceptible and resistant mice is manifest during the first 3–5 days after inoculation and before the generation of an adaptive immune response. Second, athymic nude (nu/nu) mice have a congenital lack of a thymus, are deficient in T cells, and are therefore unable to generate a cell-mediated immune response.[34] If the cell-mediated immune response was required for genetic resistance to HSV-1, nude mice should be much more susceptible to this virus than the normal heterozygous controls. However, the nude mice were found to be about as resistant to the virus as the controls.[20,34] Clearly, the lack of an adaptive cell-mediated immune response did not significantly alter resistance, indicating that this system does not play a crucial role in genetic resistance to HSV-1. However, some studies have suggested that suppression of T cell-mediated immunity with antithymocyte antiserum results in diminution of genetic resistance.[35,36] The difficulty with these studies, however, is that the antithymocyte antisera can have nonspecific suppressive effects on cells other than T cells. For example, Schlabach et al.[37] showed that antithymocyte antiserum, in addition to its long-term suppressive activity against T cells, had a short-term suppressive effect on host macrophage function which was specifically responsible for increased susceptibility to HSV-1 in treated mice.

A model of genetic resistance to HSV-1 latent infection of the peripheral nervous system in mice has been established by Kastrukoff et al.[38] Inbred strains of mice were inoculated in the lip with various concentrations of HSV-1, and the establishment of latency was evaluated 21 days later by culture of trigeminal ganglia with susceptible cells. Inbred strains of mice fell into three major categories: resistant, moderately susceptible, and very susceptible to latent infection. Genetic resistance to the development of latency was shown to be a dominant trait that was not H-2-linked. Although resistance to lethal infection correlated with resistance to latency in many strains of mice, there were some discrepancies as well. These results suggest that the resistance mechanisms operative in IP challenged mice are also operative in resistance to latency but that, in addition, other mechanisms contribute to resistance to the development of latency. The studies of Abghari et al.[39] indicate that genetic resistance to the establishment of latency requires controlling HSV replication at the site of infection and preventing the spread of virus to neurologic tissues.

Attempts have been made to determine whether a reduced ability of HSV-1 to replicate in cell cultures generated from resistant strains of mice might reflect genetic resistance in the animal.[40–42] In these studies HSV-1 was shown to replicate better in mouse embryo fibroblasts (MEF) from genetically susceptible mice than in cells from genetically resistant mice. Virus yields from resistant MEFs were lower than those from susceptible MEFs, and the efficacy of plaque formation was significantly lower. However, MEFs from resistant F_1 progeny of crosses between resistant and susceptible strains of mice replicated HSV-1 as well as the cells from the susceptible parent, indicating that this function did *not* segregate with genetic resistance.

2.2. Genetic Resistance to HSV-2-Induced Hepatitis in the Mouse

Resistance of mice to the development of focal, necrotizing hepatitis caused by HSV-2 was found by Mogenson et al.[19,43,44] to be under genetic control. Histologically, the lesions found in genetically susceptible animals were characterized by areas of degenerating liver cells, which were infiltrated with polymorphonuclear leukocytes causing central necrosis.[44] Using mature, resistant, and susceptible strains of mice, Mogenson[19] showed that resistance to HSV-2-induced hepatitis was governed by one X-linked dominant gene.

2.3. Role of Macrophages in Resistance to HSV-1 and HSV-2

Macrophages are among the first cells to be encountered by a virus upon invasion of a host.[2] These cells are large, phagocytic cells distributed throughout the liver, lungs, lymph nodes, and spleen.[45]

The first studies to demonstrate a role for macrophages in resistance to HSV-1 came out of a comparison of the susceptibility of newborn to adult mice. Andervont[46] first showed that newborn mice were much more susceptible to HSV-1 infection than adult mice. These studies were followed many years later by the observations of Johnson[48,74] that the age-related susceptibility correlated with the inability of macrophages of newborn mice to diminish replication of HSV-1 in culture. Other studies have shown that macrophage poisons such as silica particles or carrageenan killed these effector cells and resulted in marked diminution of resistance to an IP challenge with HSV-1.[49] In contrast, agents that activated macrophage function were found to augment resistance to HSV-1 infections in adult mice.[50]

Although these results suggest an important role for macrophages in resistance to HSV-1, each of the experimental approaches has a flaw, which makes the observation less than definitive. For example, natural killer (NK) cell function has also been shown to be deficient in newborn animals and may contribute to, or be responsible for, their susceptibility. Also, macrophage poisons and agents that activate the macrophages have been shown to have similar effects on the function of NK cells. Thus, resistance could be as easily attributable to NK cell function as to macrophage function.

Genetic resistance to HSV-1 was shown to be greatly diminished by treatment of mice with the various macrophage poisons.[20] Because of these results, studies were undertaken to determine whether these effector cells mediated resistance by sequestering virus replication in vitro.[51] Adherent peritoneal cells from resistant and susceptible mice replicated virus poorly when infected immediately after removing them from the mice but replicated HSV-1 to high titer when inoculated after 7 days of culture. The macrophages from genetically resistant mice consistently yielded lower peak titers of virus than did cells from susceptible mice, but this function did not segregate with genetic resistance, since cells from resistant F_1 mice replicated virus as well as those from the susceptible parents. Although these results indicate that this in vitro assay fails to correlate with genetic resistance in vivo, they do not rule out the possibility that macrophages play an important role in defense against this infection in mice. Thus, other studies suggest that macrophages are activated in vivo after infection with HSV-2 and might be required for resistance.[52]

Brucher et al.[53] have also evaluated replication of HSV-1 in macrophages from genetically susceptible and resistant mice. They found no virus replication with freshly isolated macrophages cultured in Dulbecco's modified Eagle medium, but they did find virus replication in macrophages cultured in RPMI 1640. Peak titers of HSV-1 replication were much lower with macrophage cultures from resistant mice than from susceptible mice. Although the numbers of infectious centers were equal, the plaques generated in monolayers of macrophages from resistant mice were considerably smaller than those found on monolayers from susceptible mice. The latter appeared to be due to the higher titers of interferon found in the supernatant fluids of HSV-1-infected, resistant macrophages. Studies were not carried out to determine whether this function segregated with genetic resistance to HSV-1.

The results of Mogenson[54,55] gave convincing evidence that macrophages play a crucial role in genetic resistance to HSV-2-induced hepatitis in mice. Macrophage poisons were found to greatly diminish genetic resistance, and nude mice, with activated macrophages, demonstrated greater resistance. Macrophages from genetically resistant mice inhibited the spread of virus in culture, whereas similar preparations from genetically susceptible mice failed to demonstrate this function. Most importantly, the capacity of mouse macrophages to inhibit virus spread was found to segregate with genetic resistance to hepatitis induction.[43] These results clearly indicate that a macrophage function is required for genetic resistance to HSV-2-induced hepatitis but does not indicate what that function might be. Thus, there may be a deficiency of a humoral factor that is required for the activation of these cells, and the deficiency may be in the production of or in the response to this factor.

2.4. Role of Natural Killer Cells in Resistance to HSV-1

As noted earlier, the characteristics of genetic resistance to HSV-1 in the mouse were strikingly similar to those of genetic resistance to bone marrow allographs.[20,33] Since NK cells have been shown to mediate allogeneic resistance,[56,57] studies have been undertaken to determine whether NK cells play a significant role in genetic resistance to HSV-1 in the mouse.

The first indications that NK cells might play a role in genetic resistance to HSV-1 came from the studies of ^{89}SR-treated mice.[33] Such treatment yields mice with normal macrophage, T cell, and B cell functions but with markedly depressed NK cell function.[32,56,58] Resistant mice treated with ^{89}Sr were found to be as susceptible to HSV-1 as the genetically most susceptible strains of mice tested.[33] In ^{89}SR-treated, HSV-1-challenged mice, the virus persisted in visceral tissue and was able to travel to the spinal cord. Although virus also replicated in visceral tissues of untreated mice, much of the virus was cleared from the tissues by 3 days after inoculation, and HSV-1 could not be detected in the spinal cords of untreated mice at any time. These results indicated that genetic resistance to HSV-1 is mediated by a marrow-dependent cell function that sequesters the infection early after challenge and does not allow it to spread to the central nervous system.

Several investigators have used more direct studies to evaluate the possible role of NK cells in resistance to HSV-1 infections in mice. Treatment of genetically resistant mice with anti-asialo GM_1 resulted in marked depletion of NK cell activity and increased mortality following challenge with HSV-1.[59] However, the protocol used in these studies also abrogated interferon production by peritoneal exudate cells, so it was difficult to determine whether susceptibility was due to diminished NK cell activity or to diminished interferon-generating capacity. Bukowski and Welsh[60] carried out a dose–response study with anti-asialo GM_1 in HSV-1-infected mice. Their results indicated that high doses of antibody blocked both early interferon production and NK activity and enhanced HSV-1 synthesis in inoculated mice. However, lower doses of anti-asialo GM_1 failed to block the early interferon production and had no effect on HSV-1 synthesis in challenged mice, even though NK activity was eliminated. These results indicated that resistance to HSV-1 in this model may not be dependent on NK cell function.

Using a different model system, Rager-Zisman et al.[61] provided direct evidence for a role of NK cells in protection against development of fatal HSV-1 infections in mice. These investigators rendered resistant mice susceptible to HSV-1 infection by treatment with cyclophosphomide 24 h prior to challenge. They then transferred normal spleen cells to treated animals and demonstrated that NK cells selectively transferred resistance, since spleen cells from NK-deficient beige mice failed to transfer resistance even though they transferred the interferon-generating cells. In this model, NK cells and not the interferon-generating cells appeared to be responsible for resistance to HSV-1. These conclusions differ from those of Bukowski and Welsh,[60] but this difference is probably dependent on the two very different model systems used.

2.5. Role of Interferon in Resistance of Mice to HSV-1 and HSV-2

Thirty years ago, Isaacs and Lindenmann[62] first described interferon as a factor that was capable of converting susceptible, uninfected cells into cells resistant to virus challenge. Since those early studies, many other biological roles have been ascribed to this cytokine, all induced by the interaction of interferon with the cell membrane.[63] Three major types of interferon are known: interferon-alpha is made by leukocytes; interferon-beta is made mostly by fibroblasts; and interferon-gamma is made by sensitized lymphocytes in response to the sensitizing antigen.[64] The production of interferon-alpha and interferon-beta is an immediate reaction to virus infection and therefore constitutes a part of the natural defense system.

Because of their capacity to inhibit virus replication, interferons have always been thought to be important mediators of defense against virus infections. In addition, more recent findings suggest that interferons may have other roles in host defense. Interferons may act by augmenting the efficiency of cytolytic NK cells or by recruiting pre-NK cells to differentiate into mature, functional effector cells.[65] Interferons may also act to activate macrophage function, another aspect of the host natural resistance system.[66] In this manner, interferons may act to potentiate other aspects of natural resistance systems necessary for host defense.

Extensive studies by Gresser et al.[67] showed that administration of anti-mouse interferon globulin into mice subsequently challenged with HSV-1 resulted in an earlier onset of disease and greatly increased mortality. In treated mice, much lower doses of virus were found to be lethal when inoculated by either the IP or subcutaneous route. Comparable results have been obtained using genetically resistant mice (Lopez et al., unpublished observation). Antiinterferon pretreatment of mice greatly reduces the amount of HSV-1 causing a fatal infection. Using a slightly different system, Bukowski and Welsh[60] showed that antibody to interferon greatly diminished the ability of adoptively transferred leukocytes to protect HSV-1-challenged suckling mice. Taken together, these studies strongly indicate that endogenously produced interferon plays an important, early role in host defense against HSV-1 infection. Similar studies by Wrzos et al.[68] have shown that antibody to interferon greatly diminishes resistance to an intravaginal challenge with HSV-2 in genetically resistant mice.

A number of studies from Kirchner's group[69–77] have documented a close correlation between the ability of mice to produce an early interferon response upon challenge with HSV-1 and genetic resistance to that infection. These investigators found that 2–4 h after an IP challenge with HSV-1, genetically resistant mice generated a significantly higher interferon response in peritoneal exudate than did the genetically susceptible mice. Evaluation of a number of inbred strains of mice indicated that there was a good correlation between the ability to make this early interferon and genetic resistance to HSV-1. Although the data that have been developed appear to be overwhelming, some recent observations fail to support this conclusion. For example, we have found that CBA mice are genetically susceptible to HSV-1, even though they make a strong early interferon response to challenge with HSV-1 (Lopez, unpublished observation). Also, Cmielarczyk et al.[36] showed that anti-thy-1.2 treatment of mice failed to diminish the early interferon response to HSV-1 challenge yet abrogated genetic resistance. These mice were found to be susceptible to challenge with HSV-1 even though they were capable of generating an early interferon response to that challenge.

Interferon has also been closely linked with an autointerference phenomenon induced by challenge of resistant mice with high doses of HSV-1 or HSV-2.[73,74,78] Zawatzky et al.[73,74] first noted that challenging genetically resistant mice with 250 LD_{50} of virus resulted in more rather than fewer mice surviving infection. This phenomenon was associated with the induction of high titers of interferon at the injection site 2–4 h after inoculation. Antiserum to mouse interferon rendered mice challenged with

250 LD$_{50}$ of HSV-1 susceptible to infection. Thus, interferon induced by the challenge virus appears to protect the genetically resistant animals against high doses of HSV-1. Genetic analysis revealed that HSV-induced interferon production is governed by several loci, one of which is X-linked. In similar studies, Pedersen et al.[78] demonstrated X-linked resistance of mice to high doses of HSV-2 and showed that this too correlates with early interferon production. Mice survived IP challenge with 10^6 PFU of HSV-2, whereas mice died when inoculated with 10^5 PFU. Both the autointerference phenomena and the early interferon production were shown to be influenced by loci on the X chromosome.

2.6. Mechanisms Responsible for Genetic Resistance to HSV-1 in Mouse

Although the model of genetic resistance to HSV-1 has been explored by many groups over the past 12 years, we still do not know the mechanisms responsible for genetic resistance to HSV-1 in the mouse. Clearly, resistance is dependent on natural defense mechanisms operative early after challenge with virus. We also know that the defense mechanisms responsible for resistance to HSV-1 are marrow-dependent. Nevertheless, even though interferon production, NK activation, and macrophage function have been clearly shown in various model systems to play a role in resistance against HSV-1 infections, conclusive evidence showing that one of them is responsible for genetic resistance to HSV-1 has not been forthcoming. On the other hand, the autointerference phenomenon described above is clearly induced by interferon generation, but the genetics of this model differ markedly from the genetics of resistance to HSV-1 described earlier.[12] Studies are therefore still required to define the mechanisms responsible for genetic resistance to HSV-1 and the cellular interactions leading to that resistant state.

3. Human Studies

Studies of natural defense systems operative against HSV-1 and HSV-2 infections in man have focused on the roles of interferon-alpha generated in response to virus infection and NK cells that lyse virus-infected targets. Although some studies have been carried out evaluating the antiviral role of monocytes/macrophages in man,[79–82] the difficulty of obtaining sufficient numbers of pure cells for study has limited productive investigation in this area. In the following discussion we shall focus our attention on the biology of NK cells that lyse HSV-1-infected fibroblasts and the cells that generate interferon-alpha in response to HSV-1. We shall then consider the roles these natural effector systems might play in resistance to HSV-1 in man.

3.1. Characteristics of Natural Killer Cells That Lyse HSV-1-Infected Fibroblasts

The first suggestion that NK cells might be important in virus infections was the observation that these effector cells could preferentially lyse virus-infected targets *in vitro*.[83–85] Since these early studies, many published reports have described lysis of virus-infected targets by human effector cells. The cells mediating this lysis have been shown to be NK cells by the following criteria: (1) Effector cells from seronegative controls lysed targets, indicating that presensitization of the effector cells was not necessary; (2) lytic effector cells did not express the cell surface markers characteristic of mature T cells, B cells, or macrophages; and (3) lysis of target cells was not genetically restricted with respect to the major histocompatibility antigens of the host.

Because of our interest in the possible role of NK cells in resistance to herpesvirus infections in man, we developed an assay using HSV-1-infected fibroblasts as targets (NK(HSV-FS)).[85] The effectors of NK(HSV-FS) are similar to the effectors that lyse the commonly used K562 erythroleukemia targets (NK(K562)) in that they are large granular lymphocytes[86–88] that do not require prior sensitization to be operative. As with NK cells in the mouse,[57] we have shown that NK(HSV-FS) effectors are derived from bone marrow stem cells[89] and are marrow-dependent.[90] Although the effector cells that lyse HSV-FS are similar to the effectors that lyse K562 targets in these basic characteristics, we have found that these cells differ in other characteristics, indicating that NK cells are heterogeneous.

3.2. Heterogeneity of NK Cells

Heterogeneity of NK cells was first demonstrated for mouse effector cells. Various subpopulations of lytic cells could be differentiated by their susceptibility to ^{89}SR,[91,92] by their cell surface markers,[93] and by cold-target inhibition or adsorption studies.[94,95] The use of different target cells in ^{51}Cr-release assays can detect one or more of the subpopulations of NK effector cells.[96]

We have studied the characteristics of NK(HSV-FS) effectors and compared them with the effector cells that lyse K562 erythroleukemia tumor targets (NK(K562)). Although NK cells clearly mediate lysis of each target, the effector cells that lyse these targets represent two different subpopulations of NK cells.[88] Using monoclonal antibodies, we have shown that both the NK(K562) and the NK(HSV-FS) effectors express CD-16, a cell surface determinate found on all NK cells. However, monoclonal antibodies to other cell surface markers clearly differentiated the subpopulations of effectors that mediate lysis of these targets. Some of the NK(K562) effectors were found to be positive for CD-1 and CD-4 (pan T cell) surface markers, whereas NK(HSV-FS) effectors were negative. Also, most NK(K562) cells were found to be positive for CD-2 cell surface marker, but NK(HSV-FS) effectors were again negative.

Cold-target inhibition studies have been used to indicate that the effector cells that lyse these two targets recognize different structures on the target cells.[88] Unlabeled K562 target cells reduced the lysis of ^{51}Cr-labeled K562 targets but reduced lysis of ^{51}Cr-HSV-FS targets to a much lower extent. Conversely, infected or uninfected fibroblasts reduced the lysis of ^{51}Cr-HSV-FS targets better than did K562 cells but reduced ^{51}Cr-K562 lysis relatively poorly.

Results from the study of certain patient groups provided the best evidence for heterogeneity of human NK effector cells and also indicated that these activities are under independent regulation *in vivo*.[88,96] For example, we found that patients with Wiskott-Aldrich syndrome (WAS) consistently demonstrated low levels of NK(HSV-FS) activity but normal NK(K562) activity. Conversely, other groups of individuals were found to have consistently normal levels of NK(HSV-FS) activity but low levels of NK(K562) activity.[88,96] Since consistent and significant deficiencies have been found in both directions, these results provide strong evidence that two subpopulations of effector cells mediate lysis of these targets and that these cells are independently regulated *in vivo*.

More recent studies provide further evidence to suggest that effector cells that lyse K562 targets are different from those that lyse HSV-FS targets. Thus, Fitzgerald-Bocarsly et al.[97] have shown that lysis of HSV-FS targets requires an accessory cell function whereas lysis of K562 targets does not. In these experiments, relatively pure populations of CD-16 positive cells were found to lyse K562 targets efficiently but not to lyse HSV-FS targets. Addition of an HLA-DR-positive subpopulation of cells to the CD-16-positive effector cells restored their ability to lyse the HSV-FS targets. Populations of cells enriched for dendritic cells by centrifugation on hypertonic metrizamide gradients also restored lytic capacity to CD-16-positive cells, indicating that the accessory cell function is probably provided by dendritic cells. The accessory cell function has also been shown to be required for NK of cytomegalovirus-infected fibroblasts[98] but not for lysis of HSV-1-infected B lymphoblastoid (Raji) cell lines. The mechanism by which the lysis of HSV-FS but not HSV-Raji targets requires participation of an accessory population is under investigation. One possible explanation is that different effector subpopulations are involved in the lysis of HSV-FS and HSV-Raji targets. Indeed, cold-target inhibition studies, kinetic studies, and blocking studies have all indicated that HSV-

FS, HSV-Raji, and K562 targets are lysed by at least partially nonoverlapping populations of effectors.[99] Alternatively, the infected fibroblasts may lack an activating or triggering structure that is present on the HSV-Raji targets. Together, our results indicate that the interaction of NK effector cells and HSV-infected targets is dependent on the specific target cell used in the assay and is not directed solely by the infecting virus. This conclusion is similar to that obtained by us[100] using mouse cell lines infected with HSV-1 and splenic effector cells.

3.3. Preferential Lysis of HSV-Infected Fibroblasts

As noted above, seropositive and seronegative individuals have peripheral blood mononuclear cells capable of recognizing and lysing HSV-FS in a dose-dependent manner while not lysing the uninfected targets.[85] A number of studies have focused on trying to determine what the virus infection does to the fibroblasts to make them better targets for NK effectors.

3.3.1. Role of Interferon

Early in the study of NK against virus-infected cells, Trinchieri et al.[101,102] and Santoli et al.[84] noticed that the virus-infected NK targets were capable of inducing interferon production during routine NK assays. Because the interferon was produced by cells that appeared to share many of the characteristics of NK cells themselves, these authors suggested that the induction of interferon during the NK cell assays might lead to the nonspecific augmentation of NK activity and thus be responsible for the preferential lysis of the infected targets when compared to the uninfected targets. In more recent studies, we[103] have evaluated interferon production by human effector cells in the NK(HSV-FS) assay and have found that, although interferon-alpha is produced during this assay, it cannot account for the preferential lysis of the virus-infected targets. Interferon was produced during the NK(HSV-FS) assay by effector cells from both seropositive and seronegative individuals and was shown to have the properties of interferon-alpha. However, we could find no correlation between the levels of cytotoxicity achieved in the NK assay and the amount of interferon-alpha generated during the assay. Furthermore, when we added sufficient antiinterferon-alpha antibody to neutralize all the antiviral activity generated during the assay, cytotoxicity against the virus-infected targets was not reduced. These results have recently been confirmed by others,[104] also using HSV-1-infected targets.

Patient studies have also shown that interferon production is not required for NK(HSV-FS) and that these two functions are regulated independently in vivo.[96,103] Five patients with severe combined immunodeficiency disease were found to have normal levels of NK(HSV-FS) activity despite their inability to produce interferon-alpha during the cytotoxicity assay.[96] Conversely, we have found that patients with Wiscott-Aldrich syndrome often have abnormally low levels of NK(HSV-FS) activity but a normal capacity to generate interferon-alpha in the same assay.[96]

Taken together, these results show that interferon-alpha is normally generated during the NK(HSV-FS) assay but is not required for lysis of those target cells. Nevertheless, it is possible that interferon, made in response to a virus infection *in vivo*, could lead to the recruitment and activation of NK cells which would have a positive self-regulatory effect and increase NK cell activity at the place and time of greatest need.

3.3.2. HSV-Infection May Destabilize Target Cell

Since HSV-1 infection of fibroblasts is lytic, another possible explanation for the increased susceptibility of virus-infected targets to NK cell lysis is that the virus destabilizes the host cell membrane and impairs its ability to repair itself, leaving the infected target cell inherently more susceptible to lysis than the intact uninfected target. To evaluate this possibility, infected and uninfected FS cells were subjected to hypotonic conditions ranging from 0% to 67% distilled water in culture media during a 14-h assay period. Under these conditions, infected and uninfected cells were found to be equally sensitive to osmotic shock, indicating that they did not differ in their inherent stability. Similarly, we found that HSV-infected and uninfected targets were equally susceptible to lysis by cytotoxic T cells directed at major histocompatibility determinates (Fitzgerald-Bocarsly et al., in preparation). These results indicated that the virus-infected cells were not more easily lysed than the uninfected cells because they were less stable and suggested that NK effectors were induced to lyse HSV-FS but not FS targets.

3.3.3. Role of HSV-Replication and Gene Products

What about the virus infection that makes HSV-FS better targets for NK effectors than the uninfected cells? As with vesicular stomatitis virus-infected targets,[105] viral replication is required for the target cells to be preferentially lysed. Thus, UV-inactivated HSV-1 failed to replicate in the fibroblast cells, and inoculated targets were not preferentially lysed even though virus antigens had been adsorbed to the surface of those cells (Fitzgerald-Bocarsly et al., in preparation).

Experiments were carried out to determine whether the expression of immediate early (alpha), immediate (beta), or late (gamma) viral gene products were required for the induction of preferential lysis of targets (Fitzgerald-Bocarsley et al., in preparation). To evaluate the role of the gamma gene products in preferential lysis of HSV-FS targets, NK experiments were carried out in the presence of phosphonoacetic acid (PAA), an inhibitor of HSV-1 DNA polymerase and viral DNA replication, which, in turn, is required for gamma gene expression.[106] Since only the gamma gene products made from transcripts of input DNA are made in the presence of PAA, little or no gamma gene products are found in treated cells. However, at multiplicities of infection of 1 and 0.2, treated and untreated HSV-FS were killed equally well. These results suggest that gamma gene products of HSV-1 and, thus, viral glycoproteins are not required for the induction of NK. These results are not in agreement with the results of Bishop et al.,[104,107,108] who proposed that the expression of viral glycoproteins was required for susceptibility of infected targets to NK lysis. However, the latter studies, as well as the studies of Daher and Betz,[109] utilized targets 24 h after inoculation with HSV-1 and appear to be primarily evaluating antibody-dependent cellular cytotoxicity rather than natural kill. In these studies the level of cytotoxicity depended specifically on the serostatus of the subjects under study.[109] Earlier studies showed that antibodies to herpes simplex glycoproteins, even in very low concentrations, efficiently mediated ADCC with HSV-1-infected targets.[110] Since Bishop et al.[104,107,108] also used targets 24 h after infection, cytophilic antibody may be responsible for the "specificity" demonstrated by these investigators.

Studies were also undertaken to determine whether earlier gene products of HSV-1 might be responsible for conferring susceptibility to infected targets. A temperature-sensitive mutant of HSV-1, tsLB2, fails to make beta or gamma gene products at the nonpermissive temperature (39°C) and overproduces the alpha gene product ICP4.[111] Fibroblasts infected with tsLB2 were killed even more efficiently at the nonpermissive temperature than were targets infected with the wild-type HSV-1. These results suggest that normal production of beta or gamma gene products is not essential for conferring susceptibility of fibroblasts to lysis of NK cells and that an overproduction of the alpha gene product ICP4, correlates with augmented lysis. Similarly, Borysiewicz et al.[112] showed that an immediate early gene product of cytomegalovirus induced NK lysis of infected fibroblasts.

3.3.4. Target Binding and Triggering of NK Activity

Results of cold-target inhibition studies suggest that HSV-FS and FS targets displayed structures that are recognized and bound by the subpopulation of effector cells that lyse HSV-FS targets.[88] This conclusion was confirmed using the single-cell assay (Fitzgerald-Bocarsley et al., in preparation). Human effector cells, enriched for NK effectors, were shown to bind HSV-FS and FS targets equally well in agarose.[113] Although conjugate formation was the same, only the HSV-FS/effector cell conjugates were efficiently lysed. Thus, infected and uninfected fibroblasts express structures recognized by NK effector cells, but only the virus-infected targets trigger the NK lytic function.

3.4. Interferon-Alpha-Generating Cell

As noted earlier, interferon-alpha is generated during the normal NK assay using HSV-1-infected fibroblasts as targets.[102,103] The interferon-alpha-generating cells are similar to NK cytotoxic cells in that they are null cells,[114,115] are found in Percoll gradient fractions enriched for large granular lymphocytes,[116,116a] and require no presensitization to be operative.[102,103] More recent studies, however, have shown that the interferon-alpha-generating cells can be distinguished from the lytic NK effectors by cell surface markers as well as by patient studies.[96,103,116a,117] The interferon-alpha-generating cells lack CD2, CD15, and CD16 cell surface determinates found on all or most NK lytic effector cells. However, the interferon-alpha-generating cells were found to express HLA-DR on their cell surface, a marker also expressed on the accessory cell for NK(HSV-FS).

Patient studies also indicate that the NK cytolytic effector cells and interferon-alpha-generating cells are independently regulated functions. Thus, we have found a group of four patients with severe combined immunodeficiency disease who demonstrated normal NK(HSV-FS) but abnormally low capacity to produce interferon-alpha in the same assay.[96,103] Conversely, five patients with Wiscott-Aldrich syndrome were found to have abnormally low NK(HSV-FS) activity but normal capacity to make interferon-alpha within the same assay. These results clearly indicate that the interferon-alpha-generating cells represent a subpopulation of cells different from NK lytic effectors and that this cell function is regulated independently *in vivo*.

3.5. Role of NK Cells in Resistance to HSV Infections in Man

An indication that NK cells play an important role in defense against HSV infections in man came from the studies showing that these effector cells are capable of reducing the amount of virus produced by infected cells. The studies of Fitzgerald et al.[118] first showed an antiviral effect of NK cells. These investigators measured the amount of HSV-1 secreted into supernatant fluids of cultures containing effector cells and HSV-1-infected targets at the termination of the NK assay and compared these results to HSV-1 produced by infected fibroblasts in the absence of effectors. Cytotoxic effector cells reduced the virus in a dose-dependent manner, with effector-to-target cell ratios of 200:1 reducing yield by 90% or more. Reduction in virus yield was mediated by NK effector cells, since complement elimination with monoclonal antibodies to NK-specific cell-surface determinate abrogated this function. Although interferon-alpha was generated during these assays, this cytokine was shown not to be responsible for the reduction of HSV-1 replication in experiments in which anti-interferon-alpha antibody neutralized all the antiviral activity generated during the assay. These data indicate that the cytotoxic effector cells themselves were directly able to inhibit virus replication, probably by lysing target cells prior to the production of infectious progeny.

Using an assay that required a longer incubation period, Leibson *et al.*[119] showed that CD16-positive mononuclear cells suppressed HSV replication *in vitro*. Interferon-alpha and -gamma were produced during the 3-day culture, and neutralization of either reduced the ability of mononuclear cells to suppress virus replication. Thus, NK cells were shown to suppress HSV-1 replication in an interferon-dependent manner. Of interest was their subsequent finding[120] that NK cells in cord blood mononuclear cells were deficient in both NK cell activity against HSV-infected fibroblasts and the ability to inhibit HSV-1 replication. These deficient functions correlated well with the newborn's known susceptibility to severe disease when infected with HSV-1. In other studies, Yasukawa and Kobayashi[121] used human NK clones that lysed HSV-1-infected lymphoid cell lines to show that the NK cells were capable of suppressing HSV-1 replication in these target cells.

Patient studies have been undertaken to determine whether NK effectors play an important role in resistance to HSV infections in man. We have studied the NK(HSV-FS) responses of patients known to be highly susceptible to unusually severe disease caused by HSV infections.[132] Effector cells from cord bloods and newborns were evaluated because of the newborn's known susceptibility to disseminated HSV infections. Only 30% of the cord bloods tested demonstrated responses within 2 SD of the normal, adult mean. Furthermore, peripheral blood mononuclear cells from nine premature infants were also tested and found to have responses > 3 SD below the normal mean[132] (Lopez *et al.*, unpublished observations). These observations have been confirmed by Kohl *et al.*,[123] who also showed that lysis by cord blood cells could not be augmented by pretreatment with interferon-alpha.

We have also studied the NK cell capacity of patients with Wiscott-Aldrich syndrome (WAS), an X-linked, primary immunodeficiency disorder characterized by eczema, thrombocytopenia, and recurrent infections.[96,122] Only one of the 14 WAS patients studied to date demonstrated normal NK(HSV-FS) activity, and the low NK(HSV-FS) activity correlated with a history of persistent or recurrent infections in these patients. In contrast, many of the WAS patients demonstrated normal NK(K562) activity in the face of deficient NK(HSV-FS) activity. The normal NK(K562) activity is consistent with the observations of Lipinski *et* al.[124] and suggests that the heterogeneity of NK cells may reflect the different biological roles of the subpopulations of effector cells. Alternatively, low NK(HSV-FS) may also reflect a deficiency of an accessory cell function rather than a malfunction of the NK effector.

Newborns and patients with WAS are known to have other immunodeficiencies which might account for or contribute to their susceptibility to HSV infections.[125,126] We have therefore evaluated NK capacity of patients suffering from unusually severe virus infections but with no known underlying primary or secondary cellular immunodeficiency.[122] This group of individuals also demonstrated NK(HSV-FS) responses > 2 SD below the normal mean. Two of the patients studied were also evaluated between bouts of reactivated infection, and each demonstrated depressed responses, suggesting that the low NK(HSV-FS) activity might predispose such patients to unusually severe disease caused by HSV rather than be a result of the infections. Again, some of the patients with greatly diminished NK(HSV-FS) activity were found to have normal NK(K562) capacity, suggesting that the function of the subpopulation of effector cells that lyse the HSV-FS targets gave the best correlation with resistance and may be more relevant when evaluating resistance mechanisms against virus infections.

3.6. Role of Interferon-Alpha Generation in Resistance to HSV Infections in Man

Patients with acquired immunodeficiency syndrome (AIDS) are susceptible to a variety of intracellular pathogens which usually cause severe disease only in patients with primary or secondary immunodeficiency disorders.[129] Early in the study of AIDS, we noted that these patients were also susceptible to chronic, ulcerative HSV infections rather than the self-limited lesions found in otherwise normal individuals.[128] Patients with AIDS and opportunistic infections have been shown to have a wide variety of immunological perturbations affecting both adaptive and natural defense systems. However, many of these deficiencies have also been found in homosexual controls who do not develop AIDS or in non-AIDS patients with acute virus infections. Many of the immunological abnormalities found in AIDS patients appear to be the result of infections.[127]

In a study of over 400 patients with AIDS or at risk of developing AIDS, we have found that the best correlates of susceptibility to opportunistic infections were absolute numbers of CD4-positive helper cells below 250/mm^3 and the deficiency of interferon-alpha generation by mononuclear cells challenged with HSV-FS.[129,130] These deficiencies were synergistic, and using this combination we were able to predict the individuals who would develop opportunistic infections and who were at greatest risk of dying from those infections. Thus, a deficiency of interferon-alpha generation, especially when found in the context of deficiency of CD4-positive helper cells, was associated with marked susceptibility to unusually severe disease caused by HSV.

4. Conclusions

Natural resistance mechanisms clearly play a decisive role early during a primary HSV infection. These systems are immediately available to respond to an invading microorganism and inhibit local replication and systemic spread of that infection. Failure of the natural defense system would theoretically result in widely disseminated infection long before an adaptive immune response could be induced and therefore capable of clearing that infection. For HSV infections, control of the primary infection probably also results in a diminution in the amount of virus that becomes latent in the host. The latter could result in a reduction of reactivated virus and, thus, less recrudescent disease. In addition, the natural defense mechanisms may have a direct effect of diminishing the replication and spread of reactivated latent HSV infection.

Study of the biology of the NK(HSV-FS) has led to a better understanding of the heterogeneity of these effector cells, the cellular interactions that result in the lysis of target cells, the target cell structures that bind NK effectors, and the determinants that trigger NK lysis of targets. Further understanding of the basic biology of these effector cells and the interactions required for the normal response is a necessary first step toward developing new modalities of treatment that either augment these defense mechanisms or replace necessary humoral factors in order to circumvent deficiencies. As noted frequently in this contribution, however, the heterogeneity of NK effector cells must be considered when any of these studies are being undertaken.

References

1. Notkins, A. L., 1974, Commentary: Immune mechanisms by which the spread of viral infection is stopped, *Cell Immunol.* **11**:478–83.
2. Allison, A. C., 1974, Interactions of antibodies, complement components and various cell types in immunity against virus and pyogenic bacteria, *Transplant. Rev.* **19**:3–55.
3. Nahmias, A. J. and Roizman, B., 1973, Infection with herpes-simplex viruses 1 and 2, *N. Engl. J. Med.* **299**:667–674, 719–725, 781–789.
4. Roizman, B., 1974, Herpesviruses, latency and cancer, *J. Reticuloendothel. Soc.* **15**:312–321.
5. Corey, L., 1982, The natural history of genital herpes simplex virus. Perspectives on an increasing problem, in: *The Herpesviruses* (B. Roizman, C. Lopez, eds.), Plenum Press, New York, pp. 1–35.
6. Dowdle, W. R., Nahmias, A. J., Harwell, R. W., and Pauls, F. P., 1967, Association of antigenic type of herpesvirus hominis with site of viral recovery, *J. Immunol.* **99**:974–980.

7. Stevens, J. G., 1975, Latent herpes simplex virus and the nervous system, *Curr. Top. Micro. Immunol.* **70:**31–50.
8. Hill, T. J., Field, H. J., and Blyth, W. A., 1975, Acute and recurrent infection with herpes simplex virus in the mouse: A model for studying latency and recurrent disease, *J. Gen. Virol.* **28:**341–353.
9. Rouse, B. T., 1984, Cell-mediated immune mechanisms, in: *Immunobiology of Herpes Simplex Virus Infection* (B. T. Rouse, C. Lopez eds.), CRC Press, Boca Raton, FL, pp. 107–120.
10. Abghari, S. Z., Stulting, R. D., Nigida, S. M., Downer, D. N., Kindle, J. C., and Nahmias, A. J., 1986, Spread of HSV and establishment of latency after corneal infection in inbred mice, *J. Invest. Ophthalmol.* **27:**77–82.
11. Bang, F. B., 1978, Genetics of resistance of animals to viruses: Introduction and studies in mice, *Adv. Virus Res.* **23:**269–347.
12. Lopez, C., 1975, Genetics of natural resistance to herpesvirus infections in mice, *Nature* **258:**152–153.
13. Kirchner, H., Kochen, M., Hirt, H. M., and Munk, K., 1978, Immunological studies of HSV infection of resistant and susceptible inbred strains of mice, *Z. Immun. Forsch.* **154:**147–154.
14. Kirchner, H., Hirt, H. M., Rosenstreich, D. L., and Mergenhagen, S. E., 1978, Resistance of C3H/HeJ mice to lethal challenge with herpes simplex virus, *Proc. Soc. Exp. Biol. Med.* **157:**29–32.
15. Zawatzky, R., Hilfenhaus, J., Marucci, F., and Kirchner, H., 1981, Experimental infection of inbred mice with herpes simplex virus type 1. I. Investigation of humoral and cellular immunology and of interferon induction, *J. Gen. Virol.* **43:**31–38.
16. Caspary, L., Schindling, B., Dundarov, S., and Falke, D., 1980, Infections of susceptible and resistant mouse strains with herpes simplex virus type 1 and 2, *Arch. Virol.* **65:**219–227.
17. Shellam, G. R., and Flexman, J. P., 1986, Genetically determined resistance to murine cytomegalovirus and herpes simplex virus in newborn mice, *J. Virol.* **58:**152–156.
18. Lopez, C., 1980, Resistance to HSV-1 in the mouse is governed by two major, independently segregating non-H-2 loci, *Immunogenics* **11:**87–92.
19. Mogensen, S. C., 1979, Role of macrophages in natural resistance to virus infections, *Microbiol. Rev.* **43:**1–26.
20. Lopez, C, 1978, Immunological nature of genetic resistance of mice to herpes simplex virus type 1 infection, in: *Oncogenesis and Herpesviruses III* (G. de The, W. Henle, F. Rapp, eds.), IARC, Lyon, France, pp. 775–778.
21. Rajcani, J., Gajdosova, E., and Mayer, V., 1974, Pathogenesis of herpesvirus hominis infection in immunosuppressed mice, *Acta Virol.* **18:**135–142.
22. Rager-Zisman, B., and Allison, A. C., 1976, Mechanisms of immunologic resistance to herpes simplex virus 1 (HSV-1) infection, *J. Immunol.* **116:**35–40.
23. Hough, V., and Robinson, T. W. E., 1975, Exacerbation and reactivation of herpesvirus hominis infection in mice by cyclophosphamide, *Arch. Virol.* **48:**75–83.
24. Armerding, D., Scriba, M., Hren, A., and Rossiter, H., 1982, Modulation by cyclosporin A of murine natural resistance against herpes simplex virus infection. I. Interference with the susceptibility to herpes simplex virus infection, *Antivir. Res.* **2:**3–11.
25. Armerding, D., and Rossiter, H., 1981, Induction of natural killer cells by herpes simplex virus type 2 in resistant and sensitive inbred mouse strains, *Immunobiology* **158:**369–379.
26. Schneweis, K. E., and Saftig, V., 1981, The vaginal herpes simplex virus infection of resistant (C57B1) mice, *Int. Herpesvirus Workshop*, p. 144.
27. Lopez, C., 1981, Resistance to herpes simplex virus-type 1 (HSV-1), in: *Natural Resistance to Tumors and Viruses* (O. Haller, ed), Springer-Verlag, New York, pp. 15–24.
28. Price, R. W., and Schmitz, J., 1978, Reactivation of latent herpes simplex virus infection of the autonomic nervous system by post-ganglionic neurectomy, *Infect. Immun.* **19:**523–532.

29. Cudkowicz, G., 1975, Genetic control of resistance to allogeneic and zenogeneic bone marrow grafts in mice, *Transplant Proc.* **7:**155–159.
30. Cudkowicz, G., and Bennett, M., 1971, Peculiar immunobiology of bone marrow allografts. I. Graft rejection by irradiated responder mice, *J. Exp. Med.* **134:**83–102.
31. Bennett, M., 1973, Prevention of marrow allograft rejection with radioactive strontium: Evidence for marrow-dependent effector cells, *J. Immunol.* **110:**510–516.
32. Bennett, M., Baker, E. E., Eascott, J. W., Kumar, V., and Yonkosky, D., 1976, Selective elimination of marrow precursors with the bone-seeking isotope [89]Sr: Implications for hemopoesis, lymphopoesis, viral leukemogenesis and infection, *J. Reticuloendothel. Soc.* **20:**71–87.
33. Lopez, C., Ryshke, R., Bennett, M., 1980, Marrow-dependent cells depleted by ^{89}Sr mediate genetic resistance to herpes simplex virus type 1 infections in mice, *Infect. Immun.* **28:**1028–1032.
34. Zawatzky, R., Hilfenhaus, J., and Kirchner, H., 1979, Resistance of nude mice to herpes simplex virus and correlation with *in vitro* production of interferon, *Cell. Immunol.* **47:**424–428.
35. Mori, R., Takeya, K., Minamishima, Y., and Tasiki, T., 1965, Effect of thymectomy on experimental viral infections of mice. I. Herpes simplex virus and Coksaki B5 virus, *Proc. Jpn. Acad.* **41:**975–982.
36. Chmielarczyk, W., Engler, H., Ernst, R., Optiz, U., and Kirchner, H., 1985, Injection of anti-thy-1.2 serum breaks genetic resistance of mice against herpes simplex virus, *J. Gen. Virol.* **66:**1087–1094.
37. Schlabach, A. J., Martinez, D., Field, A. K., and Tytell, A. A., 1979, Resistance of C57 mice to primary systemic herpes simplex virus infection: Macrophage dependence and T-cell independence, *Infect. Immun.* **26:**615–620.
38. Kastrukoff, L. F., Lau, A. S., and Puterman, M. L., 1986, Genetics of natural resistance to herpes simplex virus type 1 latent infection of the peripheral nervous system in mice, *J. Gen. Virol.* **67:**613–621.
39. Abghari, S. Z., Stulting, R. D., Nigida, S. M., Downer, D. N., Kindle, J. C., and Nahmias, A. J., 0000, Spread of HSV and establishment of latency after corneal infection in inbred mice, *Invest. Ophthalmol.* **27:**77–82.
40. Harnett, G. B., and Schellam, G. R., 1982, Variation in murine cytomegalovirus replication in fibroblasts from different mouse strains *in vitro*: Correlation with *in vivo* resistance, *J. Gen. Virol.* **62:**39–47.
41. Collier, L. H., Scott, Q. J., and Pani, A., 1983, Variation in resistance of cells from inbred strains of mice to herpes simplex virus type 1, *J. Gen. Virol.* **64:**1483–1490.
42. Abghari, S. Z., Stulting, R. D., Nigida, S. M., Downer, D. N., Kindle, J. C., and Nahmias, A. J., 1986, Replication of herpes simplex virus in fibroblast cells from inbred mice, *J. Invest. Ophthalmol.* **27:**57–82.
43. Mogensen, S. C., 1976, Biological conditions influencing the focal necrotic hepatitis test for differentiation between herpes simplex virus type 1 and 2, *Acta Pathol. Microbiol. Scand. (B)* **84:**154–158.
44. Mogensen, S. C., Teisner, B., and Andersen, H. K., 1974, Focal necrotic hepatitis in mice as a biological marker for differentiation of herpesvirus hominis type 1 and 2, *J. Gen. Virol.* **25:**151–155.
45. Hirsch, J. G., and Fedorko, M. E., 1970, Morphology of mouse mononuclear phagocytes, in: *Mononuclear Phagocytes* (R. von Furth, ed.), Blackwell, Oxford, U.K., p. 7.
46. Andervout, H. B., 1927, Activity of herpetic viruses in mice, *Am. J. Hyg.* **14:**383–393.
47. Johnson, R. T., 1964, The pathogenesis of herpesvirus encephalitis. I. Virus pathways to the nervous system of suckling mice demonstrated by fluorescent antibody staining, *J. Exp. Med.* **119:**343–358.
48. Johnson, R. T., 1964, The pathogenesis of herpesvirus encephalitis. II. A cellular basis for the development of resistance with age, *J. Exp. Med.* **120:**359–373.
49. Hirsch, M. S., Zisman, B., and Allison, A. C., 1970, Macrophages and age-dependent resistance to herpes simplex virus in mice, *J. Immunol.* **104:**1160–1165.

50. Halpern, B., Frey, A., Crepin, O., Platica, O., Lorinet, A. M., Rabourdin, A., Sparros, L., and Isac, R., 1973, *Corynebacterium parvum*, a potent immunostimulant in experimental infections and in malignancies, in: *Immunopotentiation*, Vol. 18, Ciba Foundation Symposium, Associated Scientific Publishers, New York p. 217.

51. Lopez, C., and Dudas, G., 1979, Replication of herpes simplex virus type 1 in macrophages from resistant and susceptible mice, *Infect. Immun.* **23:**432–437.

52. Armerding, D., Mayer, P., Scriba, M., Hren, A., and Rossiter, H., 1981, *In vivo* modulation of macrophage functions by herpes simplex virus type 2 in resistant and sensitive inbred mouse strains, *Immunobiology* **160:**217–227.

53. Brucher, J., Domke, I., Schroder, C. H., and Kirchner, H., 1984, Experimental infection of inbred mice with herpes simplex virus. VI. Effect of interferon on *in vitro* virus replication in macrophages, *Arch. Virol.* **82:**83–93.

54. Mogensen, S., 1977, Role of macrophages in hepatitis induced by herpes simplex virus types 1 and 2, *Infect. Immun.* **15:**686–691.

55. Mogensen, S. C., 1978, Macrophages and age-dependent resistance to hepatitis induced by herpes simplex virus type 2 in mice, *Infect. Immun.* **19:**46–59.

56. Kiessling, R., Hochman, P. S., Haller, O., Shearer, G. M., Wigzell, H., and Cutkowicz, G., 1977, Evidence for a similar or common mechanism for natural killer cell activity and resistance to hemopoietic grafts, *Eur. J. Immunol.* **7:**663–669.

57. Haller, O., Kiessling, R., Orn, A., and Wigzell, H., 1977, Generation of natural killer cells: An autonomous function of the bone marrow, *J. Exp. Med.* **145:**1411–1416.

58. Morahan, P. S., Coleman, P. H., Morse, S. S., and Volkman, A., 1983, Resistance to infections in mice with defects in the activities of mononuclear phagocytes and natural killer cells: Effects of immunomodulaters in beige mice and ^{89}Sr-treated mice, *Infect. Immun.* **37:**1079–1085.

59. Habu, S., Akamatsu, K., Tamaoki, N., and Okumura, K., 1984, *In vivo* significance of NK cell on resistance against virus (HSV-1) infections in mice, *J. Immunol.* **133:**2743–2747.

60. Bukowski, J. F., and Welsh, R. M., 1986, The role of natural killer cells and interferon in resistance to acute infection of mice with herpes simplex virus type 1, *J. Immunol.* **136:**3481–3485.

61. Rager-Zisman, B., Quan, P.-C., Rosner, M., Moller, J. R., and Bloom, B. R., 1987, Role of NK cells in protection of mice against herpes simplex virus-1 infection, *J. Immunol.* **138:**884–888.

62. Isaacs, A., and Lindenmann, J., 1957, Virus interference. I. The interferons, *Proc. R. Soc.* **147:**258–267.

63. Gresser, I., 1977, Commentary: On the varied biologic effects of interferon, *Cell. Immunol.* **34:**406–415.

64. Stewart, W. E., 1979, *The Interferon System*, Springer, Vienna.

65. Gidlund, M., Anders, O., Wigzell, H., Senik, A., and Gresser, I., 1979, Enhanced NK cell activity in mice injected with interferon and interferon inducers, *Nature* **273:**759–761.

66. Huang, K. Y., Donahoe, R. M., Gordon, F. B., and Dressler, H. R., 1971, Enhancement of phagocytosis by interferon-containing preparation, *Infect. Immun.* **4:**581–588.

67. Gresser, I., Tovey, M. G., Maury, C., and Bandu, M.-T., 1976, Role of interferon in the pathogenesis of virus diseases in mice as demonstrated by the use of anti-interferon serum. II. Studies with herpes simplex virus, Maloney's sarcoma, vesicular stomatitis, Newcastle disease and influenza viruses, *J. Exp. Med.* **144:**1316–1323.

68. Wrzos, H., Murasko, D. M., and Rapp, F., 1986, Effect of antibody to interferon on genital herpesvirus infection in mice, *Micro. Pathol.* **1:**71–78.

69. Zawatzky, R., Hilfenhaus, J., and Kirchner, H., 1979, Resistance of nude mice to herpes simplex virus and correlation with *in vitro* production of interferon, *Cell. Immunol.* **47:**424–428.

70. Zawatzky, R., Hilfenhaus, J., Marcucci, F., and Kirchner, H., 1981, Experimental in-

fection of inbred mice with herpes simplex virus type 1. I. Investigation of humoral and cellular immunity and of interferon induction, *J. Gen. Virol.* **53:**31–38.

71. Engler, H., Zawatzky, R., Goldbach, A., Schroder, C. H., Weyand, C., Hammerling, G. J., and Kirchner, H., 1981, Experimental infection of inbred mice with herpes simplex virus. II. Interferon production and activation of natural killer cells in peritoneal exudate, *J. Gen. Virol.* **55:**25–30.

72. Engler, H., Zawatzky, R., Kirchner, H., and Armerding, D., 1982, Experimental infection of inbred mice with herpes simplex virus. IV. Comparison of interferon production and natural killer cell activity in susceptible and resistant adult mice, *Arch. Virol.* **74:**239–247.

73. Zawatzky, R., Gresser, I., DeMaeyer, E., and Kirchner, H., 1982, The role of interferon in resistance of C57BL/6 mice to various doses of herpes simplex virus type 1, *J. Infect. Dis.* **146:**405–410.

74. Zawatzky, R., Kirchner, H., DeMaeyer-Guignard, Q., and DeMaeyer, E., 1982, X-linked locus influences the amount of circulating interferon induced in the mouse by herpes simplex virus type 1, *Virology* **63:**325–332.

75. Kirchner, H., Engler, H., Schroder, C. H., Zawatzky, R., and Storch, E., 1983, Herpes simplex virus type 1–induced interferon production and activation of natural killer cells in mice, *J. Gen. Virol.* **64:**437–441.

76. Chmielarczyk, W., Engler, H., Brucher, J., and Kirchner, H., 1983, Herpes simplex virus–induced interferon production and activation of natural killer cells in SM/J mice, relation to antiviral resistance, *Antiviral Res.* **3:**325–333.

77. Chmielarczyk, W., Domke, I., and Kirchner, H., 1985, Role of interferon in the resistance of C3H/HeJ mice to infection with herpes simplex virus, *Antiviral Res.* **5:**55–59.

78. Pederson, E. B., Haahr, S., and Mogensen, 1983, X-linked resistance of mice to high doses of herpes simplex virus type 2 correlates with early interferon production, *Infect. Immun.* **42:**740–746.

79. Daniels, C. A., Kleinerman, E. S., and Snyderman, R., 1978, Abortive and productive infections of human mononuclear phagocytes by type 1 herpes simplex virus, *Am. J. Pathol.* **91:**119–136.

80. Trofatter, K. F. Jr., Daniels, C. A., Williams, R. J. Jr., and Gall, S. A., 1979, Growth of type-2 herpes simplex virus in newborn and adult mononuclear leukocytes, *Intervirology* **11:**117–123.

81. Linnavuori, K., and Hovi, T., 1981, Herpes simplex virus infection in human monocyte cultures: Dose-dependent inhibition of monocyte differentiation resulting in abortive infection, *J. Gen. Virol.* **52:**381–385.

82. Grogan, E., Miller, G., Moore, T., Robinson, J., and Wright, J., 1981, Resistance of neonatal human lymphoid cells to interferon by herpes simplex virus overcome by aging cells in culture, *J. Infect. Dis.* **144:**547–556.

83. Diamond, R. D., Keller, R., Lee, G., and Finkel, D., 1977, Lysis of cytomegalovirus-infected human fibroblasts and transformed human cells by peripheral blood lymphoid cells from normal human donors, *Proc. Soc. Exp. Biol. Med.* **154:**259–263.

84. Santoli, D., Trinchieri, G., and Lief, F. S., 1978, Cell-mediated cytotoxicity against virus-infected target cells in humans. I. Characterization of the effector lymphocyte, *J. Immunol.* **121:**526–531.

85. Ching, C., and Lopez, C., 1979, Natural killing of herpes simplex virus type-1 infected target cells: Normal human responses and influence of anti-viral antibody, *Infect. Immun.* **26:**49–56.

86. Timonen, T., and Saksela, E., 1980, Isolation of human NK cells by density gradient centrifugation, *J. Immunol. Methods* **36:**285–291.

87. Timonen, T., Ortaldo, R. R., and Herberman, R. B., 1981, Characteristics of human granular lymphocytes and relationship to natural killer cells, *Fed. Proc.* **40:**2705–2710.

88. Fitzgerald, P. A., Evans, R., Kirkpatrick, D., and Lopez, C., 1983, Heterogeneity of

human NK cells: Comparison of effectors that lyse HSV-1-infected fibroblasts and K562 erythroleukemia targets, *J. Immunol.* **130**:1663–1668.

89. Lopez, C., Kirkpatrick, D., Sorell, M., O'Reilly, R. J., and Ching, C., 1979, Association between pretransplant natural kill and graft versus host disease following stem cell transplantation, *Lancet* **2**:1103–1106.

90. Sorell, M., Kapoor, N., Kirkpatrick, D., Rosen, J. F., Chaganti, R. S., Lopez, C., Dupont, B., Pollack, M., Terrin, B. N., Harris, M. B., Vine, D., Rose, J. S., Goosen, C., Lane, J., Good, R. A., and O'Reilly, R. J., 1981, Marrow transplantation for juvenile osteopetosis, *Am. J. Med.* **70**:1280–1287.

91. Kumar, V., Ben-Ezra, J., Bennett, M., and Sonnenfeld, G., 1979, Natural killer cells in mice treated with [89]Sr: Normal target-binding cell numbers but inability to kill even after interferon administration, *J. Immunol.* **123**:1832–1838.

92. Lust, J. A., Kumar, V., Burton, R. C., Barlett, S. P., and Bennett, M., 1981, Heterogeneity of natural killer cells in the mouse, *J. Exp. Med.* **154**:306–317.

93. Ault, K. A., and Springer, T. A., 1981, Cross-reaction of a rat-anti-mouse phagocyte-specific monoclonal antibody (anti-Mac-1) with human monocytes and natural killer cells, *J. Immunol.* **126**:359–364.

94. Stutman, O., Paige, C. J., and Figarella, E., 1978, Natural cytotoxic cells against solid tumors in mice. I. Strain and age distribution and target cell susceptibility, *J. Immunol.* **121**:1819–1826.

95. Stutman, O., Lattime, E. C., and Figarella, E. F., 1981, Natural cytotoxic cells against solid tumors in mice: A comparison with natural killer cells, *Fed. Proc.* **40**:2699–2703.

96. Messina, C., Kirkpatrick, D., Fitzgerald, P. A., O'Reilly, R. J., Siegal, F. P., Cunningham-Rundles, C., Blaese, M., Oleske, J., Pahwa, S., and Lopez, C., 1986, Natural killer cell function and interferon generation in patients with primary immunodeficiencies, *Clin. Immunol. Immunopathol.* **39**:394–404.

97. Feldman, M., Curl, S., and Fitzgerald-Bocarsly, P., 1987, The accessory cell function of HLA-DR positive cells in NK-mediated lysis of HSV-1-infected fibroblasts, *Fed. Proc.* **46**:483.

98. Bandyopadhyay, S., Perussia, B., Trinchieri, G., Miller, D. S., and Starr, S. T., 1986, Requirement for HLA-DR+ accessory cells in natural killing of cytomegalovirus-infected fibroblasts, *J. Exp. Med.* **164**:180–195.

99. Fitzgerald-Bocarsly, P., Feldman, M., Curl, S., Tehrani, S., and Denny, T., 1987, Cooperation between CD-16-positive NK cells and DR-positive accessory cells in the lysis of HSV-infected fibroblasts, *J. Leuk. Biol.* **42**:154.

100. Colmenares, C., and Lopez, C., 1986, Enhanced lysis of herpes simplex virus type-1–infected mouse cell lines by NC and NK effectors, *J. Immunol.* **136**:3473–3480.

101. Trinchieri, G., Santoli, D., and Koprowski, H., 1978, Spontaneous cell-mediated cytotoxicity in humans: Role of interferon and immunoglobulins, *J. Immunol.* **120**:1849–1855.

102. Trinchieri, G., and Santoli, D., 1978, Anti-viral activity induced by culturing lymphocytes with tumor-derived or virus-transformed cells. Enhancement of natural killer cell activity by interferon and antagonistic inhibition of susceptibility of target cells to lysis, *J. Exp. Med.* **147**:1314–1333.

103. Fitzgerald, P. A., Von Wussow, P., and Lopez, C., 1982, Role of interferon in natural kill of HSV-1-infected fibroblasts, *J. Immunol.* **129**:819–825.

104. Bishop, G. A., Glorioso, J. C., and Schwartz, S. A., 1983, Relationship between expression of herpes simplex virus glycoproteins and susceptibility of target cells to human natural killer activity, *J. Exp. Med.* **157**:1544–1561.

105. Moller, J. R., Rager-Zisman, B., Quan, P.-C., Schattner, A., Panush, D., Rose, J. K., and Bloom, B. R., 1985, Natural killer cell recognition of target cells expressing different antigens of vesicular stomatitis virus, *Proc. Natl. Acad. Sci. USA* **82**:2456–2459.

106. Honess, R., and Watson, D., 1977, Unity and diversity in the herpesviruses, *J. Gen. Virol.* **37**:15–37.

107. Bishop, G. A., Manlin, S. D., Schwartz, S. A., and Glorioso, J. C., 1984, Human natural killer cell recognition of herpes simplex virus type-1 glycoproteins: Specificity analysis with the use of monoclonal antibodies and antigenic variants, *J. Immunol.* **133:**2206–2214.

108. Bishop, G. A., Kumel, G., Schwartz, S. A., and Glorioso, J. C., 1986, Specificity of human natural killer cells in limiting dilution culture for determinants of herpes simplex virus type-1 glycoproteins, *J. Virol.* **57:**294–300.

109. El Daher, N., and Betts, R. F., 1985, New observations regarding killing of fibroblasts infected with herpes simplex virus: Cooperation between illutable factor and peripheral mononuclear cells, *J. Infect. Dis.* **152:**1197–1205.

110. Shore, S. L., Black, C. M., Melewicz, F. M., Wood, P. A., and Nahmias, A. J., 1976, Antibody-dependent cell-mediated cytotoxicity to target cells infected with type 1 and type 2 herpes simplex virus, *J. Immunol.* **116:**194–201.

111. Dixon, R., and Schaffer, P. A., 1980, Fine-structure mapping and functional analysis of temperature-sensitive mutants in the gene encoding the herpes simplex virus type 1 immediate early protein VP175, *J. Virol.* **36:**189–203.

112. Borysiewicz, L., Rodger, B., Morris, S., Graham, S., and Sissons, J., 1985, Lysis of human cytomegalovirus-infected fibroblasts by natural killer cells: Demonstration of an interferon-independent component requiring expression of early viral proteins and characterization of effector cells, *J. Immunol.* **134:**2695–2701.

113. Bradley, T., and Bonavida, B., 1981, Mechanism of cell-mediated cytotoxicity at the single cell level. IV. Natural killing and antibody-cellular cytotoxicity can be mediated by the same human effector cell as determined by the 2-target conjugate assay, *J. Immunol.* **129:**2260–2265.

114. Peter, H. H., Dallugge, H., Zawatzky, R., Eular, S., Leibold, W., and Kirchner, H., 1980, Human peripheral null lymphocytes. II. Producers of type-1 interferon upon stimulation with tumor cells, herpes simplex virus and *Corynebacterium parvum, Eur. J. Immunol.* **10:**547–555.

115. Kirchner, H., Peter, H. H., Hirt, H. M., Zawatzky, R., Dalluge, H., and Bradstreet, P., 1979, Studies of the producer cell of interferon on human lymphocyte cultures, *Immunobiology* **156:**6575.

116. Djeu, Y., Stocks, N., Zoon, K., Stanton, G. J., Timonen, T., and Herberman, R. B., 1982, Positive self-regulation of cytotoxicity in human natural killer cells by production of interferon upon exposure to influenza and herpesviruses, *J. Exp. Med.* **156:**1222–1234.

116a. Fitzerald-Bocarsley, P. A., Feldman, M., Mendelsohn, M., Curl, S., and Lopez, C., 1988, Human mononuclear cells which produce interferon-alpha during NK(HSV-FS) assays are HLA-DR positive cells distinct from cytolytic natural killer cells, *J. Leuk. Biol.* **43:**323–334.

117. Perussia, B., Fanning, V., and Trinchiere, G., 1985, A leukocyte subset bearing HLA-DA antigens is responsible for *in vitro* alpha-interferon production in response to viruses, *Nat. Immun. Cell Growth Regul.* **4:**120–137.

118. Fitzgerald, P. A., Mendelsohn, M., and Lopez, C., 1985, Human natural killer cells limit replication of herpes simplex virus type-1 *in vitro, J. Immunol.* **134:**2666–2672.

119. Leibson, P. J., Hunter-Laszlo, M., and Hayward, A. R., 1986, Inhibition of herpes simplex virus type 1 replication in fibroblast cultures by human blood mononuclear cells, *J. Virol.* **57:**976–982.

120. Leibson, P. J., Hunter-Laszlo, M., Douvas, G. S., and Hayward, A. R., 1986, Impaired neonatal natural killer-cell activity to herpesvirus: Decreased inhibition of viral replication and altered response to lymphokines, *J. Clin. Immunol.* **6:**216–224.

121. Yasukawa, M., and Kobayashi, Y., 1985, Inhibition of herpes simplex virus replication *in vitro* by human cytotoxic T-cell clones and natural killer cell clones, *J. Gen. Virol.* **66:**2225–2229.

122. Lopez, C., Kirkpatrick, D., Read, S., Fitzgerald, P. A., Pitt, J., Pahwa, S., Ching, C. Y.,

and Smithwick, E. M., 1983, Correlation between low natural kill of HSV-1-infected fibroblasts, NK(HSV-1) and susceptibility to herpesvirus infections, *J. Infect. Dis.* **147**:1030–1035.

123. Kohl, S., Frazier, J. J., Greenberg, S. B., Pickering, L. K., and Loo, L.-S., 1981, Interferon induction of natural killer cytotoxicity in human neonates, *J. Pediatr.* **98**:379–384.

124. Lipinski, M., Virelizier, J.-L., Tursz, T., and Griscelli, C., 1980, Natural killer cell activities in patients with primary immunodeficiencies or defects in immune interferon production, *Eur. J. Immunol.* **10**:246–249.

125. Stiehm, R. E., 1980, The human neonate as an immunocompromised host, in: *Infections in the Immunocompromised Host–Pathogenesis, Prevention and Therapy*, Volume 11 (J. Verhoef, P. K. Peterson, P. G. Quie, eds.), Elsevier/North-Holland, New York, p. 77.

126. Blaese, M., Strober, W., and Waldman, T. A., 1975, Immunodeficiency in the Wiscott-Aldrich Syndrome, in: *Immunodeficiency in Man and Animals* (D. Bergsman, R. A. Good, and J. Finstad, eds.), Sinauer, Sunderlind, MD., p. 250.

127. Fauci, A. S., Macher, A. M., Longo, D. L., Lane, H. C., Rook, A. H., Masur, H., and Gelmann, E. P., 1984, Acquired immune deficiency syndrome: Epidemiologic clinical immunologic and therapeutic considerations, *Ann. Intern. Med.* **100**:92–106.

128. Siegal, F. P., Lopez, C., Hammer, G. S., Brown, A. E., Kornfeld, S. J., Gold, J., Hassett, J., Hirshman, S. Z., Cunningham-Rundels, C., Adelsberg, B. R., Parham, D. M., Siegal, M., Cunningham-Rundels, S., and Armstrong, D., 1981, Severe acquired immunodeficiency in male homosexuals, manifested by chronic perianal ulcerative herpes simplex lesions, *N. Engl. J. Med.* **305**:1439–1444.

129. Lopez, C., Fitzgerald, P. A., and Siegal, F. P., 1983, Severe acquired immune deficiency syndrome in male homosexuals: Diminished capacity to make interferon-alpha *in vitro* associated with severe opportunistic infections, *J. Infect. Dis.* **148**:962–966.

130. Siegal, F. P., Lopez, C., Fitzgerald, P. A., Shah, K., Baron, P., Leiderman, I. Z., Imperato, D., and Landesman, S., 1986, Opportunistic infections in acquired immune deficiency syndrome result from synergistic defects of both the natural and adaptive components of cellular immunity, *J. Clin. Invest.* **78**:115–123.

5

Factors Influencing the Control of Virus Infections by Natural Killer Cells

RAYMOND M. WELSH, ROBERT J. NATUK,
KIM W. McINTYRE, HYEKYUNG YANG,
CHRISTINE A. BIRON, and JACK F. BUKOWSKI

1. Introduction

It has long been speculated that natural killer (NK) cells play a role in natural resistance to and regulation of virus infections.[1] Reviewed elsewhere in this volume is evidence that suggests a role for NK cells in cytomegalovirus (CMV), herpes simplex virus, and influenza virus infections. Proving definitively that NK cells provide resistance to viruses has been virtually impossible in man and difficult in animal models, many of which are currently beset with contradictory studies. Perhaps the most accurate conclusion from the available evidence is that NK cells regulate some but not all virus infections and that virus dose and inoculation routes and host species, age, and target organ may all influence the relative importance of the NK cells.

2. NK-Resistant and NK-Sensitive Viruses

In our experience the lymphocytic choriomeningitis virus (LCMV) and murine CMV (MCMV) infections of mice represent extremes for NK-

RAYMOND M. WELSH, ROBERT J. NATUK, KIM W. McINTYRE, HYEKYUNG YANG, CHRISTINE A. BIRON, and JACK F. BUKOWSKI • Department of Pathology, University of Massachusetts Medical School, Worcester, Massachusetts 01655.

resistant and NK-sensitive viruses. LCMV is a relatively noncytopathic virus which establishes a lifelong persistent infection of mice infected *in utero* or an immunizing infection regulated by cytotoxic T cells in mice infected as adults.[2] Depletion of NK cell activity in adult mice by antiserum to asialo GM$_1$[3] or by cyclophosphamide[4] does not influence LCMV synthesis. Homozygous beige mice, which have an NK cell defect, synthesize amounts of LCMV comparable to their NK-sufficient heterozygous littermates early in infection.[5] Reduced clearance of the virus is seen later in infection, correlating with a defective CTL response.[5,6] Adoptive transfers of NK cell-containing adult leukocytes into 5-day-old suckling mice that have low NK cell activity do not protect the suckling mice from LCMV.[7] Furthermore, there are no age- or strain-dependent resistance factors to LCMV that correlate with NK cell activity.

The situation with MCMV is entirely different. This is a relatively cytopathic virus which can establish a persistent infection in the salivary gland and a latent infection in leukocytes. Antibody to asialo GM$_1$ enhances MCMV infection in both acutely and persistently infected adult mice.[3,8] The biological response modifier OK432 enhances NK cell activity and induces a resistance to MCMV which can be abrogated by antiserum to asialo GM$_1$.[9] Beige mice are very sensitive to MCMV, and in bone marrow chimeras, resistance to infection is correlated with donor bone marrow cells from heterozygous but not homozygous (NK cell-deficient) mice.[10] There is a general correlation between NK cell activity and resistance to MCMV in a variety of strains of mice,[11] and suckling mice are very sensitive to MCMV, with resistance developing with age in parallel to the maturation of the NK cell response.[12] Adoptive transfer of adult leukocytes into baby mice protects the recipients from MCMV. Depletion of NK cells but not other leukocyte populations eliminates the protective effect, and partial purification of the NK cells enriches for the protective effect.[7] Furthermore, a cloned large granular lymphocyte (LGL) cell line mediating NK cell activity protected both suckling and irradiated adult mice from MCMV but not from LCMV.[7] LCMV is therefore NK-resistant, and MCMV appears to be NK-sensitive, but the NK-sensitivity of MCMV may be unusually profound, compared to other viruses.[13–18]

3. Age-Dependent Resistance

Resistance to virus infections often increases with age and may sometimes depend on the maturation of the NK cell response. However, in our hands the adoptive transfers of adult leukocytes into suckling mice, which were used to demonstrate an antiviral role for NK cells in the MCMV infection, have not shown a profound role for NK cells in other infections. Adult leukocytes did not protect recipients at all from the NK-resistant LCMV infection (Table I) but did protect against several other vi-

TABLE I
Protection of Suckling Mice from Viruses by Adult Leukocytes[a]

Virus	Donor cell number	Recipient organ	PFU/organ		
			No leukocytes (control)	Adult leukocytes	Adult NK-depleted leukocytes
LCMV	5×10^7	Spleen	4.7 ± 0.1	4.4 ± 0.1	ND[b]
MCMV	5×10^7	Spleen	4.6 ± 0.1	3.2 ± 0.1	4.1 ± 0.1
HSV-1	1.7×10^7	Spleen	4.5 ± 0.2	3.0 ± 0.1	3.1 ± 0.2
	5.6×10^6		4.5 ± 0.2	3.2 ± 0.2	3.1 ± 0.3
	1.8×10^6		4.5 ± 0.2	4.6 ± 0.2	4.5 ± 0.3
VSV	5×10^7	Spleen	3.0 ± 0.4	<2	<2
	5×10^7	Liver	4.0 ± 0.9	<2	<2
	5×10^7	Brain	7.7 ± 0.5	<2	<2
PV	5×10^7	Spleen (Exp 1)	4.1 ± 0.1	3.3 ± 0.4	4.0 ± 0.1
		(Exp 2)	3.7 ± 0.1	3.5 ± 0.1	3.7 ± 0.1
		(Exp 3)	3.7 ± 0.1	3.2 ± 0.1	3.7 ± 0.1

[a]Spleen leukocytes from 4- to 10-week-old C57BL/6 mice were injected IP into 4- to 6-day-old mice 1 day prior to IP infection with virus. Organs were titrated by plaque assay for virus 2–3 days postinfection. Mice injected with antibody to asialo GM_1 served as donors for NK cell-depleted leukocytes. Results are expressed as $\log_{10}$ plaque-forming units $\pm$ SE. See references 3 and 19 for further details.
[b]ND = Not determined.

ruses, including MCMV, herpes simplex virus (HSV) type 1, vesicular stomatitis virus (VSV), and, to a lesser extent, Pichinde virus (PV). Depletion of NK cell activity in the donor leukocytes ablated the protective effect against the NK-sensitive MCMV and inhibited the very modest protective effect against PV. This suggests that PV may be sensitive to NK cells but much less so than MCMV. Consistent with this observation are data indicating that antibody to asialo GM_1 causes a modest but significant two- to fourfold enhancement of PV synthesis in adult mice, but a 10- to 1000-fold enhancement of MCMV synthesis and no enhancement of LCMV synthesis[3] (Table II). Work suggesting that NK cells play a role in regulating murine HSV[13,14] and VSV[16–18] infections was not supported by this adoptive transfer model (Table I).

Donor leukocytes protected mice against HSV-1, but depletion of NK cell activity in the donor cells did not inhibit the protective effect (Table I). In addition, whereas 5×10^7 adult leukocytes were required to protect suckling mice from the NK-sensitive MCMV, one-tenth that number protected against HSV-1. Further investigation of the HSV-1 system revealed that virtually any tested adult leukocyte or even cultured cell lines, such as L-929 cells, mediated protection.[19] Earlier reports had indicated that adult but not suckling mouse macrophages protected suckling mice from

TABLE II
Replication of Viruses in NK Cell-Depleted Mice[a]

Virus	Mouse	Anti-asialo GM$_1$	PFU/spleen	% Lysis vs. YAC-1 cells
PV (Exp 1)	Adult	−	4.7 ± 0.1	24
		+	5.3 ± 0.1	−1.3
(Exp 2)		−	4.0 ± 0.2	20
		+	4.9 ± 0.2	−1.0
LCMV	Adult	−	5.7 ± 0.2	63
		+	5.6 ± 0.2	0.5
MCMV	Adult	−	1.7 ± 0.2	30
		+	4.6 ± 0.1	0.2
HSV-1	Adult	−	5.2 ± 0.3	38
		+	5.2 ± 0.3	1.1
MHV	Adult	−	1.7 ± 0.3	48
		+	3.4 ± 0.1	16
MCMV	5-day-old beige	−	3.8 ± 0.3	22
	5-day-old normal	−	3.6 ± 0.3	19
	5-day-old normal	+	3.7 ± 0.2	−1.5
MCMV	Adult beige	−	3.9 ± 0.2	9.5
	Adult beige/+	−	1.4 ± 0.1	51

[a]C57BL/6 normal (unless otherwise stated) or beige mice were injected IP with 20 μl anti-asialo GM$_1$ followed 4–6 h later by an IP challenge with virus. Spleens were titrated for virus 2–3 days later, and spleen cells were assayed for NK cell activity. Results are expressed as log$_{10}$ pfu/spleen ± SE. Effector-to-target ratios of 100:1 or 50:1 were employed in cytotoxicity assays.

HSV-1,[20] but in our hands spleen cells protected even after depletion of macrophages. A series of investigations by Kirchner and co-workers has suggested that the IFN made during the first round of viral replication (early IFN) is crucial to the outcome of the infection.[21–23] Resistant strains of mice tended to make higher levels of early IFN than did susceptible strains.[21] The peritoneal fluid of suckling mice receiving adoptive transfers of leukocytes generated higher levels of early IFN after HSV-1 infection than did fluid from mice not receiving leukocytes.[19] Antibody to IFN completely abrogated the protective effect of the leukocytes. Furthermore, high doses of antibody to asialo GM$_1$ in vivo abrogated the early IFN response and prevented the protective effect of donor leukocytes. Lower levels of this antibody, which still depleted NK cell activity but not early IFN production, had no effect on virus titers[19] (Table II). Prophylactic IFN treatment was highly effective in preventing HSV-1 infection of suckling mice.[19] It therefore seems that the primary mechanism for adult leukocytes to provide resistance to HSV-1 in suckling mice is by enhancing the early IFN response. It is possible that a similar mechanism could be occurring with VSV, a very IFN-sensitive virus.

Age-dependent resistance of mice to at least one strain of mouse hepatitis virus (MHV), MHV3, is consistent with an NK cell role, as it develops at 3 weeks of age.[15] With MHV3 several factors seem to be involved in resistance shown in this adoptive transfer system: a thy-1.2.-bearing (T) cell, a plastic adherent cell (macrophage), and a bone marrow cell from a mouse at least 3 weeks of age. This bone marrow cell shares similarity to NK cells in a variety of properties including culture lability, ^{89}Sr sensitivity, and stimulation by IFN inducers.[15] This strongly suggests a role for NK cells in age-dependent resistance to MHV3. Depletion of NK cell activity in adult mice with antibody to asialo GM_1 enhances the synthesis of MHV strains A-59[3] (Table II) and MHV-Y,[26] consistent with an NK cell role.

Thus, the mechanism of age-dependent resistance is complex, involving a variety of factors, of which NK cells may be only one. MCMV may be unusual in that NK cells are the only age-dependent factor with great significance. This may explain why the adoptive transfer results are so clear. Even in this model, however, there are perplexing inconsistencies. For instance, we have been unable to detect any differences in the synthesis of MCMV between suckling normal and homozygous C57BL/6 beige mice, which have an NK cell defect (Table II). In contrast, the growth of MCMV in adult beige mice is much higher than in normal adult mice[8,10] (Table II). Suckling mice have very low NK cell activity but develop significant levels of NK cell activity upon viral infection. A possible explanation for the lack of difference in MCMV synthesis in normal and beige suckling mice is that there is less of a differential between beige and normal suckling mouse NK cell activity after virus infection, whereas adult mice have a major difference[8] (Table II). However, to our great surprise, depletion of NK cell activity in suckling mice with antibody to asialo GM_1 did *not* influence MCMV titers, even though it ablated all detectable NK cell activity (Table II).

We are therefore left with the paradox that transfer of adult NK cells into baby mice renders protection against MCMV, but depletion of the suckling mouse's own NK cells has no effect on MCMV synthesis. It is difficult to find a suitable explanation for this. It seems that NK cells from the suckling mice are inadequate to control MCMV infection, even though they get activated. Perhaps their numbers are too few, their level of activation is not high enough, or there is an age-dependent defect in another function contributing to their antiviral effects.

4. Site of Virus Replication

The injection route and site of viral replication may play a role in the relative importance of NK cells in mediating antiviral effects. Although

NK cell activity is normally present predominantly in the spleen and peripheral blood, potent responses are seen in other organs that are sites of virus infection. For instance, NK cell activity is found in the cerebrospinal fluid after intracranial inoculation,[27] in the lung after intranasal[28] or intratracheal[29] inoculation, and in the peritoneal cavity after intraperitoneal inoculation.[30] Hepatotropic viruses stimulate high levels of NK cell activity in the liver,[31] and viruses that grow to high levels in the bone marrow stimulate NK cell activity therein.[32]

The increased levels of NK cell activity in organs are often associated with significant increases in NK cell number. The IFN induced during a virus infection *in vivo* stimulates the blastogenesis and proliferation of NK cells.[33–35] This is accompanied by significant increases in the number of LGLs[31,33,36] and the frequency of LGL-bearing blast morphology[37] in virus-infected organs. NK/LGLs respond chemotactically to extracts from virus-infected organs,[36] and the LGLs with blast morphology respond chemotactically much better than do resting LGLs.[38] It thus appears that virus-infected organs stimulate an accumulation of NK/LGLs by releasing factors chemotactic for blast LGLs and by synthesizing IFN, which induces NK cell blastogenesis.

This accumulation and enhanced activity of NK cells at virtually any examined site of virus infection should seemingly allow NK cells to mediate antiviral effects in most of the body's organs. Again, however, the situation may not be so simple. When the injection route is IV or IP, the titers of MCMV are greatly elevated in beige mice or in normal mice depleted of NK cell activity with antibody to asialo GM_1.[3,8,10] Dissemination of the virus into the lung is enhanced by NK cell depletion.[8] In contrast, upon instranasal inoculation with MCMV, there are no differences in virus titers between normal, beige, and NK cell-depleted mice.[8] The reason for this is not known. In this model there is substantial replication of MCMV within the lung before the virus disseminates elsewhere. It would seem, then, that whereas NK cells may inhibit the spread of MCMV *to* the lung, they do not inhibit MCMV replication *within* the lung. An obvious explanation would be that there is a defect in the activity or number of lung NK cells, but several reports indicate that they do get activated during virus infections.[28,29] This has been well documented with influenza virus and is reviewed elsewhere in this volume. Antibody to asialo GM_1 enhances the synthesis of influenza in the lung following intratracheal inoculation,[39] but other reports indicate that influenza virus pathogenesis is normal in lungs from beige mice[28] and from NK cell-depleted normal mice[40] after intranasal inoculation. Perhaps whether the inoculation route is intranasal versus intratracheal is significant.

In another study, ^{125}IUDR-labeled L-929 cells infected with LCMV were examined for clearance from the lung after intravenous injection.[41] Cells injected intranasally are initially trapped in the lung, and clearance of radiolabeled cells from the lung has been taken as an indication of *in vivo* cytotoxicity. Several studies have indicated that NK cells mediate *in vivo* lysis of radiolabeled YAC-1 cells, the prototype mouse NK-sensitive target.[42–44] The virus-infected cells were lysed *in vivo* more rapidly than uninfected cells, and hydrocortisone and cyclophosphamide abrogated the lysis. However, this lysis was not blocked by antibody to asialo GM_1, which markedly inhibited the *in vivo* lysis of YAC-1 cells.[41] This suggests the existence of an antiviral cytotoxic cell within the lung that may not be an NK cell. Whether this represents an NC- (natural cytotoxic) type of killing mechanism[45] is unclear. However, one lot of antiserum to asialo GM_1 that was highly cytotoxic to macrophages did inhibit the virus-associated *in vivo* lysis, suggesting that alveolar macrophages may be involved (Biron and Welsh, unpublished). We are left to conclude that the issue of antiviral natural cytotoxicity associated with the lung may be very complex.

Another potential complication in organ-dependent NK cell accumulation is that there may be a finite number of NK cells for which different organs compete. For example, intraperitoneal injection of mice with MHV results in a high number of NK cells within the peritoneal cavity and an intermediate number in the liver (Fig. 1). Intravenous infection stimulates very high levels of NK cells within the liver but few in the peritoneal cavity. Inoculation by both routes also results in high levels of NK cells in the liver but not the peritoneal cavity. Thus the liver and

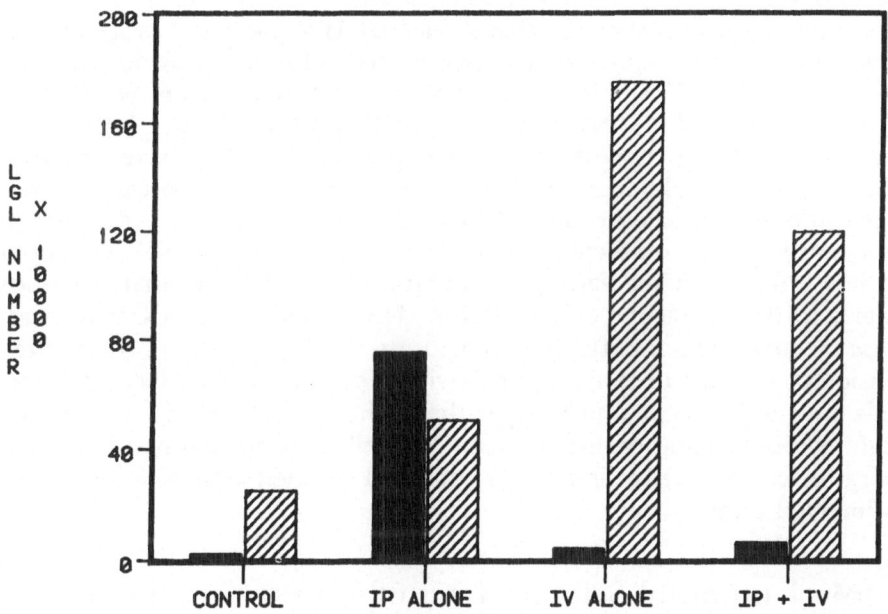

FIGURE 1. *In vivo* compartmentalization of NK/LGL responses. C57BL/6J mice were infected with 4×10^4 PFU of mouse hepatitis virus (MHV) either IP or IV or with 2×10^4 PFU MHV both IP and IV. At day 3 postinfection, peritoneal exudate cells (PEC) and liver leukocytes were prepared, and total LGL at each site (PEC, solid bars; liver, hatched bars) were enumerated as described.[31,36]

perhaps other visceral organs appear to compete with the peritoneal cavity for the attraction of NK cells. Similar results have been seen with virus-specific cytotoxic T cells in the LCMV infection.[31] The significance of this is that NK cells may not be able to mediate antiviral effects in certain organs if they cannot accumulate within them because of competition from other infected organs.

5. Interferon-Mediated Effects on Target Cells

Interferon protects target cells from NK cell-mediated lysis,[46] presumably by inhibiting the target cell from triggering the release of cytotoxic factors from bound NK cells.[47–49] As a virus infection progresses *in vivo*, cells in the body become progressively more resistant to NK cells[44,50,51] and, probably because of IFN-mediated up-regulation of class I major histocompatibility antigens, progressively more sensitive to cytotoxic T cells.[51] Infection of target cells by cytopathic viruses often renders these cells resistant to IFN-mediated effects, probably because of virus-induced inhibition of cellular RNA and protein synthesis, which IFN requires to stimulate most of its effects, including protection against NK cells.[44,46] Protection of target cells by IFN may be an important mechanism of homeostasis that prevents the IFN-mediated activation and proliferation of NK cells from causing a potent autodestructive effect.

It has been hypothesized that a selective IFN-mediated protective effect against uninfected but not virus-infected cells may provide a mechanism by which NK cells selectively lyse virus-infected cells *in vivo*.[46,47,52] *In vitro* studies with the NK-resistant LCMV and NK-sensitive MCMV are consistent with this hypothesis, as IFN protects LCMV-infected but not MCMV-infected cells from lysis by activated NK cells.[53] Within a given organ the degree of efficiency of this antiviral selectivity may depend on how sensitive the uninfected cells in the tissue are to IFN. NK cells could be less efficient in mediating an antiviral effect in tissue that responds poorly to the protective effect of IFN. This would cause NK cells to release cytotoxic factors after binding to uninfected targets. A further extension of this line of thinking deals with the demonstrated fact that some cells bind to NK cells much better than other cells.[54] In tissue in which high avidity binding occurs between NK cells and normal tissue, the antiviral effects of NK cells may be inhibited by their adsorption to those uninfected targets.

6. Mechanisms of Selective Lysis of Virus-Infected Target Cells by NK Cells

Numerous publications attest that cultures of virus-infected cells are lysed more readily than uninfected cells when exposed to NK cells *in vi-*

TABLE III
Binding of Purified NK Cells to Virus-Infected Targets[a]

Target	Infection	% Lysis	% Bound targets
Mouse embryo fibroblasts (MEF)	None	28	58
	VV-1d	32	54
	VV-2d	9.2	29
	MCMV-1d	15	36
	MCMV-2d	5.6	22
L-929	None	30	50
	VV-1d	26	42
	VV-2d	16	17
	MCMV-1d	6.5	32
	MCMV-2d	4.9	15

[a] Target cells were infected with MCMV or VV at an MOI of 3 or left untreated. These cells were then used in ^{51}Cr-release assays and target cell–binding assays[56] using purified poly I:C–activated spleen NK cells[79] as effectors at an effector-to-target ratio of 5:1. Binding assays were run at 4°C for 0.5–3 h; cytotoxicity assays were run for 4 h.

tro.[55] This has led many to speculate that NK cells selectively "recognize" virus-infected targets. Actually, the lysis of virus-infected cells is quite complex and can be the result of either of two equally complex phenomena—the activation of NK cells by virus-dependent mechanisms, and the innate sensitivity of virus-infected cells to lysis. This is an important distinction, as we have commonly found that virus-infected targets are often more resistant than uninfected targets to lysis when exposed to already activated NK cell populations. We have seen this effect with HSV-1, vaccinia virus (VV), Sendai virus, Sindbis virus, VSV, and MCMV,[53,56] and this phenomenon has also been seen by other laboratories.[11,57] Furthermore, adding to the complexity is a recent observation in our laboratory that the susceptibility of virus-infected cells to NK cell-mediated lysis changes with the duration of the infection, such that a virus such as VV may enhance sensitivity to lysis early after infection but may inhibit sensitivity to lysis late in infection, possibly because of the loss of NK cell receptors on the cell membrane (unpublished; Table III).

6.1. NK Cell Activation

Two mechanisms have been proposed for the activation of NK cells on incubation with virus-infected targets. The first involves a protein synthesis-independent triggering of NK or NK-like cells via viral glycoproteins. The second is associated with a protein synthesis-dependent step involving NK cell activation by virus-induced IFN. These will be discussed separately.

Purified glycoproteins from mumps,[58] Sendai,[59] measles,[60] LCMV,[60] and influenza[61] viruses have all been shown to activate human NK or NK-like cells. These proteins share a common property of binding to cell membranes, but they differ in their enzymatic activities. For example, both the hemagglutinin and the neuraminidase of influenza virus stimulate activity.[61] The glycoproteins act on the effector cell but are presumed to direct killing to virus-infected target cells that express them. There is controversy over the nature of the effector cell, with some evidence that a CD3+ T cell mediates much of the killing, with some variation caused by the target cell type.[62,63] The mechanism behind this phenomenon has not been elucidated. Curiously, viral glycoprotein activation of mouse NK cells has not yet to our knowledge been shown.

The original observations on selective lysis of virus-infected cells correlated the degree of lysis with the amount of IFN secreted into the culture fluid.[64] The source of IFN was either the virus-infected target cell, leukocytes responding to virus infection, or leukocytes responding directly to virus-infected cell membranes. Recent evidence has suggested that a DR (IA)-antigen expressing cell may be an accessory cell for NK cell activation, perhaps by producing IFN in this system.[65] In some systems antibody to IFN blocks lysis. An interesting dichotomy in lysis was shown with measles virus-infected targets.[66] Virus-selective lysis seen in a 4 h assay occurred in the absence of detectable IFN and was not blocked by antibody to IFN. However, elevated lysis seen in an overnight assay did correlate with IFN production and was blocked by antibody to IFN. In several other systems, an antibody to IFN did not block lysis, even in an overnight assay.[63,67–70] This does not, however, rule out a possible role for IFN, particularly if there is close contact between an NK effector cell and an IFN-producing accessory cell.[65] A more convincing experiment to rule out a role for IFN is the addition of RNA or protein synthesis inhibitors to the assay. If these fail to block the elevated killing, then IFN mediation is unlikely. This is because RNA and protein synthesis are required both for IFN production and for IFN-induced effects. In some systems such inhibitors have been used without blocking the virus-selective lysis.[68,69]

6.2. Innate Sensitivity of Virus-Infected Cells to Lysis

Although we often find that virus-infected cells are frequently more resistant to lysis, it is clear that in several systems, virus-infected cells have enhanced sensitivity to lysis. This could be mediated at several levels: (1) recognition or binding, (2) triggering of or stimulating the release of cytolysins from bound NK cells, and (3) susceptibility to released cytolysins. These will be discussed separately.

6.2.1. Recognition or Binding

Much has been written about virus-infected cells preferentially being "recognized" by NK cells. This is an attractive hypothesis to explain virus-selective killing, because viral glycoproteins on the target cell surface usually bind well to cell membranes and can often mediate cell–cell adhesion or fusion. There is little evidence that "recognition" is a major mechanism for virus-selective lysis, however. NK cells bind well to most cell types, so, except for the rare target that normally does not bind to NK cells (such as P815 in the mouse system),[54,71] enhanced binding is unlikely to make the difference between sensitivity and resistance.

Enhanced binding in target binding cell assays has been seen in several viral systems,[56,71] often correlating with enhanced killing, but reduced killing has also been seen in the presence of enhanced binding.[56] For example, Sendai virus-infected L-929 cells bind well to NK cells and can be used to deplete NK cells from leukocyte populations, but these targets are quite resistant to lysis.[56] The best correlation that can be made is the invariably reduced killing one sees when virus infections inhibit the binding of NK cells. Reduction in binding may occur as a result of obscuring or reducing the concentration of the target-binding structure. We have observed such reduced binding late in infections with HSV-1,[56] MCMV, and VV (Table III). Alternatively, a virus infection may enhance the binding of non-NK cells to targets. An example of this is the MHV-A59 infection of 3T3 cells, which causes these cells to rosette B cells,[72] which in this system mediate lysis. Lysis of these MHV-infected cells by NK cells is low until B cells are removed from the leukocyte population.

The conclusion is often made on the basis of cold-target competition assays that NK cells bind preferentially to virus-infected cells. This is an inadequate technique to measure binding, as this assay is more a measure of the inactivation of NK cells resulting from the release of cytolysins stimulated by the signal transduction or "triggering" event.[47–49] IFN-treated target cells bind to NK cells[44,47,48] but fail to cold-target-inhibit,[44,46,47,56] presumably because their cytolysins are not released.[47,49] This is discussed in the next section.

6.2.2. Triggering and Release of Cytolysin

This may be an important mechanism for the selective lysis of virus-infected cells, and an obvious mechanism may involve a stimulation or signal transduction mediated by viral glycoproteins. Other virus-induced membrane changes could occur as well. For example, vesicular stomatitis virus (VSV)-infected cells, depending on the system, may be selectively lysed by NK cells. Antibody to VSV does not block this elevated lysis, and target cells transfected with and expressing the gene for the VSV G protein are *not* more sensitive to lysis.[71] Cells infected with mutants in VSV G, M, or L genes are also not sensitive to lysis. This suggests that part of the enhanced sensitivity of infected cells may involve cellular alterations resulting from a productive virus infection. In contrast, it is reported that the lysis of HSV-1-infected target cells can be blocked by an antibody to a surface glycoprotein,[73] so different results are found in different systems.

6.2.3. Susceptibility to Released Cytolysins

Virus infections themselves cause cytopathic effects, and it is likely that the cytotoxic effects of the virus may act synergistically with the cytotoxic effect of cytolysins released from NK cells. It is becoming apparent that NK cells may be involved in the regulation of picornavirus infections and that cells infected with Coxsackie B or encephalomyocarditis viruses can be quite sensitive to NK cells[63,74] (E. Godeny, C. Gauntt, L. White, and R. Smith, personal communications). Picornaviruses do not insert glycoproteins into the plasma membrane, and, although they may mediate some membrane alterations,[75] the increased sensitivity to NK cells could be happening at the intracellular level. These viruses are potent inhibitors of cellular protein synthesis, and metabolic inhibitors of RNA and protein synthesis have been shown to enhance the sensitivity of uninfected target cells to NK cell-mediated lysis.[76,77] Since these inhibitors can inhibit membrane repair, it has been suggested but not proved that membrane repair alterations by inhibitors of protein synthesis such as these drugs or virus infections could be the cause of increased sensitivity to NK cell-mediated lysis.[76]

7. Use of NK Cells to Treat Virus Infections

Recent results in our laboratory indicate that culture-derived NK cells may be useful in controlling virus infections. Lymphokine-activated killer (LAK) cells were generated by incubating mouse spleen leukocytes in culture for 5 days with 100–1000 units of human recombinant interleukin 2. This protocol has been used in murine antitumor studies.[79] Injection of as few as 5×10^5 LAK cells into suckling mice protected them against MCMV, the NK-sensitive virus, but not against LCMV, the NK-resistant virus (Table IV). This protective effect was dramatic and required far fewer LAK cells than that reported for antitumor effects.[78] Furthermore, in contrast to the tumor studies, no *in vivo* IL-2 injections were required to produce the antiviral effect. *In vitro* cytotoxicity against YAC-1 targets was mediated predominantly by NK-1.1$^+$ leukocytes, which represented 14% of the LAK cell preparation. Injection of NK-1.1$^+$ LAK cells into suckling mice protected them against MCMV. These experi-

TABLE IV
Prophylaxis against Virus Infections by LAK Cells[a]

	Inhibition of virus synthesis		
Exp.	Donor cells	Virus	Log_{10} reduction in spleen titers
1	5×10^7 Adult leukocytes	LCMV	0
	4×10^6 LAK cells	LCMV	0
2	5×10^7 Adult leukocytes	MCMV	1.8
	4×10^6 Adult leukocytes	MCMV	0.6
	4×10^6 LAK cells	MCMV	>3.8
	1×10^6 LAK cells	MCMV	>2.6
	5×10^5 LAK cells	MCMV	>1.9
	2.5×10^5 LAK cells	MCMV	0
3	2×10^5 NK-1.1$^+$ LAK cells	MCMV	2.5
	2×10^5 Unseparated LAK cells	MCMV	2.4

[a] LAK cells were generated by *in vitro* culture of 1.5×10^7 spleen cells in 5 ml RPMI-1640 for 5 days with 100 U/ml human recombinant IL-2. Adoptive transfers of LAK cells and adult leukocytes into suckling mice were performed as described in Table I.

ments suggest that it may be possible to culture autologous NK cells in IL-2 and use them to treat human virus infections.

That human virus infections are regulated by NK cells can be surmised from the murine studies, although definitive experiments to prove this hypothesis are lacking. The most convincing data available are with the herpes group of viruses, some of which are summarized elsewhere in this volume. Whether human CMV (HCMV) is as uniquely sensitive to NK cells as is its murine counterpart is uncertain. Several reports suggest that human NK cells may selectively lyse HCMV-infected target cells *in vitro*.[65,80,81] HCMV infection is generally most severe under conditions associated with NK cell deficiency, such as congenital or neonatal infections, or in adults immunosuppressed by other viral infections, cancer, or drugs.[82] In one study with bone marrow transplant recipients, the severity of HCMV infection inversely correlated with the levels of NK activity before onset of disease.[83]

HCMV pneumonitis is a frequent, if not the major cause of infectious disease death in transplant recipients and in patients with AIDS. If HCMV is as susceptible to NK cells as MCMV, one could predict a possible therapeutic effect of LAK (activated NK) cells on that infection. Intravenous injection would deliver these cells into the lung, where they might mediate an antiviral effect.

Understanding the complexities behind the mechanisms of the antiviral effects of NK cells should help in designing appropriate regimens for the ultimate treatment of virus-infected patients with NK cells.

ACKNOWLEDGMENTS. Parts of this work were supported by USPHS research grants AI-17672, CA-34461, and AM-35506. We thank Ms. Dottie Walsh for preparation of the manuscript.

References

1. Welsh, R. M., 1978, Mouse natural killer cells: Induction, specificity, and function, *J. Immunol.* **121:**1631–1635.
2. Buchmeier, M. J., Welsh, R. M., Dutko, F. J., and Oldstone, M. B. A., 1980, The virology and immunobiology of lymphocytic choriomeningitis virus infection, *Adv. Immunol.* **30:**275–331.
3. Bukowski, J. F., Woda, B. A., Habu, S., Okumura, K., and Welsh, R. M., 1983, Natural killer cell depletion enhances virus systhesis and virus-induced hepatitis *in vivo, J. Immunol.* **131:**1531–1538.
4. Welsh, R. M., Biron, C. A., Bukowski, J. F., McIntyre, K. W., and Yang, H., 1984, Role of natural killer cells in virus infections of mice, *Surv. Synth. Pathol. Res.* **3:**409–431.
5. Welsh, R. M., and Kiessling, R. W., 1980, Natural killer cell response to lymphocytic choriomeningitis virus in beige mice, *Scand. J. Immunol.* **11:**363–367.
6. Biron, C. A., Pedersen, K. F., and Welsh, R. M., 1987, Aberrant T cells in beige mutant mice, *J. Immunol.* **138:**2050–2056.
7. Bukowski, J. F., Warner, J. F., Dennert, G., and Welsh, R. M., 1985, Adoptive transfer studies demonstrating the antiviral effects of NK cells *in vivo, J. Exp. Med.* **161:**40–52.
8. Bukowski, J. F., Woda, B. A., and Welsh, R. M., 1984, Pathogenesis of murine cytomegalovirus infection in natural killer cell-depleted mice, *J. Virol.* **52:**119–128.
9. Ebihara, K., and Minamishima, Y., 1984, Protective effect of biological response modifiers on murine cytomegalovirus infection, *J. Virol.* **51:**117–122.
10. Shellam, G. R., Allan, J. E., Papadimitriou, J. M., and Bancroft, G. J., 1981, Increased susceptibility to cytomegalovirus infection in beige mutant mice, *Proc. Natl. Acad. Sci. USA* **78:**5104–5108.
11. Bancroft, G. J., Shellam, G. R., and Chalmer, J. E., 1981, Genetic influences on the augmentation of natural killer (NK) cells during murine cytomegalovirus infection: Correlation with patterns of resistance, *J. Immunol.* **126:**988–994.
12. Boos, J., and Wheelock, E. F., 1975, Correlation of survival from murine cytomegalovirus infection with spleen cell responsiveness to concanavalin A, *Proc. Soc. Exp. Biol. Med.* **149:**443–446.
13. Lopez, C., Ryshke, R., and Bennett, M., 1980, Marrow-dependent cells depleted by ^{89}Sr mediate genetic resistance to herpes simplex virus type 1 infections in mice, *Infect. Immun.* **28:**1028–1032.
14. Habu, S., Akamatsu, K., Tamaoki, N., and Okumura, K., 1984, *In vivo* significance of NK cell on resistance against virus (HSV-1) infections in mice, *J. Immunol.* **133:**2743–2747.
15. Tardieu, M., Hery, C., and Dupuy, J. M., 1980, Neonatal susceptibility to MHV3 infection in mice. II. Role of natural effector marrow cells in transfer of resistance, *J. Immunol.* **124:**418–423.
16. Reid, L. M., Minato, N., Gresser, I., Holland, J., Kadish, A., and Bloom, B. R., 1981, Influence of anti-mouse interferon serum on the growth and metastasis of tumor cells persistently infected with virus and of human prostatic tumors in athymic nude mice, *Proc. Natl. Acad. Sci. USA* **78:**1171–1175.
17. Jones, C. L., Spindler, K. R., and Holland, J. J., 1980, Studies on tumorigenicity of cells persistently infected with vesicular stomatitis virus in nude mice, *Virology* **103:**158–166.

18. Vandepol, S. B., and Holland, J. J., 1986, Tumorigenicity of persistently infected tumors in nude mice is a function of both the virus and the host cell type, *J. Virol.* **58**:914–920.
19. Bukowski, J. F., and Welsh, R. M., 1986, The role of natural killer cells and interferon in resistance to acute infection of mice with herpes simplex virus type 1, *J. Immunol.* **136**:3481–3485.
20. Johnson, R. T., 1964, The pathogenesis of herpes virus encephalitis. II. A cellular basis for the development of resistance with age, *J. Exp. Med.* **120**:359–373.
21. Engler, H., Zawatzky, R., Kirchner, H., and Armerding, D., 1982, Experimental infection of inbred mice with herpes simplex virus. IV. Comparison of interferon production and natural killer cell activity in susceptible and resistant adult mice, *Arch. Virol.* **74**:239–247.
22. Zawatzky, R., Gresser, I., Demayer, F., and Kirchner, H., 1982, The role of interferon in the resistance of C57BL/6 mice to various doses of herpes simplex virus type 1, *J. Infect. Dis.* **146**:405–410.
23. Kirchner, H., Engler, H., Schroder, C. H., Zawatzky, R., and Storch, E., 1983, Herpes simplex virus type-1-induced interferon production and activation of natural killer cells in mice, *J. Gen. Virol.* **64**:437–441.
24. Falke, D., and Row, W. P., 1965, Die erkrangkung der Maus durch das virus der stomatitis vesicularis. I. Die Ausbreitung des virus in Abhangigkeit vom alter der Maus, *Arch. Gesamte Virusforsch.* **17**:549–559.
25. Holland, J. J., and Villarreal, L. P., 1975, Purification of defective interfering T particles of vesicular stomatitis and rabies viruses generated *in vivo* in brains of newborn mice, *Virology* **67**:438–449.
26. Carman, P. S., Ernst, P. B., Rosenthal, K. L., Clark, D. A., Befus, A. D., and Bienenstock, J., 1986, Intraepithelial leukocytes contain a unique subpopulation of NK-like cytotoxic cells active in the defense of gut epithelium to enteric murine coronavirus, *J. Immunol.* **136**:1548–1553.
27. Doherty, P. C., and Korngold, R., 1983, Characteristics of poxvirus-induced meningitis: Virus-specific and non-specific cytotoxic effectors in the inflammatory exudate, *Scand. J. Immunol.* **18**:1–7.
28. Leung, K. N., and Ada, G. L., 1981, Induction of natural killer cells during murine influenza virus infection, *Immunobiology* **160**:352–366.
29. Stein-Streilein, J., Bennett, M., Mann, D., and Kumar, V., 1983, Natural killer cells in mouse lung: Surface phenotype, target preference, and response to local influenza virus infection, *J. Immunol.* **131**:2699–2704.
30. Welsh, R. M., and Zinkernagel, R. M., 1977, Hetero-specific cytotoxic cell activity induced during the first three days of acute lymphocytic choriomeningitis virus infection in mice, *Nature* **268**:646–648.
31. McIntyre, K. W., and Welsh, R. M., 1986, Accumulation of natural killer and cytotoxic T large granular lymphocytes in the livers of virus-infected mice, *J. Exp. Med.* **164**:1667–1681.
32. Thomsen, A. R., Pisa, P., Bro-Jorgensen, K., and Kiessling, R., 1986, Mechanisms of lymphocytic choriomeningitis virus-induced hemopoietic dysfunction, *J. Virol.* **59**:428–433.
33. Biron, C. A., Turgiss, L. R., and Welsh, R. M., 1983, Increase in NK cell number and turnover rate during acute viral infection, *J. Immunol.* **131**:1539–1545.
34. Biron, C. A., and Welsh, R. M., 1982, Blastogenesis of natural killer cells during viral infection *in vivo*, *J. Immunol.* **129**:2788–2795.
35. Biron, C. A., Sonnenfeld, G., and Welsh, R. M., 1984, Interferon induces natural killer cell blastogenesis *in vivo*, *J. Leuk. Biol.* **35**:31–37.
36. Natuk, R. J., and Welsh, R. M., 1987, Accumulation and chemotaxis of large granular lymphocytes at sites of virus replication, *J. Immunol.* **138**:877–883.

37. McIntyre, K. W., Natuk, R. J., Biron, C. A., Kase, K., Greenberger, J., and Welsh, R. M., 1988, Blastogenesis and proliferation of large granular lymphocytes in non-lymphoid organs, *J. Leuk. Biol.* (in press).

38. Natuk, R. J., and Welsh, R. M., 1987, Chemotactic effect of interleukin-2 on mouse-activated large granular lymphocytes, *J. Immunol.* **139:**2737–2743.

39. Stein-Streilein, J., and Guffee, J., 1986, *In vivo* treatment of mice and hamsters with antibodies to asialo GM_1 increases morbidity and mortality to pulmonary influenza infection, *J. Immunol.* **136:**1435–1441.

40. Wells, M. A., Daniel, S., Kiley, S. C., Burlington, D. B., and Ennis, F. A., 1985, Absence of natural killer cell effects in murine influenza virus pneumonia, *J. Leuk. Biol.* **38:**124.

41. Biron, C. A., Habu, S., Okumura, K., and Welsh, R. M., 1984, Lysis of uninfected and virus-infected cells *in vivo*: A rejection mechanism in addition to that mediated by natural killer cells, *J. Virol.* **50:**698–707.

42. Riccardi, C., Puccetti, P., Santoni, A., and Herberman, R. B., 1979, Rapid *in vivo* assay of mouse natural killer (NK) cell activity, *JNCI* **63:**1041–1045.

43. Riccardi, C., Barlozzari, T., Santoni, A., and Herberman, R. B., 1981, Transfer to cyclophosphamide-treated mice of natural killer (NK) cells and *in vivo* natural reactivity against tumors, *J. Immunol.* **126:**1284–1289.

44. Welsh, R. M., Karre, K., Hansson, M., Kunkel, L. A., and Kiessling, R. W., 1981, Interferon-mediated protection of normal and tumor target cells against lysis by mouse natural killer cells, *J. Immunol.* **126:**219–225.

45. Lattime, E. C., Pecoraro, G. A., and Stutman, O., 1981, Natural cytotoxic cells against solid tumors in mice. III. A comparison of effector cell antigenic phenotype and target cell recognition structures with those of NK cells, *J. Immunol.* **126:**2011–2014.

46. Trinchieri, G., and Santoli, D., 1978, Anti-viral activity induced by culturing lymphocytes with tumor-derived or virus-transformed cells. Enhancement of natural killer activity by interferon and antagonistic inhibition of susceptibility of target cells to lysis, *J. Exp. Med.* **147:**1314–1333.

47. Trinchieri, G., Granato, D., and Perussia, B., 1981, Interferon-induced resistance of fibroblasts to cytolysis mediated by natural killer cells: Specificity and mechanisms, *J. Immunol.* **126:**335–340.

48. Perussia, B., and Trinchieri, G., 1981, Inactivation of natural killer cell cytotoxic activity after interaction with target cells, *J. Immunol.* **126:**754–758.

49. Wright, S. C., and Bonavida, B., 1983, Studies on the mechanism of natural killer cell-mediated cytotoxicity. III. Interferon-induced inhibition of NK target cell susceptibility to lysis is due to a defect in their ability to stimulate release of natural killer cytotoxic factors (NKCF), *J. Immunol.* **130:**2960–2964.

50. Hansson, M., Kiessling, R., Andersson, B., and Welsh, R. M., 1980, Effect of interferon and interferon inducers on the NK sensitivity of normal mouse thymocytes, *J. Immunol.* **125:**2225–2231.

51. Bukowski, J. F., and Welsh, R. M., 1986, Enhanced susceptibility to cytotoxic T lymphocytes of target cells isolated from virus-infected or interferon-treated mice, *J. Virol.* **59:**735–739.

52. Santoli, D., and Koprowski, H., 1979, Mechanisms of activation of human natural killer cells against tumor and virus-infected cells, *Immunol. Rev.* **44:**125–163.

53. Bukowski, J. F., and Welsh, R. M., 1985, Inability of interferon to protect virus-infected cells against lysis by natural killer (NK) cells correlates with NK cell-mediated antiviral effects *in vivo*, *J. Immunol.* **135:**3537–3541.

54. Roder, J. C., Kiessling, R., Biberfeld, P., and Andersson, B., 1978, Target-effector interaction in the natural killer (NK) cell system. II. The isolation of NK cells and studies on the mechanism of killing, *J. Immunol.* **121:**2509–2517.

55. Welsh, R. M., 1986, Regulation of virus infections by natural killer cells, *Nat. Immun. Cell Growth Regul.* **5:**169–199.

56. Welsh, R. M., and Hallenbeck, L. A., 1980, Effect of virus infections on target cell susceptibility to natural killer cell-mediated lysis, *J. Immunol.* **124:**2491–2497.
57. Munoz, A., Carrasco, L., and Fresno, M., 1983, Enhancement of susceptibility of HSV-1-infected cells to natural killer lysis by interferon, *J. Immunol.* **131:**783–787.
58. Harfast, B., Orvell, C., Alsheikhly, A., Andersson, T., Perlmann, P., and Norrby, E., 1980, The role of viral glycoproteins in mumps virus-dependent lymphocyte-mediated cytotoxicity *in vitro, Scand. J. Immunol.* **11:**391–400.
59. Alsheikhly, A., Orvell, C., Harfast, B., Andersson, T., Perlmann, P., and Norrby, E., 1983, Sendai-virus-induced cell-mediated cytotoxicity *in vitro*. The role of viral glycoproteins in cell-mediated cytotoxicity, *Scand. J. Immunol.* **17:**129–138.
60. Casali, P., Sissons, J. G. P., Buchmeier, M. J., and Oldstone, M. B. A., 1981, *In vitro* generation of human cytotoxic lymphocytes by virus. Viral glycoproteins induce nonspecific cell-mediated cytotoxicity without release of interferon, *J. Exp. Med.* **154:**840–855.
61. Arora, D. J. S., Houde, M., Justewicz, D. M., and Mandeville, R., 1984, *In vitro* enhancement of human natural cell-mediated cytotoxicity by purified influenza virus glycoproteins, *J. Virol.* **52:**839–845.
62. Alsheikhly, A. R., Andersson, T., and Perlmann, P., 1985, Virus-dependent cellular cytotoxicity *in vitro*. Mechanisms of induction and effector cell characterization, *Scand. J. Immunol.* **21:**329–335.
63. Kurane, I., Hebblewaite, D., Brandt, W. E., and Ennis, F. E., 1984, Lysis of Dengue virus–infected cells by natural cell-mediated cytotoxicity and antibody-dependent cell-mediated cytotoxicity, *J. Virol.* **52:**223–230.
64. Santoli, D., Trinchieri, G., and Koprowski, H., 1978, Cell-mediated cytotoxicity against virus-infected target cells in humans. II. Interferon induction and activation of natural killer cells, *J. Immunol.* **121:**532–538.
65. Bandyopadhyay, S., Perussia, B., Trinchieri, G., Miller, D. S., and Starr, S. E., 1986, Requirement for HLA-DR⁺ accessory cells in natural killing of cytomegalovirus-infected fibroblasts, *J. Exp. Med.* **164:**180–195.
66. Casali, P., and Oldstone, M. B. A., 1982, Mechanisms of killing of measles virus-infected cells by human lymphocytes: Interferon-associated and unassociated cell-mediated cytotoxicity, *Cell Immunol.* **70:**330–344.
67. Fitzgerald, P. A., Von Wussov, P., and Lopez, C., 1982, Role of interferon in natural kill of HSV-1 infected fibroblasts, *J. Immunol.* **129:**819–823.
68. Lee, G. D., and Keller, R., 1982, Natural cytotoxicity to murine cytomegalovirus-infected cells mediated by mouse lymphoid cells: Role of interferon in the endogenous natural cytotoxicity reaction, *Infect. Immun.* **35:**5–12.
69. Bishop, G. A., Glorioso, J. C., and Schwartz, S. A., 1983, Role of interferon in human natural killer activity against target cells infected with HSV-1, *J. Immunol.* **131:**1849–1853.
70. Fitzgerald, P. A., Mendelsohn, M., and Lopez, C., 1985, Human natural killer cells limit replication of herpes simplex virus type 1 *in vitro, J. Immunol.* **134:**2666–2672.
71. Moller, J. R., Rager-Zisman, B., Quan, P., Schattner, A., Panush, D., Rose, J. K., and Bloom, B. R., 1985, Natural killer cell recognition of target cells expressing different antigens of vesicular stomatitis virus, *Proc. Natl. Acad. Sci. USA* **82:**2456–2459.
72. Welsh, R. M., Haspel, M. V., Parker, D. C., and Holmes, K. V., 1986, Natural cytotoxicity against mouse hepatitis virus-infected cells. II. A cytotoxic effector cell with a B lymphocyte phenotype, *J. Immunol.* **136:**1454–1460.
73. Bishop, G. A., Marlin, S. D., Schwartz, S. A., and Glorioso, J. C., 1984, Human natural killer cell recognition of herpes simplex virus type 1 glycoproteins: Specificity analysis with the use of monoclonal antibodies and antigenic variants, *J. Immunol.* **133:**2206–2214.
74. Godeny, E. K., and Gauntt, C. J., 1986, Involvement of natural killer cells in Coxsackievirus B3-induced murine myocarditis, *J. Immunol.* **137:**1695–1702.

75. Lutton, C. W., and Gauntt, C. J., 1986, Coxsackievirus B3 infection alters plasma membrane of neonatal skin fibroblasts, *J. Virol.* **60:**294–296.
76. Kunkel, L. A., and Welsh, R. M., 1981, Metabolic inhibitors render "resistant" target cells sensitive to natural killer cell-mediated lysis, *Int. J. Cancer* **27:**73–79.
77. Collins, J. L., Patek, P. Q., and Cohn, M., 1981, Tumorigenicity and lysis by natural killers, *J. Exp. Med.* **153:**89–106.
78. Mule, J. J., Shu, S., and Rosenberg, S. A., 1985, The anti-tumor efficacy of lymphokine-activated killer cells and recombinant interleukin-2 *in vivo*, *J. Immunol.* **135:**646–652.
79. Biron, C. A., Pedersen, K. F., and Welsh, R. M., 1986, Purification and target cell range of *in vivo*-elicited blast natural killer cells, *J. Immunol.* **137:**463–471.
80. Diamond, R. D., Keller, R., Lee, G., and Finkel, D., 1977, Lysis of cytomegalovirus-infected human fibroblasts and transformed human cells by peripheral blood lymphoid cells from normal human donors, *Proc. Soc. Exp. Biol.* **154:**259–263.
81. Borysiewicz, L. K., Rodgers, B., Morris, S., Graham, S., and Sissons, J. G. P., 1985, Lysis of human cytomegalovirus infected fibroblasts by natural killer cells: Demonstration of an interferon-independent component requiring expression of early viral proteins and characterization of effector cells, *J. Immunol.* **134:**2696–2701.
82. Rook, A. H., and Quinnan, G. V., 1983, Cell-mediated immunity to human cytomegalovirus, in: *Human Immunity to Viruses* (F. E. Ennis, ed.), Academic Press, New York, pp. 241–256.
83. Quinnan, G. V., Kirmani, N., Rook, A. H., Manischewitz, J. F., Jackson, L., Moreschi, G., Santos, G. W., Saral, R., Burns, W. H., 1982, Cytotoxic T cells in cytomegalovirus infection: HLA-restricted T lymphocyte cytotoxic responses correlate with recovery from cytomegalovirus infection in bone marrow transplant recipients, *N. Engl. J. Med.* **307:**7–13.

6

Natural Killer Cell-Mediated Resistance to Animal Parasites

JULIA W. ALBRIGHT and JOSEPH F. ALBRIGHT

1. Introduction

The expression "natural resistance," or even "natural cell-mediated resistance," can be applied to a variety of mechanisms including those that involve macrophages, polymorphonuclear cells, and, in a broad sense, platelets, and that may involve humoral substances such as components of the complement system.[1] By inserting the word "killer" in the title of this article we hope to convey an understanding that this article will deal with a reasonably proscribed subject—viz., the role of natural killer cells in resistance to parasites. The term "natural killer" will be considered imprecise by some workers who will be correct in raising an objection to its use. We might borrow the expression "non-MHC-restricted cytotoxicity,"[20] but that also suffers from lack of precision. Because at the present time, there is no satisfactory term, we will continue with the term "natural killer" (NK) and stress the fact that, in much of the work that will be reviewed, it was an activity rather than a well-defined cell that was under investigation. The expression "NK cells" will not be used as a synonym for large granular lymphocyte (LGL). Even in those investigations to be discussed where concerted effort was made to associate phenotypic markers with NK activity, the precise identity of the active cell(s) must remain in doubt. That is because, by any set of criteria, NK cells are heterogeneous,[3–5] and there are insufficient data to allow a decision about the number of different subsets of NK cells, let alone their classification.

The expression "animal parasites" refers to eukaryotic organisms. This

JULIA W. ALBRIGHT and JOSEPH F. ALBRIGHT • Department of Microbiology, George Washington University School of Medicine, Washington, D.C. 20037.

article will be limited to host resistance to protozoan and metazoan parasites and will not consider viruses or bacteria. Natural resistance to those life forms will be considered in other chapters in this volume. Furthermore, we will not consider the fungi, because they, too, will be discussed elsewhere in this volume.

In assimilating and evaluating many publications concerned with NK activity that relate to survival and growth of parasites in their hosts, our goals have been (1) to determine whether or not, and how generally, parasites are susceptible to NK activity; (2) to estimate the contribution to NK activity to the limitation and control of infections by parasites in their hosts; (3) to consider possible approaches to elevating the NK activity of the host in order to curtail or eliminate parasitic infections; and (4) to review the evidence for, and consider the significance of, the rather marked effect parasitic infections have on the activity of NK cells. The properties, features, and functions of NK cells will be reviewed in other chapters of this volume or have been discussed elsewhere[6–8]; therefore, we will not attempt a general discussion of the characteristics of NK cells.

2. Interactions between Natural Killer Cells and Parasites

2.1. Descriptive Features

2.1.1. *Plasmodia* and *Babesia*

The earliest studies on the possible involvement of NK cells in resisting parasitic infections were concerned with intracellular protozoan parasites. Eugui and Allison[9] showed that there was a strong correlation between spontaneous NK activity and relative resistance to infection with *Plasmodium chabaudi* and *Babesia microti*. They evaluated the severity (magnitude) of infections by those parasites among different strains of mice and showed that the most resistant strains (C57BL/6, CBA) were those that displayed the highest NK cell cytotoxicity toward YAC-1 tumor cells. The most resistant strains were those in which the serum IFN concentrations were most elevated in response to infection. A subsequent study[10] cast doubt on the likelihood that NK cell activity was related to relative resistance of mice to *B. microti* or *Plasmodium vinckei petteri*. It was found that NK activity rose early during the course of infection and declined to a low level when the parasitemia reached a maximum. Mice pretreated with ^{89}Sr or the 17β-estradiol, treatments that severely depleted the NK cells,[11,12] displayed characteristic profiles of infection with both organisms. Furthermore, there was no significant alteration in the course of severity of *B. microti* infections in mice homozygous for the beige mutation; homozygosity of this mutant gene is associated with severe reduction in the activity of NK cells as well as changes in the functions of certain other leukocytes.[13] The rise in NK activity that occurs within a few days following inoculation of the malaria parasite into mice was observed by Hunter et al.[14] and correlated with a similar early rise in serum IFN levels. The early rise to a maximum was soon followed by a decline in NK activity to a level below normal. We will see that this pattern of early rise and subsequent decline in NK activity occurs in virtually every parasitic infection, including infections with bacteria and viruses.

A very clear correlation between *Plasmodium* infections, elevated serum IFN titers, and enhanced NK activity was presented by Nigerian children suffering from acute infections with *P. falciparum*.[15] The high level of antiviral activity in the serum of these children was attributed to α-type IFN. Unlike the peripheral blood leukocytes of normal children, the leukocytes of infected children responded weakly, if at all, to addition of exogenous IFN.

A recent study[16] was designed to evaluate spontaneous cell-mediated cytotoxicity (CMC) and antibody-dependent cell-mediated cytotoxicity (ADCC) against parasitized erythrocytes of the rat. The magnitude of parasitemia in rats infected with *P. berghei* was much greater in 30-day-old than in 50-day-old rats. It was found that the reduced magnitude of infection in the older rats correlated with an overall fourfold increase in the capability of the spleen of normal rats to lyse parasite-infected erythrocytes. In infected rats there was a dramatic increase in the splenic NK activity during the first 8 days of infection (assessed by an 18-h ^{51}Cr-release assay).

The rate of increase was substantially greater in 30-day-old than in 50-day-old rats, such that on day 8 of infection, the total splenic CMC was nearly equivalent in rats of the two different ages. Using the same ^{51}Cr-release assay and a rat antiserum against *P. berghei*, it was found that the ADCC activities of spleen cells from 30-day-old and 50-day-old rats were equivalent. However, within 8 days after initiation of infection, the ADCC activity per spleen rose 33-fold in the young rats and only sixfold in the 50-day-old rats.

Both the CMC and ADCC activities of the spleen cells were associated with a nonadherent cell. Both activities were eliminated when the spleen cells were treated with monoclonal antibody (MRC-OX7) against rat Thy-1.1. Thus, the identity of the effector cell remains in doubt. An additional study by Solomon[17] involved an infectivity assay, rather than a ^{51}Cr-release assay, to evaluate CMC and ADCC against *P. berghei*-parasitized rat erythrocytes. Again, the spleen cells from 50-day-old rats were about four times more effective in reducing infection of recipient animals than were 30-day-old rat spleen cells. The results of this investigation provided evidence for a role of NK cells in cytotoxic killing of parasitized erythrocytes. In this case, the effective spleen cells were both nonadherent and only slightly affected by anti-Thy-1.1 serum and complement.

An earlier study of cytotoxicity directed at parasitized erythrocytes of

Plasmodium-infected mice was performed by Coleman *et al.*[18] This study provided evidence that spleen cells, nonadherent to glass or nylon wool, were capable of lysing parasitized erythrocytes. Cells from mice previously immunized with parasite-infected erythrocytes ("immune" spleen cells) were considerably more cytolytic than cells from noninfected animals. The immune spleen cells were particularly active in the presence of immune serum. The identity of the effector cell(s) responsible for lysis of parasitized erythrocytes remains unknown.

The most definitive work concerning the possible importance of high NK activity in the resistance to *Plasmodium* infections was that of Skamene *et al.*[19] These investigators, who earlier had shown that a single gene controls resistance to murine malaria in mice of strain A compared to B10.A,[71] studied the segregation of resistance to *P. chabaudi* and of splenic NK activity in backcross offspring derived from matings of A/J (malaria-susceptible, low NK activity) and B10.A (malaria-resistant, high NK activity) parents. The results were clear: NK activity and resistance to *P. chabaudi* segregated independently. Thus, with the strains of mice and the malaria organism employed in the original studies,[9] there was no causal relationship between high NK activity and resistance to malaria.

2.1.2. *Toxoplasma*

Inoculation of mice with *Toxoplasma gondii* resulted in a rapid rise in NK activity in spleen, marrow, and peritoneal space (PS).[20,21] Peak NK activity was reached by the third day postinoculation and was followed by rapid decline of activity in all organs and body compartments. NK activity was assessed by use of standard target tumor cells such as YAC-1. Cytotoxicity was associated with NK cells by showing that activity was unaffected by removal of adherent cells and but little affected by treatment with antiserum against Thy-1 plus C. The rise of NK activity required the participation of silica-sensitive cells,[20] presumably macrophages, the need for which has been demonstrated by other investigators (e.g., Tracey[22]).

The elevated NK activity was also associated with cells that were removed by treatment with antiserum against asialo GM_1 and with cells that could be stimulated in nude mice by *T. gondii*.[20] It was clear that the elevated activity was associated with phenotypically distinct cells in the spleen and PS. For example, chronic infections with *T. gondii* caused prolonged elevation of NK activity in the PS but not in the spleen.[20] A sonicate of *T. gondii* induced elevated NK cytotoxicity toward tumor cells in both the spleen and PS[23]; however, only the particulate fraction of the sonicate could induce splenic NK activity, whereas both the soluble and particulate fractions induced elevated activity of PS cells.

A complex investigation was conducted with infected and uninfected beige (bg/bg) mutant mice and their heterozygous littermates (bg/+).[24] Both YAC-1 tumor cells and allogeneic thymocytes were employed as targets. Unimpaired cytotoxicity for thymocyte (THY) targets was found in both spleen and PS of bg/bg mice, whereas, as expected, there was very little activity toward YAC targets in bg/bg compared to bg/+ mice. Cytotoxicity toward both targets was associated with cells that were phenotypically NK-1.2$^+$, Thy-1.2$^+$, asialo GM_1^+, asialo GM_2^+ in both the spleen and PS of uninfected mice.

Infection of bg/+ mice with *T. gondii* induced elevated cytotoxicity toward YAC targets in the spleen and elevated activity toward both the THY and YAC in the PS. Infection of bg/bg mice caused no change in spleen cell cytotoxicity but induced significant elevation of activity toward both THY and YAC among the PS cells. Cytotoxicity was associated with PS cells that displayed the phenotype NK 1.2$^-$, Thy 1.2$^+$, asialo GM_1^+, asialo GM_2^+. By other criteria these activated PS cells appeared to be NK cells. The results of this investigation stress the fact that NK activity may be associated with diverse cells. Thus, there was a difference between effector cells for THY and YAK targets, between splenic and PS cytotoxic cells, and between effector cells in uninfected and infected mice. The finding of elevated NK activity toward both THY and YAC targets among the PS cells of infected bg/bg mice deserves thorough investigation.

The activated, splenic NK cells of *T. gondii*-infected mice were found to be capable of direct cytotoxic attack on isolated tachyzoites of *T. gondii*.[25] Spleen cells collected from mice on the third day of infection were capable of substantial killing of tachyzoites over a wide range of effector-to-target ratios. Under optimum conditions, better than 50% of the target tachyzoites were killed in a 4-h assay period. Killing appeared to be dependent on Ca^{2+}, which suggests the involvement of a known cytolytic mechanism.[8] Thus, both NK cells and macrophages may be important in limiting *T. gondii* infections.

Both *T. gondii* and *Trypanosoma cruzi* are intracellular parasites capable of replicating in macrophages, and in the case of both parasites, macrophages play an important role in host resistance.[26,27] Much remains to be learned about macrophages in resistance to *T. gondii* and *T. cruzi* as well as about the survival and multiplication of the parasites in macrophages. A recent study was devoted to the effects of tumor necrosis factor (TNF or "cachectin") on intracellular and extracellular forms of *T. gondii* and *T. cruzi*.[28] TNF is a product of macrophages[29] but has recently been found to be associated with the cytolytic granules of NK cells.[30] There was no effect of TNF on extracellular *T. gondii* or *T. cruzi* and no effect on intracellular *T. gondii*. However, TNF did inhibit the multiplication of *T. cruzi* inside murine macrophages. It was suggested that the differential effects of TNF on the intracellular replication of *T. gondii* and *T. cruzi* could be explained by the fact that the former exist in parasitophorous vacuoles inside the cell and are therefore unavailable to TNF. *T. cruzi*, on the other hand, resides outside vacuoles in the cytosol and may be accessible to TNF. However, it was noted[28] that TNF had no effect on *T. cruzi* inside human fibroblasts, in contrast to murine macrophages. Studies of the effect of TNF on parasites have been limited to date. TNF-containing serum has been demonstrated to destroy certain species of *Plasmodium in vitro*[31,32] and to aid mice in resisting infection with *Plasmodium*.[33,34]

The finding that activated NK cells are capable of significant destruction of extracellular tachyzoites *in vitro*[25] begs further investigation. It will be important to learn whether there is a specific moiety on the parasites that serves as a recognition site for NK cells and to investigate the mechanism for killing the tachyzoites. The effect on killing efficiency of treating the NK cells with additional activating substances such as interleukin 2(IL-2) must be evaluated.

2.1.3. Trypanosomes

The first evidence that NK cells might be involved in trypanosome infections was reported by Hatcher and Kuhn and their associates.[35,36] Two elevations of nonspecific, cell-mediated cytotoxicity toward YAC-1 tumor cells appeared after inoculating mice with blood-form trypomastigotes of *Trypanosoma cruzi*. The first peak in activity occurred on day 2 after parasite inoculation and was apparent among both spleen and PS cells. The second peak occurred later in the course of *T. cruzi* infection, the time being determined in part by the number ("dose") of parasites used to establish infection. The first peak of lytic activity was associated with cells that were nonadherent, killed by antiserum against NK-1.2 plus C, and only slightly susceptible to antiserum against Thy-1.2 plus C. There was no evidence that a similar rise in NK activity occurred in either the spleen or PS of bg/bg mice inoculated with *T. cruzi*.

The possibility that NK cells might be significant in *T. cruzi* infections was strengthened by the finding that pretreatment of mice with tilerone, an IFN inducer, substantially increased the survival of mice subsequently inoculated with the parasite.[37] High levels of IFN were present in the serum of mice 18–24 h after tilerone injection, and there was a significant rise in splenic NK activity. Roughly half of C57BL/6 mice so treated survived indefinitely following *T. cruzi* infection. Beige (bg/bg) mice, on the other hand, appeared to be partially protected from death by the tilerone treatment, although they were much more susceptible than their bg/+ counterparts. In addition to enhancing NK activity in mice, tilerone treatment significantly increased macrophage destruction of *T. cruzi*; this was true especially in the inherently resistant strain (C57BL/6) in comparison to a susceptible strain (A/J).

The blood forms of *T. cruzi* were found to be killed directly by NK cells.[38] The epimastigote stage was moderately more susceptible than the trypomastigote stage. Spleen cells and PS cells either from poly I:C-stimulated mice or from mice infected for 48 h with *T. cruzi* were considerably more efficient than normal cells in killing the parasites. The parasiticidal cells were eliminated from spleen cell preparations by treatment with anti-serum against NK-1.2 plus C and only moderately reduced by antiserum against Thy-1.2 plus C.

It appears, therefore, that *T. cruzi* is a strong candidate as a target for NK cell cytotoxicity, at least in the extracellular condition. It should be particularly informative to identify the NK cell recognition structure on *T. cruzi* and to elucidate the mechanism by which NK cells destroy the parasites. In view of the fact that IL-2 production is sharply curtailed in *T. cruzi*-infected mice,[39] it would be interesting to learn whether IL-2 (or other lymphokine) can be used to maintain the early rise in NK activity to the detriment of the parasite. There has apparently been no investigation of the effect, if any, of activated NK cells on parasitized macrophages or other host cells.

In contrast to the susceptibility of *T. cruzi* to NK cell destruction, other trypanosomes appear to be less sensitive. The rodent trypanosomes (mouse-specific, *Trypanosoma musculi*, and rat-specific, *Trypanosoma lewisi*) have been studied in some detail.[40–42] Inoculation of mice with *T. musculi* resulted in a rapid increase in NK activity toward YAC tumor targets to a peak 3 days after inoculation; this was followed by a decline in activity to a subnormal level by day 10 after parasite inoculation. During the decline and subsequent subnormal phases of NK activity it proved impossible to stimulate activity by injecting poly I:C. Although there was a fleeting, modest rise in serum IFN concentration on days 2 and 3 after parasite inoculation, it was not clear that IFN was responsible for the enhanced NK activity (see below).

The explanation of the decline and subnormal plateau in NK activity after the peak was uncertain. No suppressor cell activity was detected in the spleens of infected mice. It was demonstrated, however, that both live *T. musculi* and extracts of *T. musculi* were quite effective at inhibiting NK cell lysis of target YAC cells. In contrast *T. lewisi* failed to inhibit murine NK cells. This latter result was correlated with the finding that *T. musculi* (the murine parasite) was insensitive to murine NK cells, whereas *T. lewisi* (the rat parasite) was moderately susceptible to murine NK cells. *T. cruzi* was found to be even more susceptible than *T. lewisi* to murine cells. The adoptive transfer of activated NK cells (present in 5×10^7 donor spleen cells) failed to provide any protection against *T. musculi* infection to irradiated, syngeneic recipient mice.

The studies on *T. cruzi* and the rodent trypanosomes suggested that the ability of certain parasites (e.g., *T. musculi*) to inhibit or interfere with NK cell target recognition or with the lytic process might explain the insusceptibility of those parasites to NK cytotoxicity. These studies also provided the suggestion that such inhibition might be particularly evident in the effects of parasites on the NK cells of their natural hosts. Unfortunately, no further attention has been paid to those points.

There has been relatively little attention to the possibility that NK cells affect African trypanosomes. A study[43] of NK cell activity against *Trypanosoma brucei brucei* revealed no early activation of splenic NK cells such as has been seen in most other parasitic infections. Instead, there was a gradual decline in activity until the sixth day after parasite inoculation, followed by a rapid decline between days 6 and 12 to a markedly subnormal level. The phase of rapid decline coincided roughly with the first wave of parasitemia. The subnormal splenic level of NK activity could not be elevated by administering IFN/B or poly I:C. Cure of the trypanosome infection was followed by restoration of the NK cell activity.

In contrast to their variable susceptibility to NK cells, all trypanosomes appear to be susceptible to killing by an ADCC process. It will not be profitable to discuss the studies that have been performed at any length, because, unfortunately, in most cases the effector cells have not been clearly identified. *T. cruzi* was shown to be killed efficiently by murine spleen cells in the presence of murine antibody.[44,45] In addition, evidence that *T. cruzi*-infected mouse fibroblasts may be killed by antibodies against the parasite together with murine spleen cells has been provided.[46] Similarly, the rodent trypanosomes were found to be readily susceptible to killing by murine spleen cells facilitated by murine antibody.[47] One analysis, in which it appeared likely that lymphocytes were the effector cells, revealed efficient killing of *Trypanosoma dionisii* by human lymphocytes in the presence of parasite-specific rabbit antibodies.[48] Recently, we have obtained evidence that a cloned rat LGL tumor cell line is capable of reasonably efficient killing of *T. lewisi* when facilitated by specific murine antibodies (J. W. Albright *et al.* unpublished). The ADCC immune mechanism in parasitic infections merits extensive investigation. In the case of the rodent trypanosomes (as an example), it has been shown recently[49] from studies *in vivo* that precocious cure of infection can be achieved by an antibody-promoted, cell-mediated process. An unusually labile antibody is involved. The identity of the effector cells has not been reported, but they could include NK cells functioning in the ADCC mode.

2.1.4. *Leishmania*

The most complete analysis of the contribution of NK cells to protecting hosts from a parasite has been accomplished in the case of *Leishmania*. Initial investigations were concerned with the course of infection of two species, *L. tropica* and *L. donovani*, in beige (bg/bg) mice and their normal counterparts (bg/+ and C57BL/6).[50] No differences were found in comparing bg/bg and control mice infected with *L. tropica*; the course of infection, the serum antibody titers, and the delayed hypersensitive responses to *L. tropica* antigens were quite similar. The responses of bg/bg and control mice were quite different in the case of *L. donovani*, which causes visceral leishmaniasis. The numbers of parasites found in spleen and liver, over a 56-day period of observation, were much higher in bg/bg than in control mice, especially in the female sex.

The titers of serum antibodies and the delayed footpad response of infected mice to antigens of *L. donovani* were at least as strong in bg/bg mice as they were in controls. These results suggested that NK cell activity might be important in the control of *L. donovani* infections. *L. donovani* inoculation induced an early rise in splenic NK activity (peak around day 2) as assessed against YAC target cells in bg/+ but not in bg/bg mice.[51] A similar, early rise in activity occurred in inoculated C57BL/6, CBA, and BALB/c mice, but only a slight elevation occurred in A/J mice. The early rise in activity was followed, in short order, by a decline to a subnormal activity, except in the case of BALB/c mice, in which the decline was much more gradual.

There was no correlation between the magnitude of the early elevation in NK activity against YAC tumor cells and the severity of the parasitic infection as assessed by splenic parasite burdens. The elevated cytolytic activity toward YAC target cells was associated with NK cells as judged by the finding that treatment of spleen cells with antiserum against Qa-5 and asialo $GM_1 + C$ eliminated the cytolytic activity. Two findings of considerable interest were the absence of elevated IFN concentrations in the serum of infected mice and the unimpaired ability of cells from infected mice to produce IFN in response to stimulation with poly I:C.

To resolve the uncertainty concerning the involvement of NK cells in resistance to visceral leishmaniasis, Kirkpatrick et al.[52] conducted studies with a cloned line of murine NK cells (NKB61B10[53]). Beige mice (C57BL/6 bg/bg) were infused with 3×10^6 cloned NK cells and 5 days later inoculated with *L. donovani* amastigotes. The course of the parasitic infection in these NK cell-restored mice was compared to the infection in untreated bg/bg and in bg+ mice. The course of infection was about the same in NK cell-restored and bg/+ mice and was much less severe than in the untreated bg/bg animals. A comparison of the protection afforded to bg/bg recipient mice indicated that 3×10^6 cloned NK cells were considerably more effective than 1×10^7 normal spleen cells.

Throughout these admirable studies of *L. donovani* infections, it was observed that when the evidence suggested a positive, controlling effect of NK cells, the effect was manifested at a time when humoral antibodies against the parasites were present. Kirkpatrick and associates have concluded that NK cells contribute to the control of *L. donovani* during the acquired resistance phase of the infection. They note that clone NKB61B10 cells possess Fc receptors and cite the report[54] that macrophages infected with *L. donovani* display parasite antigens on their surfaces. It seems likely that in contributing to the control of *L. donovani* infections, the NK cells functioned in an ADCC mode.

2.1.5. *Giardia*

The gut-associated lymphoid tissue (GALT) displays abundant large lymphocytes with cytoplasmic granules.[55] It is, therefore, reasonable to inquire whether those LGL play a role in controlling intestinal parasitic infection. There is strong evidence that LGLs of the GALT are capable of spontaneous antibacterial activity toward *Salmonella typhimurium*[56] and *Shigella* (hybrid strain X16).[57] This spontaneous activity is exerted by cells that are very similar to NK cells, although they differ slightly in phenotypic characteristics from splenic NK cells. Within the GALT there is regional variation in the characteristic of these effector cells. For example, intestinal intraepithelial lymphocytes were subject to enhancement of antibacterial activity by specific serum antibodies (largely IgG), whereas mesenteric lymph node lymphocytes were only slightly enhanced by the antibodies, and the spontaneous activity of Peyer's patch lymphocytes was not enhanced at all by the antibodies.[56] An important result of these studies was the finding that the spontaneous antibacterial activity of the LGLs from all regions of the GALT was significantly enhanced by the presence of specific secretory IgA antibodies developed against the cognate bacteria.[57] This secretory IgA-mediated ADCC mechanism deserves extensive investigation.

Whether the spontaneous activity of GALT LGLs toward bacteria is effective against animal parasites, and protozoa in particular, has not been determined. A recent study[58] of the course of *Giardia muris* infections in C57BL/6J mice and their bg/bg counterparts, defective in NK activity, provided no evidence for NK involvement in the cure of this protozoan infection. It would be useful to know whether GALT LGLs are capable of spontaneously killing *G. muris* and, especially, whether antibody-facilitated LGL play an important role in the cure of giardiasis. It appears quite likely (ref. 59, e.g.) that antibody-facilitated macrophage killing of *G. muris* is an essential component of the defense against the parasite. It seems that there is much to be learned about natural killer cells and antibody-facilitated killer cells in enteric parasitic infections.

2.1.6. Metazoa

Very little interest has been devoted to the possibility that NK cells might influence metazoan parasitic infections. We are aware of studies only with *Schistosoma mansoni*. Infection of mice with cercariae resulted, 4 days later, in elevated splenic NK activity toward YAC tumor cells and in enhanced ADCC activity toward chicken erythrocyte targets.[60] It was suggested that both NK and ADCC might significantly influence the course of *S. mansoni* infections were it not for the substantial levels of circulating immune complexes; i.e., immune complexes might very well bind to effector cells through the Fc receptors and block the attack of effector cells on target parasites.

Analysis of splenic NK activity over a period of several months in mice suffering from chronic *S. mansoni* infection showed no significant elevation of activity during the first few weeks of infection. In the 12th and subsequent weeks of infection, there was a substantial (two- to four-fold) increase in the mean total lytic units per spleen of infected mice as evaluated with YAC tumor target cells.[61] Furthermore, the NK cell activity per unit number of spleen cells of the chronically infected mice could be stimulated by injection of poly I:C.

The findings in mice were not duplicated by studies of NK activity among peripheral blood cells of chronically infected humans.[62] In a group of human patients, some from Egypt and some from Brazil, no significant alterations in peripheral blood NK cell activity were found based on assays against the K562 target tumor cell line. It is worth pointing out that, in our experience, NK activity of murine peripheral blood cells is not a mirror of the status of NK cells in the organized lymphoid tissue; this may be the case, also, in humans. Thus, it remains possible that in chronically infected patients splenic NK activity may be elevated as in the schistosome-infected mice.

2.2. Mechanistic Features

2.2.1. Parasite Activation of Natural Killer Cells

In the preceding sections, we have discussed numerous examples of the rise in NK activity that soon follows the introduction of parasites. This early increase in NK activity has been reported after infections with viruses, bacteria, and both protozoan and metazoan parasites. In cases where it has been studied, the rise in activity occurs in both the spleen and peritoneal space. In most cases the increase in activity reaches a peak within 3–5 days following parasite inoculation, and peak activity is followed by a gradual, or even rapid, decline to a plateau of subnormal activity.

No satisfactory explanation of the rise and fall in NK activity has been given. It cannot be taken as an indication that NK activity is involved in restricting parasite growth or in altering the course of infection. It is possible that the change in NK activity is fortuitous and merely reflects the parasite-induced elaboration of IFN. There are, however, several examples (refs. 40,51,63) of NK activation in the absence of significant elevation of plasma IFN concentration. The early rise in activity is not associated with significant changes in tissue or organ cellularity, although at a later time it might be, as in the case of chronic schistosomiasis.[61]

It is likely that the decline in NK activity after the early peak is associated with growth of the tissue or organ. Lymphoid tissue hyperplasia is a common occurrence in parasite-infected animals and humans. In this regard, the changes that occur in the spleen of the *T. musculi*-infected mouse may be considered prototypic. A very similar course of events occurs in *Toxoplasma*-infected mice.[63] The NK activity of spleen cells toward YAC tumor targets increases about threefold by the third day following

inoculation of the parasite (Fig. 1). It then declines rather quickly, and by
the 12th day of infection, splenic NK activity is less than half of normal.
The decline in activity is associated with the marked growth of the spleen
to a peak on day 14–15 of infection (Fig. 2), at which time the total number of spleen cells is 10–20 times greater than normal. There is no indication that suppressor cells are responsible for the decline in NK activity.
Rather, the evidence indicates that the initial threefold increase in NK
activity is due to a corresponding increase in active NK cells and that this
number of cells remains more or less constant but is diluted, as the remainder of the spleen experiences considerable growth.

The initial rise in NK activity seems to be due to activation of preex-
isting cells rather than to proliferation of precursor cells. Evidence for
this in the case of *T. musculi*-induced NK cells came from studies *in vitro*.
The addition of live *T. musculi* to cultures of normal spleen cells resulted
in a rapid, fourfold increase in NK activity in a period of about 14 h. To
have been generated by proliferation, a doubling time (presumably, the
cell cycle time) of about 7 h would have been required. This seems unlikely. Similarly, in the case of *Toxoplasma* it was concluded that proliferation of existing NK cells or precursors was not responsible for the early
four- to fivefold increase in splenic NK activity.[63] A direct analysis of the
rise of NK activity following inoculation of human peripheral blood leukocyte cultures with glutaraldehyde-fixed *Salmonella* bacteria has been
performed[64] and is instructive. Activation of NK cells was prevented by
blocking protein synthesis with cycloheximide but not by blocking prolif-

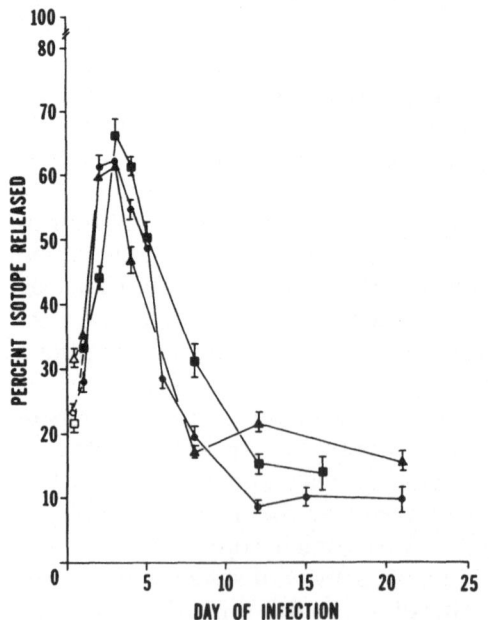

FIGURE 1. Relative NK activity of mouse spleen cells collected at various times during the course of infection of C3H (●), C57BL/6 (▲), and CBA (■) mice (corresponding open symbols are normal levels of activity of cells from uninfected mice). Ordinate shows percent of ^{51}Cr released from target YAC-1 cells after 4 h of incubation at an effector–target ratio of 200:1.

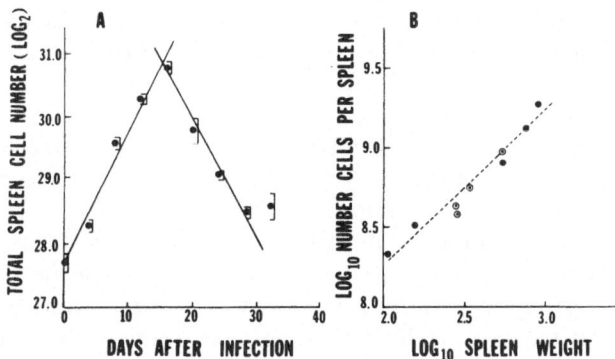

FIGURE 2. The course of splenomegaly during the course of murine infection with *T. musculi*. (A) An approximate 10-fold increase in the total number of cells in the spleen occurs during the 2 weeks following inoculation of the parasites. Subsequently, and in concert with the cure of infection, the cellularity of the spleen declines to a level about twice normal. (B) The number of cells per unit wet weight of the spleen remains constant over the rise (●) and subsequent decline (○) in total number of spleen cells.

eration with mitomycin C. Activation was unaffected by exposing the cells to ionizing radiation. Virtually all of the increased NK activity was associated with the subset of lymphocytes that displayed the Leu-19 and CD-16 antigen marker substances.

Whether there is any benefit to the host of the early activation of NK cells remains uncertain. It might represent an early step in the development of the host's defense. It is conceivable that early NK cell activation serves to the benefit of the parasite and the detriment of the host. For example, we suspect, and are attempting to demonstrate, that the activated NK cells inhibit the host's immune response to parasite antigens in a manner similar to that already demonstrated in model systems.[65,66] It seems possible that the NK cells may inhibit the initial processing/presentation of parasite antigens by inhibiting or killing antigen-presenting cells and that the resultant delay in the development of an immune response affords the parasite the opportunity to establish a vigorous infection.

2.2.2. Parasite Evasion and Inhibition of Natural Killer Cells

Those parasites that appear to be insusceptible to the action of NK cells may either (1) fail to be recognized by NK cells owing to the lack of a suitable surface ligand, or (2) be refractory to, or inhibit, the NK lytic mechanism. The insusceptibility of *T. musculi* to murine NK cells was found to be due, in part, to inhibition of NK lysis of target cells.[42] The addition either of live *T. musculi* or of cell-free extracts of *T. musculi* to mixtures of NK effector cells and YAC tumor target cells resulted in dose-dependent inhibition of target cell lysis. Preparations of *T. lewisi*, in contrast, did not inhibit murine NK cell lysis of YAC tumor target cells. As was discussed previously, *T. lewisi* was moderately susceptible to the action of murine NK cells.

Recently, we have investigated the susceptibility of trypanosomes to the isolated cytolytic granules prepared from several lines of rat large granular lymphocytic tumors (J. W. Albright *et al.*).[72] In the presence of Ca^{2+}, which is required for target lysis by the mechanism that involves cytolysin (perforin),[8] the granules were only slightly active in lysis of trypanosomes *T. musculi*, *T. lewisi*, and the African trypanosome *T. congolense*, even at very high concentration. This was true even when the trypanosomes had been treated briefly with trypsin to remove the external, surface layer of glycoprotein.

Thus, the trypanosomes appeared to be 10–50 times less susceptible to cytolysin-mediated lysis than the least susceptible nucleated mammalian cell lines that have been tested.[67] However, evidence was obtained that the parasites can actively remove cytolytic activity, either by absorbing it or, more likely, by destroying it. When trypanosome lysis by cytolytic granules was studied in the virtual absence of divalent cations (i.e., in excess EGTA), lysis was much more efficient. Although high concentrations of granules were required, the lysis of target trypanosomes was rapid and complete. Therefore, it appeared that the known lytic mechanism that involves cytolysin[8] was not effective against trypanosomes and that some other factor(s), present in low concentration in cytolytic granules, were responsible for lysis of the parasites. Characterization of this (these) factor(s) is under way.

Although *T. musculi* is insusceptible and *T. lewisi* only moderately susceptible to murine NK cell lysis, both species of trypanosome are readily killed by the ADCC mechanism. We have demonstrated this with a cloned line of rat NK tumor cells. Whether ADCC-promoted killing is attributable to the Ca-dependent cytolysin mechanism or to other substances present in cytolytic granules is unknown.

3. Parasites and Antibody-Facilitated Killer Cells

3.1. Current Paucity of Information

Very little is known about the control of parasites by antibody-facilitated killer cells—i.e., by an ADCC mechanism in which NK cells function in the killer cell mode. Most NK cells display Fc receptors and thus are capable of collaborating with specific antibodies to kill target cells. A few studies[44,45,47] have demonstrated that ADCC is an effective method of killing parasites, but in none of those studies was the effector cell(s) identified. The most informative study[52] to date involved the use of cloned murine NK cells to confer resistance to *L. donovani* on bg/bg mice. It appeared that the transferred NK cells influenced the course of infection only in concert with antibody produced by the host mice.

3.2. A Plan for Future Experimentation

The early, three- to fivefold elevation in NK activity stands out as an attractive time in which to attempt to influence the course of a parasitic infection by use of passively transferred specific antibody. In most if not all infections, strong inhibition of antibody formation is apparent by the third to fifth day after parasite inoculation, and, thus, production of facilitating antibody that might collaborate with the elevated NK cells does not occur. Whatever antibody of the IgM isotype that might appear would be of little significance, because the IgM isotype is not efficient in promoting ADCC.[68,69] It should prove interesting, and possibly quite important from a therapeutic perspective, to determine whether the introduction of enriched NK cells and specific (preferably monoclonal) antibodies of the appropriate isotype can terminate early parasitic infections. Perhaps, in the case of the enteric parasites, the administration of the NK cell-activating agent together with parasite epitope-specific, monoclonal IgA would prove to be of therapeutic benefit.

4. Conclusions

In general, it appears that direct NK cell attack on parasites is not an important mechanism for control of parasitic infections. There appear to be important exceptions, however, in the case of *T. gondii* and *T. cruzi*. There have been far too few studies of possible NK attack on parasitized host cells. In several cases the display of parasite antigens at the surface of parasitized cells has been shown, and such findings certainly suggest that parasitized cells might be as susceptible to NK cells as virus-infected cells.

The reasons for the relative insusceptibility of parasites to NK cells lie in failure of complementary molecular recognition and/or in failure of the NK lytic mechanism to inflict damage on target parasites. Considering that recognition by NK cells of even susceptible target tumor cells has not yet been elucidated, there is little that can be said about NK cell–parasite recognition at the present time. With regard to the susceptibility of parasites to known NK cytolytic attack mechanisms, the small amount of information at hand suggests that parasites may be refractory to damage by cytolysin (perforin), possibly because they inactivate it. However, there may be other substances associated with cytolytic granules in small quantities that are potent parasitolysins.

In nearly all cases, parasite infection results in an early, substantial elevation of NK cell activity in the spleen and peritoneal space and probably in other sites (e.g., blood, marrow, GALT) as well. There is the strong possibility that the activated NK cells might be very effective against the invading parasites if they could function, in collaboration with specific antibody, in the ADCC mode. That does not happen in the course of a typical parasitic infection owing to the early onset of a rather profound state of immunodepression. The NK cells themselves may contribute to this state of anergy by attacking parasite-antigen presenting cells. It may well be a matter of considerable importance to explore the combined use of enriched, activated NK cells (e.g., IL-2-activated or LAK cells [70]) and monoclonal antibody of the appropriate epitope specificity and isotype as a means of early therapy following parasitic infection.

ACKNOWLEDGMENTS. The original work described in this publication was supported by grants from the National Science Foundation (DCB-8417637) and the National Institute on Aging (RO1AGO6278).

References

1. Albright, J. F., and Albright, J. W., 1984, Natural resistance to animal parasites, in: *Immunobiology of Parasites and Parasitic Infections* (*Contemporary Topics in Immunobiology*, Vol. 12) (J. J. Marchalonis, ed.), Plenum Press, New York, pp. 1–52.
2. Lanier, L. L., Phillips, J. H., Hackett, J. Jr., Tutt, M., and Kumar, V., 1986, Natural killer cells: Definition of a cell type rather than a function, *J. Immunol.* **137**:2735–2739.
3. Lust, J. A., Kumar, V., Burton, R. C., Bartlett, S. P., and Bennett, M., 1981, Heterogeneity of natural killer cells in the mouse, *J. Exp. Med.* **154**:306–317.
4. Minato, N., Reid, L., and Bloom, B. R., 1981, On the heterogeneity of murine natural killer cells, *J. Exp. Med.* **154**:750–762.
5. Bartlett, S. P., and Burton, R. C., 1982, Studies on the natural killer (NK) cells. III. The effects of *in vitro* culture on spontaneous cytotoxicity of murine spleen cells, *J. Immunol.* **128**:1070–1075.
6. Herberman, R. B., and Ortaldo, J. R., 1981, Natural Killer cells: Their role in defenses against disease, *Science* **214**:24–30.
7. Herberman, R. B., and Callewaert, D. M. (eds.), 1985, *Mechanisms of Cytotoxicity by NK Cells*, Academic Press, New York.
8. Henkart, P., 1985, Mechanism of lymphocyte-mediated cytotoxicity, *Annu. Rev. Immunol.* **3**:31–58.
9. Eugui, E. M., and Allison, A. C., 1980, Differences in susceptibility of various mouse strains to haemoprotozoan infections: Possible correlation with natural killer activity, *Parasite Immunol.* **2**:277–292.
10. Wood, P. R., and Clark, I. A., 1982, Apparent irrelevance of NK cells to resolution of infections with *Babesia microti* and *Plasmodium vinckei petteri* in mice, *Parasite Immunol.* **4**:319–327.
11. Kumar, V., Ben-ezra, J., Bennett, M., and Sonnenfeld, G., 1979, Natural killer cells in mice treated with [89]strontium: Normal target-binding cell numbers but inability to kill even after interferon administration, *J. Immunol.* **123**: 1832–1838.
12. Seaman, W. E., and Talal, N., 1980, The effects of 17β-estradiol on natural killing in the mouse, in: *Natural Cell-Mediated Immunity Against Tumors* (R. B. Herberman, ed.), Academic Press, New York, pp. 765–777.

13. Roder, J. C., 1979, The beige mutation in the mouse. I. A stem cell predetermined impairment in natural killer cell function, *J. Immunol.* **123:**2168–2173.

14. Hunter, K. W. Jr., Folks, T. M., Sayles, P. C., and Strickland, G. T., 1981, Early enhancement followed by suppression of natural killer cell activity during murine malarial infections, *Immunol. Lett.* **2:**209–214.

15. Ojo-Amaize, E., Salimonu, L. S., Williams, A. I. O., Akinwolere, O. A. O., Shabo, R., Alm, G. V., and Wigzell, H., 1981, Positive correlation between degree of parasitemia, interferon titers, and natural killer cell activity in *Plasmodium falciparum*–infected children, *J. Immunol.* **127:**2296–2300.

16. Orago, A. S. S., and Solomon, J. B., 1986, Antibody-dependent and -independent cytotoxic activity of spleen cells for *Plasmodium berghei* from susceptible and resistant rats, *Immunology* **59:**283–288.

17. Solomon, J. B., 1986, Natural cytotoxicity for *Plasmodium berghei in vitro* by spleen cells from susceptible and resistant rats, *Immunology* **59:**277–281.

18. Coleman, R. M., Rencricca, N. J., Stout, J. P., Brissette, W. H., and Smith, D. M., 1975, Splenic mediated erythrocyte cytotoxicity in malaria, *Immunology* **29:**49–54.

19. Skamene, E., Stevenson, M. M., and Lemieux, A., 1983, Murine malaria: dissociation of natural killer (NK) cell activity and resistance to *Plasmodium chabaudi*, *Parasite Immunology* **5:**557–565.

20. Hauser, W. E. Jr., Sharma, S. D., and Remington, J. S., 1982, Natural killer cells induced by acute and chronic *Toxoplasma* infection, *Cell. Immunol.* **69:**330–346.

21. Kamiyama, T., and Hagiwara, T., 1982, Augmented followed by suppressed levels of natural cell-mediated cytotoxicity in mice infected with *Toxoplasma gondii*, *Infect. Immun.* **36:**628–636.

22. Tracey, D. E., 1979, The requirement for macrophages in the augmentation of natural killer cell activity by BCG, *J. Immunol.* **123:**840–845.

23. Hauser, W. E. Jr., Sharma, S. D., and Remington, J. S., Augmentation of NK cell activity by soluble and particulate fractions of *Toxoplasma gondii*, *J. Immunol.* **131:**458–463.

24. Kamiyama, T., 1984, *Toxoplasma*-induced activities of peritoneal and spleen natural killer cells from beige mice against thymocytes and YAC-1 lymphoma targets, *Infect. Immun.* **43:**973–980.

25. Hauser, W. E. Jr., and Tsai, V., 1986, Acute *Toxoplasma* infection of mice induces spleen NK cells that are cytotoxic for *T. gondii in vitro*, *J Immunol.* **136:**313–319.

26. Krahenbuhl, J. L., and Remington, J. S., 1982, The immunology of *Toxoplasma* and toxoplasmosis, in: *Immunology of Parasitic Diseases* (S. Cohen and K. S. Warren, eds.), Blackwell Scientific, Oxford, U.K., pp. 356–387.

27. Kuhn, R. E., 1981, Immunology of *Trypanosoma cruzi* infections, in: *Parasitic Diseases*, Volume I; *The Immunology* (J. M. Mansfield, ed.), Marcel Dekker, New York, pp. 137–166.

28. De Titto, E. H., Catterall, J. R., and Remington, J. S., 1986, Activity of recombinant tumor necrosis factor on *Toxoplasma gondii* and *Trypanosoma cruzi*, *J. Immunol.* **137:**1342–1345.

29. Beutler, B., and Cerami, A., 1986, Cachectin and tumour necrosis factor as two sides of the same biological coin, *Nature* **320:**584–588.

30. Young, J. D.-E., and Cohn, Z. A., 1986, Role of granule proteins in lymphocyte-mediated killing, *J. Cell. Biochem.* **32:**151–167.

31. Taverne, J., Dockrell, H. M., and Playfair, J. H. L., 1981, Endotoxin induced serum factor kills malarial parasites *in vitro*, *Infect. Immun.* **33:**83–89.

32. Haidaris, C. G., Haynes, J. D., Meltzer, M. S., and Allison, A. C., 1983, Serum containing tumor necrosis factor is cytotoxic for the human malaria parasite *Plasmodium falciparum*, *Infect. Immun.* **42:**385–393.

33. Clark, I. A., Virelizier, J. L., Carsell, E. A., and Wood, P. R., 1981, Possible importance of macrophage-derived mediators in acute malaria, *Infect. Immun.* **32:**1058–1066.

34. Taverne, J., Depledge, P., and Playfair, J. H. L., 1982, Differential sensitivity *in vivo* of lethal and nonlethal malarial parasites to endotoxin-induced serum factor, *Infect. Immun.* **37**:927–934.

35. Hatcher, F. M., and Kuhn, R. E., 1981, Spontaneous lytic activity against allogeneic tumor cells and depression of specific cytotoxic responses in mice infected with *Trypanosoma cruzi, J. Immunol.* **126**:2436–2442.

36. Hatcher, F. M., Kuhn, R. E., Cerrone, M. C., and Burton, R. C., 1981, Increased natural killer cell activity in experimental American trypanosomiasis, *J. Immunol.* **127**:1126–1130.

37. James, S. L., Kipnis, T. L., Sher, A., and Hoff, R., 1982, Enhanced resistance to acute infection with *Tyrpanosoma cruzi* in mice treated with an interferon inducer, *Infect. Immun.* **35**:588–593.

38. Hatcher, F. M., and Kuhn, R. E., 1982, Destruction of *Trypanosoma cruzi* by natural killer cells, *Science* **218**:295–296.

39. Tarleton, R. L., and Kuhn, R. E., 1984, Restoration of *in vitro* immune responses of spleen cells from mice infected with *Trypanosoma cruzi* by supernatants containing interleukin 2, *J. Immunol.* **133**:1570–1575.

40. Albright, J. W., Huang, K.-Y., and Albright, J. F., 1983, Natural killer activity in mice infected with *Trypanosoma musculi, Infect. Immun.* **40**:869–875.

41. Albright, J. W., and Albright, J. F., 1983, Age-associated impairment of murine natural killer activity, *Proc. Natl. Acad. Sci. USA* **80**:6371–6375.

42. Albright, J. W., Hatcher, F. W., and Albright, J. F., 1984, Interaction between murine natural killer cells and trypanosomes of different species, *Infect. Immun.* **44**:315–319.

43. Askonas, B., and Bancroft, G. J., 1984, Interaction of African trypanosomes with the immune system, *Phil. Trans. R. Soc. Lond.* **B307**:41–50.

44. Olabuenaga, S. E., Cardoni, R. L., Segura, E. L., Riera, N. E. and De Bracco, M. M. E., 1979, Antibody-dependent cytolysis of *Trypanosoma cruzi* by human polymorphonuclear leucocytes, *Cell. Immunol.* **45**:85–93.

45. Kierszenbaum, F., and Gharpure, H. M., 1983, Killing of circulating forms of *Trypanosoma cruzi* by lymphoid cells from acutely and chronically infected mice, *Int. J. Parasitol.* **13**:377–381.

46. Kuhn, R. E., and Murnane, J. E., 1977, *Trypanosoma cruzi*: Immune destruction of parasitized mouse fibroblasts *in vitro, Exp. Parasitol.* **41**:66–73.

47. Albright, J. W., and Albright, J. F., 1982, The decline of immunological resistance of aging mice to *Trypanosoma musculi, Mech. Ageing Dev.* **20**:315–330.

48. Mkwananzi, J. B., Franks, D., and Baker, J. R., 1976, Cytotoxicity of antibody-coated trypanosomes by normal human lyphoid cells, *Nature* **259**:403–404.

49. Wechsler, D. S., and Kongshavn, P. A. L., 1985, Characterization of antibodies mediating protection and cure of *Trypanosoma musculi* infection in mice, *Infect. Immun.* **48**:787–794.

50. Kirkpatrick, C. E., and Farrell, J. P., 1982, Leishmaniasis in beige mice, *Infect. Immun.* **38**:1208–1216.

51. Kirkpatrick, C. E., and Farrell, J. P., 1984, Splenic natural killer-cell activity in mice infected with *Leishmania donovani, Cell. Immunol.* **85**:201–214.

52. Kirkpatrick, C. E., Farrell, J. P., Warner, J. F., and Dennert, G., 1985, Participation of natural killer cells in the recovery of mice from visceral leishmaniasis, *Cell. Immunol.* **92**:163–171.

53. Warner, J. F., and Dennert, G., 1982, Effects of a cloned cell line with NK activity on bone marrow transplants, tumour development and metastasis *in vivo, Nature* **300**:31–34.

54. Berman, J. D., and Dwyer, D. M., 1981, Expression of *Leishmania* antigen on the surface membranes of infected human macrophages *in vitro, Clin. Exp. Immunol.* **44**:342–348.

55. Tagliabue, A., Befus, A. D., Clark, D. A., and Bienenstock, J., 1982, Characteristics of natural killer cells in the murine intestinal epithelium and lamina propria, *J. Exp. Med.* **155**:1785–1796.

56. Nencioni, L., Villa, L., Boraschi, D., Berti, B., Tagliabue, A., 1983, Natural and anti-body-dependent cell-mediated activity against *Salmonella typhimurium* by peripheral and intestinal lymphoid cells in mice, *J. Immunol.* **130:**903–907.

57. Tagliabue, A., Nencioni, L., Villa, L., Keren, D. F., Lowell, G. H., and Boraschi, D., 1983, Antibody-dependent cell-mediated antibacterial activity of intestinal lymphocytes with secretory IgA, *Nature* **306:**184–186.

58. Heyworth, M. F., Kung, J. E., and Eriksson, E. C., 1986, Clearance of *Giardia muris* infection in mice deficient in natural killer cells, *Infect. Immun.* **54:**903–904.

59. Kaplan, B. S., Uni, S., Aikawa, M., and Mahmoud, A. F., 1985, Effector mechanism of host resistance in murine giardiasis: Specific IgG and IgA cell-mediated toxicity, *J. Immunol.* **134:**1975–1981.

60. Attallah, A. M., Lewis, F. A., Urritia-Shaw, A., Folks, T., and Yeatman, T. J., 1980, Natural killer cells (NK) and antibody-dependent cell-mediated cytotoxicity (ADCC) components of *Schistosoma mansoni* infection, *Int. Arch. Allergy Appl. Immunol.* **63:**351–354.

61. Abe, T., Forbes, J. T., and Colley, D. G., 1983, Natural killer cell activity during murine Schistosomiasis mansoni, *J. Parasitol.* **69:**1001–1005.

62. Barsoum, I. S., Freeman, G. L. Jr., Habib, M., El Alamy, M. A., Rocha, R. S., Katz, N., Gazzinelli, G., and Colley, D. G., 1984, Evaluation of natural killer activity in human schistosomiasis, *Am. J. Trop. Med. Hyg.* **33:**451–454.

63. Kamiyama, T., and Hagiwara, T., 1982, Augmented followed by suppressed levels of natural cell-mediated cytotoxicity in mice infected with *Toxoplasma gondii*, *Infect. Immun.* **36:**628–636.

64. Tarkkanen, J., Saksela, E., and Lanier, L. L., 1986, Bacterial activation of human natural killer cells. Characteristics of the activation process and identification of the effector cell, *J. Immunol.* **137:**2428–2433.

65. Abruzzo, L. V., and Rowley, D. A., 1983, Homeostasis of the antibody response: Immunoregulation by NK cells, *Science* **222:**581–585.

66. Abruzzo, L. V., Mullen, C. A., and Rowley, D. A., 1986, Immunoregulation by natural killer cells, *Cell. Immunol.* **98:**266–276.

67. Henkart, P. A., Millard, P. J., Reynolds, C. W., and Henkart, M. P., 1984, Cytolytic activity of purified cytoplasmic granules from cytotoxic rat LGL tumors, *J. Exp. Med.* **160:**75–93.

68. Ralph, P. and Nakoinz, I., 1983, Cell-mediated lysis of tumor targets directed by murine monoclonal antibodies of IgM and all IgG isotypes, *J. Immunol.* **131:**1028–1031.

69. Kipps, T. J., Parham, P., Punt, J., and Herzenberg, L. A., 1985, Importance of immunoglobulin isotype in human antibody-dependent, cell-mediated cytotoxicity directed by murine monoclonal antibodies, *J. Exp. Med.* **161:**1–17.

70. Ettinghausen, S. E., Lipford, E. H. III, Mule, J. J., Rosenberg, S. A., 1985, Recombinant interleukin 2 stimulates *in vivo* proliferation of adoptively transferred lymphokine-activated killer (LAK) cells, *J. Immunol.* **135:**3623–3634.

71. Stevenson, M. M., Lyanga, J. J., and Skamene, E., 1982, Murine malaria: Genetic control of resistance to *Plasmodium chabaudi*, *Infect. Immun.* **38:**80–88.

72. Albright, J. W., Munger, W. E., Henkart, P. A., and Albright, J. F., 1988, The toxicity of rat large granular lymphocyte tumor cells and their cytoplasmic granules for rodent and African trypanosomes, *J. Immunol.* **140:**2774–2778.

7

Natural Host Resistance Mechanisms against Systemic Mycotic Agents

JUNEANN W. MURPHY

1. Introduction

The different fungi responsible for diseases in man are quite diverse in their characteristics. Some fungi are "true" pathogens in that they appear to infect normal individuals, whereas others are opportunistic and establish an infection only in an altered or compromised host. The natural portal of entry into the host varies among the fungi, but they usually fall within two major groups—those that enter by introduction into the skin, and those that enter through the respiratory route. The disease-causing fungi differ in their tissue preferences as well as in their morphological characteristics and size. These various features are important factors and must be considered in a discussion of the effectiveness of the natural host resistance mechanisms in eliminating the mycotic agents.

1.1. Diseases Caused by Fungi

Fungal diseases are frequently categorized on the basis of anatomic areas affected, such as superficial, cutaneous, subcutaneous, or systemic infections.[1,2] Although superficial infections such as pityriasis versicolor and cutaneous diseases such as dermatophytosis or ringworm and acute mucocutaneous and cutaneous candidiasis are extremely common, espe-

JUNEANN W. MURPHY • Department of Botany and Microbiology, University of Oklahoma, Norman, Oklahoma 73019.

cially in certain groups of individuals, they are relatively mild diseases and do not pose a threat to life. A notable exception is chronic mucocutaneous candidiasis, which can be a serious disease. Typically, the superficial and cutaneous mycotic diseases range from asymptomatic to chronic infections of the skin, hair, or nails. Host defenses against the etiological agents that are confined to the nonliving layers of tissue are not well understood. However, it is clear that natural resistance mechanisms, especially chemical and physical barriers, appear to be important protective mechanisms. In contrast, the subcutaneous and systemic mycotic diseases can be more serious, and many of the systemic fungi may cause life-threatening diseases. Considerably more attention has been given to assessing both natural and immune defenses against fungi causing the more serious mycotic diseases.

The subcutaneous mycotic diseases are sporotrichosis, chromomycosis, mycetomas, and candidiasis. The portal of entry into the host for the agents of subcutaneous mycotic infections is through the skin or mucous membranes, usually by traumatic implantation. In subcutaneous mycotic diseases, the organisms generally remain localized in an area or spread slowly to surrounding tissues. The organisms occasionally become disseminated, resulting in a more serious systemic disease. The etiological agents of subcutaneous mycotic diseases are not considered to be invasive; rather, they seem to take advantage of impairments in the host's defenses. The natural resistance mechanisms affecting the etiological agents of subcutaneous mycotic diseases are also not clearly defined.

A number of different fungi are capable of causing systemic mycotic disease: *Blastomyces dermatitidis, Candida albicans, Coccidioides immitis, Cryptococcus neoformans, Histoplasma capsulatum, Paracoccidioides brasiliensis,* certain *Aspergillus* species, and some organisms in the order Mucorales such as *Rhizopus oryzae.* To focus this review, emphasis will be on the natural host resistance mechanisms that function in clearance of mycotic organisms associated with disseminated or systemic forms of disease.

1.2. General Characteristics of Systemic Mycotic Diseases

With some exceptions, such as *C. albicans,* the infectious particles of the pathogenic and opportunistic fungi causing systemic disease generally gain entrance into the body via the respiratory tract. Therefore, soon after infection, the host may show no symptoms or may have a primary pulmonary form of disease ranging from an acute to chronic infection with mild to severe pulmonary distress. In the majority of cases, the disease is limited to the lungs, but in a small percentage of primary pulmonary infections, the organisms disseminate to extrapulmonary organs or tissues, depending on tissue preference of the etiologic agent. For example, after dissemination, *H. capsulatum* is found predominantly in the reticuloendothelial system; *B. dermatitidis* localizes primarily in osseous and cutaneous tissues; *C. immitis* extends to visceral organs, bones, and skin; and *Cr. neoformans* has a predilection for the central nervous system.

1.3. Characteristics of the Organisms That Cause Systemic Mycotic Diseases

There are a number of characteristics of the fungi that distinguish them from other microbial pathogens and that are important in understanding host defenses against them. First, the fungi are relatively large organisms compared to bacteria, and their size can affect entrance into the host and the way they are attacked by the host's natural cellular defenses. For example, the spherules of *C. immitis* that are found in infected tissues can be 60μm in diameter, and the yeastlike cells of *Cr. neoformans* and *P. brasiliensis* are as large as 15–20 μm.[3,4] The large size of these fungi makes it difficult for phagocytic natural effector cells to engulf them. Second, the fungi are eukaryotic organisms with cell walls containing chitin plus glucans and mannans that may be complexed to proteins.[5] The cell wall can be antiphagocytic as well as resistant to degradation by the host's cellular defenses.[3] Capsules have also been shown to inhibit phagocytosis.[6] *Cr. neoformans* is an excellent example of a highly encapsulated organism that is resistant to phagocytosis. Frequently, the outer constituents of the pathogenic fungi are released into the environment in which the organism is growing, and those soluble fungal products have the potential to modulate certain host defenses.[7-9] Third, fungal morphology, varying from yeast (blastospore) to mycelial forms that can produce an array of spore types, can influence the effectiveness of host defense mechanisms. Most of the pathogenic fungi exist as saprophytic molds in their natural habitat, and there, they produce the particles that are infectious to man. Upon entering the host, the infectious particles convert to the parasitic phase, which is the yeast form in the cases of *H. capsulatum*, *B. dermatitidis*, *Sporothrix schenckii*, and *P. brasiliensis*. On the other hand, *C. immitis* is unique in that the infectious arthroconidia produced during mycelial growth in the soil, once in the tissue, convert to immature spherules that subsequently develop into large endospore-containing structures— i.e., mature spherules. In contrast, *Candida albicans*, the *Candida* species most often associated with human disease, is generally found in the yeast phase as normal flora of the gastrointestinal tract, but in establishing infection, yeast cells multiply by budding as well as by developing into pseudomycelia and hyphae. The saprophytic form of *Cr. neoformans*, thought to be the yeast phase, is also the parasitic form. Other opportunistic fungi, such as *Aspergillus* and *Rhizopus*, grow in the mycelial phase both inside and outside the host. As the fungi convert from one form to another or adapt to the *in vivo* environment, some of their surface constituents and

their size may change. These alterations in the organism may impede natural host defenses. Several of the features identified above also affect the
induction and modulation of the immune responses, which may ultimately have an effect on certain of the natural cellular defenses.

1.4. Natural Host Resistance Mechanisms Effective against Systemic Mycotic Agents

Natural or innate host resistance mechanisms are important early after
infection with a mycotic agent, as well as late in the disease process, after
they have been augmented by immune components. Although the natural
resistance system consists of physical, chemical, and cellular defenses, this
discussion will focus on the natural cellular mechanisms of defense against
mycotic agents. The natural resistance system is considered very effective
in preventing establishment of mycotic diseases. This concept is based on
the fact that mycotic pathogens are prevalent in nature, so exposure is
frequent, yet the incidence of disease is relatively low. Mechanical and
chemical barriers in the upper respiratory tract prevent the majority of
the fungal particles inhaled from entering the alveolar spaces. However,
those fungal particles that overcome these natural defensive barriers and
get into the lungs immediately encounter the natural cellular defensive
line. An indicator of the effectiveness of the natural effector cells is suggested by the observation that individuals with defective natural cellular
defense mechanisms are more susceptible than normal individuals to certain mycotic diseases, such as invasive aspergillosis and candidiasis.[10]

With other systemic mycotic agents, the protective role of the natural
effector cells is not so evident; however, numerous investigations have
demonstrated that one or more of the natural effector cell populations
afford a degree of protection. The roles of the phagocytic effector cells,
i.e., monocyte/macrophages and polymorphonuclear leukocytes (PMNL),
in protection against the fungi have been studied for many years. There
are several detailed reviews on the effects of phagocytic cells on the fungi,
[11-15] so the discussion on monocytes/macrophages and PMNL in this
chapter will not be a complete review but will be limited to major points.
On the other hand, only recently has attention been given to the effects
of nonphagocytic cells, like macrophage precursors and natural killer (NK)
cells, on the fungal pathogens, so more detail concerning these cells in
host defense against fungal targets will be presented.

A general scenario that may occur in the natural establishment of a
systemic mycotic disease could be described as follows. When the host is
exposed to aerosolized fungal particles, virtually all of the large particles
($>4-5$ μm) and many of the smaller particles are eliminated through the
action of the mucociliary system. However, those infectious particles that
do gain entry into the alveolar spaces are immediately attacked by the
resident alveolar macrophages and NK cells. Most of the fungi, either

directly or indirectly through the alternative complement cascade, are chemotactic for PMNL, so within a few hours these professional phagocytic cells are recruited to the site of infection. In most situations, clearance or limiting of the fungal elements in the lungs, or for that matter in other tissues after dissemination, is not the result of a single natural effector cell population but rather the result of a complex array of interactions between effector cells and soluble factors. The mechanism(s) by which the natural effector cells destroy or inhibit the fungal elements may be slightly different depending on the mycotic agent, even though the same populations of natural effector cells are involved.

2. Role of Monocytes/Macrophages in Host Defense

In general, monocytes and macrophages have been considered to function in host defense by phagocytizing the pathogen prior to imposing the lethal hit. With regard to the fungi, this generalized situation is one means by which the monocytes and macrophages eliminate the organisms; however, nonphagocytic killing attributed to the monocyte/macrophage series has also been described.[17-21] For the most part, antifungal defense has been considered to be carried out by mature monocytes and macrophages, but using a mouse model, macrophage precursors found in liver,[21] spleen,[19] and bone marrow[19] have recently been described as being cytotoxic to *C. albicans*. The macrophage precursors have some characteristics in common with, but other characteristics that distinguish them from, mature monocytes and macrophages as well as natural killer cells.[18-21] For example, the macrophage precursors have large granular lymphocyte (LGL) morphology, are nylon wool nonadherent, are nonphagocytic, and are cytotoxic to YAC-1 but not EL-4 or P815 tumor target cells, characteristics that make them like NK cells; however, they lack asialo GM_1 on their membranes, a characteristic that distinguishes them from NK cells.[18,22] These immature precursor cells are considered to be in the macrophage lineage because (1) they have the macrophage surface markers, F4/80 and M143, which are identified by monoclonal antibodies; (2) they can respond with a strong proliferative response to macrophage colony-stimulating factor, CSF-1; and (3) they convert to mature macrophages after 7–10 days in tissue culture.[19] Baccarini *et al.*[19] have proposed that these macrophage precursors contribute to the natural resistance mechanisms. It will be interesting to learn, as additional studies are done, if similar effector cells can be found in humans and other animal species, if these macrophage precursors are lethal to other fungal targets, and if the mechanisms by which the organisms are killed are akin to those of the macrophages.

The contributions of mature mononuclear phagocytic cells to natural resistance against the fungi are not completely understood at this time. There are several reasons for this. The mononuclear phagocytic cells are a heterogeneous group consisting of blood monocytes and tissue macrophages. The cells within this group have the potential to change. For example, blood monocytes convert to macrophages after entering tissues or after being maintained in tissue culture, and in the conversion, there are substantial changes in the cells' antifungal properties.[23–29] Tissue macrophages vary significantly in their phagocytic and antimicrobial characteristics depending on their tissue source and their state of activation. For instance, resident alveolar macrophages differ from resident peritoneal macrophages, and resident splenic macrophages are not necessarily comparable to either of the former. Peritoneal macrophages from unstimulated animals may have different antifungal capabilities from peritoneal macrophages elicited with agents such as oil, thioglycolate broth, or glycogen.[30–32] Another factor greatly affecting the antimicrobial activities of macrophages is the immune status of the animal from which the macrophages have been obtained.[33,34] In immune animals, the macrophages may be activated by lymphokines such as macrophage-activating factor or gamma interferon, and after activation they are much more effective in killing fungal agents than are resident macrophages from normal animals.[35–37]

Besides the great degree of variability within the group of mononuclear phagocytic cells, investigators studying these cells have used a broad array of techniques for assessing their effects on the fungi. For instance, the mononuclear cell populations studied have been obtained from diverse species of animals. Moreover, some investigations focus only on phagocytic abilities or killing activities, whereas others determine phagocytic indices plus the ability of the phagocytic cells to kill or inhibit the growth of the organisms using different means of assessing the fungicidal effects. Another important variable has been the effector-to-target ratios employed.

The fungi that cause systemic mycoses are variable in size, morphological and physiological characteristics, and mode of reproduction; therefore, one effector cell type would not be expected to necessarily have the same effects on all of the fungal organisms. Given all the variables involved, one can readily see that it is difficult to make general or unqualified statements concerning the antifungal defense of mononuclear phagocytes as a group. Therefore, the approach here will be to briefly consider the fungicidal activities of the subgroups of mononuclear phagocytes that have been most studied, still recognizing the difficulties in making comparisons because of the differences in organisms and the varying experimental designs used in assessing the activities.

2.1. Alveolar Macrophages

In most systemic mycotic infections, the alveolar macrophages are among the first natural cellular defenses against the inhaled fungal elements. The primary means of evaluating the effectiveness of alveolar macrophages on mycotic agents have been through *in vitro* assays in which the viability of the organisms has been determined by vital staining, by culturing the organisms and counting the numbers of colony-forming units (CFUs), or by assessing the amount of radioisotopically labeled DNA or protein precursors incorporated. Representative data shown in Table I indicate that normal alveolar macrophages, irrespective of animal source, can phagocytize or associate with most fungal spores or hyphal elements (20–89%); however, the alveolar macrophage's abilities to kill or limit the growth of most of the fungal agents are not impressive (0–33%).

Blastomyces dermatitidis and the opportunistic organisms *Candida albicans, Rhizopus oryzae,* and *Aspergillus fumigatus* are the most susceptible to the alveolar macrophages. Waldorf *et al.*[47] suggest that normal bronchoalveolar macrophages play a central role in host defense against the two opportunistic filamentous fungi *R. oryzae* and *A. fumigatus.* They have demonstrated that alveolar macrophages from normal mice kill the conidia of *A. fumigatus* and inhibit *R. oryzae* spore germination, which is a critical step in conversion to the mycelial or tissue invasive form.[47] Additional data emphasizing the point that alveolar macrophages are important in protection against *R. oryzae* have been presented by the same investigators. They have shown that bronchoalveolar macrophages from diabetic mice do not inhibit *R. oryzae* spore germination, thereby allowing the establishment of infection.[46] These findings tend to explain why mucormycosis caused by *Rhizopus oryzae* is common in individuals with uncontrolled diabetes mellitus.[48]

In contrast to protection offered by the alveolar macrophages against *R. oryzae* is the fact that the ingestion of *H. capsulatum* by macrophages does not adversely affect the organism but rather provides the essential environment for the *Histoplasma* to reproduce.[44] It has also been proposed that the *H. capsulatum*-infected mononuclear phagocytic cells may aid in the dissemination of the organism to extrapulmonary sites.[49] When phagocytized by macrophages, *H. capsulatum* yeast cells do not trigger the respiratory burst.[50,51] Thus, by preventing the release of reactive oxygen metabolites, the organisms avoid being destroyed. *H. capsulatum* is not the only fungus listed in Table I that has been repeatedly shown to resist killing by alveolar macrophages. *C. immitis* is another. Both the arthrospores and endospore of *C. immitis* escape the lethal mechanisms of the macrophages, and they appear to do this by inhibiting fusion of the *C. immitis*-containing phagosomes with the lysosomes.[42]

In several studies, investigators have attempted to modify the activities of the natural effector cells. Fungicidal activities of alveolar macrophages could not be augmented by immunization[38,44] or by treating animals with agents such as endotoxin which tend to activate other macrophage populations[43] prior to collection of effector cells. The addition of immune serum, complement, or lung lining material to the *in vitro* assay mixtures of fungal cells and alveolar macrophages does not appear to

TABLE I

Antifungal Activities of Alveolar Macrophages

Animal source of alveolar macrophage	Target organism	Spore type	Percent phagocytosis or association	Percent killing or reduction in germination	Killing assay method (E:T ratio)	Ref.
Mouse	B. dermatitidis	Blastospore	—	21–33	CFU (100:1)	38
Human	C. albicans	Blastospore	—	0	Dye excl. (1:30–40)	39
Rabbit	C. albicans	Blastospore	48	71–93	^{3}H-leucine incorp. (1:4)	30
Rabbit	C. albicans	Blastospore	75	16	Stain (1:5)	31
Mouse	C. albicans	Blastospore	89	7	Stain (1:4)	41
Mouse	C. albicans	Blastospore	—	19	CFU (100:1)	38
Monkey	C. immitis	Endospore	20	0	CFU (10:1)	42
		Arthrospore	—	0	CFU (10:1)	42
Guinea pig	Cr. neoformans	Blastospore	52	0	CFU (2:1)	43
Rabbit	H. capsulatum	Blastospore	32	0	CFU (1:4.5)	44
Mouse	H. capsulatum	Blastospore	37	0	— (1:5)	45
		Mycelial	35	0	— (1:5)	45
Mouse	R. oryzae	Spores	70	29	Spore germination (5:1)	46
Mouse	A. fumigatus	Conidia	—	23	Spore germination (5:1)	47

increase the fungicidal effects.[42,44] However, by treating animals prior to collection of cells with complete Freund's adjuvant[31] or with *Mycobacterium bovis*,[45] the fungicidal activities of alveolar macrophages are enhanced when compared to the activities of alveolar macrophages from untreated animals. Additional studies with these models will be required to explain the mechanism of augmentation.

The means by which the alveolar macrophages kill the fungi have not been clearly elucidated, but both oxygen-dependent and -independent mechanisms could potentially function in destroying fungal organisms.[14,15] Peterson and Calderone[52] have demonstrated that lysosomal extracts from rabbit alveolar macrophages inhibit incorporation of certain amino acids by *C. albicans*. Furthermore, Lehrer and co-workers[53] have shown that low-molecular-weight cationic peptides isolated from lysosomes of resident or Freund's adjuvant elicited alveolar macrophages of rabbits are effective in killing many microorganisms, including such fungi as *C. albicans*, *C. parapsilosis*, and *Cr. neoformans*. These studies suggest that nonoxidative mechanisms may make a significant contribution to fungicidal activities of alveolar macrophages.

2.2. Blood Monocytes and Monocyte-Derived Macrophages

The antifungal activity of peripheral blood monocytes or monocyte-derived macrophages has been the subject of several investigations. Table II is a list of representative results from such studies. With *C. albicans* and *Cr. neoformans*, it has been reported that human peripheral blood monocytes could effectively kill the organisms *in vitro* (27% and 46%, respectively); however, the macrophages derived from similar monocyte populations were without effect on the organisms.[23–26] The opposite effect was observed with *B. dermatitidis*. Monocytes were slightly less effective against the yeast phase of *B. dermatitidis* (35–38% killing) than were the monocyte-derived macrophages (48–55% killing).[27] When killing of *B. dermatitidis* conidia was assessed, the monocyte-derived macrophages were considerably more effective (90% killing) than the monocytes (35% killing).[28]

Normal monocytes kill fungal organisms through oxygen-dependent mechanisms, with the myeloperoxidase-H_2O_2-halide system generally being most effective,[61,62] and by means of oxygen-independent mechanisms such as lysozyme.[63] Macrophages typically lack myeloperoxidase; therefore, the fungicidal activity is dependent on the generation of superoxide anion, hydrogen peroxide, and other active molecular species derived from oxygen as well as oxygen-independent means—e.g., lysozyme, cationic proteins. Discussions of both oxygen-dependent and -independent mechanisms by which mononuclear phagocytes and polymorphonuclear leukocytes inhibit the fungi are presented in several reviews.[14,15]

TABLE II
Fungicidal Effects of Human Peripheral Blood Monocytes or Monocyte-Derived Macrophages

Cell type	Target organism	Spore type	Percent killing	Killing assay method (E:T ratio)	Ref.
Monocyte	*B. dermatitidis*	Blastospore	35–38	CFU (400:1)	27
Monocyte	*B. dermatitidis*	Blastospore	5	Dye excl. (10:1)	28
Monocyte	*B. dermatitidis*	Conidia	35	CFU (10:1)	28
Macrophage	*B. dermatitidis*	Blastospore	48–55	CFU (400:1)	27
Macrophage	*B. dermatitidis*	Blastospore	40	Dye excl. (10:1)	28
Macrophage	*B. dermatitidis*	Blastospore	0	CFU (5:1)	29
Macrophage	*B. dermatitidis*	Conidia	90	CFU (10:1)	28
Monocyte	*C. albicans*	Blastospore	27	Stain (1:4–5)	26
Monocyte	*C. albicans*	Blastospore	44	Stain (1:2)	54
Monocyte	*C. albicans*	Blastospore	60	Stain (1:1)	55
Monocyte	*C. albicans*	Blastospore	50	CFU (300:1)	56
Monocyte	*C. albicans*	Pseudohyphae	50	Stain (1:1)	55
Monocyte	*C. albicans*	Hyphae	45–55	Uptake (10:1)	57
Macrophage	*C. albicans*	Blastospore	0	— (1:4)	25
Monocyte	*C. immitis*	Arthrospore	0	CFU (300:1)	56
Monocyte	*Cr. neoformans*	Blastospore	46	CFU (1:1)	23
Macrophage	*Cr. neoformans*	Blastospore	0	CFU (1:1)	24
Monocyte	*H. capsulatum*	Blastospore	52	CFU (5:1)	58
Monocyte	*A. fumigatus*	Hyphae	40	Uptake (10:1)	59
Monocyte	*R. oryzae*	Hyphae	41	Uptake (10:1)	60

2.3. Peritoneal Macrophages

Peritoneal macrophages from several species of laboratory animals have been assessed for their fungicidal capabilities using a wide variety of experimental approaches. The common findings have been that peritoneal macrophages from normal animals are not as effective in killing mycotic agents as monocytes or PMNL; therefore, resident macrophages are not considered to play a prominent role in natural resistance.[12] However, macrophages are extremely important in host defense against fungi but only after they have been activated by immune lymphokines, such as gamma interferon.[33–37] Considering that cell-mediated immunity is the most effective host resistance mechanism against many of the systemic mycotic organisms, one might expect the sensitized T cells to mediate protection by enhancing the activity of natural effector cells, and this has proved to be the case. Not only is macrophage activity enhanced by soluble factors from T cells, but their activity can also be down-regulated by T cell-derived factors. Recent reports indicate that macrophage activity against *Cryptococcus neoformans* and *Saccharomyces cerevisiae* can be suppressed with supernatants prepared by *in vitro* crytococcal-antigen stimulation of lymphocytes obtained from *Cr. neoformans*-infected mice or from mice given suppressive doses of soluble crytococcal antigen.[64,65]

3. Role of Polymorphonuclear Leukocytes in Host Defense

3.1. Chemotaxis of PMNL in Response to Mycotic Organisms

Polymorphonuclear leukocytes (PMNL) are usually not found in high numbers in most tissues until they are mobilized and attracted by the processes that follow introduction of a foreign substance. Many of the systemic mycotic agents either produce chemotractants[66–69] or are chemotactic to PMNL by virtue of triggering the alternative complement cascade,[70–78] thereby causing the production of the strongly chemotactic attractants for PMNL like C5a and the C5,6,7 complex. The mobilization and migration of PMNL into the infected tissue is a critical step in host defense, and if for some reason chemotactic factors are not produced or the PMNL are not responsive to chemotractants, then the host's natural resistance mechanisms may be compromised to such an extent that the infecting mycotic agent can gain control. Two studies with laboratory animals emphasize the importance of the alternative complement cascade in host defense.[74,79] Guinea pigs treated with cobra venom factor to deplete late complement components are more susceptible to an infection with *Cr. neoformans* than are normal animals.[74] Similarly, mice genetically deficient in C5 succumb to an infection with crytococci more readily than do

C5-sufficient controls.[79] It is most probable that the increased susceptibility of the complement-deficient animals to infection is not solely the
result of the lack of chemotactic factors but rather is the consequence of
combined factors, including the paucity in opsonization by C3b or iC3b.[80]

3.2. Killing of Fungi by PMNL

Once in the tissues, the PMNL are considered to be relatively effective in killing most fungal cells. Our understanding of the fungicidal activities of PMNL against the systemic fungal pathogens has been derived
primarily from *in vitro* studies, some of which are summarized in Table
III. If one compares the percent killing of the various fungi by alveolar
macrophages (Table I) and monocytes/monocyte-derived macrophages
(Table II) with the percent killing of the respective organisms by PMNL
(Table III), it is evident that PMNL and monocytes are ordinarily the
most effective in killing the fungal organisms. There are two notable exceptions, however. The arthroconidia of *C. immitis*[3] and the resting conidia of *A. fumigatus*,[88] which are the infectious stages of these organisms,
are resistant to elimination by PMNL. These situations will be discussed
later in this review.

The mechanisms by which PMNL inhibit the growth of microbial agents
are generally divided into two primary types: (1) oxidative mechanisms,
which depend on the metabolites, such as superoxide anions, hydrogen
peroxide, singlet oxygen, hydroxyl radicals, hypochlorite, and hypoiodite;
and (2) oxygen-independent components, such as lysozyme, cationic proteins, lactoferrin, etc.

Of the oxidative mechanisms, the myeloperoxidase-hydrogen peroxide-halide system seems to have the broadest antifungal effects. *B. dermatitidis*,[89,90] *C. albicans*, [61,84] *Cr. neoformans*,[23] *H. capsulatum*, [91,92] *P. brasiliensis*, [93] *A. fumigatus*,[87] and *R. oryzae*[87] have been shown to be sensitive
to the myeloperoxidase-hydrogen peroxide-halide system. Potassium iodide in conjunction with myeloperoxidase and hydrogen peroxide has
stronger fungicidal effects than the same system having chloride substi-
tuted as the halide.[91–93] Hydrogen peroxide alone kills some fungal organisms; however, in general, hydrogen peroxide is not as effective as the
myeloperoxidase system, and furthermore, concentrations of hydrogen
peroxide above physiological concentrations are required for killing.[91–93]
Levitz and Diamond[94] have described another system which effectively
kills *A. fumigatus* spores and *C. albicans* blastospores *in vitro*. The system is
composed of iron, hydrogen peroxide, and halide.

Constituents of the PMNL that do not require oxygen for killing have
also been shown to be quite effective against certain of the pathogenic
mycotic agents. For instance, lysozyme has been reported to kill *C. immitis*[63] and *Cr. neoformans*.[95] Lactoferrin can inhibit the *in vitro* growth of
C. albicans. [96] In early work, Gadebusch and Johnson[95] showed that ca-

TABLE III
Antifungal Effects of Human Polymorphonuclear Leukocytes

Target organism	Spore type	Percent killing	Killing assay method (E:T ratio)	Ref.
B. dermatitidis	Blastospore	0	CFU (1000:1)	27
		29	CFU (1:1)	81
		20	Dye excl. (10:1)	28
	Conidia	50	CFU (10:1)	28
C. albicans	Blastospore	29	Stain (1:1)	82
		35	Stain (1:1)	55
		11–18	Stain (1:1)	83
		25	Dye excl. (10:1)	28
	Pseudohyphae	58	Radioisotope uptake (1:1)	84
		10	Stain (1:1)	55
C. immitis	Arthroconidia	5	— (10:1)	3
	Endospore	10–20	— (10:1)	3
Cr. neoformans	Blastospore	81	CFU (2:1)	85
		59	CFU (1:1)	23
H. capsulatum	Blastospore	46	CFU (5:1)	58
P. brasiliensis	Blastospore	44–63	Dye excl. (5:1)	83
A. fumigatus	Hyphae	42–60	Radioisotope uptake (10:1)	86, 87
	Resting conidia	0	Germination (1:1)	88
	Swollen conidia	31	Germination (1:1)	88
B. oryzae	Hyphae	41–60	Radioisotope uptake (10:1)	86, 87

tionic proteins from granulocytes are effective in killing *Cr. neoformans.* More recently, Lehrer and co-workers[97–103] have isolated from neutrophils a family of small cationic peptides that are potent fungicides. These microbicidal peptides have been called defensins.[100,101] The defensins are 20–30 amino acids in length; are rich in cystine, arginine, and aromatic amino acids; lack free sulfhydryl groups, and are not glycosylated.[100,101] Defensins have been shown to be active against certain bacteria and viruses as well as the following fungi: *C. albicans,*[101] *C. immitis,* [102] and *Cr. neoformans.*[101] Similar cationic peptides kill *R. oryzae.*[87] The nonoxidative mechanisms probably function along with the oxidative systems in PMNL to kill fungal organisms. The reader is directed to a review by Fleischmann and Lehrer in which fungicidal mechanisms of PMNL have been discussed in some detail.[15]

3.3. Means by Which Fungi Avoid PMNL Effects

It has been noted that certain forms or structures of the same fungus are more resistant to PMNL-mediated killing than others, and the more resistant form may protect the organism to some extent from the host's natural cellular defense mechanisms. For example, as mentioned above, the resting spores of *A. fumigatus* are not killed by PMNL.[3,88] However, when these resting conidia are allowed to swell in culture, the swollen conidia become more vulnerable to PMNL killing (31% killing).[88] This example is in contrast to the conidia of *B. dermatitidis,* which are much more susceptible to the fungicidal properties of the PMNL—i.e., 50–60% killing of the ingested *B. dermatitidis* conidia[28] than are the blastospores or tissue form (0–30% killing[27,28,81]). These data indicate that the conversion of the infectious conidia to the yeast phase offers some protection to *B. dermatitidis* from the cidal effects of the PMNL.

Also worthy of mention is that different levels of killing or stimulation of growth of the fungi by PMNL have been observed among different isolates of the same organism.[27,83,92] Thus, the more virulent isolates may be so because they can evade one or more of the host defense mechanisms, such as PMNL killing, or they are stimulated to grow by the phagocytic cells or cell products. Blastospores of different isolates of *H. capsulatum* have been reported to vary in their level of sensitivity to the inhibitory effects of hydrogen peroxide, a fungicidal product of the PMNL; however, the level of resistance of each isolate to hydrogen peroxide does not correlate with the level of catalase produced by the isolate.[92] Diamond *et al.*[104] have identified an anionic, low-molecular-weight product released by hyphal forms of *C. albicans* that inhibits chemotaxis of PMNL, prevents PMNL from attaching to the surface of *C. albicans* hyphae, and inhibits the respiratory burst of the PMNL. In general, virulence factors of the fungi are not clearly defined, so considerably more attention should be given to this area of investigation.

A number of other factors influence how successful the PMNL are in reducing the numbers of viable fungal organisms in an *in vivo* environment. Phagocytic cells, such as PMNL, must first bind to their targets prior to phagocytosis and killing. As mentioned earlier in this review, capsules or capsular material[6,23] or the fungal cell walls[3] may inhibit or prevent phagocytosis. An example in which the capsule allows the organism to escape killing by preventing phagocytosis is the highly encapsulated *Cr. neoformans* yeast cell. Anticryptococcal immune serum and complement have been shown to enhance phagocytosis of *Cr. neoformans* by PMNL, thus overcoming the antiphagocytic effects of the capsule.[105] The outer cell wall layer of *C. immitis* arthroconidia limits phagocytosis of the arthroconidia.[3] Although removal of the outer wall of the arthroconidia increases phagocytosis, it has no effect on killing of the intracellular conidia.[3] Serum containing anti-*C. immitis* antibodies only slightly enhances phagocytosis but significantly improves killing of the arthroconidia by PMNL.[3] The mechanisms involved in the enhanced killing of the opsonized arthroconidia by PMNL have not been elucidated.

The size of the fungal elements frequently dictates the way PMNL proceed in the killing process. *Cr. neoformans* yeast cells are frequently as large as the PMNL, especially when the cryptococci are heavily encapsulated, making it difficult for the PMNL to phagocytize the organism. Extracellular killing of cryptococci by PMNL has been noted by several investigators.[4,77,106–108] The PMNL tend to form rings around the large encapsulated cryptococci, and histochemical studies suggest that exocytosis of granules occurs, releasing the granule contents into the area between the cryptococci and the PMNL.[106,107] Large yeast cells are not the only structures that prohibit phagocytosis; fungal organisms that grow *in vivo* in the pseudohyphal or hyphal form present similar problems to the phagocytic effector cells. Diamond and co-workers[84,87] have clearly shown that PMNL attach to the surface of the pseudohyphae or hyphae of *C. albicans, A. fumigatus,* and *R. oryzae* and then inflict damage on the hyphae through the myeloperoxidase-hydrogen peroxide-halide system. Data demonstrating that neutrophils from patients with chronic granulomatous disease do not damage the *Candida* pseudohyphae or hyphae of *A. fumigatus* or *R. oryzae* confirm the importance of oxygen-dependent mechanisms in killing these organisms and possibly explain why patients with defective production of oxygen metabolites by PMNL are so susceptible to certain opportunistic fungi such as *Candida* and *Aspergillus*.[10]

3.4. Augmentation of the Fungicidal Activity of PMNL

Drawing on information acquired from *in vitro* studies, it appears that there is the potential to enhance PMNL activity against the fungi by specific antibody[3,17,109] and by lymphokines that are induced by both specific[110,111] and nonspecific stimuli.[112] As one would certainly expect, specific antibodies directed toward fungal surface antigens can enhance phagocytosis by PMNL, although the antibodies may or may not affect killing of the organisms.[81,113] Recently, there has been a growing amount of data published demonstrating that soluble factors derived from leukocytes can modulate the antifungal activities of PMNL. Soluble supernatants from concanavalin A-stimulated murine spleen cells have been shown to augment PMNL killing of *B. dermatitidis*.[110] In addition, treatment of PMNL with culture supernatants produced by stimulating spleen cells from *B. dermatitidis*-immunized mice with *B. dermatitidis* antigens results in more effective killing of *B. dermatitidis* (22–25% killing) than is seen when PMNL are pretreated with control supernatants prior to being cocultured with *B. dermatitidis* (11–14% killing).[110]

One might assume that T cells in these two situations were responsible for the PMNL-activating lymphokines. In the first case, a nonspecific or mitogenic stimulation triggered lymphokine production, whereas in the latter situation the lymphokine was produced by immune T cells upon restimulation with specific antigen. T cells may not be the only cells with the capability of producing PMNL-activating factors. Djeu[114] has reported that NK cells can be triggered with nonviable *C. albicans* cells to produce supernatants that will significantly enhance the candidicidal activity of PMNL. This same investigator[112] has demonstrated that PMNL activity against *C. albicans* can be augmented by tumor necrosis factor and gamma interferon. Considering these reports together, one might predict that gamma interferon, which may have been a common factor in all of the PMNL-activating supernatants, was responsible for enhancing the fungicidal effects of the PMNL. These studies emphasize the potential for interactions between not only natural effector cells but also immune cells and natural effector cells to enhance the function of the natural effector cells in clearance of fungi from tissues. This should be a productive area for future research.

4. Natural Killer Cells and Other Nonphagocytic Effector Cells in Host Defense

Natural killer cells have only recently been shown to have the potential to play a role in host defense against the fungi. Studies done thus far indicate that some fungal organisms are susceptible to direct attack by NK cells, meaning that NK cells bind to the fungal targets[115–119] as they do to tumor or virus-infected cells, before inflicting the lethal hit. In other cases, NK cells do not seem to directly kill the organisms but function indirectly by producing lymphokines that are capable of augmenting the antifungal activity of other natural effector cells such as PMNL.[114] Both direct and indirect NK cell activities against mycotic agents are discussed below.

4.1. Direct Activity against the Organisms

From the very limited amount of data on NK cell effects on different fungi, it appears that NK cells are not effective against all of the pathogenic fungi. Furthermore, sufficient information has not been obtained to indicate if the animal species from which the NK cells are isolated dictates whether or not the fungal organisms are killed. To date, mouse and rat NK cells have been shown to bind and directly affect *Cr. neoformans* cells without phagocytizing the cryptococci.[115–119] Furthermore, there is evidence that NK cells limit the growth of cryptococci *in vivo*.[120–124] The yeast phase cells of *P. brasiliensis* are inhibited in their *in vitro* growth by mouse NK cells.[125] There is some evidence that human mononuclear cells obtained from peripheral blood[126] and mononuclear cells from mouse spleens[127] can inhibit proliferation of *H. capsulatum* yeast cells *in vitro*. In other preliminary reports, freshly isolated human peripheral blood lymphocytes have been described as being cytotoxic to *C. immitis* spherules and endospores *in vitro*.[128] In the latter three studies, the effector cells were not as completely characterized as one would like before labeling them as NK cells; however, they may well be NK cells. In contrast to these positive findings, several groups of investigators[112,114,129,130] using either mouse or human NK cells have demonstrated that typical NK cells do not directly kill *C. albicans* yeast cells. As mentioned earlier in this review, mouse promonocytes, which have some characteristics of NK cells such as morphology and anti-YAC-1 activity, can effectively kill *C. albicans*,[17,21] so there may be other natural effector cells, similar to NK cells, with the potential to defend the host against certain fungi.

Although the direct action of NK cells against several different mycotic agents has been assessed to varying degrees by employing *in vitro* assay systems, studies in our laboratory have examined in detail the interactions of NK cells with *Cr. neoformans*.[115–122,131] The principal approach we have taken throughout the investigations on NK cell activity against cryptococci has been to assess the NK cell activity of the various effector cell populations in parallel with evaluating the effects of the same cell populations on cryptococci. In the studies concerned with functionality of rat and mouse effector cell populations, NK cell activity was routinely determined with a standard 4-h ^{51}Cr-release assay against YAC-1 lymphoma cells, which are highly sensitive to rat and mouse NK cells,[132,133] and, concomitantly, anticryptococcal activity of the effector cell populations was measured with an 18-h growth inhibition assay.[115,118] Through the use of such parallel assays with an array of mouse effector cell populations, we have demonstrated that anticryptococcal activity copurifies with NK cell activity[115,118,119] and that the effector cells that lyse YAC-1 targets and inhibit the growth of cryptococci have the same phenotypic characteristics.[115,119] Mouse NK cells are not unique in their abilities to bind and inhibit cryptococci; rat NK cells perform similarly.[116] Agents that modulate NK cell activity, such as polyinosinic-polycytidylic acid (poly I:C),[134,135] Formalin-killed *Corynebacterium parvum*,[136,137] and cyclophosphamide (Cy)[138] have parallel effects on natural, nonphagocytic, cellular anticryptococcal activity.[115,122] It is well established that murine NK cell activity varies with strain, age, and organ from which effector cells are obtained, and we have documented concomitant variations in anticryptococcal activity.[115,122]

Using nylon wool nonadherent cell populations from two different strains of mice, we assessed the anticryptococcal activity on seven different isolates of *Cr. neoformans*.[115] Among the isolates used in these studies, all four of the cryptococcal serotypes were represented, and the isolates varied in the degree of encapsulation ranging from weak to heavy. There was no correlation between the level of anticryptococcal activity of the NK cell-enriched effector cell populations and the serotype or degree of encapsulation of the cryptococcal targets.[115] The cryptococcal capsule has been considered a virulence factor, and one reason has been that heavily encapsulated cryptococcal cells are not phagocytized by PMNL and macrophages as effectively as weakly encapsulated cells.[6,139] The capsule cannot be regarded in the same manner when considering host defenses mediated by natural effector cells, such as NK cells, which inhibit the cryptococcal targets through a direct contact mechanism rather than through a phagocytic mechanism.[115]

NK cells have Fc receptors on their surface and have been shown to be effector cells in antibody-dependent cell-mediated cytotoxicity (ADCC) against IgG-coated tumor cells and zenogenic erythrocytes.[140–144] We have found that NK cells also function in an antibody-dependent manner against cryptococci targets. Murine splenic nylon wool nonadherent (NWN) cells and the large granular lymphocyte (LGL) cell fractions from Percoll discontinuous gradients are more effective in inhibiting cryptococcal growth in the presence of rabbit anticryptococcal antibody than in the presence of normal rabbit serum or tissue culture medium.[131] These findings indicate that NK cells have the potential to function *in vivo* against *Cr. neoformans* not only as a first-line defense mechanism but also later in the course of disease, after antibodies have been produced.

The demonstration that NK cells are effective against *Cr. neoformans in vitro* prompted studies to assess their effects *in vivo*. Three different mouse models have provided data that support the contention that NK cells may contribute to early clearance of cryptococci from infected mice.[120–122,145] First, BALB/c nude mice, which have activated macrophages and high levels of NK cell activity, clear cryptococci during the first week of disease more effectively than normal heterozygous nu/+ littermates.[120,145] Second, mice homozygous for the beige mutation (C57Bl/6 bg/bg), which have defective NK cell activity but normal PMNL and macrophage activities against *Cr. neoformans*, have higher numbers of cryptococci in lungs and spleens 3 days after infection than phenotypically normal (C57B1/6 bg/+) mice.[121] In addition, using this same animal model, Marquis et al.[146] and Salkowski et al.[147] reported that bg/bg mice have significantly reduced survival rates after being given *Cr. neoformans* intravenously when compared to similarly infected bg/+ animals. Third, Cy-suppressed CBA/J mice adoptively transferred with syngeneic NK-enriched cell populations eliminate crypotococci from their tissues more efficiently early after infection than Cy-treated mice reconstituted with NK-depleted cell populations or than Cy-treated mice that were not reconstituted.[122] Lipscomb et al.[123] have reported that after intravenously infecting Cy-treated CBA/J mice with 10^4 viable cryptococci, the Cy-treated, infected group had $45 \pm 16\%$ of the original dose of cryptococci remaining in the lungs 24 h after infection, whereas mice similarly treated but reconstituted with splenic nylon wool nonadherent cells, which contained NK cells and T lymphocytes, reduced the organism count in the lungs to 26% of the original inoculum.

Although it is difficult to demonstrate the importance of NK cells *in vivo* definitively, the combined data from the various animal models and laboratories strongly indicate that NK cells play a role, at least in the mouse, in early clearance of cryptococci especially from tissues that have relatively high populations of NK cells.[122,123] Natural cellular defense mechanisms considered independently, such as NK cell defense, can be easily overwhelmed by large numbers of organisms. This is evident from the observations that mice reconstituted with NK-enriched populations could not completely eliminate the cryptococci from the tissues with the highest NK cell activities—i.e., spleens and lungs.[121] However, as we have pointed out before, the natural cellular defenses function to clear the organisms from tissues as the organisms enter the body, and in the case of natural entry of cryptococci into a host, only a few organisms would enter the lungs at the time of exposure. The natural cellular defenses in the lungs are most likely able to clear the usual numbers of organisms that gain entrance; it is only under extreme conditions of exposure to large numbers of organisms that the natural host defenses would become overwhelmed.

A similar argument could be made with regard to the natural cellular defenses in organs other than the lungs, such as spleen and liver, which come into play only after the organisms have become blood-borne. It seems unlikely that 10^4 cryptococci, which is the intravenous dose used in the experiments discussed above, would be released from the infected lung tissue into the bloodstream in one bolus but that considerably lesser numbers of organisms would be gradually sloughed off into the blood to seed other tissues. In the latter situation, the natural effector cells in the tissues would probably be able to effectively clear the incoming microbes. Unfortunately, technical limitations of the procedures used to study clearance of fungal organisms from tissues do not allow the demonstration of removal of doses of organisms lower than about 10^4.

Cr. neoformans has been reported to stimulate in the host an increase in NK cell activity above the normal level, a situation that could enhance clearance of the organism. Bartizal *et al.*[148] have demonstrated that viable cryptococci given either intravenously or via the gastrointestinal tract to germfree BALB/c nude or heterozygous mice will augment splenic NK cell activity. Increases in NK cell activity are also noted after intravenous injection of viable or heat-killed cryptococci into flora defined BALB/c animals.[148] Viable cryptococci were shown to stimulate a higher level of NK cell activity in germfree nude mice than in the heterozygous littermates, whereas opposite effects were reported to occur in the flora-defined animals.[148] Unpublished data from our laboratory indicate that not all strains of laboratory mice respond to cryptococci with increased levels of NK cell activity. Considerably more work must be done in this area to confirm these observations and to determine if various isolates of *Cr. neoformans* differentially affect the modulation of NK cell activity. If indeed cryptococci do augment NK cell activity, then one might expect enhanced natural resistance and better clearance of the organisms under these conditions.

We have begun to compare the binding characteristics of NK cells for YAC-1 targets and for cryptococci. At this early stage in these investigations, we can provide only preliminary information. With tumor cell targets, it has been clearly demonstrated that there are three distinct sequential stages that precede NK cell-, or for that matter cytotoxic T lymphocyte (CTL)-, mediated lysis of the target cells. These are binding, programming or activation, and lysis.[149,150] In both models, once the sequence of stages is completed, the effector cells detach from the target cell, and the process is repeated when another target cell is encountered.[150] It appears that a similar sequence of events occurs when NK cells encounter cryptococcal targets; however, we do not, at the present time, have any evidence for or against recycling of NK cells in the NK cell–*Cr. neoformans* model. We have observed that NK cells bind to cryptococci, but the time required for maximal conjugate formation is about 2 h, in contrast to the 20 min needed for maximal conjugate formation with YAC-1 cells as targets.[151] Roder *et al.*[151] have shown that binding of NK cells to YAC-1 targets precedes the lytic event by 5–10 min; however, a 4-h interval between binding of NK cells to cryptococci and the detection of growth inhibition seems to be required in the cryptococcus–NK cell interaction. These findings are not surprising when one considers the structural differences in the surfaces of the two different target cells; i.e., the YAC-1 cell surface is a lipid-bilayer membrane versus the polysaccharide capsule and cell wall enclosing the cytoplasmic membrane of the cryptococcal cell.

The binding event in the NK cell–cryptococci interaction is dependent on Mg^{2+}, as is the binding step of the NK cell to tumor targets and the CTL to their specific target.[152] We have unpublished data that suggest Ca^{2+} is also required for cryptococcal growth inhibition by NK cells, but conjugate formation occurs in the absence of Ca^{2+}.

Using scanning and transmission electron microscopy, we have found that the NK cells appear to focus microvilli toward the cryptococcal targets and attach to the cryptococci via the focused extended microvilli.[117,119] The actual contact points with the surface of the cryptococcus seem to be at the tips of these extended appendages of the effector cell (Fig. 1). The effector cells attached to the *Cr. neoformans* cells used in these studies were from populations of mouse splenic nylon wool nonadherent cells that had been further enriched for NK cells by Percoll fractionation, so we are confident that we are observing NK cells in conjugate formation with the cryptococcal targets. Moreover, confirmatory data were provided when we demonstrated that the effector cells in association with the cryptococci immunolabel with a concentration of anti-asialo GM_1 antibody that is cytolytic in the presence of complement for NK cells but not for macrophages or PMNL.[119] An ultrathin section through the contact site of the NK cell and the cryptococcus is shown in Fig. 2. In accordance with our observations from scanning electron micrographs, we have noted that the tips of the microvilli are the point of direct contact. This situation is considerably different from what we (unpublished data) and others[153] have observed regarding the association of NK cells with tumor targets. In NK cell–tumor cell conjugates, there is intimate contact over a large surface area between the membranes of the effector cell and the target cell.

It has been proposed that NK cells lyse their tumor targets by exocytosis of granules into the space between the NK cell and the target.[154] Furthermore, the contents of NK cell granules have been shown to contain a component (cytolysin) that is lytic to tumor cells and sheep red blood cells in a Ca^{2+}-dependent manner.[155] We have obtained from Craig Reynolds and Pierre Henkart at the National Cancer Institute, rat NK cell granule and cytolysin preparations and tested the effects of these NK cell components on cryptococci *in vitro*. The rat NK cell granules inhibit the growth of cryptococci in a dose-dependent fashion, and they act more rapidly on the crypotococcal cells than do intact NK cells. We observed maximal growth inhibition of crypotococci (65% with 20 U/ml rat LGL granules) 1 h after mixing the granule preparation with cryptococci.[156] This is in contrast to the 6 h required before inhibition of cryptococci can be detected (26% inhibition) after mixing the organisms with intact mouse NK cells. The cytolysin fraction of the rat LGL granules also limits the growth of cryptococci, and as with the intact granules, Ca^{2+} is required for cryptococcal growth inhibition.[156] Cytolysin may not be the only component in the NK cells or their granules that is effective against cryptococci; further investigations are necessary to completely define the active compounds and the mechanisms by which NK cells inhibit this yeastlike organism.

If one studies the mycology literature published prior to the time NK cells were generally recognized, one finds that the cells reported to be involved in extracellular killing of *Cr. neoformans* have some characteris-

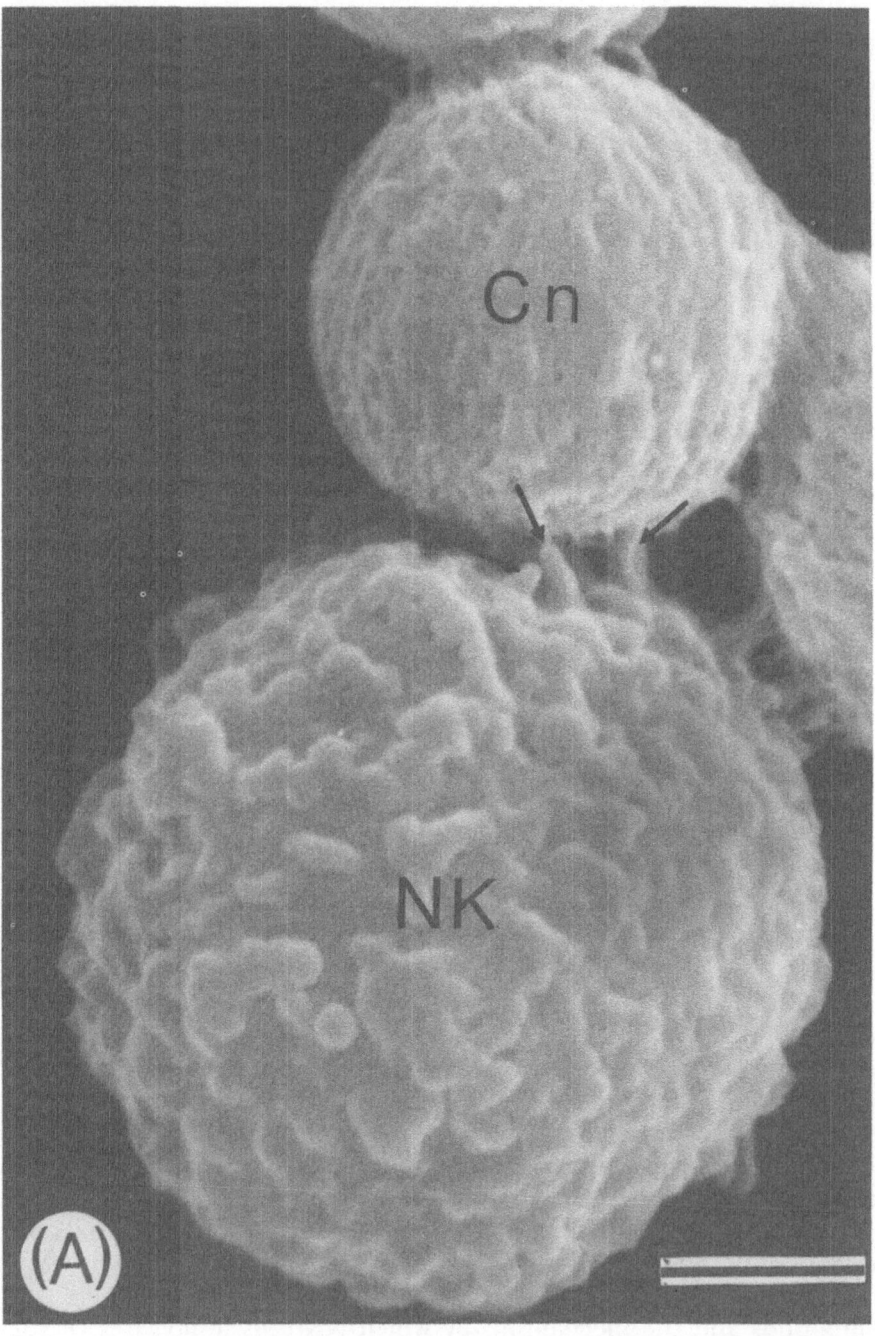

Figure 1. (A) Scanning electron micrograph of an NK cell–*Cr. neoformans* conjugate. Bar = 1 μm. (B) Higher magnification of contact area. Bar = 0.5 μm. The arrows indicate microvilli of the NK cell (NK) extended toward the cryptococcal target cell (Cn).

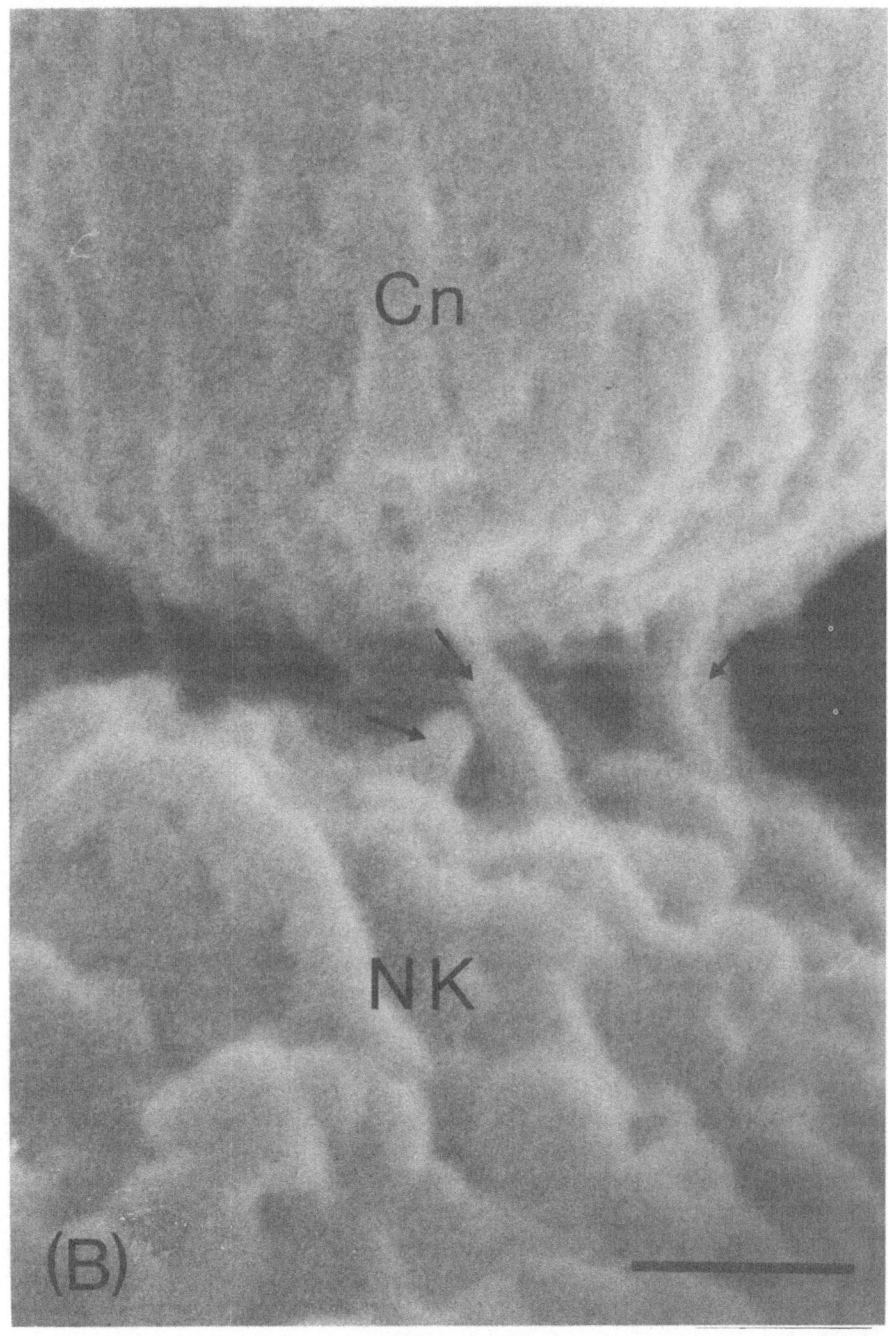

FIGURE 1. (*Continued*)

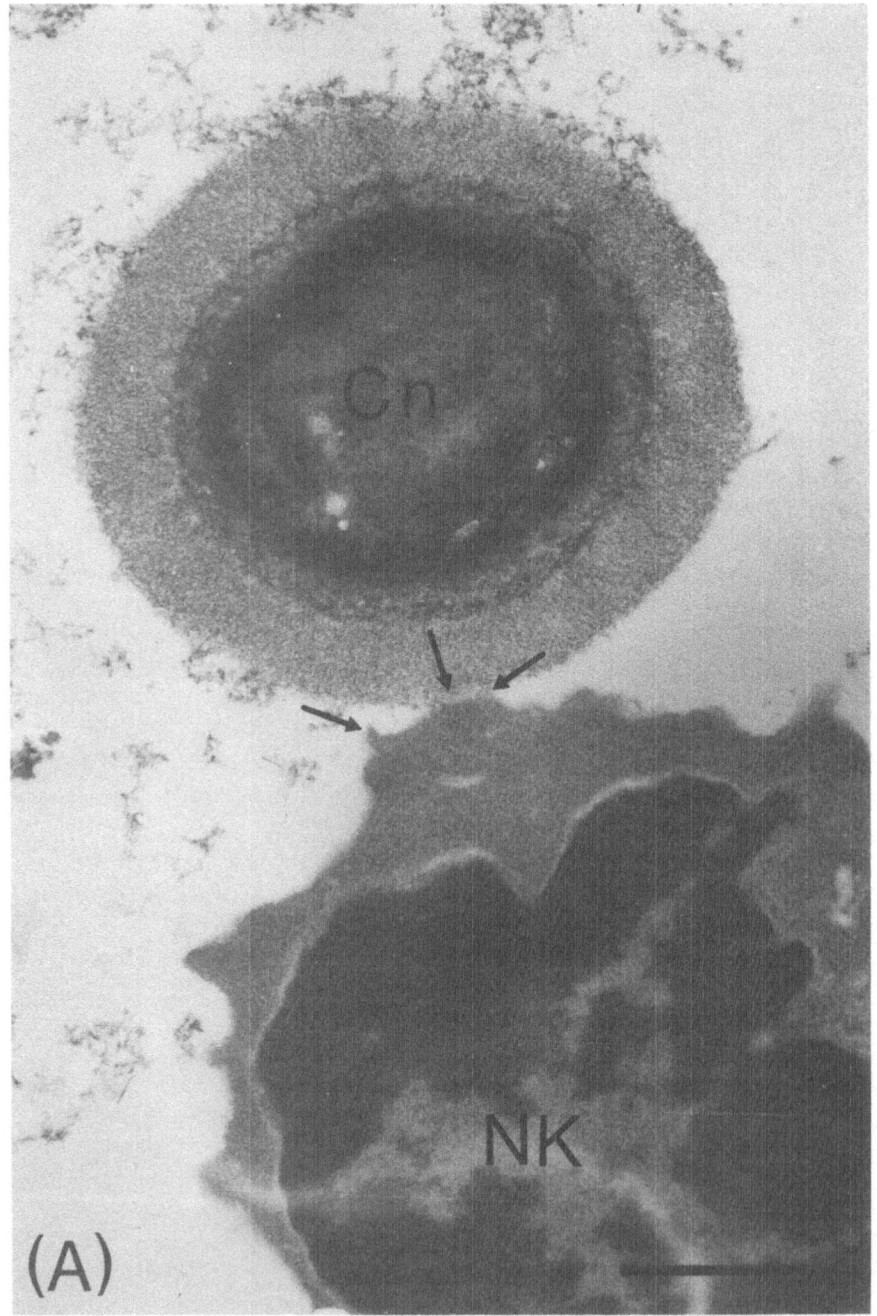

Figure 2. (A) Transmission electron micrograph of a section through the contact point of an NK cell–*Cr. neoformans* conjugate. Bar = 0.25 μm. (B) Higher magnification of contact area. Bar = 67 nm. The arrows indicate microvilli of the NK cell (NK) and in some cases are in contact with the cryptococcal target cell (Cn).

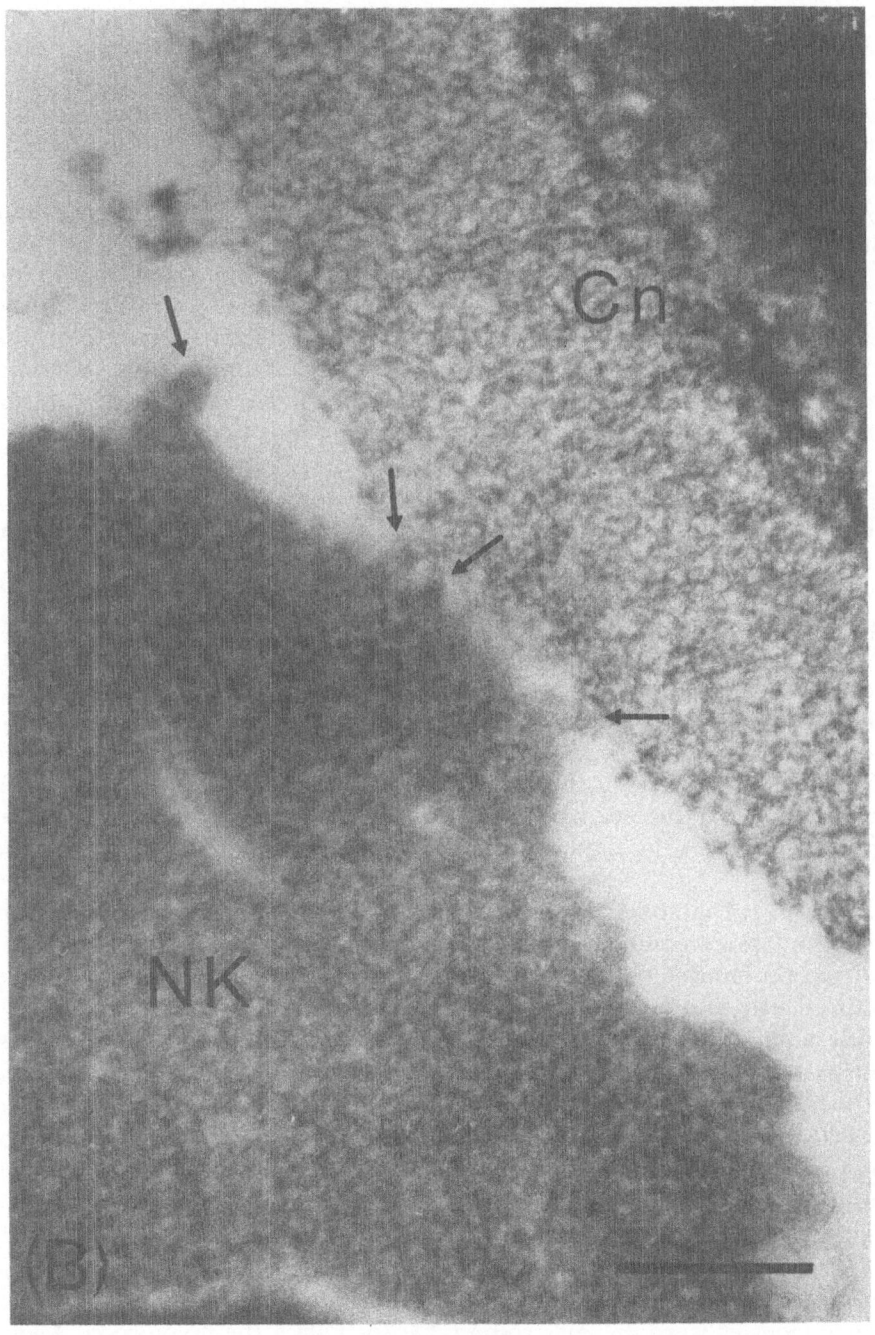

FIGURE 2. (*Continued*)

tics, but not all, in common with NK cells. As early as 1965, Scheerson-Porat et al.[4] reported the observation in both in vivo and in vitro studies of murine mononuclear cells, which they refer to as histiocytes, attaching to rabbit anticryptococcal antibody-opsonized, encapsulated cryptococci and forming rings of effector cells around the organisms. The appearance and staining characteristics described for the "histiocytes" surrounding the cryptococci are more compatible with characteristics of large granular lymphocytes than histiocytes.[4,108]

In addition, these investigators noted that, on the basis of staining characteristics, the cryptococci in the rings occasionally appeared to be destroyed. It was not until 1973 and 1974 that follow-up studies on the extracellular killing phenomena were reported. Using rabbit mononuclear cells, ring formation around large encapsulated cryptococci was observed, and electron micrographs of the formations reveal penetration of the capsule of the yeast cell by microvilli of the mononuclear cells.[107] Histochemical studies of the ring formations indicated that mononuclear cells secrete enzymes such as acid phosphatase, β-glucuronidase, and nonspecific esterase into the area of the enclosed cryptococcal yeast cell.[107] From the information available, one cannot be certain of the effector cell type in the rings surrounding the yeast cells. But considering the knowledge that we have today concerning natural effector cell types, it seems the effector cells in these early studies could have been (1) monocytes, as the investigators reported; (2) NK cells; or (3) other cytotoxic effector cells such as promonocytes similar to those described by Bistoni and Baccarini and co-workers to be effective killers of C. albicans.[18–21] Careful examination of the available information provides arguments for and against all of the suggested possibilities making it impossible to resolve the question.

In 1972, Diamond et al.[23] reported that normal human mononuclear cell populations containing 20–42% monocytes and 1–3% granulocytes and the remainder being lymphocytes killed 45.9% of the Cr. neoformans cultured with the mononuclear cells for 4 h. Capsule size did not significantly alter killing of cryptococci by mononuclear cells in their studies. This is a situation that is more akin to NK cell inhibition of Cr. neoformans growth[115] than to killing by phagocytic cells, which have been reported to effectively ingest and kill only the weakly encapsulated cryptococci.[6,139] When similar mononuclear cell populations were cultured in tissue culture dishes either with or without the nonadherent cell component, the resulting macrophages were unable to kill cryptococci.[24] Hindsight would lead one to consider NK cells as possible effector cells in these studies, although at the time the studies were done, the interpretation was that monocytes were the cells responsible for killing the cryptococci. Again, it is impossible to resolve the dilemma.

A year or so after these reports, Diamond[15] reported that human

mononuclear cells in the presence, but not in the absence, of anticryptococcal antibodies kill cryptococci by a nonphagocytic mechanism in an *in
vitro* assay. In 1976, additional data were presented by Diamond and co-
workers[17] demonstrating that mixed populations of mononuclear cells
(30% monocytes; 70% lymphocytes) obtained from human peripheral blood
kill cryptococci in the presence of anticryptococcal antibodies more effectively than phagocyte-depleted lymphoid cells, granulocytes, or nylon wool
nonadherent cells. Killing of the cryptococci was attributed to nonphagocytic lymphoid cells.[17] Unlike results from earlier studies,[23] the cell populations tested in these later studies did not kill cryptococci in the absence
of specific antibody during a 4-h incubation period.[16,17]

From what we know concerning the effects of mouse NK cells on
cryptococci, if NK cells contributed to the killing of cryptococci in these
studies with human effector cells, one would expect some killing of the
organisms in the absence of specific antibody and increased levels of killing in the presence of anticryptococcal antibody.[131] The data presented
by Diamond and Allison[17] do not completely conform to these expecta-
tions, but their data do not rule out the possibility that cells similar to NK
cells were the effector cells. Considering that the killing of cryptococci
reported by Diamond and co-workers[16,17] was attributed to nonphagocytic mononuclear cells and was independent of capsule size, there seems
to be reason to speculate that NK cells may have been the effector cells in
the antibody-dependent, cell-mediated killing of the cryptococci; however, additional investigations with human NK cells must be done to confirm this speculation.

In 1976, Walters *et al.*[157] published scanning electron micrographs
of mouse peritoneal exudate mononuclear cells attached to cryptococci
suggesting that NK cells may be the effector cells rather than peritoneal
macrophages, which was the designation given to the effector cells by the
authors. In their photomicrographs, the mononuclear cells associated with
the cryptococcal cells look very much like the NK cells we have shown in
conjugate formation with cryptococci (Fig. 1). The effector cells in their
pictures are attached to the cryptococci by microvilli and do not display
the veil-like pseudopodia of the typical macrophage or the macrophages
phagocytizing Brewer's yeast or *Staphylococcus aureus* cells shown in the
same publication.[157]

Additional photomicrographs of guinea pig mononuclear cells con-
jugated to cryptococci were published the following year by Walters in
collaboration with others.[158] Many (but not all) of the effector cells associated with the cryptococci in their photographs have the characteristics
we have been observing using Percoll fractions of murine spleen cells that
are highly enriched for NK cells. Those characteristics include attachment
of the effector cells to the cryptococci by microvilli focused toward the
cryptococci target cells and the lack of phagocytosis by pseudopodia sur-

rounding the yeast cells; focusing or polarization of the Golgi apparatus
of the effector cells in the direction of the cryopotoccus; and many micro-
villuslike processes penetrating the capsule of the associated cryptococcal
cell. The point in discussing this older literature is to emphasize that one
must consider that natural effector cells other than monocytes could have
been involved in the killing of cryptococci.

4.2. Indirect Effects of NK Cells on Mycotic Agents

Little work has been done concerning the indirect effects of NK cells
on mycotic organisms; however, because there is some evidence that NK
cells can produce cytokines, which enhance antifungal activity of other
effector cell types, studies on indirect activities of NK cells should be mentioned, if for no other reason, to stimulate additional investigations. There
are reports that certain fungi such as Cr. neoformans[148] and C. albicans[159]
when injected into animals, stimulate increased levels of NK cell activity.
Since in the case of Cr. neoformans NK cells have been shown to directly
inhibit the organisms, one would consider an increase in NK cell activity
beneficial to the host. On the other hand, NK cells do not appear to kill
C. albicans directly,[112,114,129,130] so what role could the increased NK cell
activity play in candidiasis? Djeu[114] has reported that C. albicans yeast
cells stimulate human NK cells to produce lymphokines which significantly increase the anti-Candida activity of PMNL. In addition, she has
shown that gamma interferon, which can be produced by NK cells under
certain conditions, will augment PMNL activity against C. albicans.[112] Although these studies are in preliminary stages, they indicate that there
may be a broader role for NK cells in host defense than has been previously considered—i.e., regulation of other effector cell activities.

5. Summary

Natural resistance mechanisms are generally very effective first-line
defenses against the usual dose of systemic mycotic pathogens. A single
component is not usually responsible for clearance of fungal organisms,
but rather the various cells and factors that comprise the natural resistance system tend to function together to eliminate systemic mycotic agents.
In the normal or unprimed state, the natural effector cells usually display
a low level of antifungal activity which can be augmented by soluble products from both the natural and the immune systems. Thus, once the secondary or immune responses come into play, the first-line defenses acquire an elevated level of activity and continue to be the major means of
eliminating the fungal pathogens.

References

1. Emmons, C. W., Binford, C. H., Utz, J. P., and Kwon-Chung, K. J., 1977, *Medical Mycology*, 3d Ed., Lea and Febiger, Philadelphia.
2. Rippon, J. W., 1982, *Medical Mycology*, 2d Ed., W. B. Saunders, Philadelphia.
3. Drutz, D. J., and Huppert, M., 1983, Coccidioidomycosis: Factors affecting the host–parasite interaction, *J. Infect. Dis.* **147**:372–390.
4. Schneerson-Porat, S., Shahar, A., and Aronson, M., 1965, Formation of histiocyte rings in response to *Cryptococcus neoformans* infection, *J. Reticuloendothel. Soc.* **2**:249–255.
5. Farkas, V., 1979, Biosynthesis of cell walls of fungi, *Microbiol. Rev.* **43**:117–144.
6. Bulmer, G. S., and Sans, M. D., 1968, *Crypotococcus neoformans*. III. Inhibition of phagocytosis, *J. Bacteriol.* **95**:5–8.
7. Artz, R. P., and Bullock, W. E., 1979, Immunoregulatory responses in experimental disseminated histoplasmosis: Depression of T-cell-dependent and T-effector responses by activation of splenic suppressor cells, *Infect. Immun.* **23**:893–902.
8. Murphy, J. W., and Moorhead, J. W., 1982, Regulation of cell-mediated immunity in cryptococcosis. I. Induction of specific afferent T suppressor cells by cryptococcal antigen, *J. Immunol.* **128**:276–283.
9. Rivas, V., and Rogers, T. J., 1983, Studies on the cellular nature of *Candida albicans*-induced suppression, *J. Immunol.* **130**:376–379.
10. Cohen, M. S., Isturiz, R. E., Malech, H. L., Root, R. K., Wilfert, C. M., Gutman, L., and Buckley, R. H., 1981, Fungal infection in chronic granulomatous disease, *Am J. Med.* **71**:59–66.
11. Howard, D. H., 1975, The role of phagocytic mechanisms of defense against *Histoplasma capsulatum*, in: *Mycoses*, Pan American Health Organization, Washington, pp. 50–59.
12. Howard, D. H., 1981, Mechanisms of resistance in the systemic mycoses, *Compr. Immuno.* **8**:475–494.
13. Diamond, R. D., 1981, Immunology of invasive fungal infections, *Compr. Immunol.* **8**:585–633.
14. Lehrer, R. I., and Fleishchmann, J., 1982, Antifungal defense by macrophages, in: *Microbiology—1982*, (D. Schlessinger, ed.), American Society for Microbiology, Washington, pp. 385–387.
15. Fleishchmann, J., and Lehrer, R. I., 1983, Phagocytic mechanisms in host response, in: *Fungi Pathogenic for Humans and Animals*, Part B, Pathogenicity and Detection, II (D. Howard, ed.), Marcel Dekker, New York, pp. 123–149.
16. Diamond, R. D., 1974, Antibody-dependent killing of *Cryptococcus neoformans* by human peripheral blood mononuclear cells, *Nature* **247**:148–150.
17. Diamond, R. D., and Allison, A. C., 1976, Nature of the effector cells responsible for antibody-dependent cell-mediated killing of *Cryptococcus neoformans*, *Infect. Immun.* **14**:716–720.
18. Baccarini, M., Bistoni, F., and Lohmann-Matthes, M., 1985, *In vitro* natural cell-mediated cytotoxicity against *Candida albicans*: Macrophage precursors as effector cells, *J. Immunol.* **134**:2658–2665.
19. Baccarini, M., Kiderlen, A. F., Decker, T., and Lohmann-Matthes, M., 1986, Functional heterogeneity of murine macrophage precursor cells from spleen and bone marrow, *Cell. Immunol.* **101**:339–350.
20. Decker, T., Lohmann-Matthes, M., and Baccarini, M., 1986. Heterogeneous activity of immature and mature cells of the murine monocyte-macrophage lineage derived from different anatomical districts against yeast-phase *Candida albicans*, *Infect. Immun.* **54**:477–486.
21. Decker, T., Baccarini, M., and Lohmann-Matthes, M., 1986, Liver-associated macrophage precursors as natural cytotoxic effectors against *Candida albicans* and YAC-1 cells, *Eur. J. Immunol.* **16**:693–699.

22. Baccarini, M., Bistoni, F., and Lohmann-Matthes, M., 1986, Organ-associated macrophage precursor activity: Isolation of candidacidal and tumoricidal effectors from the spleens of cyclophosphamide-treated mice, *J. Immunol.* **136**:837–843.
23. Diamond, R. D., Root, R. K., and Bennett, J. E., 1972, Factors influencing killing of *Cryptococcus neoformans* by human leukocytes *in vitro*, *J. Infect. Dis.* **125**:367–376.
24. Diamond, R. D., and Bennett, J. E., 1973, Growth of *Cryptococcus neoformans* within human macrophages *in vitro*, *Infect. Immun.* **7**:231–236.
25. Lehrer, R. I., 1970, The fungicidal activity of human monocytes: A myeloperoxidase-linked mechanism, *Clin. Res.* **18**:408.
26. Lehrer, R. I., 1975, The fungicidal mechanisms of human monocytes. I. Evidence for myeloperoxidase-linked and myeloperoxidase-independent candidacidal mechanisms, *J. Clin. Invest.* **55**:338–346.
27. Brummer, E., and Stevens, D. A., 1982, Opposite effects of human monocytes, macrophages and polymorphonuclear neutrophils on replication of *Blastomyces dermatitidis in vitro*, *Infect. Immun.* **36**:297–303.
28. Drutz, D. J., and Frey, C. L., 1985, Intracellular and extracellular defenses of human phagocytes against *Blastomyces dermatitidis* conidia and yeasts, *J. Lab. Clin. Med.* **105**:737–750.
29. Bradsher, R. W., Ulmer, W. C., Marmer, D. J., Townsend, J. W., and Jacobs, R. F., 1985, Intracellular growth and phagocytosis of *Blastomyces dermatitidis* by monocyte-derived macrophages from previously infected and normal subjects, *J. Infect. Dis.* **151**:57–64.
30. Swenson, F. J., and Kozel, T. R., 1978, Phagocytosis of *Cryptococcus neoformans* by normal and thioglycolate-activated macrophages, *Infect. Immun.* **21**:714–720.
31. Lehrer, R. I., Ferrari, L. G., Patterson-Delafield, J., and Sorrell, T., 1980, Fungicidal activity of rabbit alveolar and peritoneal macrophages against *Candida albicans*, *Infect. Immun.* **28**:1001–1008.
32. Granger, D. L., Perfect, J. R., and Durack, D. T., 1986, Macrophage-mediated fungistasis *in vitro*: Requirements for intracellular and extracellular cytotoxicity, *J. Immunol.* **136**:672–680.
33. Howard, D. H., and Otto, V., 1977, Experiments on lymphocyte-mediated cellular immunity in murine histoplasmosis, *Infect. Immun.* **16**:226–231.
34. Beaman, L., Benjamini, E., and Pappagianis, D., 1981, Role of lymphocytes in macrophage-induced killing of *Coccidioides immitis in vitro*. *Infect. Immun.* **34**:347–353.
35. Wu-Hsieh, B., Zlotnik, A., and Howard, D. H., 1984, T-cell hybridoma-produced lymphokine that activates macrophages to suppress intracellular growth of *Histoplasma capsulatum*, *Infect. Immun.* **43**:380–385.
36. Wu-Hsieh, B., and Howard, D. H., 1984, Inhibition of growth of *Histoplasma capsulatum* by lymphokine-stimulated macrophages. *J. Immunol.* **132**:2593–2597.
37. Brummer, E., Morrison, C. J., and Stevens, D. A., 1985, Recombinant and natural gamma-interferon activation of macrophages *in vitro*: Different dose requirements for induction of killing activity against phagocytizable and nonphagocytizable fungi, *Infect. Immun.* **49**:724–730.
38. Sugar, A. M., Brummer, E., and Stevens, D. A., 1986, Fungicidal activity of murine broncho-alveolar macrophages against *Blastomyces dermatitidis*, *J. Med. Microbiol.* **21**:7–11.
39. Cohen, A. B., and Cline, M. J., 1971, The human alveolar macrophage: Isolation cultivation *in vitro*, and studies of morphologic and functional characteristics, *J. Clin. Invest.* **50**:1390–1398.
40. Peterson, E. M., and Calderone, R. A., 1977, Growth inhibition of *Candida albicans* by rabbit alveolar macrophages, *Infect. Immun.* **15**:910–915.
41. Sugar, A. M., Brummer, E., and Stevens, D. A., 1983, Murine pulmonary macrophages: Evaluation of lung lavage fluids, miniaturized monolayers, and candidacidal activity, *Am. Rev. Respir. Dis.* **127**:110–112.

42. Beaman, L., and Holmberg, C. A., 1980, *In vitro* response of alveolar macrophages to infection with *Coccidioides immitis, Infect. Immun.* **28**:594–600.
43. Bulmer, G. S., and Tacker, J. R., 1975, Phagocytosis of *Cryptococcus neoformans* by alveolar macrophages, *Infect. Immun.* **11**:73–79.
44. Sanchez, S. B., and Carbonell, L. M., 1975, Immunological studies on *Histoplasma capsulatum, Infect. Immun.* **11**:387–394.
45. Kimberlin, C. L., Hariri, A. R., Hempel, H. O., and Goodman, N. L., 1981, Interactions between *Histoplasma capsulatum* and macrophages from normal and treated mice: Comparison of the mycelial and yeast phases in alveolar and peritoneal macrophages, *Infect. Immun.* **34**:6–10.
46. Waldorf, A. R., Ruderman, N., and Diamond, R. D., 1984, Specific susceptibility to mucormycosis in murine diabetes and bronchoalveolar macrophage defense against *Rhizopus, J. Clin. Invest.* **74**:150–160.
47. Waldorf, A. R., Levitz, S. M., and Diamond, R. D., 1984, *In vivo* bronchoalveolar macrophage defense against *Rhizopus oryzae* and *Aspergillus fumigatus, J. Infect. Dis.* **150**:752–760.
48. Lehrer, R. I., Howard, D. H., Sypherd, P. S., Edwards, J. E., Segal, G. P., and Winston, D. J., 1980, Mucormycosis, *Ann. Intern. Med.* **93** (Part 1):93–108.
49. Goodwin, R. A., and DesPrez, R. M., 1978, Histoplasmosis, *Am. Rev. Respir. Dis.* **117**:929–955.
50. Wolf, J. E., Kerchberger, V., Kobayashi, G. S., and Little, J. R., 1987, Modulation of the macrophage oxidative burst by *Histoplasma capsulatum, J. Immunol.* **138**:582–586.
51. Eissenberg, L. G., and Goldman, W. E., 1987, *Histoplasma capsulatum* fails to trigger release of superoxide from macrophages, *Infect. Immun.* **55**:29–34.
52. Peterson, E. M., and Calderone, R. A., 1978, Inhibition of specific amino acid uptake in *Candida albicans* by lysosomal extracts from rabbit alveolar macrophages, *Infect. Immun.* **21**:506–513.
53. Lehrer, R. I., Szklarek, D., Selsted, M. E., and Fleischmann, J., 1981, Increased content of microbicidal cationic peptides in rabbit alveolar macrophages elicited by complete Freund adjuvant, *Infect. Immun.* **33**:775–778.
54. Kletter, Y., and Nagler, A., 1984, Function of peripheral blood and bone-marrow monocytes in preleukemic patients: Normal phagocytosis and intracellular killing of *Candida albicans, Acta Haematol.* **72**:379–383.
55. Schuit, K. E., 1979, Phagocytosis and intracellular killing of pathogenic yeast by human monocytes and neutrophils, *Infect. Immun.* **24**:932–938.
56. Beaman, L., and Pappagianis, D., 1985, Fate of *Coccidioides immitis* arthroconidia in human peripheral blood monocyte cultures *in vitro*, in: *Coccidioidomycosis*, National Foundation Infectious Diseases, Washington, pp. 170–180.
57. Diamond, R. D., and Haudenschild, C. C., 1981, Monocyte-mediated serum-independent damage to hyphal and pseudohyphal forms of *Candida albicans in vitro, J. Clin. Invest.* **67**:173–182.
58. Holland, P., 1971, Circulating human phagocytes and *Histoplasma capsulatum*, in: *Histoplasmosis*, Proceedings of the 2d National Conference (L. Ajello, E. W. Chick, and M. L. Furcolow, eds.), Charles C. Thomas, Springfield, IL, pp. 380–383.
59. Diamond, R. D., Huber, E., and Haudenschild, C. C., 1983, Mechanisms of destruction of *Aspergillus fumigatus* hyphae mediated by human monocytes, *J. Infect. Dis.* **147**:474–483.
60. Diamond, R. D., Haudenschild, C. C., and Erickson, N. F. III, 1982, Monocyte-mediated damage to *Rhizopus oryzae* hyphae *in vitro, Infect. Immun.* **38**:292–297.
61. Lehrer, R. I., and Cline, M. J., 1969, Leukocyte myeloperoxidase deficiency and disseminated candidiasis: The role of myeloperoxidase in resistance to *Candida* infection, *J. Clin. Invest.* **48**:1478–1488.
62. Lehrer, R. I., 1975, The fungicidal mechanisms of human monocytes. I. Evidence for

myeloperoxidase-linked and myeloperoxidase-independent candidacidal mechanisms, *J. Clin. Invest.* **55**:338–346.

63. Collins, M. S., and Pappagianis, D., 1974, Inhibition by lysozyme of growth of the spherule phase of *Coccidiodes immitis in vitro, Infect. Immun.* **10**:616–623.

64. Morgan, M. A., Blackstock, R. A., Bulmer, G. S., and Hall, N. K., 1983, Modification of macrophage phagocytosis in murine crytococcosis, *Infect. Immun.* **40**:493–500.

65. Blackstock, R. A., McCormack, J. M., and Hall, N. K., 1987, Induction of a macrophage-suppressive lymphokine by soluble cryptococcal antigens and its association with models of immunologic tolerance, *Infect. Immun.* **55**:233–239.

66. Cutler, J. E., 1977, Chemotactic factor produced by *Candida albicans, Infect. Immun.* **18**:568–573.

67. Weeks, B. A., Escobar, M. R., Hamilton, P. B., and Fueston, V. M., 1976, Chemotaxis of polymorphonuclear neutrophilic leukocytes by mannan-enriched preparations of *Candida albicans, Adv. Exp. Med. Biol.* **73**:161–169.

68. Chinn, R. Y. W., and Diamond, R. D., 1982, Generation of chemotactic factors by *Rhizopus oryzae* in the presence and absence of serum: Relationship to hyphal damage mediated by human neutrophils and effects of hyperglycemia and ketoacidosis, *Infect. Immun.* **38**:1123–1129.

69. Thurmond, L. M., and Mitchell, T. G., 1984, *Blastomyces dermatitidis* chemotactic factor: Kinetics of production and biological characterization evaluated by a modified neutrophil chemotaxis assay, *Infect. Immun.* **46**:87–93.

70. Ratnoff, W. D., Pepple, J. M., and Winkelstein, J. A., 1980, Activation of the alternative complement pathway by *Histoplasma capsulatum, Infect. Immun.* **30**:147–149.

71. Galgiani, J. N., Yam, P., Petz, L. D., Willims, P. L., and Stevens, D. A., 1980. Complement activation by *Coccidioides immitis: In vitro* and clinical studies, *Infect. Immun.* **28**:944–949.

72. Waldorf, A. R., and Diamond, R. D., 1985, Neutrophil chemotactic responses induced by fresh and swollen *Rhizopus oryzae* spores and *Aspergillus fumigatus* conidia, *Infect. Immun.* **48**:458–463.

73. Diamond, R. D., May, J. E., Kane, M., Frank, M. M., and Bennett, J. E., 1973, The role of late complement components and the alternate complement pathway in experimental cryptococcosis, *Proc. Soc. Exp. Biol. Med.* **144**:312–315.

74. Diamond, R. D., May, J. E., Kane, M., Frank, M. M., and Bennett, J. E., 1974, The role of the classical and alternate complement pathways in host defenses against *Cryptococcus neoformans* infection, *J. Immunol.* **112**:2260–2270.

75. Diamond, R. D., and Erickson, N. F. III, 1982, Chemotaxis of human neutrophils and monocytes induced by *Cryptococcus neoformans, Infect. Immun.* **38**:380–382.

76. Laxalt, K. A., and Kozel, T. R., 1979, Chemotaxigenesis and activation of the alternative complement pathway by encapsulated and non-encapsulated *Cryptococcus neoformans, Infect. Immun.* **26**:435–440.

77. Gadebusch, H. H., 1972, Mechanisms of native and acquired resistance to infection with *Cryptococcus neoformans*, in: *Macrophages and Cellular Immunity* (A. Laskin and H. Lechevalier, eds.), CRC Press, Cleveland, OH, pp. 3–12.

78. Ray, T. L., and Wuepper, K. D., 1976, Activation of the alternative (properdin) pathway of complement by *Candida albicans* and related species, *J. Invest. Dermatol.* **67**:700–703.

79. Rhodes, J. C., Wicker, L. S., and Urba, W. J., 1980, Genetic control of susceptibility to *Cryptococcus neoformans* in mice, *Infect. Immun.* **29**:494–499.

80. Kozel, R. R., and Pfrommer, G. S. T., 1986, Activation of the complement system by *Cryptococcus neoformans* leading to binding of iC3b to the yeast, *Infect. Immun.* **52**:1–5.

81. Sixbey, J. W., Fields, B. T., Sun, C. N., Clark, R. A., and Nolan, C. M., 1979, Interactions between human granulocytes and *Blastomyces dermatitidis, Infect. Immun.* **23**:41–44.

82. Lehrer, R. I., and Cline, M. J., 1969, Interaction of *Candida albicans* with human leukocytes and serum, *J. Bacteriol.* **98**:996–1004.

83. Goihman-Yahr, M., Essenfeld-Yahr, E., De Albornoz, M., Yarzabal, L., De Gomez, M. H., Martin, B. S., Ocanto, A., Gil, F., and Convit, J., 1980, Defect of *in vitro* digestive ability of polymorphonuclear leukocytes in paracoccidioidomycosis, *Infect. Immun.* **28:**557–566.

84. Diamond, R. D., Krzesicki, R., and Jao, W., 1978, Damage to pseudohyphal forms of *Candida albicans* by neutrophils in the absence of serum *in vitro, J. Clin. Invest.* **61:**349–359.

85. Tacker, J. R., Farhi, F., and Bulmer, G. S., 1972, Intracellular fate of *Cryptococcus neoformans, Infect. Immun.* **6:**162–167.

86. Diamond, R. D., Krzesicki, R., Epstein, B., and Jao, W., 1978, Damage to hyphal forms of fungi by human leukocytes *in vitro, Am. J. Pathol.* **91:**313–327.

87. Diamond, R. D., and Clark, R. A., 1982, Damage to *Aspergillus fumigatus* and *Rhizopus oryzae* hyphae by oxidative and nonoxidative microbicidal products of human neutrophils *in vitro, Infect. Immun.* **38:**487–495.

88. Levitz, S. M., and Diamond, R. D., 1985, Mechanisms of resistance of *Aspergillus fumigatus* conidia to killing by neutrophils *in vitro, J. Infect. Dis.* **152:**33–42.

89. Sugar, A. S., Chahal, R. S., Brummer, E., and Stevens, D. A., 1983, Susceptibility of *Blastomyces dermatitidis* strains to products of oxidative metabolism, *Infect. Immun.* **41:**908–912.

90. Brummer, E., Sugar, A. M., and Stevens, D. A., 1984, Immunological activation of polymorphonuclear neutrophils for fungal killing: Studies with murine cells and *Blastomyces dermatitidis in vitro, J. Leuk. Biol.* **36:**505–520.

91. Howard, D. H., 1973, Fate of *Histoplasma capsulatum* in guinea pig polymorphonuclear leukocytes, *Infect. Immun.* **8:**412–419.

92. Howard, D. H., 1983, Studies on the catalase of *Histoplasma capsulatum, Infect. Immun.* **39:**1161–1166.

93. McEwen, J. G., Sugar, A. M., Brummer, E., Restrepo, A., and Stevens, D. A., 1984, Toxic effect of products of oxidative metabolism on the yeast form of *Paracoccidioides brasiliensis, J. Med. Microbiol.* **18:**423–428.

94. Levitz, S. M., and Diamond, R. D., Killing of *Aspergillus fumigatus* spores and *Candida albicans* yeast phase by the iron-hydrogen peroxide-iodine cytotoxic system: Comparison with the myeloperoxidase-hydrogen peroxide-halide system, *Infect. Immun.* **43:**1100–1102.

95. Gadebusch, H. H., and Johnson, A. G., 1966, Natural host resistance to infection with *Cryptococcus neoformans.* IV. The effect of some cationic proteins on the experimental disease, *J. Infect. Dis.* **116:**551–565.

96. Kirpatrick, C. H., Green, I., Rich, R. R., and Schade, A. L., 1971, Inhibition of growth of *Candida albicans* by iron-unsaturated lactoferrin: Relation to host defense mechanisms in chronic mucocutaneous candidiasis, *J. Infect. Dis.* **124:**539–544.

97. Lehrer, R. I., and Ladra, K. M., Fungicidal components of mammalian granulocytes active against *Cryptococcus neoformans, J. Infect. Dis.* **136:**96–99.

98. Selsted, M. E., Szklarek, D., and Lehrer, R. I., 1984, Purification and antibacterial activity of antimicrobial peptides of rabbit granulocytes, *Infect. Immun.* **45:**150–154.

99. Selsted, M. E., Szklarek, D., Ganz, T., and Lehrer, R. I., 1985, Activity of rabbit leukocyte peptides against *Candida albicans, Infect. Immun.* **49:**202–206.

100. Selsted, M. E., Harwig, S. S. L., Ganz, T., Schilling, J. W., and Lehrer, R. I., 1985, Primary structures of three human neutrophil defensins, *J. Clin. Invest.* **76:**1436–1439.

101. Ganz, T. Selsted, M. E., Szklarek, D., Harwig, S. S. L., and Daher, K., Bainton, D. F., and Lehrer, R. I., 1985, Defensins—natural peptide antibiotics of human neutrophils, *J. Clin. Invest.* **76:**1427–1435.

102. Segal, G. P., Lehrer, R. I., and Selsted, M. E., 1985, *In vitro* effect of phagocyte cationic peptides on *Coccidioides immitis, J. Infect. Dis.* **151:**890–894.

103. Lehrer, R. I., Szklarek, K., Ganz, T., and Selsted, M. E., 1986, Synergistic activity of rabbit granulocyte peptides against *Candida albicans, Infect. Immun.* **52:**902–904.

104. Diamond, R. D., Oppenheim, F., Nakagawa, Y., Krzesicki, R., and Haudenschild, C.

C., 1980, Properties of a product of *Candida albicans* hyphae and pseudohyphae that inhibits contact between the fungi and human neutrophils *in vitro, J. Immunol.* **125**:2797–2804.

105. Kozel, T. R., Highison, B., and Stratton, J., 1984, Localization on encapsulated *Cryptococcus neoformans* of serum components opsonic for phagocytosis by macrophages and neutrophils, *Infect. Immun.* **43**:574–579.

106. Kalina, M., Kletter, Y., Shahar, A., and Aronson, M., 1971, Acid phosphatase release from intact phagocytic cells surrounding a large-sized parasite, *Proc. Soc. Exp. Biol. Med.* **136**:407–410.

107. Kalina, M., Kletter, Y., and Aronson, M., 1974, The interaction of phagocytes and the large-sized parasite *Cryptococcus neoformans* cytochemical and ultrastructural study, *Cell. Tissue Res.* **152**:165–174.

108. Aronson, M., and Kletter J., 1973, Aspects of the defense against a large-sized parasite, the yeast, *Cryptococcus neoformans*, in: *Dynamic Aspects of Host Parasitic Relationships* (D. Weiss and A. Zuckerman, eds.) Academic Press, New York, pp. 132–162.

109. Miller, G. P., and Kohl, S., 1983, Antibody-dependent leukocyte killing of *Cryptococcus neoformans, J. Immunol.* **131**:1455–1459.

110. Brummer, E., and Stevens, D. A., 1984, Activation of murine polymorphonuclear neutrophils for fungicidal activity with supernatants from antigen-stimulated immune spleen cell cultures, *Infect. Immun.* **45**:447–452.

111. Brummer, E., Sugar, A. M., and Stevens, D. A., 1985, Enhanced oxidative burst in immunologically activated but not elicited polymorphonuclear leukocytes correlates with fungicidal activity, *Infect. Immun.* **49**:396–401.

112. Djeu, J. Y., Blanchard, D. K., Halkias, D., and Friedman, H., 1986, Growth inhibition of *Candida albicans* by human polymorphonuclear neutrophils: Activation by interferon-α and tumor necrosis factor, *J. Immunol.* **137**:2980–2984.

113. Kagaya, K., and Fukazawa, Y., 1981, Murine defense mechanism against *Candida albicans* infection. II. Opsonization, phagocytosis, and intracellular killing of *C. albicans, Microbiology* **25**:807–818.

114. Djeu, J. Y., Blanchard, D. K., and Friedman, H., 1987, Regulation of human polymorphonuclear neutrophil activity against *Candida albicans* by large granular lymphocytes via release of interferon, tumor necrosis factor and another unidentified cytokine, *Lymphokine Res.* **6**:1414.

115. Murphy, J. W., and McDaniel, D. O., 1982, *In vitro* reactivity of natural killer (NK) cells against *Cryptococcus neoformans, J. Immunol.* **128**:1577–1583.

116. Murphy, J. W., and McDaniel, D. O., 1982, *In vitro* effects of natural killer (NK) cells on *Cryptococcus neoformans*, in: *NK Cells and Other Natural Effector Cells* (R. B. Herberman, ed.), Academic Press, New York, pp. 1105–1112.

117. Murphy, J. W., 1984, Natural killer cells against a mycotic pathogen, in: *Microbiology— 1984* (L. Leive and D. Schlessinger, eds.), American Society for Microbiology, Washington, pp. 328–332.

118. Murphy, J. W., 1985, Natural cell-mediated resistance against *Cryptococcus neoformans*. A possible role for natural killer (NK) cells, in: *Current Topics in Medical Mycology*, Volume 1 (M. R. McGinnis, ed.), Springer-Verlag, New York, pp. 135–154.

119. Nabavi, N., and Murphy, J. W., 1985, *In vitro* binding of natural killer cells to *Cryptococcus neoformans* targets, *Infect. Immun.* **50**:50–57.

120. Murphy, J. W., 1982, Natural cell-mediated resistance in cryptococcosis, in: *NK Cells and Other Natural Effector Cells* (R. B. Herberman, ed.), Academic Press, New York, pp. 1503–1511.

121. Hidore, M. R., and Murphy, J. W., 1986, Natural cellular resistance of beige mice against *Cryptococcus neoformans, J. Immunol.* **137**:3624–3631.

122. Hidore, M. R., and Murphy, J. W., 1986, Correlation of natural killer cell activity and clearance of *Cryptococcus neoformans* from mice after adoptive transfer of splenic nylon wool-nonadherent cells, *Infect. Immun.* **51**:547–555.

123. Lipscomb, M. F., Evans, Z. A., Toew, G. B., Hauff, N., Hackett, J., and Kumar, V.,

1985, A role for natural killer (NK) cells in the clearance of intravenously injected *Cryptococcus neoformans, Fed. Proc.* **44:**1084.

124. Lipscomb, M. F., Toews, G. B., Evans, Z., Tompkins, R., Alvarellos, T., and Kumar, V., 1986, A role for natural killer cells in murine cryptococcosis, *6th Int. Cong. Immunol.* **5.16.9:**611.

125. Jimenez, B. E., and Murphy, J. W., 1984, *In vitro* effects of natural killer cells against *Paracoccidioides brasiliensis* yeast phase, *Infect. Immun.* **46:**552–558.

126. Tewari, R. P., Mkwananzi, J. B., McConnachie, P, Von Behren, L. A., Eagleton, L., Kulkarni, P., and Bartlett, P. C., 1981, Natural and antibody-dependent cellular cytotoxicity (ADCC) of human peripheral blood mononuclear cells (PBMC) to yeast cells of *Histoplasma capsulatum* (HC), *ASM Abstr.* F-66:324.

127. Yamada, T., Khardori, N., and Tewari, R. P., 1982, Natural and antibody-dependent cellular cytotoxicity (ADCC) of macrophages and lymphocytes from normal and immune mice for yeast cells of *Histoplasma capsulatum* (HC), *ASM Abstr.* F-9:328.

128. Petkus, A. F., and Baum, L. L., 1986, NK cell activity against *Coccidioides immitis, Fed. Proc.* **45:**4516.

129. Zunino, S., and Hudig, D., 1986, Human NK lymphocytes do not kill *Candida albicans,* although *Candida* can block NK lysis of K562 cells, *Nat. Immun. Cell Growth Regul.* **5:**166.

130. Baccarini, M., Bistoni, F., Puccetti, P., and Garaci, E., 1983, Natural cell-mediated cytotoxicity against *Candida albicans* induced by cyclophosphamide: Nature of the *in vitro* cytotoxic effector, *Infect. Immun.* **42:**1–9.

131. Nabavi, N., and Murphy, J. W., 1986, Antibody-dependent natural killer cell-mediated growth inhibition of *Cryptococcus neoformans, Infect. Immun.* **51:**556–562.

132. Kiessling, R., Klein, E., and Wigzell, H., 1975, "Natural" killer cells in the mouse. I. Cytotoxic cells with specificity for mouse Moloney leukemia cells. Specificity and distribution according to genotype, *Eur. J. Immunol.* **5:**112–117.

133. Reynold, C. W., Timonen, T., and Herberman, R. B., 1981, Natural killer (NK) cell activity in the rat. I. Isolation and characterization of the effector cells, *J. Immunol.* **127:**282–287.

134. Djeu, J. Y., Heinbaugh, J. A., Holden, H. T., and Herberman, R. B., 1979, Role of macrophages in the augmentation of mouse natural killer cell activity by poly I:C and interferon, *J. Immunol.* **122:**182–188.

135. Djeu, J. Y., Heinbaugh, J. A., Viera, W. D., Holden, H. T., Herberman, R. B., 1979, The effect of immunopharmacologic agents on mouse natural killer cell-mediated cytotoxicity and its augmentation by poly I:C, *Immunopharmacology* **1:**231–244.

136. Savary, C. A., and Lotzova, E., 1978, Suppression of natural killer cell cytotoxicity by splenocytes from *Corynebacterium parvum*-injected, bone marrow-tolerant, and infant mice, *J. Immunol.* **120:**239–243.

137. Ojo, E., Haller, O., Kimura, A., and Wigzell, H., 1978, An analysis of conditions allowing *Corynebacterium parvum* to cause either augmentation or inhibition of natural killer cell activity against tumor cells in mice, *Int. J. Cancer* **21:**444–451.

138. Riccardi, C., Barlozzari, T., Santoni, A., Herberman, R. B., and Cesarini, C., 1981, Transfer to cyclophosphamide-treated mice of natural killer (NK) cells and *in vivo* natural reactivity against tumors, *J. Immunol.* **126:**1284–1289.

139. Kozel, T. R., and Mastroianni, R. P., 1976, Inhibition of phagocytosis by cryptococcal polysaccaride: Dissociation of the attachment and ingestion phases of phagocytosis, *Infect. Immun.* **14:**62–67.

140. Timonen, T., Ortaldo, J. R., and Herberman, R. B., 1981, Characteristics of human large granular lymphocytes and relationship to natural killer and K cells, *J. Exp. Med.* **153:**569–575.

141. Bradley, T. P., and Bonavida, B., 1982, Mechanism of cell-mediated cytotoxicity at the single cell level. IV. Natural killing and antibody-dependent cellular cytotoxicity can be mediated by the same human effector cell as determined by the two-target conjugate assay, *J. Immunol.* **129:**2260–2265.

142. Kumagai, K., Itoh, K., Suzuki, R., Hinuma, S., and Saitoh, F., 1982, Studies of murine

large granular lymphocytes. I. Identification as effector cells in NK and K cytotoxicities, *J. Immunol.* **129:**388–394.

143. Ojo, E., and Wigzell, H., 1978, Natural killer cells may be the only cells in normal mouse lymphoid cell populations endowed with cytotoxicity ability for antibody-coated tumor target cells, *Scand. J. Immunol.* **7:**297–306.

144. Wahlin, B., and Perlmann, P., 1983, Characterization of human K cells by surface antigens and morphology at the single cell level, *J. Immunol.* **131:**2340–2347.

145. Cauley, L. K., and Murphy, J. W., 1979, Response of congenitally athymic (nude) and phenotypically normal mice to a *Cryptococcus neoformans* infection, *Infect. Immun.* **23:**644–651.

146. Marquis, G., Montplasir, S., Pelletier, M., Mousseau, S., and Auger, P., 1985, Genetic resistance to murine cryptococcosis: The beige mutation (Chediak-Higashi syndrome) in mice, *Infect. Immun.* **47:**288–293.

147. Salkowski, C. A., Bartizal, K. F., and Balish, E., 1985, Susceptibility of congenitally immunodeficient mice to *Cryptococcus neoformans*, in: *Germfree Research: Microflora Control and Its Application to the Biomedical Sciences* (B. W. Wostmann, ed.), Alan R. Liss, New York, pp. 407–410.

148. Bartizal, K. F., Salkowski, C. A., and Balsih, E., 1985, Modulation of murine natural killer cells by *Cryptococcus neoformans*, in *Germfree Research: Microflora Control and Its Application to the Biomedical Sciences* (B. W. Wostmann, ed.), Alan R. Liss, New York, pp. 333–336.

149. Henkart, P. A., 1985, Mechanism of lymphocyte-mediated cytotoxicity, *Annu. Rev. Immunol.* **3:**31–58.

150. Goldfarb, R. H., Timonen, T., and Herberman, R. B., 1983, Mechanism of tumor cell lysis by natural killer cells, in: *NK Activity and Its Regulation* (T. Hoshino, H. S. Koren, and A. Uchids, eds.), International Congress Series 641, Excerpta Medica, Amsterdam, pp. 403–421.

151. Roder, J. C., Kiessling, R., Biberfeld, P., and Anderson, B., 1978, Target-effector interaction in the natural killer (NK) cell system. II. The isolation of NK cells and studies on the mechanism of killing, *J. Immunol.* **121:**2509–2517.

152. Hiserodt, J. C., Britvan, L. J., and Targan, S. R., 1982, Characterization of the cytolytic reaction mechanism of the human natural killer (NK) lymphocyte: Resolution into binding, programming and killer cell-independent steps, *J. Immunol.* **129:**1782–1787.

153. Hiserodt, J. C., and Beals, T. F., 1985, Ultrastructural analysis of human natural killer cell–target cell interactions leading to target cell lysis, in: *Mechanisms of Cytotoxicity by NK Cells* (R. B. Herberman and D. M. Callewaert, eds.), Academic Press, New York, pp. 195–204.

154. Henkart, M. P., and Henkart, P. A., 1982, Lymphocyte-mediated cytotolysis as a secretory process, in: *Mechanisms of Cell-Mediated Cytotoxicity* (W. R. Clark and R. Golstein, eds.), Plenum, New York, pp. 227–242.

155. Henkart, P., Millard, P., Yue, C., Fredrickse, P., Blumenthal, R., Bluestone, J., Reynolds, C. W., and Henkart, M. P., 1985, Biochemical and functional properties of LGL and CTL cytoplasmic granules. *Adv. Exp. Med. Biol.* **184:**121–134.

156. Murphy, J. W., Reynolds, C. W., and Henkart, P. A., 1986, Effects of NK cell granules and cytolysin on *Cryptococcus neoformans*, *Nat. Immun. Cell Growth Regul.* **5:**150.

157. Walters, M., Papadimitriou, J. M., and Robertson, T. A., 1976, The surface morphology of the phagocytosis of micro-organisms by peritoneal macrophages, *J. Pathol.* **118:**221–226.

158. Papadimitriou, J. M., Robertson, T. A., Kletter, Y., Aronson, M., and Walters, M. N.-I., 1978, An ultrastructural examination of the interaction between macrophages and *Cryptococcus neoformans*, *J. Pathol.* **124:**103–109.

159. Marconi, P., Scaringi, L., Tissi, L., Boccanera, M. Bistoni, F., Bonmassar, E., and Cassone, A., 1985, Induction of natural killer cell activity by inactivated *Candida albicans* in mice, *Infect. Immun.* **50:**297–303.

8

Role of Natural Killer Cells in Inflammation and Antibacterial Activity

ARNOLD H. GREENBERG

1. Introduction

The purpose of this chapter is to review the evidence that large granular lymphocytes (LGL) can participate in the inflammatory response, with particular reference to bacterial infections. In this discussion, LGL will refer only to those cells that are identifiably of the natural killer (NK) cell lineage with respect to their membrane phenotype, in contrast to those LGL that may be derived from the T cell lineage. In this way, we will be referring to a natural effector population which has historically been associated with a nonrestricted lytic function and acts independently of the T cell receptor.

Other chapters in this book will discuss in more detail the distinguishing features of the cell types, including the functional and phenotypic markers that characterize them. This definition is required because of the similarities between T cells and NK cells that have led to lengthy discussions about the possibilities of their common lineage and function. The same may be said of the NK cell and monocyte. Features such as their similar morphology, their ability to secrete IL-1,[1] their highly motile nature,[2,3] and the identification of a phagocytic LGL subpopulation[4] all lend weight to the notion that there may be a common biological role for these cells.

ARNOLD H. GREENBERG • Manitoba Institute of Cell Biology, University of Manitoba, Winnipeg, Manitoba, Canada R3E 0V9.

It has long been considered that monocytes and granulocytes are "first-contact" leukocytes which function both as initial effector populations for bacterial pathogens and as recruiting and immunoamplifying cells. As we have come to understand some of the nonlytic functions of the NK cell, the similarity of certain NK and monocyte functions has raised the possibility that the NK cell could also act as a monocytelike inflammatory cell. These predictions have been tested in a number of *in vitro* and *in vivo* systems with results that allow one to seriously consider this novel role for the NK cell. In this chapter, then, we will review the evidence for this concept.

2. Locomotion of Large Granular Lymphocytes

2.1. Chemotactic and Chemokinetic Responses

Locomotion is a phenomenon that is typical of many eukaryocytic and prokaryotic cells. Leukocytes exhibit the capacity for locomotion to a group of unrelated molecules including bacterial products, casein, C5a, oxidized derivatives of arachidonic acid, and some lymphokines and monokines.[5–10] Polymorphonuclear (PMN) leukocytes and monocytes have been the most widely studied[11,12]; however, lymphocyte locomotion has also been well documented.[2,3,12–18] Leukocyte locomotion can either be directional (chemotactic) or nondirected (chemokinetic). Although chemokinetic responses of both T and non-T lymphocytes are commonly found,[15] chemotactic locomotion of lymphocytes to antigen, casein, zymosan-activated serum (C5a), and lymphokines has also been reported.[2,3,13,16–19]

Recent observations indicate that LGL are also fully capable of migrating in response to a number of well-known chemoattractants. Two groups have reported this phenomenon—Mantovani and colleagues[2] and our own laboratory[3]—and the two are in close agreement about the nature of the migratory cell type in humans. The cells were identified as LGL after enrichment on discontinuous Percoll gradients.[2,3] These LGL bear the phenotypic membrane markers of human NK cells, including HNK-1 (Leu 7), B73.1 (Leu 11, CD16), and OKT11, identified by complement depletion, rosetting, and FACS sorting. They were distinguished from contaminating T cells in the population on the basis of T cell markers, T3 (CD3), T4 (CD4), and T8 (CD8), which were absent from the migrating cells, as well as from monocytes, using the markers M1 and MO2.

These observations confirmed earlier work suggesting that LGL-like cells were motile based on such evidence as time-lapse cinematographic analysis and the ability of LGL to form uropods, the cellular structure associated with motile, polarized cells.[20,21] Indeed, the suggestion that non-T peripheral lymphocytes could migrate had been made 10 years earlier by O'Neill and Parrot[22] in experiments in which they observed cells with a large cytoplasmic to nuclear ratio capable of migrating in a casein gradient.

The migratory responses of the LGL seems to be both chemotactic and chemokinetic.[2,3] The classical method of identifying directed migration is to perform the checkerboard analysis of Zigmond and Hirsch.[23] In this assay, chemoattractant is placed above or below the filter of the Boyden chamber, with the migrating cells always above the filter. If chemotaxis is directional, then migration of the cells into the filter will only occur when the concentration of chemoattractant is greater in the lower chamber. What was evident from the experiments examining migration patterns of LGL is that there was a great deal of nondirected locomotion observed when chemoattractant was placed in the upper compartment with LGL. Furthermore, random movement was detected if LGL were allowed to migrate over longer periods of time in the absence of chemoattractant, or simply by incubating LGL in tissue culture for over 24 h prior to testing in the chemotaxis assays.[3,24]

The relative biological significance of chemokinetic versus chemotactic activity is not known. However, within the context of a local environment, nondirected locomotion will likely be effective, though less efficient, in promoting LGL contact with the relevant target cells. Where directed locomotion might be more important is in attracting LGL from the circulation to specific organ sites. As will be discussed in detail in a later section, NK/LGL migration to sites of inflammation has been observed in rodents during virus infection[25,26] and following injection of biological response modifiers.[27] Since it is not yet possible to distinguish between directed and nondirected migratory responses in vivo, the relative importance of these types of locomotion remains uncertain.

2.2. Chemoattractants for LGL

There is general agreement that the LGL can migrate in casein and C5a gradients,[2,3] and our laboratory reported that LGL can respond quite vigorously to a synthetic oligopeptide bearing the formylmethionyl group at the N terminus that has been shown to be an extremely potent chemoattractant for polymorphonuclear leukocytes[3] (Fig. 1). The peptide is formylmethionylleucylphenylalanine (f-MLP) and shares structural similarities with a peptide released by E. coli. The dose–response curve to f-MLP is similar to that for PMN, and the LGL chemotactic response to f-MLP is completely inhibitable by an inactive structural analog (CBZ-Phe-Met).[3] Interestingly, we were unable to directly demonstrate f-MLP binding directly to isolated human LGL using an FITC-labeled hexapeptide (N-formyl-nle-leu-phe-nle-tyr-lys-fluorescein) that binds to f-MLP receptors on PMN.[29] However, it was possible to detect f-MLP receptors on

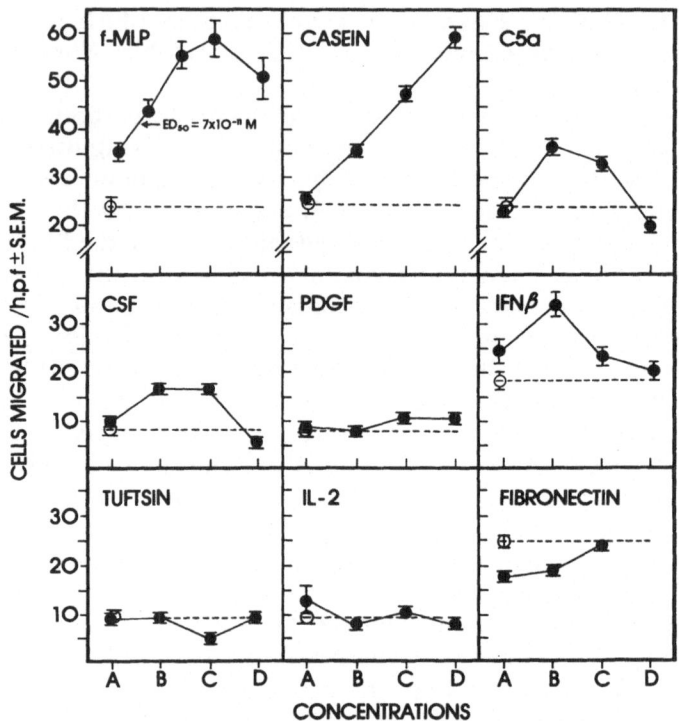

FIGURE 1. Migration of LGL to putative chemoattractants. These compounds were placed in the bottom chamber in increasing concentration as follows (A to D): f-MLP—3×10^{-12}, 10^{-11}, 3×10^{-11}, and 10^{-10} M. Casein—0.1, 0.5, and 2.0 mg/ml. Purified C5a—10, 30, 60, and 300 μg/ml. CSF—0.01, 0.1, 1, and 10 U/ml. PDGF—0.05, 0.15, 5, and 50 ng/ml. Natural IFNβ—500, 1000, 5000, and 10,000 U/ml. Tuftsin—10^{-10}, 10^{-9}, 10^{-8}, and 10^{-7} M. Natural IL-2—0.5, 5, 50, and 125 U/ml. Fibronectin—0.01, 0.2, and 0.2 mg/ml. Random migration (○) was determined from duplicate chambers that did not contain chemoattractant.

an LGL leukemia cell line, described by Reynolds et al.,[30] called RNK. These cells were also able to migrate toward f-MLP, although in a less vigorous response when compared to peripheral blood LGL, and f-MLP receptors were demonstrated by direct radioligand binding using f-ML(^{3}H)P.[3]

A number of other substances that have been shown to stimulate NK activity have also been evaluated for their chemotactic ability. Neither interleukin 2 (IL-2) nor tuftsin is capable of acting as a chemoattractant for LGL,[3] and both of these substances can activate NK lysis. Although our laboratory found that IFN-β can be a weak chemoattractant for LGL,[3] Polentarutti et al.[24] reported that neither interferon-β nor interferon-γ was a chemoattractant. It is clear, however, that both IFN and IL-2, when preincubated with LGL prior to assessing their migratory capacity, produce a general increase in locomotion in the absence of a specific chemoattractant.[3,24] This type of observation would suggest that the chemokinetic activation of LGL by substances that also activate the cytolytic capacity of these cells produces the net effect of increasing the range of activity of the effector cell within the organ site.

The conclusion that can be taken from this group of studies is that LGL-bearing typical phenotypic markers of the human NK cell can migrate unidirectionally to several common chemoattractants. However, the cell, which is spontaneously motile, can be further activated to increase locomotion in a nondirected manner upon stimulation by activators of NK lysis.

3. In Vivo Migratory Responses of Large Granular Lymphocytes/NK Cells

3.1. Tissue Distribution

To understand the migratory responses of LGL during an inflammatory response, we first must consider the normal tissue distribution and circulation of this lymphoid population. There is, in fact, little known about the migratory patterns of the LGL. These cells almost certainly originate in the bone marrow,[31] where their proliferation and differentiation occur in a manner that is not understood. The cells then enter the peripheral circulation, and, while some continue to circulate, others must migrate to the various organs where they have been identified.

Since the distribution of lymphocytes throughout the compartments of the body is not random, it raises the question of what determines the number and type of cell at each tissue site. The functional distribution of lymphoid subsets in vivo is, at least in some organs, highly represented by quite unusual cell types such as surface IgA-bearing lymphocytes in the mucosa-associated lymphoid organs and antigen-specific B and T cells in lymph nodes or spleen challenged with antigen.[22–34] The ability of lymphocytes to enter any tissue site requires that they bind selectively to endothelial cells at the site of inflammation and penetrate the vessel wall. In the case of mature B and T lymphocytes, these preferentially bind to high endothelial venules (HEV), and this selective interaction is important in controlling lymphocyte traffic.[35,36] It is unknown whether the LGL/NK cell binds to the HEV, but since most mature virgin lymphocyte populations can bind to these endothelial cells and exhibit some organ selectivity, this seems quite likely.

Evidence from a number of sources has demonstrated that LGL are widespread in their distribution,[37] having been identified in blood,[38,39] spleen,[40,41] lymph nodes,[42] thymus,[43] lungs,[26] liver,[27,44,45] and intestine.[46,47] Of interest is the apparent absence of LGL in thoracic duct effluent, unlike most mature lymphocytes, which generally recirculate leaving the bloodstream and reentering through efferent lymphatics.[39,48] This observation suggests that although LGL are present in the circulation and can enter a variety of lymphoid and nonlymphoid tissues from the blood, they may not recirculate in the same way as other lymphoid populations. Studies on homing of LGL purified from rat peripheral leukocytes after labeling with [^{3}H]uridine or [^{111}In]oxine demonstrated that after intravenous injection, the main sites of localization were the alveolar walls of the lungs and spleen red pulp.[48] They were not found in spleen white pulp and lymph nodes, the main sites of T cell traffic.[48] It would be of interest to know if LGL from specific organ sites, such as lung and spleen, preferentially home to the same sites. These cells might be a subpopulation of the circulating pool that have the ability to recognize and enter the specific organ.

The widespread distribution of LGL indicates the availability of these cells to respond to local inflammation at many tissue sites, in addition to entering the site directly from the blood. As indicated earlier, the distribution of NK cells in the various anatomical compartments is not uniform, and the number of LGL that can be isolated from both lymphoid and nonlymphoid organs of normal animals varies greatly.[42] For example, NK activities in cells isolated from human, rat, and mouse blood, spleen, lung, and liver are widely divergent.[26,27,40–42,46,47] Indeed, NK activity in some organs, such as the lung, appears to be very low in both normal rodents[26] and humans,[49,50] possibly owing to the presence of potent inhibitory macrophages.[51] Most importantly, for the purposes of present discussion, it has been demonstrated even in organs with very little endogenous activity, that in response to infectious agents including viruses[25,26,52–57] and parasites,[58] a significant and preferential augmentation of NK activity is seen.

3.2. LGL/NK Cell Response to Infection and Noninfectious Stimulation

It has been known for many years that a number of synthetic and microbial-derived substances are able to augment NK activity *in vivo* and *in vitro*. Preferential augmentation of NK activity in selective anatomical compartments has been observed with particular treatment regimens.[26,27,60] Several mechanisms may account for the means by which NK activity is enhanced within a compartment, including an increase in the activation state of mature differentiated NK cells, the redistribution of active or precursor cells from other sites, differentiation and/or proliferation of cells and their precursors within the site, or any combination of these factors.

That NK cells can be stimulated by bacterial products is well illustrated by the recent work of Tarkkanen *et al.*,[59] who found that *Salmo-*

nella can directly activate Leu-19$^+$, CD3$^-$, and CD16$^+$, or CD16$^-$ human peripheral blood NK cells for lysis of NK-sensitive and -resistant targets. The movement of NK cells into certain nonlymphoid organ sites and activation of cells within the tissues in response to active infection or stimulation with noninfectious bacterial preparations and synthetic activators has also been well documented.[59–63] For example, augmentation of peritoneal and liver NK activity by administration of *Corynebacterium parvum* and pyran copolymer (MVE-2) was observed by Wiltrout *et al.*[27] This was accompanied by a substantial increase in both LGL and a smaller increase in large agranular lymphocytes. The lytic cell was a typical NK cell in its pattern of tumor target cell lysis, and membrane phenotype, which was asialo $GM_1{}^+$, Ly5$^+$, Qa5$^+$, Ly1$^-$, Ly2$^-$ lytic population, predominated. One difference was that the LGL in the liver of stimulated animals were strongly Thy-1$^+$, whereas endogenous NK cells are low Thy-1 expressors. More recently, this observation has been confirmed using other biological response modifiers including poly ICLC, OK-432, and *Propionibacterium acnes.*[60]

The role of the NK cell in resistance to viral infection will be discussed in another chapter; therefore, we will limit our comments to the migratory responses and accumulation of NK cells at infection sites. A number of good examples of accumulation of natural killer cells at the site of virus infection have been published. Local augmentation of NK activity has been observed in the lungs of mice infected with influenza virus,[26] in the liver of mice infected with hepatitis[56] and lymphocytic choriomeningitis virus (LCMV), and in the cerebrospinal fluid of vaccinia-induced meningitis.[64,65] The NK activity in the lung and liver of virus-infected mice increases correspondingly with the numbers of LGL.[25,26]

MacIntyre and Welsh[25] reported that the hepatotropic LCMV-WE strain and nonhepatotropic LCMV-ARM strain produce quite similar NK/LGL responses, so this augmentation in activity is likely not due to virus replication in the liver but rather some systemic effect, possibly as a result of interferon release. These investigators argue that the increased cell-mediated lysis detected in liver and blood is more likely due to increases in NK cell number than state of activation. However, the evidence for this is weak and based only on the preferential loss of lysis against relatively resistant target cells.

Earlier work by Biron *et al.*[66] found that hydroxyurea, which preferentially kills cells synthesizing DNA, significantly reduced the augmented NK activity of LCMV-infected mice. These data suggested that NK cells in these mice have an increased turnover rate and that a significant portion of the increase in activity was due to NK cell division. It is unclear from their experimental design whether the cell division occurred *in situ* or dividing cells had migrated into the spleen. Since stimulated NK cells can migrate more readily than resting cells,[3,24] this is a distinct possibility. More recent work on the regulation of LGL influx following viral infection by Natuk and Welsh[67] has identified a chemotactic factor for NK/LGL in the peritoneal fluid of infected mice. This evidence provides a mechanistic explanation for increased cell number and NK activity in the organ sites. In conclusion, increased NK activity following viral infection is likely a result of NK cell influx as well as division and an enhanced state of activation.

Murine hepatitis virus and cytomegalovirus, which are highly cytopathic for hepatocytes in contrast to LCMV, stimulate high levels of NK/LGL in the liver,[25] and this may be a response to the viral infection of the hepatocyte. Interestingly, MacIntyre and Welsh[25] note that the CTL lytic activity of liver leukocytes, which peaks several days after NK/LGL infiltration, is also mediated by a cell with LGL morphology. Wiltrout *et al.*[27] noted that BRM-stimulated NK cells from liver express thy-1, although in the case of the LCMV-induced CTL/LGL, most appear to be CTL of Lyt2+ and Thy1.2+ phenotype. Thus, there are two distinguishable populations with LGL morphology appearing in livers of LCMV-treated mice—an early NK/LGL cell population, and a CTL/LGL population which appears somewhat later.[25]

3.3. Migration of LAK Cells

Although there is still some controversy surrounding the phenotype of the cell(s) that mediates lymphocyte-activated killer (LAK) cell activity, recent results suggest that the NK lineage is the origin of the vast majority of human effector cells.[68] At least for the present discussion, I will consider the LAK cell a natural effector cell, likely to be predominantly of the NK lineage. The work of Rosenberg and colleagues has documented the proliferation and migration of the LAK cell in IL-2-treated animals.[69,70] IL-2 is an NK cell activator, and mononuclear cells cultured in IL-2 will divide and generate cells that are capable of lysing autologous or syngeneic tumor cells.[71] In the experiments of Rosenberg's group, the *in vivo* administration of these cells significantly reduces the proliferation of hepatic and pulmonary metastases.[73,74]

Employing *in vivo* DNA labeling with [^{125}I]UdR, this group has demonstrated that IL-2-treated mice developed large numbers of activated lymphoid cells in the lung, liver, and kidneys, something that was not seen in preirradiated mice.[69] The same sort of observation has been made in animals receiving both LAK cells and IL-2; however, in this case, proliferation at tissue sites was increased above that seen with animals receiving only IL-2, suggesting that LAK cells had entered these tissues and proliferated under the influence of the IL-2 therapy. This was substantiated by the observation that the LAK cell proliferation was completely abrogated by preirradiation of the cells.[69] These observations were confirmed by histological examination of the hepatic and pulmonary tissue, where large numbers of lymphoid cells were found in the tissue parenchyma. Actively dividing lymphoid cells were identified in lungs, liver, kidneys, spleen, and lymph nodes.[69,70] These observations, then, suggest that the LAK cell has the capacity to infiltrate a number of organs from the circulation.

Although the absolute level of proliferation of LAK cells entering from the blood was highest in the liver, significant levels could be detected in the lungs, spleen, and kidneys.[69] The migratory pattern was not typical of T cells, since the LAK cell was able to localize in mesenteric lymph nodes, which differs from findings using cultured T cells, even after IL-2 administration.[75] It is unknown whether non-IL-2-stimulated NK cells have the same trafficking pattern as the LAK cell.

4. Can NK Cells Regulate the Inflammatory Response?

Since LGL/NK cells enter the sites of viral infection and BRM-induced inflammation very rapidly, it is tempting to speculate that they may participate in regulating the accompanying and subsequent influx of inflammatory cells. To qualify as a regulator of this sort, the cell must be capable of releasing cellular products that either can act as chemoattractants or are capable of generating chemoattractants. The NK cell is able to release a number of substances that can act in exactly this capacity.

NK cells release IFN-α in response to viruses and tumor cells,[76,77] and IFN-γ production is induced by IL-2[78] and hydrogen peroxide.[79] As mentioned earlier, the interferons enhance nondirected migration of LGL,[24] and IFN release can therefore potentially act as a mechanism of recruiting more LGL to the site of inflammation as well as activating them *in situ*.

Another important chemotactic cytokine produced by NK cells is IL-1.[1] Human NK cell IL-1 is indistinguishable from that produced by monocytes,[1] and IL-1 can induce chemotaxis of PMN[80] as well as B and T lymphocytes.[18,81] Since it has not been tested directly, one can only infer from these data that NK cells may be able to stimulate migration of leukocytes via this pathway. Similarly, this type of reasoning allows us to argue that since NK cells can produce tumor necrosis factor (TNF)[82] and TNF is a potent chemoattractant for both monocytes and PMN,[83] NK cells can likely stimulate chemotaxis of these inflammatory leukocytes via TNF release. NK cells produce colony-stimulating factor,[84] which is a mild chemoattractant for LGL,[3] and B cell growth factor,[85] which has not yet been reported to be a leukocyte chemoattractant.

The only direct evidence that NK/LGL can release a chemoattractant has come from work in our own laboratory.[86] In these experiments, we found that within 1 h of contact with gluteraldehyde-fixed tumor, human LGL of the HNK1$^+$, OKT11$^+$, OKT3$^-$, OKM$^-$ phenotype released a factor called NK leukocyte chemotactic factor (NK-LCF) into supernatant

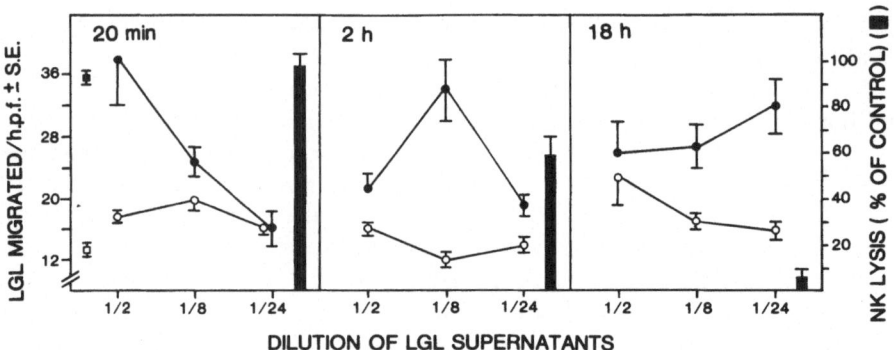

FIGURE 2. Stimulation of NK leukocyte chemotactic factor (NK-LCF) release by Sr^{2+} stimulation of LGL. Cells were incubated for 20 min (left), 2 h (center), or 18 h with 25 mM $SrCl_2$ (●) or in medium (○). Cell-free supernatants were dialyzed and tested for chemotactic activity and compared with f-MLP (■) and Hank's balanced salt solution controls (□). The remaining LGL were assayed for NK lytic activity on K562 (solid bars). A reduction in NK lysis is accompanied by increasing release of chemotactic activity (from Greenberg et al.[86]).

that was chemotactic for LGL, neutrophils, and macrophages. Of particular interest was the ability of the degranulating cation Sr^{2+} to release NK-LCF (Fig. 2), suggesting it may be located in the LGL granule. A direct test of this hypothesis, using granules of the RNK leukemia,[86] confirmed that a chemotactic factor was present in the granules and may be similar to that released by LGL, since an antigranule antibody that neutralized granule chemotactic activity also blocked activity in LGL supernatants.[86] However, it is not proved that the activity in granules is identical, because a polyclonal antibody directed at whole granule preparations was used in these experiments and could have neutralized another granule component. In addition, contamination of the immunizing granule preparation could have resulted in antibody that would react with nongranule components. Isolation and characterization of the chemotactic factor will resolve this issue. Since NK cytokines with known chemotactic activity, such as IL-1 and TNF, could also be present in the granules, we examined several granule preparations for both IL-1 and TNF biological activity but were unable to detect any significant responses. TNF, measured by a sensitive ELISA technique, was detected in one of three granule preparations, and only at very low levels; this lymphokine is therefore unlikely to be the granule chemoattractant (A. H. Greenberg and M. Palladino, unpublished data). Other granule constituents including cytolysin,[87] serine esterases,[88] and an unusual chondroitin sulfate proteoglycans[89] have yet to be examined but might also be candidates.

5. Do NK Cells Have Antibacterial Activity?

Although it is still not conclusively resolved whether NK cells have biologically important antibacterial responses, some indirect evidence suggests that this may be possible. Antibacterial activity could take two forms: direct cytotoxicity or indirect effects on activity of other cells such as monocytes and PMN. NK cell participation in both forms of bacterial cytotoxicity has been reported[5,90] and deserves some critical appraisal. The first suggestion that a natural effector cell could kill bacteria was the report of Lowell *et al.* that, in addition to monocyte and PMNs,[91,92] ADCC of *Shigella flexneri* and group C meningococci was mediated by normal human non-nylon wool adherent Fc receptor-positive mononuclear cells, as well as a non-E rosette-forming subpopulation of these cells.[92,93]

Although this phenotype was consistent with a K cell at the time of the study, it does not meet the more rigorous criteria for identifying an NK cell with ADCC activity that have since been developed. Morgan *et al.*[94,95] were also able to establish that lymphocytes could mediate anti-*Shigella* ADCC, this time isolated from either breast milk[94] or gut lymphocytes.[95] They also correlated the conversion of *Shigella*-resistant guinea pigs to *Shigella* susceptibility with a reduction in gut natural and ADCC killing of the bacteria.[95] However, the phenotype of the nonadherent gut lymphocytes was not further characterized. Lowell's idea that antibacterial ADCC may be important[93] was subsequently reexamined by Tagliabue *et al.*[96,97] Their studies concluded that gut-associated lymphocytes could mediate direct anti-*Shigella* cytotoxicity, and this was increased in the presence of antibacterial secretory IgA, presumably in an ADCC-type reaction.[96] These investigators also found that lymphocytes mediated anti–*Salmonella typhimurium* activity in the presence or absence of antibody.[97] This effector was asialo GM_1^+, Fc_r^+, Thy 1.2^-, nonadherent and nonphagocytic,[97] all characteristics that are typical of the murine NK cell.

Curiously, Tagliabue later rejected the NK role in *Salmonella* resistance *in vivo* because of the different patterns of strain distribution of early *S. typhimurium* resistance and NK activity against tumor cells *in vitro*.[97] This was an unwarranted conclusion, because we have no reason to assume that the genetics of NK antitumor activity is identical to the putative NK antibacterial response. The same study reported that the bg/bg mutant was completely unable to resist the bacterium. Although the susceptibility could have been due to a monocyte or PMN lysosomal defect in the mutant beige strain, it is certainly compatible with NK participation in antibacterial activity *in vivo*. This model therefore deserves further critical testing.

The enhanced lytic response of Leu 11^+ as well as Leu 11^- LGL to *Shigella flexneri*-infected HeLa cells[98] is also worth noting, since it is well known that NK cells can preferentially lyse virally infected cells (see Chapters

3–5). The enhanced recognition of bacterial-infected HeLa cells was dependent on bacterial invasion of the target, since neither soluble products nor noninvasive bacteria altered tumor cell sensitivity. The mechanism of this enhanced recognition is, however, unknown.

The next question to be posed is how NK cells could mediate their antibacterial activity. Direct lysis implies the activation of the NK lytic machinery. This might take two forms. One mechanism of NK kill is hypothesized to be the result of granule exocytosis and the release of the granule-associated cytolysin.[87] The other mechanism is thought to be the result of secretion of lymphotoxins such as TNF[99] and NKCF.[100] It is not known whether bacteria are susceptible to cytolysin. Similarly, the susceptibility of bacteria to kill by lymphotoxins has not been established.

Phagocytosis of *Staphylococcus aureus* by LGL is an unexpected observation and mechanism of kill.[4] Abo *et al.* reported that Leu 11$^+$ LGL that were either Leu 7$^+$ or Leu 7$^-$ could preferentially phagocytose gram-positive bacteria and subsequently secrete significant amounts of IL-1 and IFN[4] (Fig. 3). Some gram-negative bacteria such as *E. coli* and *S. minnesota* could stimulate IL-1 but were not phagocytosed. These authors argue the interesting point that Leu 11$^+$ LGL may represent a distinct lineage of LGL, more closely resembling myelomonocytes in function, since they possess both phagocytic and IL-1 secretory capacity. These cells would be distinct from the Leu 11$^-$, Leu 7$^+$ LGL, which is the phenotype of virtually all long-term IL-2-dependent cell lines.[101,102]

Whether or not the phagocytic LGL is a distinct lineage, it is clear that LGL of the Leu 11$^+$ phenotype can participate in gram-positive and gram-negative antibacterial responses by phagocytosis and/or IL-1 secretion. It is also worth noting that if an NK cell subpopulation can effectively phagocytose and kill pathogens, it raises the interesting possibility that it could act as an antigen-presenting cell. To my knowledge, this has never been tested.

Indirect bacterial killing through the release of LGL factors that can recruit and activate macrophages has also been hypothesized as an NK/LGL role in bacterial resistance.[103] In work from our laboratory, a clue to this relationship came from the observation that monocytes contaminating human LGL preparations could be triggered to a respiratory burst by a soluble factor released following tumor activation.[104] The LGL were HNK-1 (Leu 7$^+$) and OKT 11$^+$. The activation of oxidative metabolism in the monocytes raised the possibility that the factor could stimulate antimicrobial activity as well.

In a test of this hypothesis, we examined the intracellular lysis of *S. aureus* by factors released by LGL.[103] An LGL macrophage-activating factor (LGL-MAF) was detected following the stimulation of an HNK-1$^+$ LGL population by NK-sensitive tumors. The factor was rapidly released and could be blocked by monensin, a carboxylic ionophore that is an inhibitor of vesicular traffic and can block LGL degranulation and lysis.[105] The

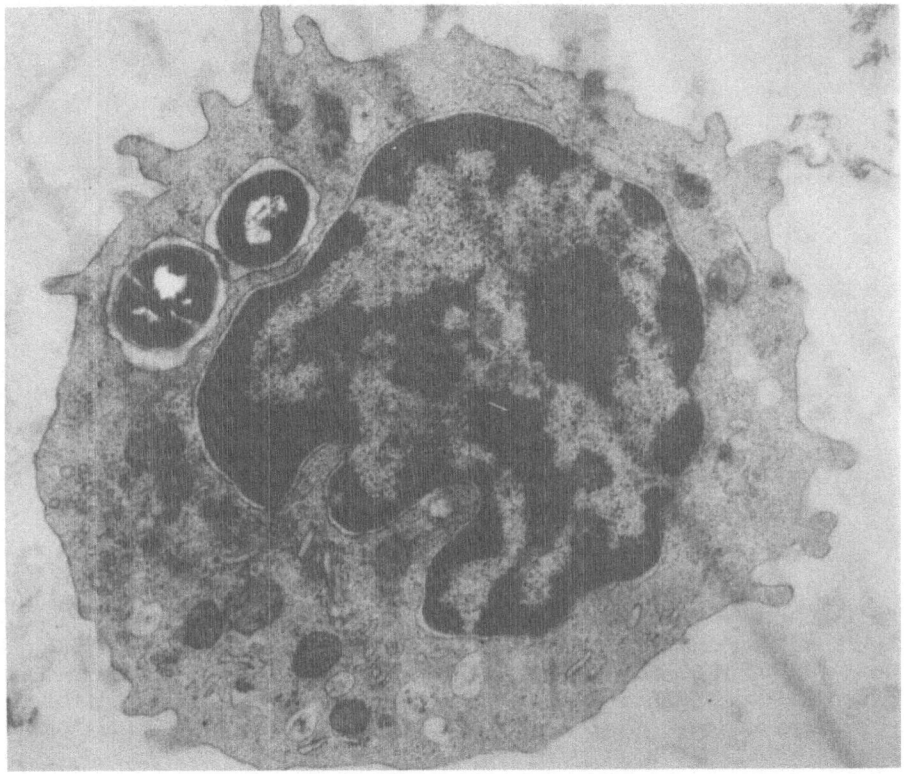

FIGURE 3. Microscopic observation of Leu-7⁺ cells phagocytosing *S. aureus* into their cytoplasm (× 15,000) (from Abo *et al.*[4]).

LGL-MAF was a low-molecular-weight protein of less than 20- kD[103] (Fig. 4). It has not yet been identified, although a MAF activity has recently been detected in the granule of the LGL leukemia RNK.[106] It should be noted that the LGL cytokines IL-1 and TNF have been shown to stimulate neutrophil oxidative metabolism,[107] neutrophil degranulation,[108] and chemotaxis.[10] The possibility that LGL-MAF is IL-1 or TNF has not been completely excluded.

6. Concluding Remarks

The NK cell appears to have many of the properties of an inflammatory cell. That is, it is highly motile and can respond by increasing its migratory activity in response to bacterial products, complement components, and lymphokines. *In vivo*, it is a widely distributed cell that can enter a specific organ site in response to bacterial and viral infection and

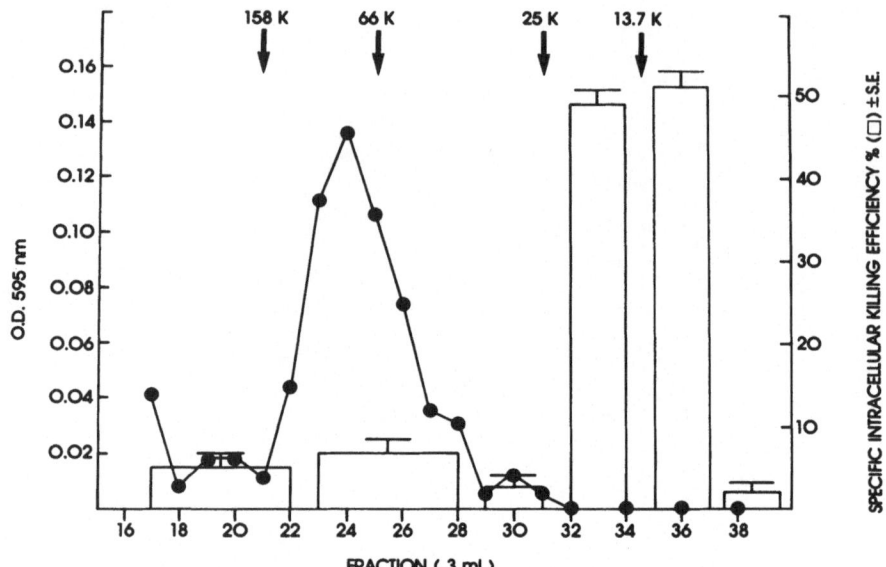

FIGURE 4. Ultragel fractionation of LGL MAF in supernatants from human LGL stimulated with K562. Six fractions were examined. Fractions <30,000 daltons were split into four pools of 10,000, 10,000–15,000, 15,000–20,000, and 20,000–30,000 daltons along with two pools of 30,000–120,000 and >120,000 daltons. All pools were then concentrated back to 3 ml and were assayed on humans alveolar macrophages. Antistaphlococcal activity was seen only in fractions less than 20,000 daltons ($p < 0.001$) (from Gomez et al.[103]).

following stimulation by noninfectious bacterial extracts. It can also be directly activated for lysis by some bacterial species. The NK cell can synthesize cytokines (IL-1, TNF) that are chemoattractants, and a chemoattractant is released from NK cells (NK-LCF) that can stimulate the migration of neutrophils, monocytes, and LGL. Some evidence suggests that the NK cell can kill *Shigella* and *Salmonella* both directly and by an ADCC mechanism. A subpopulation of NK cells can both phagocytose gram-positive bacteria and release IL-1 following contact with certain bacterial species. Finally, evidence exists for the release of an NK macrophage-activating factor that can stimulate monocyte oxidative metabolism and intracellular killing of *S. aureus* by macrophages.

Taken together, these observations argue that the NK cell could be an active participant in resistance to bacterial infection. One important question that needs to be addressed before accepting this view is whether NK antibacterial activity is biologically important. The prediction of the hypothesis is that NK deficiency should result in susceptibility to pyogenic infections. This has not been resolved. The NK-deficient bg/bg strain is not unusually predisposed to infection in a nonprotected environment, but one report suggests these mice are susceptible to bacterial challenge.[95] Humans with a similar defect, the Chediak-Higashi (CH) syndrome, appear to have a high incidence of pyogenic infection.

The CH syndrome and bg/bg strain may, however, not be good experimental models to test the question, because the degranulation defect is general to many cells with antibacterial activity, including PMN and monocytes. Furthermore, the NK deficiency is not absolute, and cytokine production is likely not affected. Other models of NK deficiency, such as chronic treatment of mice with anti-asialo GM$_1$, or anti-NK1 have not been studied. This type of experiment is also not free of interpretation problems because of the presence of the asialo–GM$_1$ antigen on other leukocytes. These problems with the animal models are obviously not limited to this area of research but reflect the failure of the field to find a model that defines only an NK cell defect. Even with these limitations, however, the null hypothesis that the absence of NK activity has no effect on bacterial resistance can be posed and, quite possibly rejected.

References

1. Scala, G., Allavena, P., Djeu, J. Y., Kasahara, T., Ortaldo, J. R., Herberman, R. B., and Oppenheim, J. J., 1984, Human large granular lymphocytes are potent producers of interleukin-1, *Nature* **309**:56–59.
2. Bottazzi, B., Introna, M., Allavena, P., Villa, A., and Mantovani, A., 1985, *In vitro* migration of human large granular lymphocytes, *J. Immunol.* **134**:2316–2321.
3. Pohajdak, B., Gomez, J., Orr, F. W., Khalil, N., and Greenberg, A. H., 1986, Chemotaxis of large granular lymphocytes, *J. Immunol.* **136**:278–284.
4. Abo, T., Sugawara, S., Amenomori, A., Itoh, H., Rikiishi, H., Moro, I., Kumagai, K., 1986, Selective phagocytosis of gram-positive bacteria and interleukin-1-like factor production by a subpopulation of large granular lymphocytes, *J. Immunol.* **136**:3189–3197.
5. Ward, P. A., Lepow, I. H., and Newman, L. J., 1968, Bacterial factors chemotactic for polymorphonuclear leukocytes, *Am. J. Pathol.* **52**:725–732.
6. Snyderman, R., Phillips, J., and Mergenhagen, S. E., 1970, Polymorphonuclear leukocyte chemotactic activity in rabbit serum and guinea pig serum treated with immune complexes: Evidence for C5a as the major chemotactic factor, *Infect. Immun.* **1**:521–527.
7. Goldman, D. W., and Goetzl, E. J., 1984, Heterogeneity of human polymorphonuclear leukocyte receptors for leukotriene B4, *J. Exp. Med.* **159**:1027–1032.
8. Luger, T. A., Charon, J. A., Colot, M., Micksche, M., and Oppenheim, J. J., 1983, Chemotactic properties of partially purified human epidermal cell-derived thymocyte activating factors (ETAF) for polymorphonuclear luekocytes and mononuclear cells, *J. Immunol.* **131**:816–820.
9. Sauder, D. N., Marinessa, N. C., Katz, S. I., Dinarello, C. A., and Gallin, J. I., 1984, Chemotactic cytokines: The role of leukocyte pyrogen and epidermal cell thymocyte activating factor in neutrophil chemotaxis, *J. Immunol.* **132**:828–832.
10. Ming, W. J., Bersoni, L., and Mantovani, A., 1987, Tumor necrosis factor is chemotactic for monocytes and polymorphonuclear leukocytes, *J. Immunol.* **138**:1469–1474.
11. Snyderman, R., and Goetzl, E. J., 1981, Molecular and cellular mechanisms of leukocyte chemotaxis, *Science* **213**:830–837.
12. Snyderman, R., and Pike, M. C., 1984, Chemoattractant receptors on phagocytic cells, *Annu. Rev. Immunol.* **2**:257–271.

13. Russel, R. J., Wilkinson, P. C., Sless, F., and Parrott, D. M. V., 1975, Chemotaxis of lymphoblasts, *Nature* **256**:646–648.
14. Schreiner, G. F., and Unanue, E. R., 1975, Anti-Ig triggered movements of lymphocytes. Specificity and lack of evidence for directional migration, *J. Immunol.* **114**:809–814.
15. Parrott, D. M. V., and Wilkinson, P. C., 1981, Lymphocyte locomotion and migration, *Prog. Allergy* **28**:193.
16. Center, D. M., and Cruikshank, W., 1982, Modulation of lymphocyte migration by human lymphokines. I. Identification and characterization of chemoattractant activity for lymphocytes from mitogen-stimulated mononuclear cells, *J. Immunol.* **128**:2563–2574.
17. Kornfeld, H., Berman, J. C., Beer, D. J., Center, D. M., 1985, Induction of human T lymphocytes motility by interleukin-2, *J. Immunol.* **134**:3887–3890.
18. Hunninghake, G. W., Glazier, A. J., Monick, M. M., and Dinarello, C. A., 1987, Interleukin-1 is a chemotactic factor for human T-lymphocytes, *Am. Rev. Respir. Dis.* **135**:66–71.
19. Wilkinson, P. C., Parrott, M. V., Russel, R. J., and Sless, F., 1977, Antigen-induced locomotor responses in lymphocytes, *J. Exp. Med.* **145**:1158–1168.
20. Muse, K. E., and Koren, H. S., 1982, The uropod as an integral and specialized structure of large granular lymphocytes, in: *NK and Other Natural Effector Cells* (R. B. Herberman, ed.), Academic Press, New York, pp. 1035–1040.
21. Uchida, A., Colot, M., and Micksche, M., 1984, Suppression of natural killer cell activity by adherent effusion cells of cancer patients. Suppression of motility, binding capacity and lethal hit of NK cells, *Br. J. Cancer* **49**:17–23.
22. O'Neill, G. J., and Parrott, D. M. V., 1977, Locomotion of human lymphoid cells. I. Effect of culture and ConA and T and non-T lymphocytes, *Cell. Immunol.* **33**:257–266.
23. Zigmond, S. H., and Hirsch, J. G., 1973, Leukocyte locomotion and chemotaxis. New methods for evaluation, and demonstration of a cell derived chemotactic factor, *J. Exp. Med.* **137**:387–410.
24. Polentarutti, N., Bottazzi, B., Balotta, C., Erroi, A., and Mantovani, A., 1986, Modulation of the locomotor capacity of human large granular lymphocytes, *Cell. Immunol.* **101**:204–212.
25. McIntyre, K. W., and Welsh, R. M., 1986, Accumulation of natural killer and cytotoxic T large granular lymphocytes in the liver during virus infection, *J. Exp. Med.* **164**:1667–1681.
26. Stein-Streilein, J., Bennett, M., Mann, D., and Kumar, V., 1983, Natural killer cells in mouse lung, surface phenotype, and response to local influenza virus infection, *J. Immunol.* **131**:2699–2704.
27. Wiltrout, R. H., Mathieson, B. J., Talmadge, J. E., Reynolds, C. W., Zhang, S. R., Herberman, R. B., and Ortaldo, J. R., 1982, Augmentation of organ-associated NK activity by biological response modifiers: Isolation and characterization of large granular lymphocytes from the liver, *J. Exp. Med.* **160**:1431–1449.
29. Sklar, L. A., and Finnay, D. A., 1982, Analysis of ligand-receptor interactions with the fluorescence activated cell sorter, *Cytometry* **3**:161–167.
30. Reynolds, C. W., Bere, E. W., and Ward, J. M., 1984, Natural killer activity in the rat. III. Characterization of transplantable large granular lymphocyte (LGL) leukemias in F344 rat, *J. Immunol.* **132**:534–540.
31. Haller, O., and Wigzell, H., 1977, Suppression of natural killer cell activity with radioactive strontium: Effector cells are marrow dependent, *J. Immunol.* **118**:1503–1506.
32. Lamm, M. E., 1976, Cellular aspects of immunoglobulin A, *Adv. Immunol.* **22**:223–290.
33. Kraal, G., Weissman, I. L., and Butcher, E. C., 1982, Germal center B cells: Antigen specificity and changes in heavy chain isotype expression, *Nature* **298**:377–379.
34. Sprent, J., 1980, Antigen-induced selective sequestration of T lymphocytes: Role of the major histocompatibility complex, *Monogr. Allergy* **16**:233–244.

35. Gowans, J. L., and Knight, E. J., 1964, The route of recirculation of lymphocytes in the rat, *Proc. R. Soc. Ser. B* **159**:257–264.
36. Butcher, E. C., 1988, The regulation of lymphocyte traffic, *Curr. Top. Microbiol. Immunol.* (in press).
37. Reynolds, C. W., and Ward, J. M., 1986, Tissue and organ distribution of NK cells, in: (E. Lotzova and R. B. Herberman eds.), *Immunobiology of Natural Killer Cells*, CRC Press, Boca Raton, FL, pp. 63–72.
38. Timonen, T., Reynolds, C. W., Ortaldo, J. R., and Herberman, R. B., 1982, Isolation of human and rat natural killer cells, *J. Immunol. Methods* **51**:269–277.
39. Fox, R. I., Fong, S., Tsoukas, S., and Vaughn, J. H., 1984, Characterization of recirculating lymphocytes in rheumatoid arthritis patients: Selective deficiency of natural killer cells in thoracic duct lymph, *J. Immunol.* **132**:2883–2887.
40. Kunagai, K., Itoh, K., Suzuki, R., Hinuma, S., and Saitoh, F., 1982, Studies of murine large granular lymphocytes. I. Identification as effector cells in NK and K cytotoxicities, *J. Immunol.* **129**:388–394.
41. Itoh, K., Suzuki, R., Umezu, Y., Hanaumi, K., and Kumagai, K., 1982, Studies of murine large granular lymphocytes. II. Tissue, strain and age distribution of LGL and LAL, *J. Immunol.* **129**:395–400.
42. Ward, J. M., Argilan, F., and Reynolds, C. W., 1983, Immunoperoxidase localization of large granular lymphocytes in normal tissues and lesions of thymic nude rats, *J. Immunol.* **131**:132–139.
43. Zoller, M., Andrighetto, G., and Heyman, B., 1982, Natural and antibody-dependent killer cells in the thymus, *Eur. J. Immunol.* **12**:914–921.
44. Kaneda, K., Dan, C., and Wake, K., 1983, Pit cells as natural killer cells, *Biomed. Res.* **4**:567–576.
45. Lukomska, B., Olzewski, W. L., and Engeset, A., 1983, Rat liver contains a distinct blood-borne population of NK cells resistant to anti-asialo GM$_1$ antiserum, *Immunol. Lett.* **6**:277–281.
46. Tagliabue, A., Befus, A. D., Clark, D. A., and Bienenstock, J., 1982, Characteristics of natural killer cells in the murine intestinal epithelium and lamina propria, *J. Exp. Med.* **155**:1785–1796.
47. Leventon, G. S., Kulkarni, S. S., Meistrich, M. L., Newland, J. R., and Zanden, A. R., 1983, Isolation of murine small bowel intraepithelial lymphocytes, *J. Immunol. Methods* **63**:35–44.
48. Rolstad, B., Herberman, R. B., and Reynolds, C. W., 1986, Natural killer cell activity in the rat. V. The circulation patterns and tissue localization of peripheral blood large granular lymphocytes (LGL), *J. Immunol.* **136**:2800–2808.
49. Villa, C. B. F., Vecchi, A., Giavazzi, R., Introna, M., Avallone, R., and Mantovani, A., 1982, Natural cytotoxic activity in human lungs, *Clin. Exp. Immunol.* **47**:437–444.
50. Robinson, B. W. S., Pinkston, P., and Crystal, R. G., 1984, Natural killer cells are present in normal human lung but are functionally impotent, *J. Clin. Invest.* **74**:942–950.
51. Bordignon, C., Villa, F., Allavena, P., Intrana, M., Biondi, A., Avallone, R., and Mantovani, A., 1982, Inhibition of natural killer activity by human bronchial alveolar macrophages, *J. Immunol.* **129**:587–591.
52. Mann, D. W., Sonnenfeld, G., and Stein-Streilein, J., 1983, Pulmonary compartmentalization of interferon and natural killer cell activity, *Proc. Soc. Exp. Biol. Med.* **180**:224–230.
53. Bukowski, J. F., Woda, B. A., Habu, S., Okumura, K., and Welsh, R. M., 1983, Natural killer cell depletion enhances virus synthesis and virus-induced hepatitis *in vivo*, *J. Immunol.* **131**:1531–1538.
54. Bukowski, J. F., Biron, C. A., and Welsh, R. M., 1983, Elevated natural killer cell-mediated cytotoxicity, plasma interferon and tumor cells rejection in mice persistently infected with lymphocytic choriomeningitis, *J. Immunol.* **131**:991–996.

55. Biron, C. A., and Welsh, R. M., 1982, Proliferation and role of natural killer cells during viral infection, in: *NK Cells and Other Natural Effector Cells* (R. B. Herberman, ed.), Academic Press, New York, p. 493.

56. Welsh, R. M., 1978, Cytotoxic cells induced during lymphocyte choriomeningitis virus infection in mice. I. Characterization of natural killer cell induction, *J. Exp. Med.* **148:**163–181.

57. Stein-Streilein, J., Witte, P. L., Streilein, J. W., and Guffee, J., 1985, Local cellular defenses in influenza-infected lungs, *Cell. Immunol.* **95:**234–246.

58. Niederkorn, J. Y., Brieland, J. K., and Mayhew, E., 1983, Enhanced natural killer cell activity in experimental murine encephalitozoonosis, *Infect. Immun.* **41:**302–307.

59. Tarkkanen, J., Saksela, E., and Lanier, L. L., 1986, Bacterial activation of human natural killer cells. Characteristics of the activation process and identification of the effector cells, *J. Immunol.* **137:**2418–2433.

60. Wiltrout, R. H., Denn, A. C., and Reynolds, C. W., 1986, Augmentation of organ-associated NK activity by BRM's: Association of NK activity with mononuclear cell infiltration, *Pathol. Immunopathol. Res.* **5:**219–233.

61. Wiltrout, R. H., Herberman, R. B., Zhang, S. K., Chirigos, M. A., Ortaldo, J. R., Green, K. M. Jr., and Talmadge, J. E., 1985, Role of organ-associated NK cells in decreased formation of experimental metastases in lung and liver, *J. Immunol.* **134:**4267–4275.

62. Talmadge, J. E., Schneider, M., Collins, M., Phillips, H., Herberman, R. B., Wiltrout, R. H., 1985, Augmentation of NK cell activity in tissue specific sites by liposomes incorporating MTP-PE, *J. Immunol.* **135:**1477–1485.

63. Wiltrout, R. H., Talmadge, J. E., and Herberman, R. B., 1988, Biological response modifiers in augmentation of natural killer activity: Potential role in prevention and treatment of metastatic disease, *Adv. Immun.* Cancer Ther. (in press).

64. Doherty, P. C., and Korngold, R., 1983, Characteristics of poxvirus-induced meningitis: Virus-specific and non-specific cytotoxic effectors in the inflammatory exudate, *Scand. J. Immunol.* **18:**107–114.

65. Griffin, D. E., and Hess, J. L., 1986, Cells with natural killer activity in the cerebrospinal fluid of normal mice and athymic nude mice with acute sinbus virus encephalitis, *J. Immunol.* **136:**1841–1845.

66. Biron, C. A., Turgiss, L. R., and Welsh, R. M., 1983, Increase in NK cell number and turnover rate during acute viral infection, *J. Immunol.* **131:**1539–1545.

67. Natuk, R. J., and Welsh, R. M., 1987, Accumulation and chemotaxis of natural killer/large granular lymphocytes at sites of virus replication, *J. Immunol.* **138:**877–883.

68. Phillips, J. H., and Lanier, L. L., 1986, Dissection of the lymphokine activated killer phenomena: Relative contribution of peripheral blood natural killer cells and T lymphocytes to cytolysis, *J. Exp. Med.* **164:**814–825.

69. Ettinghausen, S. E., Lipford, E. H., Mule, J. J., and Rosenberg, S. A., 1985, Recombinant interleukin 2 stimulates *in vivo* proliferation of adoptively transferred lymphokine-activated killer (LAK) cells, *J. Immunol.* **135:**3623–3635.

70. Ettinghausen, S. E., Lipford, E. H., Mule, J. J., and Rosenberg, S. A., 1985, Systemic administration of recombinant interleukin 2 stimulates *in vivo* lymphoid proliferation in tissues, *J. Immunol.* **135:**4488–4497.

71. Henney, C. S., Kuribayashi, K., Kern, D. E., and Gillis, S., 1981, Interleukin-2 augments natural killer cell activity, *Nature* **291:**335–337.

72. Grimm, E. A. A., Mazumider, A., Zhang, H. Z., and Rosenberg, S. A., 1982, Lymphokine-activated killer phenomenon. Lysis of natural-killer resistant fresh solid tumor cells by interleukin 2–activated autologous human peripheral blood lymphocytes, *J. Exp. Med.* **155:**1823–1841.

73. Mule, J. J., Shu, S., Schwartz, S. L., and Rosenberg, S. A., 1984, Adoptive immunotherapy of established pulmonary metastases with LAK cells and recombinant interleukin-2, *Science* **255:**1487–1489.

74. Lafreniere, R., and Rosenberg, S. A., 1985, Successful immunotherapy of murine experimental hepatic metastases with lymphokine-activated killer cells and recombinant interleukin 2, *Cancer Res.* **45**:3735–3741.
75. Carroll, A. M., Palladino, M. A., Oettgen, H., and De Sousa, M., 1983, *In vivo* localization of cloned IL-2-dependent T cells, *Cell. Immunol.* **76**:69–76.
76. Djeu, J. Y., Stocks, N., Zoon, K., Stanton, G. J., Timonen, T., and Herberman, R. B., 1982, Positive self-regulation of cytotoxicity in human natural killer cells by production of interferon upon exposure to influenza and herpes virus, *J. Exp. Med.* **156**:1222–1234.
77. Timonen, T., Saksela, E., Virtanen, I., and Cantell, K., 1980, Natural killer cells are responsible for interferon-induced in human lymphocytes by tumor cell contact, *Eur. J. Immunol.* **10**:422–427.
78. Handa, K., Suzuki, R., Matsui, H., Shimizu, Y., and Kumagai, K., 1983, Natural killer (NK) cells as a responder to interleukin 2 (IL2). II. IL2-induced interferon-γ production, *J. Immunol.* **130**:988–992.
79. Munakata, K., Semba, U., Shibuya, Y., Kuwano, K., Akagi, M., and Arai, S., 1985, Induction of interferon-γ by human natural killer cells stimulated by hydrogen peroxide, *J. Immunol.* **134**:2449–2455.
80. Sauder, D., Mounessa, N. L., Katz, S. I., Dinarello, C. A., and Gallin, J. I., 1984, Chemotactic cytokines: The role of leukocytic pyrogen and epidermal cell thymocyte-activating factor in neutrophil chemotaxis, *J. Immunol.* **132**:828–832.
81. Miossec, P., Yu, C.-L., and Ziff, M., 1984, Lymphocyte chemotactic activity of human interleukin-1, *J. Immunol.* **133**:2007–2011.
82. Degliantoni, G., Murphy, M., Kobayashi, M., Francis, M. K., Perussia, B., and Trincheri, G., 1985, Natural killer (NK) cell-derived hematopoietic colony-inhibiting activity and NK cytotoxic factor. Relationship with tumor necrosis factor and synergism with immune interferon, *J. Exp. Med.* **162**:1512–1530.
83. Ming, W. J., Bersani, L., and Mantovani, A., 1987, Tumor necrosis factor is chemotactic for monocytes and polymorphonuclear leukocytes, *J. Immunol.* **138**:1469–1474.
84. Kasahara, T., Djeu, J. Y., Dougherty, S. F., and Oppenheim, J. J., 1983, Capacity of human large granular lymphocytes (LGL) to produce multiple lymphokines: Interleukin 2, interferon and colony-stimulating factor, *J. Immunol.* **131**:2379–2384.
85. Procopio, A. D. G., Allavena, P., and Ortaldo, J. R., 1985, Noncytotoxic functions of natural killer (NK) cells: Large granular lymphocytes (LGL) produce a B cell growth factor (BCGF), *J. Immunol.* **135**:3264–3271.
86. Greenberg, A. H., Khalil, N., Pohajdak, B., Talgoy, M., Henkart, P., and Orr, F. W., 1986, NK-leukocyte chemotactic factor (NK-LCF). A large granular lymphocyte (LGL) granule-associated chemotactic factor, *J. Immunol.* **137**:3224–3230.
87. Henkart, P. A., Millard, P. J., Reynolds, C. W., and Henkart, M. P., 1984, Cytolytic activity of purified cytoplasmic granules from cytotoxic rat large granular lymphocyte tumors, *J. Exp. Med.* **160**:75–93.
88. Henkart, P. A. (personal communication).
89. MacDermott, R. P., Schmidt, R. E., Caulfield, J. P., Hein, A., Bartley, G. T., Ritz, J., Schlossman, S. F., Austen, K. F., and Stevens, R. L., 1985, Proteoglycans in cell mediated cytotoxicity. Identification, localization and exocytosis of a chondroitin sulfate proteoglycan from human cloned natural killer cells during target cell lysis, *J. Exp. Med.* **162**:1771–1787.
90. Nencioni, K., Villa, L., Boraschi, D., Berti, B., and Taglibue, A., 1983, Natural and antibody-dependent cell-mediated activity against *Salmonella typhimurium* by peripheral and intestinal lymphoid cells in mice, *J. Immunol.* **130**:903–907.
91. Lowell, G. H., Smith, L. F., Griffiss, J. M., and Brandt, B. L., 1980, IgA-dependent, monocyte-mediated antibacterial activity, *J. Exp. Med.* **152**:452–457.
92. Lowell, G. H., Smith, L. F., Artenstein, M. S., Nash, G. S., and MacDermott, R. P.,

1979, Antibody-dependent cell-mediated antibacterial activity of human mononuclear cell. I. K-lymphocytes and monocytes are effective against meningococci in cooperation with human immune sera, *J. Exp. Med.* **150**:127–137.

93. Lowell, G. H., MacDermott, R. P., Summers, P. L., Reeder, A. A., Bertovich, M. J., and Formal, S. B., 1980, Antibody-dependent cell-mediated antibacterial activity: K lymphocytes, monocytes and granulocytes are effective against *Shigella*, *J. Immunol.* **125**:2778–2784.

94. Morgan, D. H., DuPont, H. L., Gonik, B., and Kohl, S., 1984, Cytotoxicity of human peripheral blood and colostral leukocytes against *Shigella* species, *Infect. Immun.* **46**:25–33.

95. Morgan, D. R., DuPont, H. L., Wood, L. V., and Kohl, S., 1984, Cytotoxicity of leukocytes from normal and *Shigella*-susceptible (opium-treated) guinea pigs against virulent *Shigella sonnei*, *Infect. Immun.* **46**:22–24.

96. Tagliabue, A., Nencioni, L., Villa, L., Keren, D. F., Lowell, G. H., and Boraschi, D., 1983, Antibody-dependent cell-mediated antibacterial activity of intestinal lymphocytes with secretory IgA, *Nature* **300**:184–186.

97. Tagliabue, A., Nencioni, L., Villa, L., and Boraschi, D., 1984, Genetic control of *in vitro* natural cell-mediated activity against *Salmonella typhimurium* by intestinal and splenic lymphoid cells in mice, *Clin. Exp. Immunol.* **56**:531–536.

98. Klimpel, G., Niesel, D. W., and Klimpel, K. D., 1986, Natural cytotoxic effector cell activity against *Shigella flexneri*-infected HeLa cells, *J. Immunol.* **136**:1081–1086.

99. Peters, P. M., Ortaldo, J. R., Shalaby, M. R., Svedersky, L. P., Nedwin, G. E., Bringman, T. S., Hass, P. E., Aggarwal, B. B., Herberman, R. B., Goedel, D. V., and Palladino, M. A. Jr., 1986, Natural killer cell-sensitive targets stimulate production of TNF α but not TNF β (lymphotoxin) by highly purified human peripheral blood large granular lymphocytes, *J. Immunol.* **137**:2592–2598.

100. Wright, S. C., and Bonavida, B., 1983, YAC-1 variant clones selected for resistance to natural killer cytotoxic factors are also resistant to natural killer cell-mediated cytotoxicity, *Proc. Natl. Acad. Sci. USA* **80**:1688–1692.

101. Krensky, A. M., Ault, K. A., Reiss, J., Stronunger, J. L., and Burakoff, S. J., 1982, Generation of long term human cytolytic cell lines with persistent natural killer cell activity, *J. Immunol.* **129**:1748–1752.

102. Van de Griend, R. J., Van Krinpen, B. A., Ronteltop, C. P., and Bolhuis, R. L. H., 1984, Rapidly expanded activated human killer cell clones have strong anti-tumor cell activity and have the surface phenotype of either T, T-non-T or null cells, *J. Immunol.* **132**:3185–3191.

103. Gomez, J., Pohajdak, B., O'Neill, S., Wilkins, J., and Greenberg, A. H., 1985, Activation of rat and human alveolar macrophage intracellular microbicidal activity by a preformed cytokine, *J. Immunol.* **135**:1194–1200.

104. Pohajdak, B., Gomez, J. C., Wilkins, J. A., Greenberg, A. H., 1984, Tumor-activated NK cells trigger oxidative metabolism, *J. Immunol.* **133**:2430–2436.

105. Tortakoff, A. M., and Vassalli, P., 1978, Comparative studies of nitrocellular transport of secretory proteins, *J. Cell. Biol.*, **79**:694–699.

106. Roussel, E., Talgoy, M., Henkart, P. A., and Greenberg, A. H., 1986, Stimulation of macrophage (MPH) tumoricidal activity by a cytokine in granules from the rat RNK large granular lymphocyte (LGL) tumor, in: *Sixth International Congress of Immunology*, Toronto, Canada, p. 560.

107. Klempner, M. S., Dinarello, C. A., Henderson, W. R., and Gallin, J. I., 1979, Stimulation of neutrophil oxygen-dependent metabolism by human leukocytic pyrogen, *J. Clin. Invest.* **64**:996–1102.

108. Klempner, M. S., Dinarello, C. A., and Gallin, J. L., 1978, Human leukocyte pyrogen induces the release of specific granule contents from human neutrophils, *J. Clin. Invest.* **61**:1330–1336.

Summary of Part II

One of the earliest indications that natural effector cells could have important biological functions other than immune surveillance to tumors came from the observation that nonimmune lymphocytes are able to lyse virus-infected target cells. These results led to the suggestion that natural effector cells may also serve as a first line of defense against a variety of microbial infections. Even to date, much of the strongest evidence for a nontumoricidal function of the NIS has come from experiments with viruses and virus-infected target cells. The chapters by Stein-Streilein (Chapter 3), Lopez and Fitzgerald-Bocarsly (Chapter 4), and Welsh *et al.* (Chapter 5) each review different aspects of the evidence for and against a role of the NIS in the host's protection against virus infections.

In the review by Stein-Streilein, it is clear that a strong association can be made between natural killer (NK) activity and a susceptibility to influenza infections. Influenza virus infections in the lungs induce high levels of interferon (IFN) and an increase in pulmonary NK activity. Depression of pulmonary NK activity results in no change in IFN levels but markedly increased virus titers and a decrease in survival. In addition, this association between NK activity and viral pathogenicity is not limited to the lungs but can also be seen within the cerebral spinal fluid following the intracranial injection of virus. Similar results have also been reported in the development of acute viral encephalitis,[1] suggesting that the NIS may play a very important role in limiting early viral infections of the central nervous system.

The review by Lopez and Fitzgerald-Bocarsly also points out a decisive role for the NIS in the early response to primary herpes simplex virus (HSV) infections. As in the influenza virus system, experimental animal models have also been used to successfully examine the role of the NIS in this virus system. Although it is clear that the host's overall resistance to virus is dependent on natural effector cells which are operative early after challenge with virus, we still do not know what contributions are provided to this resistance by NK cells, macrophages, PMN, or other effector cells that may produce IFN.

The review by Welsh *et al.* also points out that natural effector cells (especially NK cells) are active against some, but not all, types of experimental virus infections. The data clearly demonstrate that murine cytomegalovirus (MCMV) is effectively controlled in the early stages of infection by the NIS, whereas lymphocytic choriomeningitis virus (LCMV) is poorly or not at all affected by natural effector cells. The reason(s) for this differential effect of the NIS on various viruses are not yet clear but may relate to the susceptibility of different viruses or virus-infected target cells to natural effector mechanisms or differences in the route or kinetics of viral spread. Further study on this important question is clearly necessary. The review by Welsh *et al.* also points out that depression of NK activity prior to virus inoculation abrogates the early IFN response. Unlike with influenza virus, this early IFN response by the NIS seems to play a major role in inhibiting the development of some virus infections.

Recent experiments have also demonstrated that cells with NK activity can also effectively lyse HTLV-1 and HTLV-3 infected target cells.[2] These data suggest that natural effector cells may also play an important role in controlling the spread of HTLV-3 virus during AIDS, but the clinical relevance of this finding is still unknown.

There is also strong evidence to suggest that the NIS plays an important role in inhibiting the development of microbial infections other than viruses. In a chapter by Murphy (Chapter 7), evidence is presented that macrophages, PMN, and cells with NK activity can affect the development of some fungal infections. Recent data have also demonstrated that cells with NK activity are capable of directly inhibiting the *in vitro* growth of young spherules and endospores from *Coccidioides immitis*.[3] Although more work clearly needs to be done in this area, the available evidence suggests that the synthesized cytokines (TNF, IFN-α, Ab) may act alone or in synergy with macrophage, PMN, or NK effector cells to produce a significant antifungal response in the host. Recent *in vivo* studies in NK-deficient mice are consistent with this conclusion.[4] However, these studies further suggest that cells with NK activity play a role only in the early resistance to *Cryptococcus neoformans* infections in the blood but not at extravascular sites of infections.

The review by Albright and Albright (Chapter 6) suggests that natural effector cells may also be effective against some forms of protozoan and other extracellular and intracellular parasites. As with virus infections, infection with parasites results in a rapid synthesis of IFN and an augmentation of both NK and macrophage activities. It is fairly clear that these activated natural effector cells can kill some parasites directly (e.g., *Trypanosoma gondei*) or indirectly via the synthesis of molecules that adversely affect infected target cells (TNF, IFN-α, etc.). There is also evidence that some cytokines may affect some parasites directly (e.g., *Plasmodium* by TNF). Experiments have also demonstrated that the pretreatment of hosts with some activators of the NIS (IFN) prior to parasite inoculation can increase survival. Other organisms, including *T. musculi* and *Schistosoma*,[5] are also known to depress the NIS (NK activity), which may account for the resistance of these organisms to natural effector mechanisms.

In contrast to the above microorganisms, there is little evidence that lymphoid cells within the NIS play any direct effector role in controlling the development of bacterial infections. Some experimental data with *Shigella* and *Salmonella* do suggest that natural effector cells may contribute to the resistance against these organisms by functioning as antibody-dependent effector cells (ADCC). However, the review by Greenberg (Chapter 8) also points out the indirect immunoregulatory involvement of natural effector cells in a variety of inflammatory responses, including bacterial infections. It has been known for some time that macrophages and PMN are major components of the early inflammatory response. It is now clear that large granular lymphocytes (LGL), the major NK effector cells, are also found at many sites of inflammation. These cells chemotax to a variety of signals and are themselves a major source of chemotactic signals for macrophages and PMN. Once activated, LGL can produce a variety of cytokines (IL-1, IL-2, TNFα, IFN, NK-LCF, NK-MAF) that can markedly regulate the local development of both specific and nonspecific immune responses to microbial infections. It seems likely that this involvement of the LGL in early inflammatory responses to a variety of microorganisms is a major function of these cells.

Although it seems fairly clear that natural effector cells play an important role in the host's defense against a variety of microorganisms, many questions remain to be answered. Specifically, for most microbial infections it is not yet clear which of the components of the NIS (macrophages, NK cells, NC cells, PMN, etc.) are critical elements in the host's response. It is also not yet clear how these various effector cells are able to recognize microbial organisms or even what types of antigens are being recognized. As pointed out earlier, it is also not clear why some classes of virus, fungi, and parasites are efficiently handled by the NIS while others are completely resistant to these effector cells. And lastly, the exact mechanism(s) by which natural effector cells mediate their protection have not yet been clearly defined. The availability of established animal models and the recent advances in our ability to specifically identify and selectively modulate various arms of the NIS should help to clarify these issues.

In addition to the above issues, the question remains whether various cells within the NIS can be effectively stimulated to treat established infections. However, as in the case with tumors, most natural effector cells are seemingly ineffective against established and rapidly progressing infections. Instead, it appears that these effector cells are primarily responsible for limiting the early development of infections until more potent antigen-stimulated specific immune responses can develop (i.e., CTL, Ab).

Recently, however, it has been observed that the addition of IL-2 to unstimulated natural effector cells can result in a marked increase in the antitumor efficiency of these lymphokine-activated killer (LAK) cells (see review by Lefor *et al.*, Chapter 2). These promising results with LAK cells in tumor systems raise the question of whether such an approach using IL-2 or other lymphokines for the treatment of severe or chronic infections like CMV or *Coxsackievirus* infections[6] might be feasible.

Considering the promising potential of IL-2 therapy against tumors, this approach seems logical. However, because of the difficulties and severe side effects previously seen with IL-2 in this treatment approach, serious consideration of such adverse side effects must be considered first. As an initial step, it is imperative that the effectiveness of LAK cell therapy for microbial infections be proved in the well-established animal models. If the results in these models are positive, then trials of this potentially beneficial new forms of immunotherapy for severe or chronic infections should be considered for patients refractory to established means of therapy.

Although IL-2 or lymphokine therapy may provide a novel and effective form of therapy for established infections, the ideal scenario would be to better use the NIS to prevent the development of these infections. Since most evidence suggests a role for the NIS in the early resistance to microbial organisms and since a great deal is now known about the regulation of the NIS by biological response modifiers (BRM), we should now begin too think about the use of these agents prophylactically in high-risk patients (such as bone marrow transplant recipients, patients given high-dose chemotherapy, or individuals with a variety of immunodeficiency diseases) with the intent to prevent the development of infections. It seems possible that these immunocompromised patients, at high risk for the development of viral, fungal, or parasitic infections, may be beneficially affected by the prophylactic use of BRM to increase their levels of natural immunity. Once again, such an approach should be taken with caution and the benefits proved in animal models prior to the initiation of clinical trials.

In any case, it now seems likely that the near future may prove to be a very exciting period with regard to the NIS and microbial infections. At the very least, we should soon know much more regarding the genetics, specificity, and mechanisms by which natural effector cells control microbial infections. Optimistically, we may also use this knowledge of the NIS to help prevent or cure a variety of diseases in patients with severe or chronic microbial infections.

References

1. Griffin, D. E., and Hess, J. L., 1986, Cells with natural killer activity in the cerebrospinal fluid of normal mice and athymic nude mice with acute sindbis virus encephalitis, *J. Immunol.* **136:**1841–1845.

2. Ruscetti, F. W., Mikovits, J. A., Kalyanaraman, V. S., Overton, R., Stevenson, H., Stromberg, K., Herberman, R. B., Farrar, W. L., and Ortaldo, J. R., 1986, Analysis of effector mechanisms against HTLV-I and HTLV-III/LAV-infected lymphoid cells, *J. Immunol.* **136:**3619–3624.
3. Petkus, A. F., and Baum, L. L., 1987, Natural killer cell inhibition of young spherules and endospores of *Coccidioides immitis, J. Immunol.* **139:**3107–3111.
4. Lipscomb, M. F., Alvarellos, T., Toews, G. B., Tompkins, R., Evans, Z., Koo, G., and Kumar, V., 1987, Role of natural killer cells in resistance to *Cryptococcus neoformans* infections in mice, *Am. J. Pathol.* **128:**354–361.
5. Gastl, G. A., Feldmeier, H., Kortmann, C., Daffalla, A. A., and Peter, H. H., 1986, Human schistosomiasis: Deficiency of large granular lymphocytes and indomethacin-sensitive suppression of natural killing, *Scand. J. Immunol.* **23:**319–325.
6. Godeny, E. K., and Gauntt, C. J., 1987, Murine natural killer cells limit coxsackievirus B3 replication. *J. Immunol.* **139:**913–918.

PART III
IMMUNOREGULATORY
PROPERTIES OF THE NIS

9
Role of the Natural Immune System in the Antibody Response

Regulatory Effect of NK Cells

KATSUO KUMAGAI, SATSUKI SUZUKI, AND RYUJI SUZUKI

1. Introduction

Natural killer (NK) cells were originally described as effector cells capable of *in vitro* lysis of certain tumor targets.[1] However, concurrent with the revelation of NK cell antitumor potential, it became evident that these cells are also involved in the regulation of the growth and differentiation of hematopoietic systems.[2] This was initially demonstrated by the evidence that histoincompatible bone marrow cells would grow in the NK cell-depleted, NK cell-deficient, or NK cell–immature mice but not in the NK cell-stimulated mice.[3,4] Later, the direct involvement of NK cells in murine bone marrow transplantation was demonstrated by abrogation of resistance to parental and allogeneic bone marrow grafts in mice depleted of NK cells by NK 1.1 monocloned antibody and by restoration of bone marrow graft resistance in NK cell-depleted mice by transfer of cloned NK cells.[5]

The role of NK cells in regulation of the growth of hemopoietic tissues in man was further suggested by the evidence that NK cells displayed

KATSUO KUMAGAI, SATSUKI SUZUKI, and RYUJI SUZUKI • Department of Microbiology, To-hoku University School of Dentistry, Sendai, Japan.

inhibitory activity on the growth of granulocytic cells and erythroid precursors.[6–10] NK cells are also suggested to control the proliferation and
differentiation of thymocytes.[11–13] It also appears that proliferation and
differentiation of the peripheral T and B cells may be regulated by NK
cells. For instance, *in vitro* colony formation by T cells has been shown to
be potentiated by NK cells.[14] Murine NK cell mediated the enhancing
effect on induction of allogeneic cytotoxic T lymphocytes.[15] Human DR$^+$
NK cells have also been shown to act as effective antigen-presenting cells
(APC) for T lymphocyte activation.[16] In contrast, NK cells have been
shown to inhibit immunoglobulin (Ig) secretion and differentiation of B
cells.[17,18]

From these observations, it appears that one of the important biological functions of NK cells may be the regulation of the growth and differentiation of hemopoietic and lymphoid cells. Therefore, the NK cells may
play an important regulatory role in a variety of T cell- and B cell-mediated immune responses. This paper will present the most recent data
on the NK cell role in the regulation of B cell differentiation and antibody
response.

2. B Cell Differentiation and Antibody Response

To analyze the effect of NK cells on the B cell differentiation and the
antibody response, an understanding of the activation, proliferation, and
differentiation of B cells that lead to antibody production is necessary.
The antibody response is B cell reaction to the antigen, coordinated by T
cells and APC. The major APCs are dendritic cells and macrophages, which
process the antigen entering the body and present it in a highly immunogenic form to the T (helper) cells and B cells.[19,20] There is evidence
that the presentation of antigen to responding helper T cells is major
histocompatibility complex (MHC)-restricted—that is to say, responding
T cells only recognize antigen on the APC surface provided that both the
T cell and the APC share determinant of the MHC.[21,22] In the antibody
response system, besides helper T cells, there exist T cells that specifically
suppress antibody responses—T suppressor cells, which also recognize the
antigen in an MHC-restricted fashion.[23] However, there also exist non-
antigen-specific mechanisms of regulation of the antibody response. Some
activated T cells, for example, which can be induced by certain immune
reaction or nonspecific stimulus such as concanavalin A, suppress the
pokeweed mitogen-driven Ig secretion and antibody production to T cell-
dependent and -independent antigen.[23,24] Certain T cells activated by
nonspecific stimulus also release a factor, interleukin 2 (IL-2), that has a
nonspecific effect in amplifying the proliferation of other T cells.[25] Both
helper and suppressor T cells are affected. Macrophages can also be stim-

ulated by nonspecific stimuli such as lipopolysaccharide (LPS) and secrete a T cell-activating factor, interleukin 1 (IL-1), which acts on helper T cells and B cells. [26,27]

The helper T cells, when being stimulated with antigen and IL-2, produce soluble factors that mediate the proliferation and differentiation of B cells. These soluble factors (B cell-acting lymphokines) are functionally divided into two groups: B cell growth factor (BCGF), thought to be involved in B cell proliferation; and B cell differentiation factor (BCDF), responsible for maturation of activated B cells into Ig-secreting cells.[28,29] At least three distinct B cell-acting lymphokines have been isolated; B cell growth factor I (BCGF-I) (or B cell stimulatory factor 1, BSF-1, or interleukin 4, IL-4)[30]; B cell growth factor II (BCGF-II) (or T cell-replacing factor, TRF, or interleukin 5, IL-5)[31]; and B cell-differentiating factor (BCDF or BSF-2), [32] although their mechanisms of action have yet to be established.

The potential for regulation of antibody response exists at multiple points during the above process of B cell differentiation: antigen presentation by APCs, activation and differentiation of helper and suppressor T cells, or B cell growth and differentiation into antibody-producing cells, all of which are mediated by the antigen stimulation and the following production of APC, T, and B cell-derived soluble mediators.

3. Evidence for Regulation of Antibody Response by NK Cells

The first indication that NK cells may have a regulatory effect on the antibody response arose from the finding that Fc receptor-bearing non-T, non-B human lymphocytes (L cells) mediated the enhancing effect on antigen (keyhole limpet hemocyanin, KLH)-induced lymphocyte blastogenesis.[33] This L cell population, of which a large portion consists of NK cells, when stimulated with IgG antibody complexed to KLH, augmented blastogenesis of lymphocytes induced by KLH. On the other hand, it has been reported that although L cells when cocultured with peripheral blood mononuclear cells (PBMC) enhanced Ig synthesis and cell proliferation, those pretreated with IgG antibody-sensitized erythrocytes inhibited Ig synthesis despite promoting cell proliferation.[34] Nable et al.[35,36] also showed that a cloned NK cell line was able to lyse LPS-activated B cell blasts. This NK cell clone was also capable of suppressing LPS-induced antibody production.

Subsequently, the convincing evidence for NK cells' influence on the differentiation, proliferation, and activity of B cells in mouse and human has been reported. For instance, in the murine system, induction of high NK activity by polyinosinic and cytidylic acid in mice promotes early termination of the ongoing primary IgM antibody response.[17] In human system, purified NK cells, after activation with immune complexes, were shown to mediate suppression of pokeweed mitogen (PWM)-induced Ig secretion.[37] Our group[18] also provided the evidence showing that HNK-1 (Leu7)$^+$ large granular lymphocyte (LGL) fraction was able to suppress PWM-induced Ig secretion without overt activation. It has later been shown by Clement *et al.*[38] that the vast majority of Leu2$^+$ (suppressor/cytotoxic) T cells that suppress PWM-induced B cell differentiation have the Leu2$^+$15$^+$ phenotype and are morphologically LGL. The cells are shown to coexpress the Leu7 antigen, which is detected only on granular lymphocytes. More recently, it has been shown that human LGL that express type 3 complement receptors (CR3) and coexpress Leu2 antigens but no IgG Fc receptor (FCR) (CF3$^+$Leu2$^+$FCR$^-$) have suppressive effect on PWM-induced B cell proliferation and Ig production.[39] Purified human endogenous NK cells (LGL) were also shown to be suppressive for Ig production by lymphoblastoid B cells[39] or for Epstein-Barr virus (EBV)-induced Ig synthesis of peripheral B cells.[41]

It has since been repeatedly demonstrated in mice that endogenous NK cells are able to mediate the suppression of B cell differentiation and Ig secretion.[17,42–44] We showed that depletion of endogenous NK activity from C3H/He mice could be accomplished with intravenous (IV) injections of the rabbit antiserum to asialo GM_1 ($ASGM_1$).[42,44] NK activity was rapidly depleted *in vivo* within 12 h after administration of anti-$ASGM_1$ antiserum, almost completely disappeared for 2 days, and gradually returned to near normal levels by 72 h. Corresponding to selective depletion of $ASGM_1$$^+$ cells and NK activity by a minute amount of 1 μl antiserum, the spleen cells showed increased number of B cells and augmented mitogenic responses to B cell but not T cell mitogens (Fig. 1). No changes in the percentages of Thy1$^+$, Lyt1$^+$, or Lyt2$^+$ cells or phagocytic cells were induced in the spleen of mice injected with anti-$ASGM_1$ even at a maximum amount used, 10 μl per mouse.

Treatment of spleen cells with this antiserum in the presence of complement *in vitro* also resulted in the complete removal of NK activity, without any deterioration of T cells or their helper and suppressor subsets. These NK-depleted spleen cells showed production of PWM-driven plaque-forming cell (PFC) to the levels much higher than that of control spleen. When T cells isolated from these NK-depleted spleens were also examined for their helper activity on PFC response of purified B cells to PWM, a much higher helper activity on PWM-driven PFC production than with NK-containing T cells was obtained. In addition, when the NK cells (LGL) purified from the spleen to the levels greater than 90% by Percoll gradient centrifugation and complement-dependent killing of T cells (Thy1$^+$, Lyt1$^+$) were added into NK-depleted spleen cells, the PFC response was suppressed depending on the number of cells added. These results indicate that NK cells in the spleen of mice exhibit a suppressor property on B cells stimulated *in vitro* by PWM.

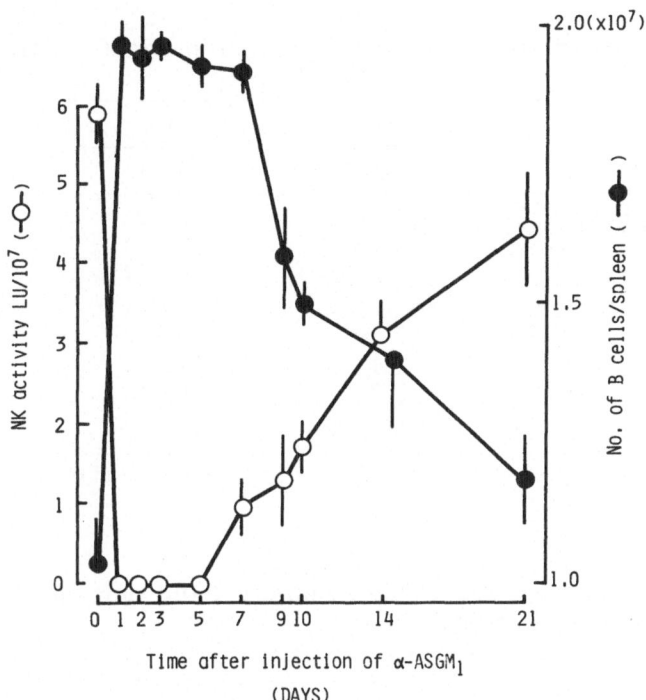

FIGURE 1. Time course of NK depletion and increase in the number of SIg^+ B cells in the spleens of mice after administration of anti-$ASGM_1$. C3H/He mice were injected intravenously with 10 μl of anti-$ASGM_1$ at the indicated time prior to sacrifice. The isolated spleen mononuclear cells were then examined for the number of SIg^+ cells by FACS analyzer and NK cytotoxicity to YAC-1 targets (LU/10^7 cells).

Antigen-specific PFC responses to sheep blood cells were also found to increase after *in vivo* depletion of NK cell activity with an alloantiserum anti-NK1.1.[43,45] NK activity is rapidly depleted *in vivo* within 4 h after administration of anti-NK1.1 antiserum and gradually returns to near normal levels by 72 h. Four hours after administration of anti-NK1.1 antiserum to deplete NK cells, animals were immunized with sheep red blood cells (SRBC). Four and 5 days later, the spleens were harvested, and the number of antigen-specific PFC was assessed. As a result, depletion of NK activity prior to immunization led to a twofold increase in the number of antigen-specific PFC. Using the mice treated with anti-$ASGM_1$, similar results were obtained (Table I).

There is also evidence from immunochemical studies that cells that stain with NK cell-specific monoclonal HNK1 or Leu7 are preferentially localized in the B cell area of lymphoid tissue.[46-52] This close juxtaposition of NK cells and B cells in lymphoid tissue, together with the inhibi-

TABLE I
Augmentation of Antibody (PFC) Production to SRBC in Mice Treated with Anti-ASGM₁ᵃ

Mice	No. of mice	No. of PFC/10^6 spleen cells	
		IgM	IgG
Anti-ASGM₁-treated	1	575 ± 75	393 ± 123
	2	500 ± 50	2557 ± 165
	3	725 ± 125	410 ± 83
	Mean ± SD	600 ± 94	1120 ± 1016
Control	1	250 ± 50	228 ± 69
	2	363 ± 88	341 ± 41
	Mean ± SD	307 ± 57	285 ± 57

ᵃMice were injected IV with 10 μl of anti-ASGM₁ and 24 h later immunized IP with SRBC (1×10^7). Assays were carried out on day 7.

tory effects discussed above, lends support to the view that NK cells may play some regulatory role *in vivo*.

4. Cellular Interaction in NK Regulation of B Cell Differentiation and Ig Secretion

The means whereby NK cells recognize B cells and subsequently modify their function has yet to be established. Their influence might be direct or indirect or both. Direct mechanisms would depend on NK cells' being able to recognize B cells, and a number of suggestions have been made about how they might do so. For example, Brieva et al.[40] showed that human (Leu7⁺) NK cells might act directly on human lymphoblastoid B cells producing antitetanus toxid antibody to suppress the antibody production. It has also been suggested that lymphoblastoid B cells express the transferrin receptor at a stage of the *in vitro* culture and that this structure may be involved in the NK-mediated inhibition of antibody response.[53]

With respect to this point, Storkus and Dawson[54] demonstrated that the expression of an NK-susceptible phenotype occurs at a late stage of B cell development, using B cell lines at different stages of B cell differentiation. It has also been shown that the intensity of a transferrin receptor antigen 4F2 expression appeared to correlate well with NK sensitivity on both resting and differentiated B cell lines. On the other hand, Ritchie et al.[49] proposed that NK cells might interact with Ig D molecules on B cell surfaces and that this interaction may influence the maturation of antibody affinity. Massucci et al.[55] and others[56] showed that human NK cells inhibited outgrowth of autologous Epstein-Barr virus (EBV)-infected B lymphocytes. Targan et al.[57] also strongly suggested that the NK lytic process itself is involved in the suppression of lymphoblastoid B cell function. Kuwano et al.[41] showed that human NK cells suppressed both IgM and IgG synthesis induced by EBV. Interferon-α (IFNα) produced was required for the NK-mediated suppression of Ig synthesis, suggesting that NK cells display an interaction with EBV-infected B cells and produce IFNα, which in turn activates NK cells, and these activated NK cells suppress the Ig synthesis by B cells, which undergo transformation induced by EBV.

If NK cells' influence on B cell differentiation or activity is indirect, two possible mechanisms have been proposed. One postulate, Arai and our group proposed, is that unactivated NK cells interfere with T helper activity.[18] Human peripheral NK cells suppressed PWM-driven Ig production by B cell in the presence of helper T4$^+$ T cells; however, they showed no inhibition of soluble factor-induced B cell differentiation assayed in the absence of helper T cells.[18] An alternative suggestion is that NK cells induced after immunization may regulate both the primary and the anamnestic response by interacting with a target that is expressed on accessory cells after their association with antigen.[17] The authors also showed that dendritic cells (DC) that have interacted with antigen are targets for NK cells, and only Thy1$^-$ ASGM$_1^+$ NK cells suppress the antibody response by suppressing or eliminating these antigen-pulsed DC.[58,59]

5. Production of Cytokines by NK Cells and Their Role in the Regulation of Antibody Response

Both murine and human NK cells are heterogeneous and include cytotoxic cells together with other subsets with different functions.[60–63] A number of specialized NK (LGL) subsets can be identified through the expression of different surface markers. For example, studies in our laboratories have shown that a murine LGL population, with a strong NK activity, on which the ASGM$_1$ and Ly5 antigens were expressed at higher quantities, responded directly to IL-2 and produced IFNγ, whereas another subset of LGL, on which ASGM$_1$ and Ly5 were expressed at lower quantities, responded to various stimuli and secreted IL-2[64–66] (unpublished observation). In the human LGL, the T3$^-$Leu7$^-$Leu11$^+$ subset with a high NK activity could produce IFNγ in response to IL-2.[67,68] The same subset could also produce IL-1 and IFNγ in response to gram-positive bacteria.[69] On the other hand, a major population that produced IL-2 in response to mitogens was another Leu11$^-$ subset without NK activity (unpublished observation). Studies in other laboratories have also shown that IL-1-producing LGL are HLA-DR$^+$M$_1^+$B83.1$^+$ (Leu11$^+$) with an NK activity[69] and those that produce IL-2 are HLA-DR$^+$T11$^{+(71)}$. They have also shown that highly purified human LGL, depleted of any detectable contaminant T and B cells or monocytes, were found to be potent producers *in vitro* of BCGF able to sustain proliferation of B cells activated by anti-μ.[72,73] The BCGF (probably IL-4)-producing cells are LGL with a phenotype of Leu-7$^+$M$_1$$^+$HLA-DR$^-E^-$ Leu11$^-$T3$^-$ and without NK activity[72] or 3G8$^+$, HNK1$^+$/OKT11$^+$, DR$^-$, OKT3$^-$, Leu-M$_1$$^-$.[73] These findings indicate that the LGL with the ability to produce cytokines may not necessarily be identical to the ones that have the NK activity.

It has been suggested that the activity of NK cells to produce various cytokines may be subject to autoregulation. Thus, NK subset(s) responding to some stimuli, produce cytokines, which in turn stimulate another NK (LGL) subset(s) to become activated cells. Our recent investigations have revealed that the human Leu7$^+$Leu11$^-$ LGL subset responds to PHA and IL-1 and produces IL-2, which can, in turn, stimulate a T3$^-$ Leu7$^-$Leu11$^+$ subset to become activated killer (AK) or lymphokine-activated killer (LAK) cells, which have the ability to kill a variety of freshly isolated tumor cells and tumor cell lines insensitive to NK lysis[67,74,75] (unpublished observations). Murine NK cells in response to IL-2 also develop into activated cells cytotoxic for tumor cell targets either sensitive or insensitive to NK lysis of syngeneic and allogeneic tumor origins. In these cases, IFNγ produced by NK cells played a role as a triggering signal in differentiation of NK cells to AK or LAK cells.[67,74] Activated NK (LAK) cells were found to have the ability to kill mitogen-induced activated T cells, in addition to tumor cells.[75]

These activated NK cells may play a critical role in the regulation of T and B cell-mediated immune responses, mediating through cytocidal effect on different target cells. Poly I:C-activated NK cells were found to have the ability to suppress T lymphocyte proliferation in mixed lymphocyte reaction (MLF) cultures, acting on dendritic antigen-presenting cells.[58] It is also possible that B cells at late stages of differentiation are killed by these activated NK cells, as discussed by Storkus and Dawson.[54] On the other hand, it seems likely that autoactivation of NK cells or production of cytokines such as IL-1, IL-2, or IFN by NK cells may influence the processes of antibody response, mediating through their influence on production of cytokines by T and B cells during immune responses. We have examined the effect of NK depletion by anti-ASGM$_1$ in mice on cytokine production by spleen cells in response to PWM stimulation. Spleen cells from anti-ASGM$_1$-treated mice were cultured in the presence of 10 μl PWM and tested for production of IL-2, IL-3,[76] and BSF-1 (IL-4) in the culture supernatants (Fig. 2). Spleen cells from mice treated with anti-ASGM$_1$, in which PWM-driven Ig secretion was increased, generated IL-2 to the much lower levels than that of controls. IFN and IL-3 productions were also impaired. However, these spleen cells responded to PWM with production of IL-4 to levels similar to those of controls. These NK-

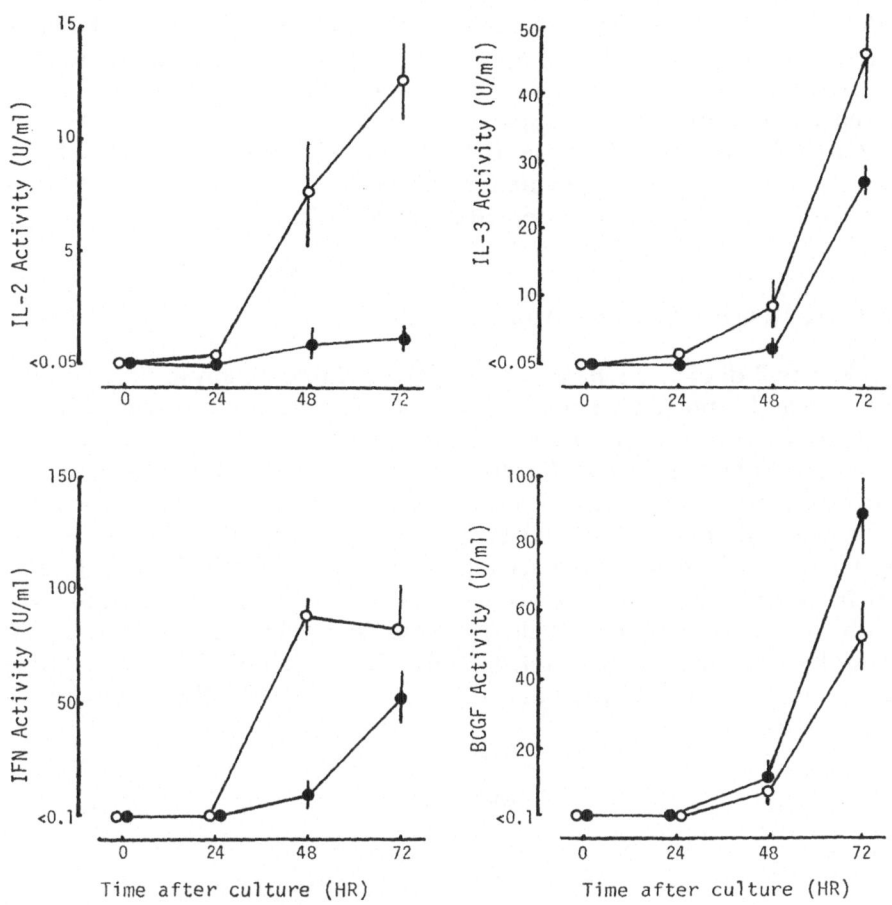

Figure 2. Suppression of PWM-induced production of IL-2 and IFN in NK-depleted spleen cells. C3H/He mice were injected intravenously with 10 μl of ASGM$_1$ antiserum (treated group) and saline containing 20% normal rabbit sera (control group). Twenty-four hours later, spleen cells were isolated from the treated (●) and control (○) groups, cultured in the presence of 1% PWM for 72 h, and then examined for productions of IL-2, IFN, Il-3, and BCGF (IL-4) during culture in the culture fluids.

depleted spleen cells also exhibited the lower responses to Con A in production of IL-2, IFN, and IL-3 but not IL-4.

The mechanisms by which NK depletion resulted in reduced production of IL-2, IFN, and IL-3 but not IL-4 among the spleen lymphocyte cultures remain unclear. In the antibody response, however, production of BSF-1 (IL-4) may be a critical requirement for helper T cell function because of its multiple functions to stimulate growth of B cells and their differentiation like surface Ia antigen expression.[77,78] Although other soluble factors, IL-2, or IFN may also be required for T cell-dependent B cell activation or differentiation,[25,79,80] both factors are known to be involved in the induction of suppressor T cells in a variety of immune responses.[25,81] Selective decrease of IL-2 and IFN, but not IL-4, might be causally related to the augmented B cell proliferation and Ig production induced by PWM in the NK-depleted spleen cells.

6. Conclusions and Summary

Figure 3 depicts a scheme for a cell-to-cell interaction in the antibody response and a possible regulatory role of NK cells in the reaction. In the antibody response there exists a linear cell interaction between antigen-presenting cells (APC) and T cells, between T cells and T cells, and also between T cells and B cells. APCs process the antigens entering the body and present them to the T and B cells. APCs also produce T cell-activating factor, IL-1, to stimulate activation of helper T cells. Helper T cells, which recognize antigens, are activated to secrete T cell growth factor, IL-2, which in turn stimulates helper T cells themselves to produce B cell-acting factors such as BSF-1 (IL-4), TRF (IL-5), or BSF-2 (BCDF). These factors mediate proliferation and differentiation of B cells, which are co-

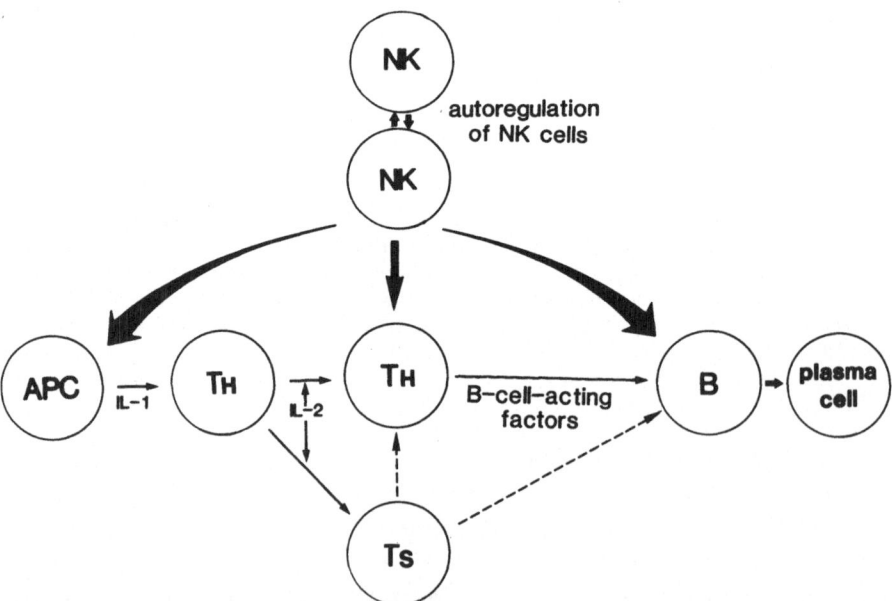

FIGURE 3. Schematic representation of suppression of B-cell differentiation by NK cells. APC, antigen-presenting cell; TH, helper T cell; Ts, suppressor T cell; B, B cell; ————→NK cell- mediated suppression; ---------→ Ts-mediated suppression.

stimulated by the antigens. NK cells appear to suppress the antibody response at the beginning of the reaction such as antigen presentation by the APC to the helper T cells or subsequent activation of helper T cells. It also seems likely that B cells, which are differentiating into the antibody-producing cells, are a target for NK suppression. Thus, NK cells recognize B cells during maturation in some way but are not antigen-specific, and they directly suppress their growth or differentiation. These both indirect and direct influences on the B cell functions may be involved in NK regulation of antibody response.

The mechanisms by which NK cells affect the APC, T cell, or B cell functions remain to be established. Cytokines secreted from NK cells and activation of NK cells mediated by the cytokines produced may be involved in both the direct and indirect influences of NK cells on the antibody response. Thus, NK subset(s) stimulated during antibody response reaction produce IL-1, IL-2, IL-4, or IFN, which in turn stimulate another NK subset to become activated cells. These activated NK cells play a major role in the NK regulation of antibody response. The cytokines produced by NK cells may also play a role in the regulation of antibody response, acting on APCs, T cells, or directly on B cells.

It is clear from the work reviewed here that NK cells are an important lymphoid cell population capable of influencing the antibody response of B cells. Now that this regulatory effect of NK cells has been firmly established, the next step will be to analyze the further mechanisms by which NK cells or their soluble products may control APC, T cell, and B cell functions involved in the regulation of antibody response. Several autoimmune diseases such as active lupus erythematosus (SLE) or rheumatoid arthritis, in which suppressor function on antibody production is impaired, show reduced NK activity.[82–85] Decreased NK activity may be causally related to the development of such immune disorders with abnormal Ig production. Further studies on this effect of NK cells in the regulation of the antibody response may provide valuable information on the immunopathogenesis of autoimmune diseases.

References

1. Herberman, R. B., Djeu, J. Y., Kay, H. D., Ortaldo, J. R., Riccardi, C., Bonnard, G. D., Holden. H. T., Tagnani, R., Santoni, A. S., and Puccetti, P., 1979, Natural killer cells: Characteristics and regulation of activity, *Immunol. Rev.* **44**:43–70.
2. Lotzová, E., 1986, NK cell role in regulation of the growth and functions of hematopoietic and lymphoid cells, in: *Immunobiology of Natural Killer Cells*, Volume II (E. Lotzová and R. B. Herberman, eds.), CRC Press, Boca Raton, FL, pp. 89–105.
3. Kiessling, R., Hochman, P. S., Häller, O., Shearer, G. M., Wigzell, H., and Cudkowicz, G., 1977, Evidence for a similar or common mechanism for natural killer cell activity and resistance to hematopoietic grafts, *Eur. J. Immunol.* **7**:655–663.
4. Lotzová, E., and Savary, C. A., 1977, Possible involvement of natural killer cells in bone marrow graft rejection, *Biomedicine* **27**:341.

5. Lotzová, E., Savary, C. A., and Pollack, S. B., 1983, Prevention of rejection of allogeneic bone marrow transplantation by Nk 1.1 antiserum, *Transplantation* **35:**490–494.
6. Mangan, K. F., Chikkappa, G., Bieler, L. Z., Scharfman, W. B., and Parkinson, D. R., 1982, Regulation of human blood erythroid burst forming unit (BFU-E) proliferation by T lymphocyte subpopulations defined by Fc receptors and monoclonal antibodies, *Blood 59:*990–996.
7. Mangan, K. F., Chikkappa, G., and Farley, P. C., 1982, T gamma (Tγ) cells suppress growth of erythroid colony-forming units *in vitro* in the pure red cell aplasia of B cell chronic lymphocytic leukemia, *J. Clin. Invest.* **70:**1148–1156.
8. Barr, R. D., and Stevens, C. A., 1982, The role of autologous helper and suppressor T cells in the regulation of human granulopoiesis, *Am. J. Hematol.* **12:**323–326.
9. Hansson, M., Beran, M., Andersson, B., and Kiessling, R., 1982, Inhibition of *in vitro* granulopoiesis by autologous and allogeneic human NK cells, *J. Immunol.* **129:**126–132.
10. Mangan, K. F., Hartnett, M. E., Matis, S. A., Winkelstein, A., and Abo, T., 1984, Natural killer cells suppress human erythroid stem cell proliferation *in vitro*, *Blood* **63:**260–269.
11. Ono, A., Amos, D. E., and Koren, H. S., 1977, Selective cellular natural killing against human leukemic T cells and thymus, *Nature* **266:**546–548.
12. Hansson, M., Kiessling, R., Andersson, B., Kärre, K., and Roder, J., 1979, NK cell–sensitive T-cell subpopulation in thymus: Inverse correlation to host NK activity, *Nature* **278:**174–176.
13. Hansson, M., Kiessling, R., and Andersson, B., 1981, Human fetal thymus and bone marrow contain target cells for natural killer cells, *Eur. J., Immunol.* **11:**8–12.
14. Pistoia, U., Nocera, A., Ghio, R., Leprini, A., Perata, A., Pistone, H., and Ferravini, M., 1983, PHA-induced human T cell colony formation: Enhancing effect of large granular lymphocytes, *Exp. Hematol.* **11:**249–253.
15. Suzuki, R., Suzuki, S., Ebina, N., and Kumagai, K., 1985, Suppression of alloimmune cytotoxic T lymphocytes (CTL) generation by depletion of NK cells and restoration by interferon and/or interleukin 2, *J. Immunol.* **134:**2139–2148.
16. Scala, G., Allavena, P., Ortaldo, J. R., Herberman, R. B., and Oppenheim, J. J., 1985, Subsets of human large granular lymphocytes (LGL) exhibit accessory cell functions, *J. Immunol.* **134:**3049–3055.
17. Abrruzo, L. B., and Rowley, D. A., 1983, Homeostasis of the antibody response: Immunoregulation by NK cells, *Science* **222:**581–585.
18. Arai, S., Yamamoto, H., Itoh, K., and Kumagai, K., 1983, Suppressive effect of human natural killer cells on pokeweed mitogen-induced B cell differentiation, *J. Immunol.* **131:**651–657.
19. Beller, D. I., and Unanue, E. R., 1980, Ia antigens and antigen-presenting function of thymic macrophages, *J. Immunol.* **124:**1433–1440.
20. Inaba, K., Steinman, R. M., Van Voorhis, W. D., and Muramatsu, S., 1983, Dendritic cells are critical accessory cells for thymus-dependent antibody responses in mouse and in man, *Proc. Natl. Acad. Sci USA* **80:**6041–6045.
21. Beller, D. I., 1983, Functinal significance of the regulation of macrophage Ia expression, in: *Progress in Immunology* (Y. Yamamura and T. Tada, eds.), Academic Press, Tokyo, pp. 985–988.
22. Farr, A. G., Kiely, J.-M., and Unaue, E. R., 1979, Macrophage–T cell interactions involving *Listeria monocytogenes*—role of the H-2 gene complex, *J. Immunol.* **122:**2395–2404.
23. Sorensen, C. M., and Pierce, C. W., 1981, Haplotype-specific suppression of antibody responses *in vitro*. I. Generation of genetically restricted suppressor T cells by neonatal treatment with semiallogeneic spleen cells, *J. Exp. Med.* **154:**35–47.
24. Maier, T., Holda, J. H., and Claman, H. N., 1985, Graft-vs-host reactions (GVHR) across murine histocompatibility barriers. II. Development of natural suppressor cell activity, *J. Immunol.* **135:**1644–1651.
25. Farrar, J. J., Benjamin, W. R., Hilfiker, M. L., Howard, M., Farrar, W. L., and Fuller-

Farrar, J., 1982, The biochemistry, biology, and role of interleukin 2 and the induction of cytotoxic T cell and antibody-forming cell response, *Immunol. Rev.* **63:**129–166.

26. Scala, G., and Oppenheim, J. J., 1983. Antigen presentation by human monocytes: Evidence for stimulant processing and requirement for interleukin 1, *Nature* **309:**56–59.

27. Pike, B. L., and Nossal, G. J. V., 1985. Interleukin 1 can act as a B-cell growth and differentiation factor, *Proc. Natl. Acad. Sci. USA* **82:**8153–8157.

28. Swain, S., Howard, M., Kappler, J., Marrack, P., Watson, J., Booth, R., and Dutton, R., 1983, Evidence for two distinct classes of murine B cell growth factors with activities indifferent functional assays, *J. Exp. Med.* **158:**822–835.

29. Howard, M., Farrar, J., Hilfiker, M., Johnson, B., Takatsu, K., Hamaoka, K., and Paul, W. E., 1982. Identification of a T cell-derived B cell growth factor distinct from interleukin 2, *J. Exp. Med.* **155:**914–923.

30. Lee, F., Yakota, T., Otsuka, T., Meyerson, P., Villaret D., Coffman, R. L., Mosmann, T., Rennick, D., Riehen, N., Smith, C., Zlotnik, A., and Arai, K., 1986, Isolation and characterization of a mouse interleukin cDNA clone that expresses B-cell stimulatory factor 1 activities and T-cell- and mast-cell-stimulating activities, *Proc. Natl. Acad. Sci. USA* **83:**2061–2065.

31. Kinashi, T., Harada, N., Seversinson, E., Tanabe, T., Sideras, P., Konishi, M., Azuma, C., Tominaga, A., Bergstedt-Lindgvist, S., Takahashi, M., Matsuda, F., Yaoita, Y., Takatsu, K., and Honjo, T., 1986, Cloning of complementary DNA encoding T-cell replacing factor and identity with B-cell growth factor II, *Nature* **324:**70–73.

32. Hirano, T., Yasukawa, K., Harada, H., Taga, T., Watanabe, Y., Matsuda, T., Kashiwamura, S., Nakajima, K., Koyama, K., Iwamatsu, A., Tsunasawa, S., Sakiyama, F., Matsui, H., Takahara, Y., Taniguchi, T., and Kishimoto T., 1987, Complementary DNA for a novel human interleukin (BSF-2) that induces B lymphocytes to produce immunoglobulin, *Nature* **324:**73–76.

33. Carbalho, E. M., and Horwitz, D. A., 1980, Characterization of a non-T, non-B human blood lymphocyte that mediates the enhancing effects of immune complexes on lymphocyte blastogenesis, *J. Immunol.* **124:**1656–1661.

34. Lobo, P. I., 1981, Characterization of a non-T, non-B human lymphocyte (L cell) with use of monoclonal antibodies, *J. Clin. Invest.* **68:**431–438.

35. Nabel, G., Bucalo, L. R., Allard, J., Wigzell, H., and Cantor, H., 1981, Multiple activities of a cloned cell line mediating natural killer cell function, *J. Exp. Med.* **153:**1582–1591.

36. Nabel, G., Allard, W. J., and Cantor, H., 1982, A cloned cell line mediating natural killer cell function inhibits immunoglobulin secretion, *J. Exp. Med.* **156:**658–663.

37. Tilden, A. B., Abo, T., and Balch, C. M., 1983, Suppressor cell function of human granular lymphocytes identified by the NHK-1 (Leu 7) monoclonal antibody, *J. Immunol.* **130:**1171–1175.

38. Clement, L. T., Gross, C. E., and Gartland, G. L., 1984, Morphologic and phenotypic features of the subpopulation of Leu-2[+] cells that suppress B cell differention, *J. Immunol.* **133:**2461–2468.

39. Abo, W., Gray, J. D., Bakke, A. C., and Horwitz, D. A., 1987, Studies on human blood lymphocytes with iC3b (type 3) complement receptors. II. Characterization of subsets which regulate pokeweed mitogen-induced lymphocyte proliferation and immunoglobulin synthesis, *Clin. Exp. Immunol.* **67:**544–555.

40. Brieva, J. A., Targan, S., and Stevens, R. H., 1984, NK and T cell subsets regulate antibody production by human *in vivo* antigen-induced lymphoblastoid B cells, *J. Immunol.* **122:**611–615.

41. Kuwano, K., Arai, S., Munakata, T., Tomita, Y., Yoshitake, T., and Kumagai, K., 1986, Suppressive effect of human natural killer cells on Epstein-Barr virus-induced immunoglobulin synthesis, *J. Immunol.* **137:**1462–1467.

42. Suzuki, S., Suzuki, R., Onta, T., and Kumagai, K., 1984. Suppression of B cell differentiation by NK cells in mice, in: *Natural Killer Activity and Its Regulation* (T. Hoshino, H. S. Koren, and A. Uchida, eds.), Excerpta Medica, Amsterdam, pp. 296–300.

43. Robles, C. P., Pereira, P., Wortley, P., and Pollack, S. B., 1985, Regulation of the B cell response by NK cells, in: *Mechanisms of Cytotoxicity by NK Cells* (R. B. Herberman and D. M. Callewaert, eds.), Academic Press, New York, pp. 499–506.
44. Suzuki, S., Suzuki, R., Onta, T., and Kumagai, K., 1986. Suppression of B-cell differentiation by natural killer (asialo GM_1^+) cell in mice, *Nat. Immun. Cell Growth Regul.* 5:75–89.
45. Robles, C. P., and Pollack, S. B., 1986, Antibody responses and regulation, *Nat. Immun. Cell Growth Regul.* 5:64–74.
46. Banerjee, D., and Thibert, R. F., 1983, Natural killer-like cells found in B-cell compartments on human lymphoid tissues, *Nature* 304:270–272.
47. Poppema, S., Visser, L., and De Leij, L., 1983, Reactivity of presumed anti-natural killer cell antibody Leu 7 with intrafollicular T lymphocytes, *Clin. Exp. Immunol.* 54:834–837.
48. Mori, S., Mohri, N., Morita, H., Yamaguchi, K., and Shimamine, T., 1983, The distribution of cells expressing a natural killer cell marker (HNK-1) in normal human lymphoid organs and malignant lymphoma, *Virchows Arch. (Cell. Pathol.)* 43:253–263.
49. Ritchie, A. W. A., James, K., and Micklem, H. S., 1983, The distribution in human lymphoid tissue and possible significance of cells identified by the monoclonal antibody HNK-1, *Clin. Exp. Immunol.* 51:439–447.
50. Si, L., and Witeside, T. L., 1983, Tissue distribution of human NK cells with anti-Leu-7 monoclonal antibody, *J. Immunol.* 130:2149–2155.
51. Tanaka, H., Takasaki, S., Muroya, T., Suzuki, T., and Ishikawa, E., 1984, The distribution and possible significance of natural killer cells (HNK-1$^+$ cells) in the human spleen, *Acta Histochem. Cytochem.* 17:339–358.
52. Pizzolo, G., Smenzato, G., Chilosi, M., Morittu, L., Ambrosetti, A., Warner, N., Bofill, M., and Hanossy, G., 1984, Distribution and heterogeneity of cells detected by HNK-1 monoclonal antibody in blood and tissues in normal, reactive and neoplastic conditions, *Clin. Exp. Immunol.* 57:195–206.
53. Brieva, J. A., and Stevens, R. H., 1984, Involvement of the transferrin receptor in the production and NK-induced suppression of human antibody synthesis, *J. Immunol.* 133:1288–1292.
54. Storkus, W. J., and Dawson, J. R., 1986, B cell sensitivity to natural killing: Correlation with target cell stage of differentiation and state of activation, *J. Immunol.* 136:1542–1547.
55. Massucci, M. G., Bejarano, M. T., Massucci, G., and Klein, E., 1983, Large granular lymphocytes inhibit the *in vitro* growth of autologous Epstein-Barr-virus infected B cells, *Cell. Immunol.* 76:311–321.
56. Kaplan, J., and Shope, T. C., 1985, Natural killer cells inhibit outgrowth of autologous Epstein-Barr virus-infected B cells, *Nat. Immun. Cell Growth Regul.* 4:40–47.
57. Targan, S., Brieva, J., Newman, W., and Stevens, R., 1985, Is the NK lytic process involved in the mechanism of NK suppression of antibody-producing cells? *J. Immunol.* 134:666–669.
58. Shah, P. D., Gilbertson, S. M., and Rowley, D. A., 1985, Dendritic cells that have interacted with antigen are targets for natural killer cells, *J. Exp. Med.* 162:625–636.
59. Shah, P. D., Keÿ, J., Gilbertson, S. M., and Rowley, D. A., 1986, Thy-1$^+$ and Thy-1$^-$ natural killer cells. Only Thy-1$^-$ natural killer cells suppress dendritic cells, *J. Exp. Med.* 163:1012–1017.
60. Burns, G. F., Begley, C. G., Mackay, I. R., Triglia, T., and Werkmeister, J. A., 1985, Supernatural killer cells, *Immunol. Today* 6:370–373.
61. Grossman, Z., and Herbermen, R. B., 1986, Natural killer cells and their relationship to T-cells: Hypothesis on the role of T-cell receptor gene rearrangement on the course of adaptive differentiation, *Cancer Res.* 46:2651–2658.
62. Lanier, L. L., Phillips, J. H., Hackett, J., Tutt, M., and Kumar, V., 1986, Opinion—

Natural killer cells: Definition of a cell type rather than a function, *J. Immunol.* **137**:2735–2739.

63. Ortaldo, J. R., and Reynolds, C. W., 1987, Natural killer activity: Definition of a function rather than a cell type, *J. Immunol.* **138**:4545–4546.

64. Suzuki, R., Handa, K., Itoh, K., and Kumagai, K., 1983, Natural killer cells as a responder to interleukin 2 (IL2). I. Proliferative response and establishment of cloned cells, *J. Immunol.* **130**:981–987.

65. Handa, K., Suzuki, R., Matsui, H., Shimizu, Y., and Kumagai, K., 1983, Natural killer (NK) cells as a responder to interleukin 2 (IL-2). II. IL-2 induced interferon γ production, *J. Immunol.* **130**:988–992.

66. Kumagai, K., Suzuki, R., Suzuki, S., and Arai, S., 1985, Immunoregulatory effects of NK cells, In: *Mechanisms of Cytotoxicity* (R. B. Herberman and D. M. Callewaert, eds.), Academic Press, Orlando, FL, pp. 489–498.

67. Itoh, K., Shiiba, K., Shimizu, Y., Suzuki, R., and Kumagai, K., 1985, Generation of activated killer cells by recombinant interleukin 2 (rIL2) in collaboration with interferon γ (IFNγ), *J. Immunol.* **134**:3124–3129.

68. Kumagai, K., Suzuki, R., Itoh, K., Shiiba, K., Ebina, N., and Igarashi, M., 1986, Interleukin-2-induced differentiation of natural killer cells to activated killer cells, in: *Natural Immunity, Cancer and Biological Response Modification* (E. Lotzova and R. B. Herberman, eds.), Karger, Basel, pp. 131–142.

69. Abo, T., Sugawara, S., Amenomori, A., Itoh, H., Rikiishi, H., Moro, I., and Kumagai, K., 1986, Selective phagocytosis of gram-positive bacteria and interleukin 1-like factor production by a subpopulation of large granular lymphocytes, *J. Immunol.* **136**:3189–3197.

70. Scala, G., Allavena, P., Djeu, J. Y., Kasahara, T., Orgaldo, J. R., Herberman, R. B., and Oppenheim, J. J., 1984, Human large granular lymphocytes (LGL) are potent producers of interleukin 1, *Nature* **309**:56–59.

71. Kasahara, T., Djeu, Y. Y., Dongherty, S. F., and Oppenheim, J. J., 1983, Capacity of human large granular lymphocytes (LGL) to produce multiple lymphokines: Interleukin 2, interferon, and colony stimulating factor, *J. Immunol.* **131**:2379–2385.

72. Pistoia, V., Cozzolino, F., Torcia, M., Castigli, E., and Ferrarini, M., 1985, Production of B cell growth factor by a Leu-7[+], OKM₁[+] non-T cell with the features of large granular lymphocytes (LGL), *J. Immunol.* **134**:3179–3184.

73. Procopio, A. D. G., Allavena, P., and Ortaldo, J. R., 1985, Noncytotoxic functions of natural killer (NK) cells: Large granular lymphocytes (LGL) produce a B cell growth factor (BCGF), *J. Immunol.* **135**:3264–3271.

74. Itoh, K., Tilden, A. B., Kumagai, K., and Balch, C. M., 1985, Leu11[+] lymphocytes with natural killer (NK) activity are precursors of recombinant interleukin 2 (rIL2)-induced activated killer (AK) cells, *J. Immunol.* **134**:802–807.

75. Shiiba, K., Suzuki, R., Kawakami, K., Ohuchi, A., and Kumagai, K., 1986, Interleukin 2–activated killer cells: Generation in collaboration with interferon γ and its suppression in cancer patients, *Cancer Immunol. Immunother.* **21**:119–128.

76. Palacios, R., Henson, G., Steinmetz, M., and McKearn, J. P., 1984, Interleukin-3 supports growth of mouse pre-B-cell clones *in vivo*, *Nature* **309**:126–131.

77. Finkelman, F. D., Katona, I. M., Urban, J. F. Jr., Snapper, C. M., Ohara, J., and Paul, W. E., 1986, Suppression of *in vivo* polyclonal IgE responses by monoclonal antibody to the lymphokine B cell-stimulatory factor 1, *Proc. Natl. Acad. Sci. USA* **83**:9675–9678.

78. Killar, L., MacDonald, G., West, J., Woods, A., and Bottomly, K., 1987, Cloned, Ia-restricted T cells that do not produce interleukin 4 (IL-4)/B cell stimulatory factor 1 (BSF-1) fail to help antigen-specific B cells, *J. Immunol.* **138**:1674–1679.

79. Leibson, H. J., Gofter, M., Zlotnik, A., Marrack, P., and Kappler, J. W., 1984, Role of γ-interferon in antibody-producing responses, *Nature* **309**:799–801.

80. Sidman, C. L., Marxhall, J. D., Schultz, L. D., Gray, P. W., and Johnson, M. M., 1984,

γ-Interferon is one of several direct B cell-maturating lymphokines, *Nature* **309**:801–804.

81. Aune, T. M., and Pierce, C. W., 1982, Activation of a suppressor T-cell pathway by interferon, *Proc Natl. Acad. Sci. USA* **79**:3808–3812.

82. Karsh, J., Dorval, G., and Osterland, C. K., 1981, Natural cytotoxicity in rheumatoid arthritis and systemic lupus erythematosus, *Clin. Immunol. Immunopathol.* **19**:437–446.

83. Itoh, K., Saitoh, F., Kumagai, K., and Kosaka, S., 1981, Depressed natural killer activity ' in rheumatoid arthritis and its *in vitro* augmentation with interferon and N-(2-carboxyphenyl)-4-chloroanthranilic acid sodium salt (CCA), an anti-arthritis agent, *Ryumachi* **21**:69–74.

84. Goto, M., Tanimoto, K., and Horiuchi, Y., 1980, Natural cell-mediated cytotoxicity in systemic lupus erythematosus, *Arthritis Rheum* **23**:1274–1281.

85. Sibbitt, W. L., Mathews, P. M., and Bankhurst, A. D., 1983, Natural killer cells in systemic lupus erythematosus. Defects in effector lytic activity and response to interferon and interferon inducers, *J. Clin. Invest.* **71**:1230–1239.

10

Role of Natural Effector Cells in the Regulation of Cell-Mediated Immune Responses

ARABELLA B. TILDEN and LORAN T. CLEMENT

1. Introduction

The natural immune system is composed of host defense mechanisms that do not appear to require sensitization and lack antigen specificity and MHC restriction. In contrast, the immune system mounts responses to foreign materials that are specific for the antigens encountered and exhibit a memory or enhanced response upon reexposure to the agent. Although the classical immune system and the natural immune system differ fundamentally in these respects, it is now clear that these systems nonetheless interact with or influence one another. For example, a number of lymphokines are produced by T lymphocytes, in the course of an immunological response to antigen, which have significant effects on the functional capabilities of natural effector cells (this is the subject of Chapter 2). Conversely, there is now a large body of evidence that effector cells of the natural immune system exert profound effects on the immunological functions of T and B lymphocytes. The properties of natural effector cells affecting cell-mediated immune responses will be the focus of this chapter.

The growth of interest and research directed at understanding the immunoregulatory effects of natural effector cells has been accompanied

ARABELLA B. TILDEN • Department of Medicine, Division of Hematology and Oncology, University of Alabama at Birmingham, and Veterans Administration Medical Center, Birmingham, Alabama 35294. LORAN T. CLEMENT • Department of Pediatrics, University of California at Los Angeles School of Medicine, Los Angeles, California 90024.

by controversy, primarily because efforts to define the lineage and/or role of the thymus in the generation of various populations of immunoregulatory cells have failed. Because this issue remains unsettled, it is difficult to distinguish with precision or certainty the immunoregulatory cells that are derived from (or influenced by) the thymus from those produced in the bone marrow independently of the thymus. Consequently, the following review of natural effector cells that regulate cell-mediated immune responses will include discussion of the properties, tissue distribution, and functions of a variety of immunoregulatory cells that share phenotypic, morphological, or related characteristics.

2. Characteristics of Natural Immunoregulatory Effector Cells

2.1. Morphological and Physical Characteristics

The characteristic morphological feature of natural effector cells with immunoregulatory functions is the presence of lysosomes dispersed throughout the cytoplasm, which accounts for the "granular lymphocyte" terminology typically applied to these cells.[1–4] This contrasts with the intracellular lysosomal organization of classic thymus-derived T lymphocytes, in which lysosomes are clustered together in association with a lipid droplet, thereby forming a perinuclear Gall body.[1,5] The granular lymphocyte morphology of immunoregulatory natural effector cells has been demonstrated by a number of techniques. Granular lymphocytes can be identified using Wright's stain or May-Gruenwald-Giemsa staining, or they may be identified by histochemical analyses based on the activity of lysosomal enzymes such as α-naphthyl acetate esterase (ANAE), β-glucuronidase (BG), or acid phosphatase (AP).[1–4] Perhaps the most reliable analysis of such cells has been obtained with electron microscopy, which has shown that natural immune effector cells have a low nuclear to cytoplasmic ratio, an irregular cell surface with short microvilli, and an indented nucleus with Golgi apparatus located in the nuclear cleft. Electron-denser granules are dispersed in the cytoplasm as well as in the Golgi region.[1,3,4] Multivesicular bodies, felt to be precursors of the electron-dense granules, are also present in the abundant cytoplasm.

Whereas the ultrastructural characteristics of immunoregulatory natural effector cells generally resemble those described for other cells of the natural immune system, particularly natural killer cells, there is some evidence suggesting that these cells may differ morphologically and/or histochemically from natural cytotoxic cells. For example, granular lymphocytes expressing the CD3 T cell antigen (which, as discussed later, have potent immunoregulatory functions) appear to have fewer cytoplasmic granules than are present in natural killer effector cells.[6] Histochemical comparisons of granular lymphocyte subsets expressing the CD2 antigen (the E receptor that binds to sheep erythrocytes) have revealed that the lysosomal enzyme α-naphthyl acetate esterase is present in virtually all cells that suppress cell-mediated immune responses, whereas this enzyme is frequently undetected in CD2$^+$ cells that have natural killer activity.[7-9] The reasons for the morphological and cytochemical differences among these granular lymphocyte subpopulations are unknown. They might be due to differences in the relative maturity of these natural immunity effector cells.[4] Alternatively, they could reflect differences in the lineage or differentiation pathway from which these cells are derived.

Another property of immunoregulatory natural effector cells that has been used to characterize and isolate these cells is their relative buoyant density on Percoll gradients.[10] Natural effector cells with suppressor functions are of low to medium density,[11] a property shared with other granular lymphocyte subsets and distinct from the majority of resting T cells with higher density.[11,12] Although the technique of fractionation of lymphocytes on Percoll gradients has been very useful for isolating relatively pure populations of cells with natural effector cell characteristics, it is important to note that the density of granular lymphocytes is heterogeneous, and some cells with this morphology do not copurify with the major, medium-density population.[11] Thus, cells purified by this technique do not necessarily include all immunoregulatory effector cells. Conversely, T cells activated *in vivo* may be present in lower-density cell fractions.[12]

2.2. Phenotypic Characteristics

The tremendous heterogeneity observed is the most striking feature of the cell membrane antigens expressed by natural effector cells with immunoregulatory properties. In studies of human natural effector cells (which, as a consequence of the advent of heterologous monoclonal antibodies, have been most thoroughly studied), a wide variety of cell surface antigens have been identified. These membrane markers include antigens regarded as "pan-T cell" markers, such as the CD2, CD3, CD5, and CD7 antigens.[6-14] In addition, the CD4 and CD8 membrane antigens that distinguish the helper/inducer T cell subset and the cytotoxic/suppressor subpopulation, respectively, have also been shown to be expressed by cells with morphological characteristics of cells comprising the natural immune system.[7-9,13-17] A high proportion of human natural immunoregulatory cells may also express antigens expressed by myelomonocytic-lineage cells. Prominent among these are the CD16 antigen, which is the receptor for IgG antibodies,[14,16,18,19] and the CD11 antigen, which functions as the CR3 receptor for the C3bi complement fragment.[7-9,20] Other markers expressed by cells with granular lymphocyte morphology are the Leu-7

(HNK-1) antigen and the NKH-1 (or Leu-19) antigens.[6,21,22] At present there is no identified membrane antigen that is expressed uniformly and exclusively by human natural immune effector cells.

In addition to this diverse group of membrane markers, immunoregulatory cells with natural effector cell characteristics may also express membrane molecules found on activated lymphocytes. These include class II antigen encoded by the major histocompatibility complex (such as HLA-DR) and receptors for interleukin 2 (IL-2) or transferrin.[14,23]

Immunoregulatory natural effector cells can also be recognized by virtue of their lack of expression of certain markers. For example, the membrane enzyme 5'-ectonucleotidase is not present on the population of CD8$^+$ granular lymphocytes that suppress T cell responses, but it is present on CD8$^+$ cytotoxic T lymphocytes.[24] Another marker that is useful in this regard is the antigen recognized by the 9.3 monoclonal antibody.[25] As discussed in detail later, this membrane molecule is not present on cells with a granular lymphocyte morphology, and it provides a particularly useful characteristic for distinguishing natural immune effector cells from classical T lymphocytes.

Cells comprising the natural immune system in mice also have a granulate lymphocyte morphology and express a variety of distinctive cell surface antigens. As seen in the human system, these cells may express antigens characteristic of T lymphocytes, such as the Thy-1 and Lyt-2 alloantigens, as well as certain other non-lineage-specific markers useful for their identification, including the Ly-5 antigen and asialo GM$_1$.[26–28] Recently, monoclonal antibodies reactive with antigens apparently unique to granular lymphocytes of mice have also been produced.[29]

In view of the striking phenotypic heterogeneity of cells of the natural immune system, it is not surprising that a great many phenotypically distinct subpopulations of such cells have been recognized. With regard to immunoregulatory functions, however, only a minority of these subsets have been isolated and characterized. In the discussions that follow, the specific phenotypic characteristics of cells with immunoregulatory capabilities will be discussed in the context of the functional studies that have been performed. This does not exclude the likely possibility that other phenotypically dissimilar subsets of natural effector cells are capable of mediating similar (or additional) immunoregulatory functions. However, emphasis will be placed on discussing the known functions of phenotypically identifiable natural effector cell populations.

2.3. Tissue Distribution of Natural Effector Cells

Although early studies of immunoregulatory natural effector cells focused on cells in the blood and/or secondary lymphoid tissues, it is now clear that cells with immunoregulatory properties and a granular lymphocyte morphology or phenotypic characteristics of natural immunity effector cells are widely distributed throughout the body.

2.3.1. Skin

In studies of murine epidermal cells, a novel cell with immunoregulatory function and natural effector cell characteristics has been identified. These cells express large amounts of the Thy-1 antigen as well as the Ly-5 and asialo GM_1 antigens.[30,31] These cells are of bone marrow origin and do not express other T-cell markers.[32] As discussed later, these cells appear to down-regulate contact sensitivity reactions occurring in the skin.

2.3.2. Gastrointestinal Tract

Lymphocytes with characteristics of natural immunity effector cells have been shown to be present in several distinct anatomical sites in the gastrointestinal tract. In humans, intraepithelial lymphocytes of the gut mucosa have been shown to be granular lymphocytes that express the CD8 and CD3 antigens.[33,34] Cells with comparable features have also been found in mice.[35] The lamina propria of primates also contains cells expressing the CD8 and CD11 (Leu-2 and Leu-15) antigens,[36] thus phenotypically resembling a subset of blood granular lymphocytes that suppresses cellular immune responses.[7,8] Although information about the immunoregulatory capabilities of these gut lymphocytes is limited, it is likely that these cells share at least some of the regulatory functions of the phenotypically similar cells found elsewhere in the body.

2.3.3. Lymph Node

Granular lymphocytes expressing the Leu-7 antigen are present in lymph nodes (and tonsils) and are concentrated with germinal center zones.[37–39] Interestingly, the vast majority of these cells coexpress the (CD4) (Leu-3) antigen,[39,40] a phenotypic feature uncommon for granular lymphocytes in blood or other tissues except in certain disease states.[41] Cells expressing other markers characteristic of natural effector cells (i.e., CR3 receptors) are infrequent in lymph nodes.

2.3.4. Thymus

When thymus cells are examined in suspension or in sections by immunohistologic techniques, cells with phenotypic characteristics of natural effector cells are very infrequent.[37,38] However, in recent studies of murine thymocytes, we have observed that cells with a granular lymphocyte morphology can be readily identified among cortisone-resistant thymocytes (CRT). In thymic cells obtained from normal 6-week-old mice, lymphocytes with a granular lymphocyte morphology (as determined by cytochemical analysis for acid phosphatase or ANAE) are uncommon and represent less than 3–5% of the cells present. However, when thymocytes from mice treated 2 days previously with cortisone acetate (which reduced the number of total thymocytes by >90%) were analyzed, a significant portion (40–65%) of the CRT had a granular lymphocyte morphology. Both Lyt-2$^+$ and L3T4$^+$ CRT with a granular lymphocyte morphology were seen. Morphological analysis by electron microscopy confirmed that these cells had a typical granular lymphocyte morphology indistinguishable from that of natural immunity (NK) cells in peripheral tissues. However, virtually none of these cells expressed the Mac-1 (CR3) antigen characteristically present on peripheral NK cells. Although the immunoregulatory functions of these cells have not yet been assessed, these data suggest that at least some cells with morphological features of lymphocytes comprising the natural immune system are indeed derived from the thymus. Future studies investigating whether these cells have immunoregulatory functions and/or antigen specificity should be of considerable significance in defining the ontogeny and functional properties of cells frequently considered to be natural effector cells at present.

3. Regulation of Cell-Mediated Immune Functions

3.1. Suppression of Cellular Immunity

3.1.1. Responses of T Cells to Mitogens and Nominal Antigens

Among the immunoregulatory functions of the natural immune system, suppression of cell-mediated immune responses has been noted frequently. The initial descriptions of such cells in mice and humans came from *in vitro* studies of lymphocytes expressing Fc receptors for IgG[42,43] and having a granular lymphocyte morphology.[1] The demonstration that cells with these characteristics were able to suppress a number of T cell responses, including proliferation and helper activity, has been followed by a large number of studies examining the suppressor functions of cells with related characteristics. In humans, *in vitro* T cell proliferative responses induced by polyclonal mitogens can be suppressed by cells expressing Fc-IgG receptors that are activated by immune complexes.[44] Additionally, a population of Fc receptor-negative cells with a granular lymphocyte morphology that express antigens recognized by the Leu-2 (CD8), Leu-15 (CD11), and Leu-7 antibodies can inhibit T cell proliferative responses to mitogens.[7,8,44] These cells are also able to suppress the *in vitro* proliferative responses of CD4$^+$ cells to soluble antigens such as tetanus toxoid.

The antiproliferative effects of these suppressor cells thus appeared to be relatively nonspecific with regard to the specificity or nature of the response, which is characteristic of the suppression observed with other natural effector populations. It is of interest, however, that studies of human antigen-specific suppressor cells have shown that these cells are CD8$^+$ cells that do not express the 9.3 antigen.[46,47] As previously noted, since expression of the 9.3 antigen and the CD11 antigen is reciprocal among E$^+$ cells (i.e., CD11$^+$ cells do not express the 9.3 antigen and *vice versa*),[25] it would appear that the nonspecific suppression of mitogen responses is mediated by the same subset of cells that exhibit antigen-specific suppressor activity.

Moreover, we have examined the morphology of Leu-2$^+$ 9.3$^-$ lymphocytes and have found that these cells uniformly have a granular lymphocyte morphology and may infrequently express the CD16 (Fc receptor) molecule (unpublished observations). These data demonstrate the difficulty of distinguishing natural immunity effector cells from other immunoregulatory cells with apparent immunological specificity on the basis of phenotypic or morphological criteria, and highlight the need for additional study of the lineal relationships of such cells.

In addition to the suppression of T cell proliferative responses, human natural effector cells also suppress polyclonal B cell differentiation induced by T-dependent mitogens, such as pokeweed mitogen.[45] This suppressive activity may affect the B cell directly or it may be a manifestation of the suppression of helper T cell-mediated immune responsiveness. Discussion of the possible mechanisms for suppression of antibody production by natural effector cells is the subject of another chapter in this volume.

In the mouse, natural immune effector cells that suppress antigen- or mitogen-induced proliferative responses have also been identified and characterized. Natural suppressor (NS) cells are normally present in neonatal lymphoid tissues and the bone marrow of adult mice and can also be induced in adult lymphoid tissues by total lymphoid irradiation or in graft versus host disease.[48–51] Natural suppressor cells can inhibit a broad range of cell-mediated immune responses in a non-MHC-restricted fashion, including T cell proliferation induced by mitogens,[50] and they have a number of morphological and phenotypic characteristics of cells comprising the natural immune system.[51] Another effector cell population with immunoregulatory functions, which has been termed the natural cytotoxic (NC) cell subset, is also present in neonatal tissues.[51] These cells have a number of similarities to NS and NK cells, but they appear to differ from these other natural immune effector cells in their cytotoxic activity and their response to lymphokines.[51] Although it seems likely that these populations are closely related to one another, these data again point out the heterogeneity and complexity of natural effector cells with regard to their functional capabilities and the nature of their interactions with other cell populations.

In addition to the suppression of *in vitro* T cell responses to mitogens, there is evidence that contact hypersensitivity reactions against specific antigens encountered in the epidermis are also suppressed by natural effector cells. The murine epidermis contains a population of Thy-1[+] cells that express markers present on other murine natural immunity cells[30–32] and have the ability to down-regulate contact hypersensitivity reactions.[53] Interestingly, the suppression of these responses was most notable when these cells (in conjunction with antigen) were introduced systemically, perhaps indicating the need for interactions of these cells with immunocompetent cells not normally present in the epidermis as a requirement for developing immunosuppressive activity.

3.1.2. Responses of T Cells to Alloantigens

The ability of natural effector cells to suppress cell-mediated responses to alloantigens has been demonstrated in a number of systems. In humans, granular lymphocytes that express the Leu-7 antigen can inhibit the proliferative response of T cells in an MLR.[54] The Leu-7[+] suppressor cells are found in both CD2-positive and -negative populations as determined by erythrocyte rosette separation. However, the CD2[−] Leu-7[+] fraction has slightly more efficient suppressor cell activity. The suppressor cell properties of Leu-7[+] cells are different from those of CD8[+] suppressor cells but similar to Tγ cells in requiring immune complex exposure for activation.[7,43,54] Granular lymphocytes that express the Leu-7 antigen can also suppress T lymphocyte responses to the mitogens Con A and PHA (unpublished observations).

In a related fashion, murine natural suppressor (NS) or natural cytotoxic (NC) cells that inhibit mitogen-induced T cell responses are also capable of suppressing T cell proliferation induced by alloantigens.[48–52] Natural effector cells also can block the development of alloimmune cytotoxic T cells[55,56] or the development of graft versus host disease.[56] This activity may be of special significance in maintaining the maternal-fetal barrier that prevents maternal rejection of the implanted allogeneic fetus *in utero*. Suppressor cells of non-T cell lineage are present in decidual tissue of the fetoplacental unit and block the development of cytotoxic cells that can mediate rejection of the fetus.[57,58]

3.2. Enhancement of Immune Responses

In addition to their immunosuppressive functions, natural immunity effector cells have capabilities that may enhance cellular immune responses. In general, these immunostimulatory functions have been less prominent than the suppressive actions of these cells, and, as discussed below, they often relate to the ability of natural effector cells to secrete stimulatory lymphokines.

In humans, the immunoregulatory activity of natural effector cells may vary. Hence, non-T, non-B lymphocytes that express Fc receptors, which typically suppress cell immunity, may enhance mitogen- or antigen-induced proliferation under certain circumstances.[59] Large granular lymphocytes also provide an accessory function enhancing the development of virus-specific cytotoxic T cells.[60] One unique population of human granular lymphocytes that has attracted attention as potential immunoenhancing natural effector cells is the subset expressing the CD4 antigen. Because these cells are located within the B cell germinal centers of lymph nodes and express the CD4 antigen characteristic of helper T cells, it has been an attractive hypothesis that these cells with natural effector cell characteristics[40] might enhance immune responses. However, in analyses to date, these cells do not appear to have significant effects on immune responses,[41,61] and the functions (and lineage) of these cells remain to be determined.

3.3. Mechanisms of Activation

From a wide variety of studies, it has become clear that lymphocytes must be activated by exogenous stimuli before they can exert their effector functions, and lymphoid cells of the natural immunity system are no exception. In view of the striking phenotypic and functional heterogeneity of natural effector cells, it is perhaps not surprising that these cells are also heterogeneous with regard to the activation stimuli that induce their immunoregulatory functions. For example, immune complexes composed of IgG antibodies are potent stimuli for inducing suppressor activity by Fc receptor-bearing natural effector cells.[11,42–44,54,62] Suppression of cell-mediated immune responses by natural effector cells can also be induced or enhanced by lymphokines such as IL-2,[62–64] interferon gamma,[64] or interferon alpha.[62,65] Finally, natural suppressor cells in mice[66] or human suppressor cells with phenotypic features of natural effector cells[67] can be activated by autocoids such as histamine.

3.4. Mechanisms of Immunoregulation

The mechanism by which activated natural effector cells mediate their immunoregulatory functions is somewhat controversial. Owing to the cytotoxic capabilities of natural killer cells, which comprise a functional subpopulation of the natural immune system, some investigators have proposed that the suppression of immune responses by natural effector cells may be due to a cytotoxic attack against activated T or B lymphocytes, thereby abrogating the immune response.[68] Similarly, lysis of antigen-presenting cells by NK-like cells has also been suggested as an immunosuppressive mechanism.[69] However, many natural effector cells with immunoregulatory activity have no demonstrable cytotoxic capabilities. Furthermore, even some cells with NK activity (i.e., Leu-7 cells that express CD16) can suppress mitogenic responses to PHA but not kill PHA blast cells (unpublished observation). In many instances, these cells have been shown to secrete soluble factors that suppress cellular immune responses.[51,57,70] These factors appear to be distinct from interferons, and they may exert their activity by blocking the production or responsiveness to IL-2[70,72] and/or by a direct cytostatic effect.[71] It has recently been suggested that such a factor released by human natural effector cells might be leukoregulin.[71]

Although relatively little is known about the mechanisms by which natural effector cells may enhance cell immunity, most of the available evidence suggests that these capabilities are probably mediated by soluble lymphokines secreted by natural effector cells. At least some of these cells are capable of secreting a broad variety of well-characterized factors, including IL-1, IL-2, and interferon gamma,[73–75] all of which may augment cellular immunity under certain circumstances. In addition, granular lymphocytes release other uncharacterized lymphokines that activate macrophages[76] or enhance development of cytotoxic T cells.[60] The ability of natural effector cells to produce lymphokines that enhance cell-immune reactions, on the one hand, and are capable of inducing suppressin, on the other hand, suggests that the ultimate *in vivo* immunoregulatory effects of activated natural effector cells may depend on a complex series of variables that are currently difficult to predict.

4. Relevance to Disease States

The ability to identify, purify, and analyze populations of natural effector cells has stimulated increased efforts to determine the role of these cells in the causation or mitigation of a number of disease states in which regulation of cellular immunity is of particular importance. Although many of these clinical conditions are discussed in detail elsewhere in this volume, several examples that highlight the apparent importance of the role of natural effector cells in conditions with disordered cellular immunity are introduced below.

4.1. Autoimmune Disorders

In addition to autoantibody production, systemic lupus erythematosus (SLE) is characterized by defective production of IL-2 by activated T cells. Studies of the mechanism underlying this defect have demonstrated that IL-2 production is suppressed by a soluble macromolecule secreted by cells expressing the Leu-2 and/or Leu-7 antigens.[77] These are granular lymphocytes,[45] and they may be present in increased frequency in at least some SLE patients.[78] Although considerable gaps in our understanding of the complex immunoregulatory defects in SLE remain, it is likely that quantitative or qualitative alterations in natural effector cells participate in the pathogenesis of (or inadequate regulatory responses in) this and other autoimmune diseases.

4.2. Infectious Diseases

The importance of natural killer cells in host defenses against viral infections has been well characterized in many experimental systems. However, recent studies of the phenotypic and functional properties of lymphoid cells in certain viral infections have suggested that natural effector cell subsets that regulate cell-mediated immunity may be dramatically altered in the course of these infections. This is perhaps most notable in AIDS and the lymphadenopathy syndrome, where the pronounced HIV-induced reduction of CD4$^+$ cells is accompanied by a striking increase in CD8$^+$ Leu-7$^+$ cells.[79,80] These lymphocytes are phenotypically similar to naturally occurring granular lymphocytes with immunosuppressive activity.[44] Since these cells normally have no NK activity,[9] and since AIDS patients have defective NK activity,[80] it is quite possible that the increase in the frequency of these cells represents an immunoregulatory response to some aspect of the disordered immune system in these patients rather than a cytotoxic response to HIV-infected cells. Whether this increase in cells with potent immunosuppressive capabilities contributes to the abnormal host defense mechanisms induced by (or secondarily associated with) HIV infections remains to be determined.

Prolonged increased frequencies of CD8$^+$ Leu-7$^+$ granular lymphocytes have also been observed in cytomegalovirus (CMV) infections following cardiac or bone marrow transplantation.[81,82] Since comparable changes are not seen in CMV-negative transplant patients,[81] it is again likely that a viral-induced perturbation of a suppressed host immune system evokes this natural effector cell expansion. The consequences of this alteration (whether beneficial or detrimental) are unknown.

4.3. Transplantation

Patients who have received allografts often have significant deficits in their cellular immunity. This may be due to a number of factors, including immunosuppressive drug therapy, activation of virus infections such as CMV, and/or conditions such as graft versus host disease. In bone marrow transplantation, the initial recovery of circulating lymphocytes is composed largely of cells with phenotypic characteristics of granular lymphocytes.[83] Among these are Leu-2+15+ cells, which are present in considerably increased frequencies[79,83] and which are able to suppress IL-2 production.[79] In renal transplant patients, increased numbers of granular CD4+ Leu-7+ lymphocytes are found.[84] These cells resemble those identified in germinal centers and which are increased in patients with B cell malignancies.[41] As previously noted, the functions of the cells have not been determined. Although the precise effects of natural effector cells on cell-mediated immunity after allograft transplantation are only beginning to become apparent, it is clear that the interplay of allogeneic stimuli, opportunistic infections, and/or immunosuppressive drug therapy produce profound alterations in natural effector cells with immunoregulatory capabilities, an event that is highly likely to be of significance in determining the level of cell-mediated immune competence in these patients.

5. Summary and Future Directions

The natural immunity system is composed of phenotypically and functionally heterogeneous lymphoid cells that have pronounced immunoregulatory effects on cell-mediated immune responses. Although it is clear that the natural immune system is derived from bone marrow precursor cells, the morphological, phenotypic, or functional criteria applied to the recognition of these cells are diverse, vary from laboratory to laboratory, and are not sufficiently precise to distinguish these cells from those that may be progeny of a separate differentiation pathway. Hence, many immunoregulatory cells may express characteristics of both natural immune effector cells (such as a granular lymphocyte morphology, low cell density, and/or the expression of "NK" markers) as well as certain T cell characteristics (in particular, expression of the CD3 antigen and rearrangement of T cell receptor genes). Whereas it is possible to distinguish certain bone marrow-derived natural effector cells from certain thymus-derived effector cells, there remain a large number of potent immunoregulatory cell subsets whose lineage is uncertain. Precise understanding of the role of natural effector cells in regulating cell-mediated immune response will thus not be attained until these cells can be distinguished unequivocally from thymus-derived immunoregulatory subsets.

This limitation notwithstanding, it is still clear that cells of the natural immune system are present in many tissues of the body and have potent immunoregulatory capabilities. At present, the major functions identified for these cells have been suppressive in nature; natural effector cells suppress the proliferation of a broad variety of cell types, including T cells active in cell-mediated immunity. However, these cells may also enhance certain responses by the elaboration of immunostimulatory lymphokines. Although the activation stimuli for natural effector cells are poorly understood at present, it appears that IL-2 may have a pivotal role. Thus, it is conceivable that IL-2 generated in the initial stages of an immune response may also induce nonspecific suppression characteristic of natural effector cells. However, a variety of other stimuli, such as immune complexes or histamine, can also induce nonspecific suppression by certain subsets of these cells. It is important to note that the mere presence of natural effector cells in a tissue or in blood is not necessarily predictive of their functional activity; natural effector cells with immunoregulatory potential may be greatly expanded in frequency in lymphoproliferative (or other) diseases without any evidence of immune dysfunction or down-regulation. Thus, an important area for future research is to define the activation requirements, cellular interactions, and immunoregulatory mechanisms of natural effector cells. Such information is of considerable importance for defining the potential involvement of these cells in the pathogenesis of (or defense against) a number of disease states. Phenotypic analyses have now demonstrated abnormal frequencies of a number of lymphocyte subsets with natural effector cell characteristics in a variety of diseases or clinical conditions, and it is likely that improved knowledge of the physiology and functions of natural effector cell subsets will provide a clearer understanding of these disease-related quantitative changes. There is considerable reason to believe that this will be a fruitful and exciting area of research that will bear upon virtually every clinical discipline.

ACKNOWLEDGMENTS. Studies reported herein were supported by PHS grant CA 42735, awarded by the National Cancer Institute, DHHS, the Veterans Administration Medical Research Council, and NIH grants CA27197, CA03013, and CA13148.

References

1. Grossi, C. E., Webb, S. R., Zicca, A., Lydyard, P. M., Moretta, L., Mingari, M. C., and Cooper, M. D., 1978, Morphological and histochemical analyses of two human T-cell subpopulations bearing receptors for IgM or IgG, *J. Exp. Med.* **147**:1405–1417.
2. Saksela, E., Timonen, T., Ranki, A., and Hayry, P., 1979, Morphological and functional characterization of isolated effector cells responsible for human natural killer activity to fetal fibroblasts and to cultured cell line targets, *Immunol. Rev.* **44**:71–99.
3. Ferrarini, M., Cadoni, A., Franzi, A. T., Ghigliotti, C., Leprini, A., Zicca, A., and Grossi, C. E., 1980, Ultrastructure and cytochemistry of human peripheral blood lymphocytes. Similarities between the cells of the third population and T_G lymphocytes, *Eur. J. Immunol.* **10**:562–568.
4. Grossi, C. E., Cadoni, A., Zicca, A., Leprini, A., and Ferrarini, M., 1982, Large granular lymphocytes in human peripheral blood. Ultrastructural and cytochemical characterization of the granules, *Blood* **59**:227–284.
5. Armitage, R. J., Linch, D. C., Worman, C. P., and Cawley, J. C., 1982, The morphology and cytochemistry of human T-cell subpopulations defined by monoclonal antibodies and Fc receptors. *Br. J. Haematol.* **52**:605–609.

6. Abo, T., Cooper, M. D., and Balch, C. M., 1982, Characterization of HNK-1⁺ (Leu-7) human lymphocytes. I. Two distinct phenotypes of human NK cells with different cytotoxic capability, *J. Immunol.* **129:**1752–1757.
7. Landay, A., Gartland, G. L., and Clement, L. T., 1983, Characterization of a phenotypically distinct subpopulation of Leu-2⁺ cells that suppresses T cell proliferative responses, *J. Immunol.* **131:**2757–2761.
8. Landay, A., Clement, L. T., and Grossi, C. E., 1984, Phenotypically and functionally distinct subpopulations of human lymphocytes with T cell markers also exhibit different cytochemical patterns of staining for lysosomal enzymes, *Blood* **63:**1067–1071.
9. Clement, L. T., Dagg, M. K., and Landay, A., 1984, Characterization of human lymphocyte subpopulations: Alloreactive cytotoxic T-lymphocyte precursor and effector cells are phenotypically distinct from Leu-2⁺ suppressor cells, *J. Clin. Immunol.* **4:**395–402.
10. Timonen, T., and Saksela, E., 1980, Isolation of human natural killer cells by density gradient centrifugation, *J. Immunol. Methods* **36:**285–292.
11. Brieva, J. A., Targan, S., and Stevens, R. H., 1984, NK and T cell subsets regulate antibody production by human *in vivo* antigen-induced lymphoblastoid B cells, *J. Immunol.* **132:**611–615.
12. London, L., Perussia, B., and Trinchieri, G., 1986, Induction of proliferation *in vitro* of resting human natural killer cells: IL-2 induces into cell cycle most peripheral blood NK cells, but only a minor subset of low density T cells, *J. Immunol.* **137:**3845–3854.
13. Allavena, P., and Ortaldo, J. R., 1984, Characteristics of human NK clones: Target specificity and phenotype, *J. Immunol.* **132:**2363–2369.
14. Lanier, L. L., and Phillips, J. H., 1986, A map of the cell surface antigens expressed on resting and activated human natural killer cells, in: *Leukocyte Typing II*, Volume 3 (E. L. Reinherz, B. F. Haynes, L. M. Nadler, and I. D. Bernstein, eds.), Springer-Verlag, New York, pp. 157–170.
15. Perussia, B., Fanning, V., and Trinchieri, G., 1983, A human NK and K cell subset shares with cytotoxic T cells expression of the antigen recognized by antibody OKT8, *J. Immunol.* **131:**223–229.
16. Titus, J. A., Sharrow, S. O., and Segal, D. M., 1983, Analysis of Fc(IgG) receptors on human peripheral blood leukocytes by dual flourescence flow microfluoremetry. II. Quantitation of receptors on cells that express the OKM1, OKT3, OKT4, and OKT8 antigens, *J. Immunol.* **130:**1152–1158.
17. Velardi, A., Grossi, C. E., and Cooper, M. D., 1985, A large subpopulation of lymphocytes with T helper phenotype (Leu-3⁺/T4⁺) exhibits the property of binding to NK cell targets and granular lymphocyte morphology, *J. Immunol.* **134:**58–64.
18. Trinchieri, G., and Perussia, B., 1984, Human natural killer cells. Biological and pathological aspects, *Lab. Invest.* **50:**489–513.
19. Rumpold, H., Kraft, D., Obexer, G., Bock, G., and Gebhart, W., 1982, A monoclonal antibody against a surface antigen shared by human large granular lymphocytes and granulocytes, *J. Immunol.* **129:**1458–1463.
20. Ceuppens, J. L., Gualde, N., and Goodwin, J. S., 1982, Phenotypic heterogeneity of the OKM1-positive lymphocyte population: Reactivity of OKM1 monoclonal antibody with a subset of the suppressor/cytotoxic T-cell population, *Cell Immunol.* **69:**150–165.
21. Hercend, T., Griffin, J. D., Bensussan, A., Schmidt, R. E., Edson, M. A., Brennan, A., Murray, C., Daley, J. F., Schlossman, S. F., and Ritz, J., 1985, Generation of monoclonal antibodies to a human natural killer clone. Characterization of two natural killer-associated antigens, NKH1A and NKH2, expressed on subsets of large granular lymphocytes, *J. Clin. Invest.***75:**923–943.
22. Lanier, L. L., Le, A. M., Cwirla, S., Federspeil, N., and Phillips, J. H., 1986, Antigenic, functional, and molecular genetic studies of human natural killer cells and cytotoxic T lymphocytes not restricted by the major histocompatibility complex, *Fed. Proc.* **45:**2823–2828.
23. Trinchieri, G., Matsumoto-Kobayashi, M., Clark, S. V., Seehra, J., London, L., and Pe-

russia, B., 1984, Response of resting human peripheral blood natural killer cells to interleukin-2, *J. Exp. Med.* **160**:1147–1169.

24. Dianzani, U., Massaia, M., Pileri, A., Grossi, C. E., and Clement, L. T., 1986, Differential expression of ecto-5′-nucleotidase activity by functionally and phenotypically distinct subpopulations of human Leu-2+/T8+ lymphocytes, *J. Immunol.* **137**:484–489.

25. Yamada, H., Martin, P. J., Bean, M. A., Braun, M. P., Beatty, P. G., Sadamoto, K., and Hansen, J. A., 1985, Monoclonal antibody 9.3 and anti-CD11 antibodies define reciprocal subsets of lymphocytes, *Eur. J. Immunol.* **15**:1164–1168.

26. Brooks, G. G., Kuribayashi, K., Sale, G. E., and Henny, C. S., 1982, Characterization of five cloned murine cell lines showing high cytolytic activity against YAC-1 cells, *J. Immunol.* **128**:2326–2332.

27. Kasai, M., Yoneda, T., Habu, S., Maruyama, Y., Okumura, K., and Tokunaga, T., 1981, *In vivo* effect of anti-asialo GM₁ antibody on natural killer activity, *Nature* **291**:5813–5816.

28. Kedar, E., Ikejiri, B. L., Srendi, B., Bonavida, B., and Herberman, R. B., 1982, Propagation of mouse cytotoxic clones with characteristics of natural killer (NK) cells, *Cell. Immunol.* **69**:305–311.

29. Koo, G. C., and Peppard, J. R., 1984, Establishment of monoclonal anti-NK-1 antibody, *Hybridoma* **3**:301–306.

30. Bergstresser, P. R., Tigelaar, R. E., Dees, J. H., and Streilein, J. W., 1983, Thy-1 antigen-bearing dendritic cells populate murine epidermis, *J. Invest. Dermatol.* **81**:286–291.

31. Tschachler, E., Schuler, G., Hutterer, J., Leibl, H., Wolff, K., and Stuigl, G., 1983, Expression of Thy-1 antigen by murine epidermal cells, *J. Invest. Dermatol.* **81**:282–285.

32. Breathnach, S. M., and Katz, S. I., 1984, Thy-1+ dendritic cells in murine epidermis are bone marrow derived, *J. Invest. Dermatol.* **83**:74–82.

33. Selby, W. S., Janossy, G., Goldstein, G., and Jewell, D. P., 1981, T lymphocyte subsets in human intestinal mucosa: The distribution and relationships to MHC-determined antigens, *Clin. Exp. Immunol.* **44**:453–460.

34. Cerf-Bensussan, N., Schneeberger, E. E., and Bhan, A. K., 1983, Immunohistologic and immunoelectron microscopic characterization of the mucosal lymphocytes of human small intestine by the use of monoclonal antibodies, *J. Immunol.* **130**:2615–2262.

35. Schrader, J. W., Scollay, R., and Battye, F., 1983, Intramucosal lymphocytes of the gut: Lyt-2 and Thy-1 phenotype of the granulated cells and evidence for the presence of both T cells and mast cell precursors, *J. Immunol.* **130**:558–564.

36. Graeff, A. S., Strober, W., and James, S. P., 1985, Intestinal lamina propria lymphocytes (LPL) in non-human primates lack cells with suppressor-inducer phenotype, *Fed. Proc.* **44**:564.

37. Hsu, S., Cossman, J., and Jaffee, E. S., 1983, Lymphocyte subsets in normal human lymphoid tissues, *Am. J. Clin. Pathol.* **7**:21–30.

38. Si, L., and Whiteside, T. L., 1983, Tissue distribution of human NK cells studies with anti-Leu-7 monoclonal antibody, *J. Immunol.* **130**:2149–2154.

39. Porwit-Ksiazek, A., Ksiazek, T., and Biberfeld, P., 1983, Leu-7+ (HNK-1+) cells. I. Selective compartmentalization of Leu-7+ cells with different immunophenotypes in lymphatic tissues and blood, *Scand. J. Immunol.* **18**:485–493.

40. Velardi, A., Tilden, A. B., Millo, R., and Grossi, C. E., 1986, Isolation and characterization of Leu-7+ germinal-center cells with the T helper-cell phenotype and granular lymphocyte morphology, *J. Clin. Immunol.* **6**:205–215.

41. Velardi, A., Clement, L. T., and Grossi, C. E., 1985, Quantitative and functional analysis of a human lymphocyte subset with T-helper (Leu-3/T4+) phenotype and NK cell characteristics in patients with malignancies, *J. Clin Immunol.* **5**:329–339.

42. Stout, R. D., and Herzenberg, L. A., 1975, The Fc receptor on thymus-derived lymphocytes. I. Detection of a subpopulation of murine T lymphocytes bearing the Fc receptor, *J. Exp. Med.* **142**:611–620.

43. Moretta, L., Webb, S. R., Grossi, C. E., Lydyard, P. M., and Cooper, M. D., 1977, Func-

tional analysis of two human T-cell subpopulations: Help and suppression of B-cell responses by T cells bearing receptors for IgM or IgG, *J. Exp. Med.* **146**:184–200.

44. Moretta, L., Mingari, M. C., Moretta, A., and Cooper, M. D., 1979, Human T lymphocyte subpopulations: Studies of the mechanism by which T cells bearing Fc receptors for IgG suppress T-dependent B cell differentiation induced by pokeweed mitogen, *J. Immunol.* **122**:984–991.

45. Clement, L. T., Grossi, C. E., and Gartland, G. L., 1984, Morphologic and phenotypic features of the subpopulation of Leu-2⁺ cells that suppresses B cell differentiation, *J. Immunol.* **133**:2461–2468.

46. Damle, N. K., Mohagheghpour, N., Hansen, J. A., and Engleman, E. G., 1983, Alloantigen-specific cytotoxic and suppressor T lymphocytes are derived from phenotypically distinct precursors, *J. Immunol.* **131**:2296–2303.

47. Damle, N. K., Mohagheghpour, N., and Engleman, E. G., 1984, Soluble antigen-primed inducer T cells activate antigen-specific suppressor T cells in the absence of antigen-pulsed accessory cells. Phenotypic definition of suppressor-inducer and suppressor-effector cells, *J. Immunol.* **132**:644–651.

48. Duwe, A. K., and Singhal, S. K., 1979, The immunoregulatory role of bone marrow. II. Characterization of a suppressor cell inhibiting the *in vitro* antibody response, *Cell. Immunol.* **43**:372–381.

49. Oseroff, A., Okada, S., and Strober, S., 1984, Natural suppressor (NS) cells found in the spleen of neonatal mice and adult mice given total lymphoid irradiation (TLI) express the null surface phenotypes, *J. Immunol.* **132**:101–110.

50. Maier, T., Holda, J. H., and Claman, H. N., 1985, Graft-versus-host reactions (GVHR) across minor murine histocompatibility barriers. II. Development of natural suppressor cell activity, *J. Immunol.* **135**:1644–1651.

51. Maier, T., Holda, J. H., and Claman, H. N., 1986, Natural suppressor (NS) cells. Members of the LGL regulatory family, *Immunol. Today* **7**:312–315.

52. Jadus, M. R., and Parkman, R., 1986, The selective growth of murine newborn-derived suppressor cells and their probable mode of action, *J. Immunol.* **136**:783–792.

53. Sullivan, S., Bergstresser, P. R., Tigelaar, R. E., and Streilein, J. M., 1986, Induction and regulation of contact hypersensitivity by resident, bone marrow-derived, dendritic epidermal cells: Langerhans cells and Thy-1⁺ epidermal cells, *J. Immunol.* **137**:2460–2467.

54. Tilden, A. B., Abo, T., and Balch, C. M., 1983, Suppressor cell function of human granular lymphocytes identified by the HNK-1 (Leu-7) monoclonal antibody, *J. Immunol.* **130**:1171–1175.

55. Dorshkind, K., Klimpel, G. R, and Rosse, C., 1980, Natural regulatory cells in murine bone marrow: Inhibition of *in vitro* proliferative and cytotoxic responses to alloantigens, *J. Immunol.* **124**:2584–2590.

56. Strober, S., Okada, S., and Oseroff, A., 1984, Role of natural suppressor cells in allograft tolerance, *Fed. Proc.* **43**:263–265.

57. Slapsys, R. M., and Clark, D. A., 1982, Active suppression of host-versus-graft reaction in pregnant mice. IV. Local suppressor cells in decidua and uterine blood, *J. Reprod. Immunol.* **4**:355–362.

58. Clark, D. A, Slapsys, R. M., Croy, B. A., Krcek, J., and Rossant, J., 1984, Local active suppression by suppressor cells in decidua: A review, *Am. J. Reprod. Immunol.* **5**:78–86.

59. Carvalho, E. M., and Horwitz, D. A., 1980, Characterization of non-T, non-B human blood lymphocyte that mediates the enhancing effects of immune complexes on lymphocyte blastogenesis, *J. Immunol.* **124**:1656–1661.

60. Burlington, D. B., Djeu, J. Y., Wells, M. A., Kiley, S. C., and Quinnan, G. V., 1984, Large granular lymphocytes provide an accessory function in the *in vitro* development of influenza A virus-specific cytotoxic T cells, *J. Immunol.* **132**:3154–3158.

61. Velardi, A., Mingari, M. C., Moretta, L., and Grossi, C. E., 1986, Functional analysis of

cloned germinal center CD4$^+$ cells with natural killer cell-related features. Divergence from typical T helper cells, *J. Immunol.* **137**:2808–2813.

62. Abo, W., Bakke, A. C., and Horwitz, D. A, 1985, The regulatory properties of the third mononuclear population on lymphocyte proliferation and immunoglobulin synthesis, *Clin. Res.* **33**:555A.

63. Thoman, M. L., and Weigle, W. O., 1984, Interleukin-2 induction of antigen-non-specific suppressor cells, *Cell. Immunol.* **85**:215–224.

64. Holda, J. H., Maier, T., and Claman, H. N., 1986, Natural suppressor activity in graft-versus-host spleen and normal bone marrow is augmented by IL-2 and interferon-γ, *J. Immunol.* **137**:3538–3543.

65. Kuwano, K., Arai, S., Munakata, T., Tomita, Y., Yoshitake, Y., and Kumagai, K., 1986, Suppressive effect of human natural killer cells on Epstein-Barr virus-induced immunoglobulin synthesis, *J. Immunol.* **137**:1462–1467.

66. Khan, M. M., Marr-Leisy, D., Verlander, M. S., Bristow, M. R., Strober, S., Goodman, M., and Melmon, K. L., 1986, The effects of derivatives of histamine on natural suppressor cells, *J. Immunol.* **137**:308–314.

67. Sansoni, P., Silverman, E. D., Khan, M. M., Melmon, K. L., and Engleman, E. G., 1985, Immunoregulatory T cells in man: Histamine induced suppressor T cells are derived from a Leu-2$^+$ (T8$^+$) subpopulation distinct from that which gives rise to cytotoxic T cells, *J. Clin. Invest.* **75**:650–656.

68. Phillips, J. H., and Lanier, L. L., 1986, Lectin-dependent and anti-CD3 induced cytotoxicity are preferentially mediated by peripheral blood cytotoxic T lymphocytes expressing Leu-7 antigen, *J. Immunol.* **136**:1579–1585.

69. Rowley, D. A., and Shah, P. D., 1986, The immunological meaning of Thy-1-negative NK cells, *Immunol. Today* **7**:196–199.

70. Clark, D. A., Chaput, A., Walker, C., and Rosenthal, K., 1985, Active suppression of host-versus-graft reaction in pregnant mice. VI. Soluble suppressor activity obtained from decidua blocks the response to IL-2, *J. Immunol.* **134**:1659–1666.

71. Sayers, T. J., Ransom, J. H., Denn, A. C., Herberman, R. B., and Ortaldo, J. R., 1986, Analysis of a cytostatic lymphokine produced by incubation of lymphocytes with tumor cells: Relationship to leukoregulin and distinction from recombinant lymphotoxin, recombinant tumor necrosis factor,and natural killer cytotoxic factor, *J. Immunol.* **137**:385–390.

72. Gebel, H. M., Kaizer, H., and Landay, A. L., 1985, Leu-2$^+$15$^+$ T cells mediate suppression of IL-2 production in bone marrow recipients, *Fed. Proc.***44**:1303.

73. Scala, G., Allavena, P., Djeu, J. D., Kasahara, T., Ortaldo, J. R., Herberman, R. B., and Oppenheim, J. J., 1984, Human large granular lymphocytes are potent producers of interleukin-1, *Nature* **309**:56–59.

74. Kasahara, T., Djeu, J. D., Dougherty, S. F., and Oppenheim, J. J., 1983, Capacity of human large granular lymphocytes (LGL) to produce multiple lymphokines: Interleukin-2, interferon, and colony-stimulating factor, *J. Immunol.* **131**:2379–2386.

75. Suzuki, R., Suzuki, S., Ebina, N., and Kumagai, K., 1985, Suppression of alloimmune cytotoxic T lymphocyte (CTL) generation by depletion of NK cells and restoration by interferon and/or interleukin-2, *J. Immunol.* **134**:2139–2148.

76. Gomez, J., Pohajdak, B., O'Neill, S., Wilkins, J., and Greenberg, A. H., 1985, Activation of rat and human alveolar macrophage intracellular microbicidal activity by a preformed LGL cytokine, *J. Immunol.* **135**:1194–1200.

77. Linker-Israeli, M., Bakke, A. C., Quismorio F. P. Jr., and Horwitz, D. A., 1985, Correction of interleukin-2 production in patients with systemic lupus erythematosis by removal of spontaneously activated suppressor cells, *J. Clin. Invest.* **75**:762–768.

78. Egan, M. L., Mendelsohn, S. L., Abo, T., and Balch, C. M., 1983, Natural killer cells in systemic lupus erythematosus. Abnormal numbers and functional immaturity of HNK-1$^+$ cells, *Arth. Rheum.* **26**:623–629.

79. Lewis, D. E., Puck, J. M., Babcock, G. F., and Rich, R. R, 1985, Disproportionate expansion of a minor T cell subset in patients with lymphoadenopathy syndrome and acquired immunodeficiency syndrome, *J. Infect. Dis.* **151:**555–559.
80. Plaeger-Marshall, S., Spina, C. A., Giorgi, J. V., Mitsuyasu, R., Wolfe, P., Gottleib, M., and Beall, G., 1987, Alterations in cytotoxic and phenotypic subsets of natural killer cells in acquired immune deficiency syndrome (AIDS), *J. Clin. Immunol.* **7:**16–23.
81. Maher, P., O'Toole, C. M., Wreghitt, T. G., Spiegelhalter, D. J., and English, T. A. H., 1985, Cytomegalovirus infection in cardiac transplant recipients associated with chronic T cell subset ratio inversion with expansion of a Leu-7$^+$ T$_{s-c}$$^+$ subset, *Clin. Exp. Immunol.* **62:**515–524.
82. Worsch, A. M., Gratama, J. W., Middeldorp, J. M., Nissen, C., Gratwohl, A., Speck, B., Jansen, J., D'Amaro, J., The T. H., and DeGast, G. C., 1985, The effect of cytomegalovirus infection on T lymphocytes after allogeneic bone marrow transplantation, *Clin. Exp. Immunol.* **62:**278–287.
83. Ault, K. A., Autin, J. H., Ginsburg, D., Orkin, S. H., Rappeport, J. M., Keohan, M. L., Martin, P., and Smith, B. R., 1985, The phenotype of recovering lymphoid cell populations following marrow transplantation, *J. Exp. Med.* **161:**1483–1502.
84. Legendre, C. M., Guttman, R. D., Hou, S. K., and Jean, R., 1985, Two-color immunoflourescent and flow cytometry analysis of lymphocytes in long-term renal allotransplant recipients: Identification of a major Leu-7$^+$/Leu-3$^+$ subpopulation, *J. Immunol.* **135:**1061–1066.

11

Control of Hematopoietic Progenitor Cells by Natural Killer Cells

GIORGIO TRINCHIERI, MARIANNE MURPHY,
MARIA CRISTINA CUTURI, IGNACIO ANEGON, and
BICE PERUSSIA

1. Introduction

In adult animals and in physiological conditions, hematopoiesis occurs only in the bone marrow and in lymphoid organs, and it is not observed in organs such as liver or spleen, sites of fetal hematopoiesis. Precursor cells committed to erythroid and myeloid hematopoietic lineages originate from pluripotent stem cells and, through several cycles of cell division, give rise to terminally differentiated cells.[1] The entire process of hematopoiesis is regulated by the equilibrium between the self-renewal capability of the stem cell and the commitment to differentiate along one or more hematopoietic lineages. The maintenance of this equilibrium underlies the hematopoietic homeostasis necessary for the continuous production of the different types of hematopoietic cells required in physiological conditions. The regulation of hematopoiesis is also sufficiently flexible to enable hematopoietic organs to respond effectively to pathological situations (e.g., during bleeding or infections) requiring a rapidly increased production of a particular blood cell type.

In the bone marrow, stem cells, committed progenitor cells, and ma-

GIORGIO TRINCHIERI, MARIANNE MURPHY, MARIA CRISTINA CUTURI, IGNACIO ANEGON, and BICE PERUSSIA ● Wistar Institute of Anatomy and Biology, Philadelphia, Pennsylvania 19104.

turing cells are intermixed with stromal cell types (endothelial cells, macrophages, adipocytes, lymphocytes, and other cell types). The differentiation of the precursor cells depends on their interactions with the stromal environment both through direct cellular contacts and through soluble factors released by the stromal cells. Hematopoietic growth factors have been purified from the medium conditioned by a variety of normal and malignant cell lines, whether of hematopoietic origin or not, and by activated lymphocytes.[2,3]

Stem and progenitor cells represent no more than 1% of total bone marrow cells and lack distinctive morphological or molecular markers. Assays of clonal growth *in vivo* and *in vitro* evaluate indirectly the number of precursor cells from the number of colonies formed. The murine spleen colony assay detects a cell, the colony-forming unit–spleen (CFU-S), with characteristics of self-renewal and pluripotency like those of very early progenitor cells but probably not those of the most primitive stem cells.[1] *In vitro* colonies of pluripotent blast cells, possibly corresponding to the more immature CFU-S, have been described.[4] In the human system, multipotential cells can be cloned *in vitro* from bone marrow or peripheral blood cells.[5] These multi-CFU form colonies containing granulocytes, eosinophils, macrophages, mast cells, megakaryocytes, and some cells with self-renewal activity. Committed precursor cells give rise to colonies containing cells of only a single lineage (or two in the case of granulocyte/macrophage CFU (CFU-GM)). The more immature CFU-GM form colonies *in vitro* mostly at 14 days of culture (day 14 or early CFU-GM), whereas the more mature ones form colonies at 7 days (day 7 or late CFU-GM). In the erythroid system, early precursor cells form large colonies at day 14 (burst-forming unit–erythroid (BFU-E)), whereas more mature precursor cells (CFU-E) form smaller colonies after only a few days of culture.

Lymphocytes, mostly T cells, represent a small but significant proportion of bone marrow cells from healthy donors. Natural killer (NK) cells originate and differentiate in the bone marrow,[6] but active mature NK cells are almost entirely absent from the bone marrow of healthy donors.[7] Alterations of T and NK cells in the bone marrow could be quantitative (increased number or change in the proportion of different subsets) or qualitative (activation of the cells). Although T and NK cells can produce both stimulating and inhibiting factors, bone marrow failure in one or more lineages is the hematopoietic condition most often associated with lymphocyte activation. In a classical study by Bagby *et al.*,[8] 234 patients with neutropenia and granulocytic hypoplasia of diverse etiologies were analyzed for evidence of lymphocyte-mediated suppression of granulopoiesis. Removal of lymphocytes from bone marrow cells resulted in increased formation of CFU-GM in 49 patients (16.6%).

The presence of inhibitory lymphocytes may represent a primary autoimmune mechanism, or they may be generated as a reaction to a pathogenic stimulus (e.g., infection or malignancy) with a secondary effect on hematopoietic cells. In some patients, the clonal or malignant expansion of a lymphocyte population with inhibitory activity is responsible for the failure of other hematopoietic cells. Lymphocytes may act directly on progenitor or stem cells, or they may affect other accessory cell types required for growth factor production. Inhibition by lymphocytes may require direct cellular contact or be mediated via soluble factors. In this review, *in vivo* and *in vitro* evidence for a role of NK cells in regulation of hematopoiesis and the mechanisms of interaction of NK cells with hematopoietic precursor cells will be discussed.

2. Experimental and Clinical *in Vivo* Evidence for a Role of NK Cells in Regulation of Hematopoiesis

A role for NK cells in hematopoietic hemeostasis was originally suggested by the pioneering studies by Cudkowicz and collaborators[9–11] on hybrid resistance to parental bone marrow transplantation in irradiated mice. Parental hematopoietic or lymphoid grafts do not survive in lethally irradiated F_1 hybrids, even though these animals are universal recipients of grafts of other types of parental tissue.[9] The genetic control of hybrid resistance contrasts with the classical transplantation studies, which show that graft compatibility rests predominantly on multiple genetic determinants of cellular antigens inherited codominantly: the histocompatibility (H) antigens. The F_1 hybrid antiparent reaction has been explained assuming the existence of a class of noncodominant genes, designated Hh for hematopoietic (or hybrid) histocompatibility, with tissue distribution restricted to hematopoietic cells.[9] By transplanting across allogeneic and xenogeneic barriers using recipients in which the T cell response has been abrogated by irradiation, it was possible to demonstrate an Hh-controlled allogeneic and xenogeneic resistance to hematopoietic cells that shares most of the properties of hybrid resistance.[12] The characteristics of the effector cells mediating hybrid resistance (e.g., radioresistance, age of maturation, bone marrow dependence, thymus independence, sensitivity to split-dose irradiation, lack of immunological memory) suggested their identity with NK cells.[10,11] In the mouse, both hybrid resistance and NK cell activity are under similar genetic control[13] and are abrogated *in vivo* by treatment with antisera recognizing NK cells[14,15]; hybrid resistance is reduced in NK-deficient beige mice,[16] and the ability to reject bone marrow in a genetically restricted way is adoptively transferred by clones with NK cell activity.[17]

The list of properties shared between the cells responsible for hybrid resistance and NK cells does not include, however, the single most pertinent property of hematopoietic resistance—its immunogenetic specificity. The genetic restriction of natural hybrid resistance has been reproduced in an *in vitro* system in which purified murine F_1 NK cells inhibit parental CFU-GM,[18] although a lower but significant suppression was also observed against syngeneic progenitor cells.[18,19] The genetic specificity has been shown by *in vivo* experiments of competitive inhibition to reside at the effector cell level.[20] A possible role of regulatory radioresistant T cells or of natural antibodies in determining genetic specificity of hybrid resistance has been proposed, but these models do not account for all properties of hybrid resistance.[21]

In vivo NK cells suppress hematopoietic progenitors in mice experimentally infected with lymphocyte choriomeningitis virus (LCMV).[22] Adult mice injected intraperitoneally with LCMV undergo a relatively mild disease followed by marked immunological and hematological dysfunction.[23,24] During the first week of infection, there is a profound suppression of CFU-S and CFU-GM.[23,24] Erythropoiesis, as measured by ^{59}Fe uptake into hematopoietic tissue, is also markedly suppressed. After day 10 of infection, CFU-S and erythropoiesis return to levels higher than normal in spleen, whereas hematopoiesis remains depressed for over 3 weeks in bone marrow. The *in vivo* infection of mice with LCMV results in interferon (IFN) production and increased NK activity in spleen and bone marrow,[25] accompanied by the appearance of NK blasts and proliferation of NK cells.[26] In the infected mice, NK activity and tissue distribution in the animals correlated with hematopoietic dysfunction,[22] although the long-lasting bone marrow defect could not be completely explained by the effect of NK cells.

NK activity was detected in the bone marrow during LCMV infection, suggesting that the depression of hematopoiesis at early times during infection might be attributed to NK cells.[22] The NK cells in bone marrow had the antigenic phenotype of immature NK cells, suggesting that either increased local production or delayed migration of NK cells from bone marrow was responsible for the increased cytotoxic activity.[22] Using an adoptive transfer system, it was shown that irradiated LCMV-infected mice reject syngeneic bone marrow and that this resistance is almost completely abolished by treatment with anti-asialo GM_1 antiserum, which abolishes NK activity.[22] These experimental observations in LCMV-infected mice demonstrated that *in vivo*-activated NK cells can suppress growth and proliferation of syngeneic hematopoietic progenitor cells and that this suppression can occur in organs, such as the bone marrow, in which NK cell-mediated cytotoxicity is normally low.

The possibility that NK cells play an important regulatory role in physiological hematopoiesis at least in extramedullary sites is strongly suggested by recent data (R. Kiessling, personal communication) showing that CFU-GM precursors in the spleen but not in the bone marrow are increased severalfold in normal mice depleted of endogenous NK cells by chronic treatment with antibody NK1.1.

In humans, several clinical situations of bone marrow depression are associated with the presence of activated lymphocytes.[8] In many cases the activated lymphocytes capable of hematopoietic suppression are T cells, mostly of the suppressor/cytotoxic CD8(+) subset,[27] which express HLA-DR and Tac activation antigens.[28] The identification of NK cells as responsible for bone marrow suppression in human pathology has been difficult because of the ambiguity of distinctive characteristics between NK cells and activated T cells. The morphology of large granular lymphocytes (LGL), typical of resting NK cells, is often presented by activated T cells, especially CD8(+) T cells. In several early studies, the antibody HNK-1 (anti-Leu7) was used as a reagent for NK cells.[29] The Leu7 antigen is present in normal peripheral blood lymphocytes on a proportion of NK cells and in a small subset of T cells,[7] and in patients with activated T cells, a large proportion of Leu7(+) T cells is often observed.

The low affinity Fc receptor (FcR) recognized by anti-CD16 antibodies[7] is also expressed on T cells from some patients. Cells bearing receptor for sheep erythrocytes (CD2 antigen) and FcR, often referred to as Tγ cells, in normal peripheral blood correspond to the NK cell subset.[7] However, in several patients with bone marrow failure (e.g., EBV infection, pure red cell aplasia during chronic lymphocytic leukemia, LGL lymphocytosis), CD2(+) cells expressing FcR and/or CD16 antigens have characteristics of T cells; i.e., they express the T cell antigen receptor (TCAR) and the TCAR-associated CD3 antigen. This cell type (CD3(+), CD16(+)) is observed in a small proportion of lymphocytes from approximately 4% of normal donors,[30] and it is expanded monoclonally in the majority of patients with LGL lymphocytosis. LGL lymphocytosis is usually manifested by granulocytopenia and/or red cell aplasia, thrombocytopenia, hypo- or hypergammaglobulinemia, and a relative or absolute increase of cells with LGL morphology.[31] In most cases the phenotype of LGL is rather homogeneous, and these cells express CD2, CD3, and CD8 antigens, but the 65,000 mol. wt. CD5 antigen present on all normal T cells and on leukemic B-CLL cells is usually absent or expressed at very low density on the LGL.[31]

All cases of LGL lymphocytosis that express CD3 antigen show rearrangement of the genes for the β and γ chains of the TCAR, indicating the T cell origin of these cells.[31] The CD3(+) LGL from all patients express FcR for IgG, as shown by rosette formation with IgG-sensitized red cells, and the FcR is usually functionally active in mediating antibody-dependent cytotoxicity, but low or no spontaneous cytotoxicity[31] is detected. Surface antigens preferentially expressed on NK cells, such as OKM1, HNK-1 (Leu7) and NKH-1 (N901 or anti-Leu19 antibodies) are expressed on the cells from some patients. Some studies,[32,33] but not others,[34,35] reported reactivity of cells from most patients with anti-CD16 (FcR) antibodies, suggesting that technical differences among laboratories more than patient heterogeneity are responsible for these contrasting results.

Approximately 10% of the patients with LGL lymphocytosis present cells with a different phenotype. These cells are CD3(−), CD16(+), CD8(+) or (−), and they usually have high spontaneous cytotoxic activity and do not show rearrangements in the genes for TCAR.[32–34,36] Evidence for monoclonality in at least two cases was presented based on chromosomal abnormality.[37,38] These cells have phenotypic and functional characteristics identical to those of peripheral blood NK cells. Chan *et al.*[35] observed that the nine patients with CD3(+) LGL presented neutropenia, whereas the two patients with CD3(−) LGL did not present abnormalities in granulopoiesis. However, other recent studies described patients with CD3(−), CD16(+) LGL lymphocytosis associated to neutropenia and anemia.[38–40] Cells from two of these patients were studied *in vitro* and shown to inhibit proliferation/differentiation of progenitor cells.[39,40]

Expansion of Leu7(+) LGL was also reported in patients with Felty's syndrome (neutropenia, arthritis, splenomegaly) and with adult-onset cyclic neutropenia.[41,42] In the patients with Felty's syndrome the Leu7(+), LGL are CD3(+) and of T cell origin, but, unlike in LGL lymphocytosis, the CD3(+) cells express the CD5 antigen and show polyclonality of TCAR gene rearrangement.[41,43] On the other hand, two out of three patients with cyclic neutropenia described by Loughran *et al.*[42] showed expansion of LGL with typical phenotype of NK cells.

Activated T cells with the ability to suppress hematopoiesis have been demonstrated in many patients with idiopathic acquired aplastic anemia.[44,45] However, rejection of bone marrow grafts from an identical twin was mediated by cells with characteristics of NK cells in at least one case of aplastic anemia.[46] This observation suggests the possibility that NK cells might be involved both in the pathogenesis of the anemia and in the mechanism of rejection of the graft.

3. Inhibition of *in Vitro* Hematopoiesis by Human NK Cells

It has been hypothesized that the *in vivo* role of NK cells might be surveillance of primitive cells.[47] The proportion of NK cells is high in blood, spleen, and liver but low in bone marrow and thymus. Normal primitive cell types with significant susceptibility to NK lysis *in vitro* can be found in bone marrow and thymus.[48–51]

Because progenitor cells represent only a very small proportion of bone marrow cells and their purification has presented serious technical difficulties, it has been very difficult to analyze directly a cytotoxic effect of NK cells on progenitor cells. Most of the *in vitro* evidence for a role of NK cells in inhibiting hematopoiesis comes from experiments testing the ability of purified NK cell preparations to suppress proliferation and differentiation of colony-forming cells. Inhibition of colony formation by CFU in 7-to-14-day assays could reflect a direct cytotoxic or cytostatic effect on the progenitor cells or any alteration in the process of proliferation and/ or differentiation that leads to the formation of colonies of differentiated cells. In these assays NK cells could act by direct cellular contact or by release of soluble factors.

An earlier suggestion that human NK cells inhibit *in vitro* hematopoiesis comes from the work of Morris *et al.*[52] showing inhibition of CFU-GM day 7 by non-B, non-T lymphocytes, and from that by Barr and Stevens,[53] who showed inhibition mediated by sheep erythrocyte rosetting (E(+)), FcR(+) cells. A better characterization of the suppressor cells was reported by Hansson *et al.*[54] using NK cells purified by discontinuous Percoll gradient centrifugation.[55] Those authors [54] showed that NK cells inhibit both autologous and allogeneic CFU-GM, that the inhibition was enhanced by pretreatment of NK cells with IFN, and that NK-sensitive target cells competed for the inhibition. The inhibitor cells were resistant to 10 Gy irradiation but required several hours of contact with the bone marrow cells before plating in semisolid medium, in order to mediate maximum inhibition. Late CFU-GM day 8 were inhibited somewhat more efficiently than the more immature CFU-GM day 14, whereas no inhibition was observed on BFU-E.

Spitzer and Verma[56] reported that both E(+) and E(−), FcR(+) lymphocytes from bone marrow were inhibitory, whereas in peripheral blood, only E(−), FcR(+) cells inhibited bone marrow CFU-GM. In these experiments,[56] no requirement for preincubation of suppressor cells with bone marrow was observed, perhaps because separation of cells through their FcR determines cellular activation. Matera *et al.*[57] also showed that Percoll-purified LGL inhibit bone marrow CFU-GM day 7 following culture with the bone marrow cells before plating, whereas FcR(+) LGL inhibited CFU-GM day 7 without requirement for preincubation. FcR(+) LGL stimulated peripheral blood CFU-GM day 14, suggesting production of a growth factor. A possible role of NK cells in inhibiting not only normal hematopoietic progenitor cells but also clonogenic growth of leukemia cells was suggested by Beran *et al.*[58]: allogeneic Percoll-purified NK cells prevented colony formation by the blasts of three patients with acute myeloid leukemia. The antileukemia cell effect of NK cells was boosted by pretreatment of effector cells with IFN. Interestingly, as observed for NK-mediated killing of target cell lines,[59,60] IFN treatment of the leukemic cells rendered them resistant to the suppressive effect of NK cells.[58]

Mangan *et al.*[61] showed that neither E(+), FcR(+), nor Percoll-enriched NK cells significantly affected BFU-E formation, whereas Percoll-enriched NK cells inhibited CFU-E growth, and this effect was enhanced by pretreatment with IFN. Purified HNK-1(+) cells suppressed both BFU-E and CFU-E, although the reactivity of antibody HNK-1 with cells other than NK prevents firm conclusions about the effect of NK cells on BFU-E.

Degliantoni *et al.*[62,63] showed that the peripheral blood cells that spontaneously suppress bone marrow hematopoietic colonies have the exact phenotype of NK cells (CD16(+), NKH-1(+), CD3(−), CD5(−), CD4(−), HLA-DR(−), mostly CD2(+), and, in part, CD8(+) and HNK-1(+)). The suppressive effect of these cells was increased by pretreatment with IFN-α. CFU-GM day 14, CFU-E, and CFU-GEMM, but neither BFU-E nor CFU-GM day 7, were suppressed by NK cells.

Hermann et al.[64] have recently analyzed the ability of human NK cell clones and CD3(+) T cell clones with NK-like cytotoxic activity to suppress in vitro hematopoiesis. The NK clones did not promote hematopoietic colony growth, and individual NK cell clones suppressed subpopulations of progenitor cells in a heterogeneous but clonally stable manner. The generation of the inhibitory effect required cell-to-cell contact, and maximum inhibition was observed after 8–18 h of preincubation.

The possible in vivo relevance of the observed in vitro reactivity of NK cells against syngeneic progenitor cells is suggested by studies with cells from a patient with aplastic anemia that twice failed to reconstitute after engraftment with bone marrow of an identical twin.[46] The patient's peripheral blood cells, with characteristics of NK cells (LGL, CD4(−), CD8(−), cytotoxic for K562 cells) caused marked inhibition of syngeneic CFU-GM colonies.[46]

Inhibition of hematopoietic colonies in vitro by cells of patients with LGL lymphocytosis have been analyzed in only a few studies. In one report,[40] cells from a patient with NK type LGL [CD3(−), CD16(+)] efficiently inhibited CFU-E and BFU-E and, less consistently, CFU-GM. In the same report, cells from six other patients with CD3(+) LGL either did not inhibit or inhibited colonies only after 24 h coculturing with the bone marrow cells. Another report[38] showed inhibition by cells from a patient with CD3(−) LGL lymphocytosis only upon stimulation with interleukin 2 (IL-2).

4. Role of Soluble Factors in the Modulation of Hematopoiesis by Lymphocytes

Lymphocytes produce an array of lymphokines with enhancing or suppressive effect on hematopoiesis. Several of these factors synergize or antagonize in their effects on progenitor cells. The types of lymphokines produced depend both on the subset of producer lymphocytes and on the stimulus inducing their production. Normal resting lymphocytes usually do not produce significant levels of lymphokines, but they do so upon activation; in pathological conditions in vivo, activated normal lymphocytes or clonally expanded malignant lymphocytes can be observed that constitutively produce lymphokines affecting hematopoiesis.[65] Lymphokines can affect hematopoiesis directly, by acting on progenitor cells, or indirectly, by inducing cytokine production by other lymphocyte subsets, mature hematopoietic cells, or stromal cells.

Lymphocytes, both T and NK cells, both *in vitro* and *in vivo*, produce lymphokines during immune or inflammatory response, and a role for these lymphokines in some of the alterations of hematopoiesis observed in pathological situations has been proposed.[65] However, it is more difficult to prove a role for lymphokines in normal hematopoietic homeostasis. *In vitro* experiments of hematopoietic colony growth have shown requirement for lymphocytes, particularly T cells, when colony-stimulating factors (CSF) are not added to the cultures.[3,66] Stimulated T cells produce (1) factors that induce growth of CFU-GM (in particular GM-CSF); (2) burst-promoting activity (BPA), which stimulates formation of BFU-E (at least three factors produced by T cells have BPA activity—GM-CSF, interleukin 3 (IL-3), and B cell-stimulating factor 1 (BSF-1)); and (3) factors that allow growth of the pluripotent CFU-GEMM (including GM-CSF, IL-3, and BSF-1). In addition, T cells can produce IFN-γ and lymphotoxin (LT)/tumor necrosis factor (TNF), which induce production of GM-CSF from monocytes and endothelial cells, respectively.[67,68] However, the absolute requirement for lymphocytes *in vitro* is in part an artifact, since most stromal cells are removed from bone marrow cell preparations, and the bone marrow microenvironment is destroyed: *in vivo* stromal cells probably provide the interactions and the factors required for hematopoiesis, and lymphocytes represent only one of the components of bone marrow microenvironment. In the absence of lymphocytes, a new equilibrium allowing hematopoietic homeostasis can be reached, as demonstrated by the normal hematopoiesis usually observed in animals or patients with severe immunodeficiencies.

NK cells also produce various types of factors, including growth factors, which affect hematopoiesis. The number of factors the production of which has been attributed to NK cells is high, but in most cases a sufficiently accurate identification of the producer cells has not been presented, and conclusive evidence that NK cells are the producer is lacking.[69] However, production of IL-1,[70] IFN-γ,[71] and B cell growth factors [72] by NK cells is supported by careful experimental data. BPA activity has also been shown to be produced by NK cells,[73,74] but the exact factor with this activity has not been characterized. The presence of B cell growth activity in NK cell supernatants suggests the possibility that NK cells produce BSF-1 (also known as IL-4, or B cell growth factor 1 (BCGF-1)), a factor with BPA functions. The possibility that NK cells produce the two other lymphokines with known BPA activity, IL-3 and GM-CSF, has not been directly tested. Production of CSF able to sustain the growth of CFU-GM day 7 in supernatants of mitogen-stimulated LGL was described; however, the type of CSF was not characterized, and NK cells were not positively identified as the producer cells.[75]

Matera *et al.*[57] reported that purified NK cells and their supernatant fluid enhance formation of CFU-GM day 14 for human peripheral blood. Although these data might suggest that production of CSF, possibly GM-CSF, an enhancing effect was described only in the presence of a source of CSF, leaving open the possibility that a costimulating factor, and not a CSF, was measured. Such a costimulating factor produced by NK cells could be IL-1α that has been shown to be equivalent with hemopoietin 1, a factor without CSF activity but synergizing with other types of CSF such as CSF-1 or IL-3.[76] Analysis of supernatant fluids from NK cell clones reported by Hermann et al.[64] did not show the presence of any CSF activity. Like T cells, NK cells produce IFN-γ and LT/TNF that induce GM-CSF production from monocytes and endothelial cells, and IL-1, which has also been demonstrated to induce GM-CSF production by endothelial cells.[77].

Inhibitory factors released by activated T cells and NK cells have also been shown to be responsible for hematopoietic suppression *in vivo* and possibly *in vitro*.[65,78] Upon virus infection, peripheral blood mononuclear cells produce IFN-α, a strong inhibitor of colony formation by CFU-GM day 7 and, less so, by CFU-GM day 14.[63,79] Although it was suggested that NK cells might be the major producer of IFN-α,[80] other studies have shown that the IFN-α producer cell is an HLA-DR(+) light-density cell type, possibly dendritic cells, clearly distinct from NK cells, but co-purifying with them on Percoll gradients.[81,82] IFN-γ, a product of both T and NK cells,[71] has been demonstrated in the serum of some patients with aplastic anemia.[78] Lymphocytes from these patients also spontaneously produced high levels of IFN-γ when cultured *in vitro*.[78] IFN-γ-containing supernatants obtained from the patients' lymphocytes or from lymphocytes from normal donor stimulated *in vitro* with PHA inhibited colony formation, and the inhibition was prevented by antibodies to IFN-γ.[78] Because preparations of natural IFN-γ also inhibited colony formation,[79,83] it was concluded that IFN-γ is responsible for the inhibition observed with supernatants from activated lymphocytes. However, other studies using homogeneous recombinant IFN-γ showed that IFN-γ was able to inhibit colonies only at concentrations much higher than those required for IFN-α or β.[63,84] Murphy et al.[84] reexamined the inhibiting activity of supernatants from PHA-stimulated T cells and demonstrated that the inhibition was mediated by a synergistic interaction between IFN-γ and LT, each factor alone being present in the supernatants at concentrations too low to account for the observed inhibition. The inhibitory activity present in the preparations of natural IFN-γ was probably due to contamination with LT, as shown in a different experimental system by Stone-Wolfe et al.[85]

Degliantoni et al.[62,63] showed that purified NK cells produce a colony-inhibiting activity (NK-CIA) when cocultured for several hours with NK-sensitive cells (such as K562 cells) or with allogeneic or autologous bone marrow cells but not with NK-insensitive cells (such as Raji cells). HLA-DR(+) bone marrow cells, highly enriched for hematopoietic progenitor cells, induce NK-CIA production, whereas HLA-DR(−) cells, depleted of precursor cells, fail to do so, suggesting that NK cells produce NK-CIA following direct interaction with the progenitor cells. The specificity of inhibition of hematopoietic colonies by NK-CIA and by NK cells was almost identical: both inhibited CFU-GEMM, CFU-E, and CFU-GM day 14, but neither inhibited BFU-E nor CFU-GM day 7.[62] NK-CIA was synergistic with IFN-γ in inhibiting CFU-GM; NK-CIA and IFN-γ, together but not separately, also inhibited CFU-GM day 7.[63] The NK-CIA concentration in the supernatant was sufficient to account for the inhibition of colony formation observed when NK cells were added directly to the bone marrow cells used for colony formation, although the contribution of a direct cytotoxic effect of NK cells on progenitor cells to the observed inhibitory effect cannot be ruled out.

Incubation of NK cells with NK-sensitive target cells was previously shown to induce production of an NK cell cytotoxic factor (NKCF), cytotoxic for NK targets such as K562, or more efficiently, the myeloid U937 cells.[86] NKCF activity was also observed in the supernatant fluid from NK cells cultured with HLA-DR(+) bone marrow cells.[63] NKCF- and NK-CIA-containing supernatants did not contain significant amounts of IFN-α or γ, and the NK-CIA activity was not inhibited by antibodies to IFN.[63] Both NK-CIA and NKCF activity in the supernatant fluids were mediated by a factor of approximately 20,000 mol. wt.[63] NK-CIA inhibition of colony formation and NKCF cytotoxicity on U937 cells were efficiently abolished by monoclonal antibodies to TNF but not to LT.[63] The NK-CIA-containing supernatants have low (0.1–10 U/ml) TNF activity as evaluated by biological assay (cytotoxicity on actinomycin D-treated mouse L cells) or by radioimmunoassay[87]: such levels of TNF were sufficient to account for the observed inhibition of colony formation as determined using recombinant TNF.[63,88,89]

The inhibition of NKCF activity on U937 cells by anti-TNF antibodies suggested a similarity between NKCF and TNF. However, NKCF-containing supernatants lysed K562 or Molt 4 cell lines, which, unlike U937 cells, are not sensitive to TNF, suggesting that not all NKCF activity could be accounted for by TNF.[86,91] A recent study by Wright and Bonavida[91] showed that two factors with different isoelectric points are responsible for NKCF activity: one probably corresponding to TNF, and one different from TNF and possibly active on K562 and Molt 4 cells. TNF alone, however, seems to be responsible for the NK-CIA activity observed in NK cell supernatants.

The ability of recombinant TNF, LT, and IFN-γ to inhibit various types of colonies was analyzed on highly enriched progenitor cell preparations, stimulated by homogeneous recombinant GM-CSF and erythropoietin.[89] Inhibition by recombinant TNF and LT was observed in these experimental conditions, suggesting that the factors directly affect the progenitor cells and do not act on other cell types (e.g., macrophages, lymphocytes) present in the bone marrow cell suspension. TNF and

LT inhibited CFU-GEMM, BFU-E, and CFU-E with similar efficiency (approximately 50% inhibition with 1 U/ml), and the inhibition was slightly augmented by IFN-γ.[89] That homogeneous TNF but not supernatants from NK cells containing TNF inhibited BFU-E colonies is probably due to the fact that NK cells produce BPA, masking the inhibition by TNF.[74]

Both TNF and LT poorly inhibited CFU-GM day 7 at doses higher than 100 U/ml, but they strongly synergized with IFN-γ in inhibiting this colony type.[89] The two cytotoxins had different effects on CFU-GM day 14: TNF alone was highly inhibitory, whereas doses of LT at least 100-fold higher were required for inhibition. Both cytotoxins, however, were strongly inhibitory for CFU-GM day 14 in synergism with IFN-γ.[84,89] This differential activity of TNF and LT explains why NK cell supernatant containing TNF were able to inhibit CFU-GM day 14 colonies in the absence of IFN-γ,[63] whereas the presence of both LT and IFN-γ in supernatants from PHA-stimulated T cells was required for inhibition.[84]

In the study by Hermann *et al.*,[64] NK clones that produced both IFN-γ and NK-CIA activity inhibited erythroid and myeloid colonies, including CFU-GM day 7. Anti-IFN-γ monoclonal antibodies prevented the inhibition of CFU-GM day 7 but not of other colony types, suggesting that the inhibition was mediated by a factor (possibly TNF) acting synergistically with IFN-γ. Cells from several LGL lymphocytosis patients have been shown to produce IFN-γ.[33,38] In one case of CD3(−), CD16(+) LGL lymphocytes, IL-2-stimulated LGL produced both IFN-γ and a colony-inhibiting activity that was only partially abolished by anti-IFN-γ antibodies, also suggesting a synergistic effect between IFN-γ and other factors, possibly cytotoxins.[38]

TNF and LT are not toxic for human myeloid cell lines, but they inhibit their proliferative potential only after inducing terminal differentiation to cells with monocyte/macrophage characteristics.[92] IFN-γ strongly synergizes with TNF and LT in inducing differentiation. The possibility that the reduction of the proliferative potential of the progenitor cells depends on a similar mechanism—i.e., induction of differentiation—is suggested by the observation that doses of TNF ineffective to inhibit CFU-GM colonies induce a significant increase in the proportion of macrophage colonies at day 7 of culture.[89]

5. Regulation of the Production of Cytotoxins by Human Lymphocytes

The finding that NK cells produce TNF, a factor originally considered a product of monocyte/macrophages, prompted us to investigate the ability of lymphocytes to produce TNF and LT. TNF and LT have almost identical biological activity and are encoded by two closely linked genes within the HLA complex on human chromosome 6.[93] The two genes have 46% homology at nucleotide level in the protein coding region and 28% homology at amino acid level.[94] The flanking regions of the two genes are different, however, suggesting separate transcriptional controls.

Monocytes produced high levels of TNF in response to LPS, whereas no TNF production was induced by LPS in lymphocyte preparations carefully depleted of monocytes.[87] However, more than 20 ng/ml of TNF was produced by purified lymphocytes by simultaneous stimulation with phorbol diesters or PHA and calcium ionophore.[87] Phorbol diesters, but not PHA, induced moderate levels of TNF from monocytes, but unlike lymphocytes, no synergistic stimulation by calcium ionophore was observed.[87] PHA alone induced only low levels of TNF from lymphocytes. LT was also produced by lymphocytes together with TNF. However, when both protein secretion and mRNA accumulation were analyzed, PHA was a better stimulus for LT, and phorbol diester together with calcium ionophore was a better stimulus for TNF production.[87]

The ability of purified T and NK cells to produce TNF was analyzed. PHA, in combination or not with phorbol diesters, induced production of TNF only from T cells, whereas upon stimulation with phorbol diester and calcium ionophore, T cells produced 5- to 10-fold more TNF than NK cells.[87] To confirm the previous observation that NK cells produce TNF only in response to NK-specific stimuli, we examined TNF production from fresh NK cell preparations or from preparations of NK cells expanded in culture for 10 days. NK cells were stimulated with anti-CD16 (FcR) antibodies or with immune complexes, in order to avoid the possible artifacts due to the presence of two different cell types as in our originally described system of coculture of NK cells with cell lines or bone marrow cells. The FcR (CD16 antigen) on NK cells is involved in their functional activity such as cytotoxicity of antibody-coated target cells. It is therefore reasonable to assume that cross-linking of the FcR on NK cells might induce modifications similar to those observed upon NK cell interaction with NK-sensitive target cells.

On fresh NK cells, Sepharose-linked anti-CD16 antibodies induced production of TNF but not of IFN-γ. On cultured NK cells, IL-2 strongly synergized with anti-CD16 antibodies in inducing TNF synthesis. IL-2 on fresh NK cells and IL-2 with anti-CD16 antibodies on cultured NK cells also induced high levels of IFN-γ production. As shown in Northern blotting analysis and nuclear run-off experiments in progress, the control of induction of synthesis of these factors is at the transcriptional level, and accumulation of mRNA reaches a maximum approximately 1 h after stimulation. These results confirm that NK cells can produce TNF and are the only cell type producing it when NK specific stimuli (anti-CD16 antibodies or target cells) are used. TNF production is not coupled to IFN-γ production unless IL-2 is present or preactivated cells are used. This observation explains why production of only TNF was observed when resting peripheral blood NK cells were used,[63] whereas IFN-γ and, probably, TNF were produced when IL-2-dependent NK cell clones[64] or

IL-2-stimulated LGL from lymphocytosis patients[38] were analyzed. These results also show that NK and T cells might affect hematopoiesis with very similar mechanisms and that the recruitment of either cell type, or their subsets, and the relative contribution of the various soluble factors depends on the type of stimulus triggering the cellular response.

6. Conclusions

Both experimental and clinical observations support the possibility that NK cells are the cellular mediators of certain types of pathological disregulation of hematopoiesis. *In vitro* models have offered insights into the mechanisms of interaction of NK cells with hematopoietic progenitor cells. NK cells probably directly interact with progenitor cells, with a mechanism of specificity still unknown, and are triggered to produce various soluble mediators. NK cells can produce factors with both enhancing and inhibiting activity on hematopoiesis. The activity of NK cells and their ability to produce cytotoxins is also regulated by other cell types through factors such as IL-2, IFN-α, etc.

The evidence for a role of NK cells in maintaining physiological hematopoietic homeostasis is much less compelling. However, if the pathological aspects of NK cell functions are interpreted, as they are for many other systems, to be an exaggeration of the physiological functions of this cell type, a role of NK cells in hematopoietic homeostasis can be assumed. The observation by R. Kiessling that depletion of NK cells *in vivo* does not affect bone marrow hematopoiesis but determines a significant increase in the number of progenitor cells in the spleen suggests the possibility that NK cells are mostly involved in the regulation of extramedullary hematopoiesis. This possibility would also be compatible with observations in the hybrid resistance system,[16] with NK cell organ distribution, and with the localization of their effect against metastatic diffusion of tumors or parasite infection.[69]

ACKNOWLEDGMENTS. We thank Marion Kaplan for typing the manuscript. The experimental data by the authors described in this chapter were obtained through support by U.S. Public Health Service grants CA10815, CA20833, CA32898, CA37155, CA40256, and CA45284. B. Perussia is a Scholar of the Leukemia Society of America.

References

1. Till, J. E., and McCulloch, E. A., 1961, A direct measurement of the radiation sensitivity of normal mouse bone marrow cells, *Radiat. Res.* **14**:213–222.
2. Metcalf, D., 1986, The molecular biology and functions of the granulocyte-macrophage colony-stimulating factors, *Blood* **67**:257–267.

3. Cline, M. J., and Golde, D. W., 1974, Production of colony-stimulating activity by human lymphocytes, *Nature* **248**:703–704.
4. Nakahata, T., and Ogawa, M., 1982, Hemopoietic colony-forming cells in umbilical cord blood with extensive capability to generate mono- and multipotential hemopoietic progenitors, *J. Clin. Invest.* **70**:1324–1328.
5. Fauser, A. A., and Messner, H. A., 1978, Granuloerythropoietic colonies in human marrow, peripheral blood and cord blood, *Blood* **52**:1243–1248.
6. Haller, O., and Wigzell, H., 1977, Suppression of natural killer cell activity with radioactive strontium: Effector cells are marrow dependent, *J. Immunol.* **118**:1503–1506.
7. Perussia, B., Starr, S., Abraham, S., Fanning, V., and Trinchieri, G., 1983, Human natural killer cells analyzed by B73.1, a monoclonal antibody blocking Fc receptor functions. I. Characterization of the lymphocyte subset reactive with B73.1, *J. Immunol.* **130**:2133–2141.
8. Bagby, G. C., Lawrence, H. J., and Neerhout, R. C., 1983, T-lymphocyte-mediated granulopoietic failure. *In vitro* identification of prednisone-responsive patients, *N. Engl. J. Med.* **309**:1073–1078.
9. Cudkowicz, G., and Stimpfling, J. H., 1964, Deficient growth of C57B1 mouse marrow cells transplanted in F1 hybrid mice. Association with the histocompatibility-2 locus, *Immunology* **7**:291–306.
10. Cudkowicz, G., and Hochman, P. S., 1979, Do natural killer cells engage in regulated reaction against self to ensure homeostasis? *Immunol. Rev.* **44**:13–41.
11. Kiessling, R., Hochman, P. S., Haller, O., Shearer, G. M., Wigzell, H., and Cudkowicz, G., 1977, Evidence for a similar or common mechanism for natural killer cell activity and resistance to hemopoietic grafts, *Eur. J. Immunol.* **7**:655–663.
12. Cudkowicz, G., and Bennett, M., 1971, Peculiar immunobiology of bone marrow allografts. I. Graft rejection by heavily "responder" mice, *J. Exp. Med.* **134**:83–102.
13. Clark, E. A., and Harmon, R. C., 1980, Genetic control of natural cytotoxicity and hybrid resistance, *Adv. Cancer Res.* **31**:227–285.
14. Okumura, K., Habu, S., and Shimamura, K., 1982, The role of asialo GM_1^+ (GA_1^+) cells in the resistance to transplants of bone marrow or other tissues, in: *Nk Cells and Other Natural Effector Cells* (R. B. Herberman, ed.), Academic Press, New York, pp. 1527–1533.
15. Lotzova, E., Pollack, S. B., and Savary, C. A., 1982, Direct evidence for the involvement of natural killer cells in bone marrow transplantation, in: *NK Cells and Other Natural Effector Cells* (R. B. Herberman, ed.), Academic press, New York, pp. 1535–1540.
16. Harrison, D. E., and Carlson, G. A., 1983, Effect of the beige mutation on natural resistance to marrow grafts, *J. Immunol.* **130**:484–489.
17. Warner, S. F., and Dennert, G., 1982, Effects of a cloned cell line with NK activity on bone marrow transplants, tumor development and metastasis *in vivo*, *Nature* **300**:31–34.
18. Bordignon, C., Daley, J. P., and Nakamura, I., 1985, Hematopoietic histoincompatibility reactions by NK cells *in vitro:* Model for genetic resistance to marrow grafts, *Science* **230**:1398–1401.
19. Holmberg, L. A., Miller, B. A., and Ault, K., 1984, The effect of natural killer cells on the development of syngeneic hematopoietic progenitors, *J. Immunol.* **133**:2933–2939.
20. Daley, J. P., and Nakamura, I., 1984, Natural resistance of lethally irradiated F1 hybrid mice to parental marrow grafts 8is a function of H-2/Hh restricted effectors, *J. Exp. Med.* **159**:1132–1148.
21. Warner, J. F., and Dennert, G., 1985, Bone marrow graft rejection as a function of antibody-directed natural killer cells, *J. Exp. Med.* **161**:563–576.
22. Randrup Thomsen, A., Pisa, P., Bro-Jorgensen, K., and Kiessling, R., 1986, Mechanisms of lymphocytic choriomeningitis virus-induced hemopoietic dysfunction, *J. Virol.* **59**:428–433.
23. Bro-Jorgensen, K., 1978, The interplay between lymphocytic choriomeningitis virus, immune function, and hemopoiesis in mice, *Adv. Virus Res.* **22**:327–369.

24. Bro-Jorgensen, K., and Knudtzon, S., 1977, Changes in hemopoiesis during the course of the acute LCM virus infection in mice, *Blood* **49:**47–57.
25. Welsh, R. M., 1978, Cytotoxic cells induced during lymphocytic choriomeningitis virus infection of mice. I. Characterization of natural killer cell induction, *J. Expl. Med.* **148:**163–181.
26. Biron, C. A., and Welsh, R. M., 1982, Blastogenesis of natural killer cells during viral infection *in vivo, J. Immunol.* **129:**2788–2795.
27. Bagby, G. C., 1981, T. lymphocytes involved in inhibition of granulopoiesis in two neutropenic patients are of the cytotoxic/suppressor (T3[+] T8[+]) subset, *J. Clin. Invest.* **68:**1597–1600.
28. Zoumbos, N. C., Gascon, P., Djeu, J., Trost, S. R., and Young, N. S., 1985, Circulating activated suppressor T lymphocytes in aplastic anemia, *N. Engl. J. Med.* **312:**257–265.
29. Abo, T., and Balch, C., 1981, A differentiation antigen of human NK and K cells identified by a monoclonal antibody (HNK-1), *J. Immunol.* **127:**1024–1029.
30. Lanier, L. L., Kipps, T. J., and Phillips, J. H., 1985, Functional properties of a unique subset of cytotoxic CD3[+] T lymphocytes that express Fc receptors for IgG (CD16/Leu11 antigen), *J. Exp. Med.* **162:**2089–2106.
31. Reynolds, C. W, and Foon, K. A., 1984, T-lymphoproliferative disorders in man and experimental animals: A review of the clinical cellular and functional characteristics, *Blood* **64:**1146–1158.
32. Van De Griend, R. J., and Bolhuis, R. L. H., 1985, *In vitro* expansion and analysis of cloned cytotoxic T cells derived from patients with chronic T lymphoproliferative disorders, *Blood* **65:**1002–1009.
33. Pistoia, V., Prasthofer, E. F., Tilden, A. B., Barton, J. C., Ferrarrini, M., Grossi, C. E., and Zuckerman, K. S., 1986, Large granular lymphocytes (LGL) from patients with expanded LGL populations acquire cytotoxic functions and release lymphokines upon *in vitro* activation, *Blood* **68:**1095–1100.
34. Rambaldi, A., Pelicci, P., Allavena, P., Knowles, D. M., Rossini, S., Bassan, R., Barbri, T., Dala-Favera, R., and Montovani, A., 1985, T cell receptor β chain gene rearrangements in lymphoproliferative disorders of large granular lymphocytes/natural killer cells, *J. Exp. Med.* **162:**2156–2162.
35. Chan, W. C., Link, S., Mawle, A., Check, I., Byrnes, R. K., and Winton, E. G., 1986, Heterogeneity of large granular lymphocyte proliferations: Delineation of two major subtypes, *Blood* **68:**1142–1153.
36. McKenna, R. W., Arthur, D. C., Gajl-Paczalska, K. J., Flynn, P., and Brunning, R. D., 1985, Granulated T cell lymphocytosis with neutropenia: Malignant or benign chronic lymphoproliferative disorder? *Blood* **66:**259–266.
37. Pistoia, V., Carroll, A. J., Prasthofer, E. F., Tilden, A. B., Zuckerman, K. S., Ferrarini, M., and Grossi, C. E., 1986, Establishment of TAC-negative, IL-2-dependent cytotoxic cell lines from large granular lymphocytes (LGL) of patients with expanded LGL populations, *J. Clin. Immunol.* **6:**457–466.
38. Koizumi, S., Seki, H., Tachinami, T., Taniguchi, M., Matsuda, A., Taga, K., Nakarai, T., Kato, E., Taniguchi, N., and Nakamura, H., 1986, Malignant clonal expansion of large granular lymphocytes with a Leu11[+], Leu-7[-] surface phenotype: *In vitro* responsiveness of malignant cells to recombinant human interleukin 2, *Blood* **68:**1065–1073.
39. Tagawa, S., Tokumine, Y., Ueda, E., Waki, K., Kanayama, Y., Taniguchi, N., Nakanishi, T., Inoue, R., and Kitani, T., 1986, Leu11[+] T cell chronic lymphocytic leukemia with partially activated natural killer function and its further activation by recombinant IL-2 *in vitro, Blood* **68:**846–852.
40. Grillot-Courvalin, C., Vinci, G., Tsapis, A., Dokhelar, M. C., Vainchenker, W., and Brouet, J. C., 1987, The syndrome of T8 hyperlymphocytosis: Variation in phenotype and cytotoxic activities of granular cells and evaluation of their role in associated neutropenia, *Blood* **69:**1204–1210.

41. Friemark, B., Lanier, L., Phillips, J., Quertermous, T., and Fox, R., 1987, Comparison of T cell receptor gene rearrangement in patients with large granular T cell leukemia and Felty's syndrome, *J. Immunol.* **138:**1724–1729.

42. Loughran, T. P. J., Clark, E. A., Price, T. H., and Hammond, W. P., 1986, Adult-onset cyclic neutropenia is associated with increased large granular lymphocytes, *Blood* **68:**1082–1087.

43. Linch,D. C., Newland, A. C., Turnbull, A. L., Knott, L. J., MacWhannel, A., and Beverley, P., 1984, Unusual T cell proliferations and neutropenia in rheumatoid arthritis: Comparison with classical Felty's syndrome, *Scand. J. Haematol.* **33:**342–350.

44. Torok-Storb, B. J., Sieff, C., Storb, R., Adamson, J., and Thomas, E. D., 1980, *In vitro* tests for distinguishing possible immune-mediated aplastic anemia from transfusion-induced sensitization, *Blood* **55:**211–215.

45. Bacigalupo, A., Podesta, M., Mingari, M. C., Moretta, L., Van Lint, M. T., and Marmont, A., 1980, Immune suppression of hematopoiesis in aplastic anemia: Activity of T lymphocytes, *J. Immunol.* **125:**1449–1453.

46. Goss, G. D., Wittwer, M. A., Bezwoda, W. R., Herman, J., Rabson, A., Seymour, L., Derman, D. P., and Mendelow, B., 1985, Effect of natural killer cells on syngeneic bone marrow: *In vitro* and *in vivo* studies demonstrating graft failure due to NK cells in an identical twin treated by bone marrow transplantation, *Blood* **60:**1043–1046.

47. Kiessling, R., and Wigzell, H., 1981, Surveillance of primitive cells by natural killer cells, *Curr. Top. Microbiol. Immunol.* **92:**107–123.

48. Hansson, M., Kiessling, R., and Andersson, B., 1981, Human fetal thymus and bone marrow contain target cells for natural killer cells, *Eur. J. Immunol.* **11:**8–12.

49. Hansson, M., Kiessling, R., Andersson, B., Karre, K., and Roder, J., 1979, Natural killer (NK) sensitive T-cell subpopulation in the thymus: Inverse correlation to NK activity of the host, *Nature* **278:**174–176.

50. Riccardi, C., Santoni, A., Barlozzari, T., and Herberman, R. B., 1981, *In vivo* reactivity of mouse natural killer (NK) cells against normal bone marrow cells, *Cell. Immunol.* **60:**136–143.

51. Gidlund, M., Nose, M., Axberg, I., Wigzell, H., Totterman, T., and Nilsson, K., 1982, Analysis of differentiation events causing changes in NK cell tumor-target sensitivity, In: *NK Cells and Other Natural Effector Cells* (R. B. Herberman, ed.), Academic Press, New York, pp. 733–741.

52. Morris, T. C. M., Vincent, P. C., Sutherland, R., and Hersey, P., 1980, Inhibition of normal granulopoiesis *in vitro* by non-B non-T lymphocytes, *Br. J. Haematol.* **45:**541–550.

53. Barr, R. D., and Stevens, C. A., 1982, The role of autologous helper and suppressor T cells in the regulation of human granulopoiesis, *Am. J. Hematol.* **12:**323–326.

54. Hansson, M., Beran, M., Andersson, B., and Kiessling, R., 1982, Inhibition of *in vitro* granulopoiesis by autologous and allogeneic human NK cells, *J. Immunol.* **129:**126–132.

55. Timonen, T., and Saksela, E., 1980, Isolation of human natural killer cells by density gradient centrifugation, *J. Immunol. Methods* **36:**285–291.

56. Spitzer, G., and Verma, D. S., 1982, Cells with Fc receptors from normal donors suppress granulocyte-macrophage colony formation, *Blood* **60:**758–766.

57. Matera, L., Santoli, D., Garbarino, G., Pegoraro, L., Bellone, G., and Pagliardi, G., 1986, Modulation of *in vitro* myelopoiesis by LGL: Different effects on early and late progenitor cells, *J. Immunol.* **136:**1260–1265.

58. Beran, M., Hansson, M., and Kiessling, R., 1983, Human natural killer cells can inhibit clonogenic growth of fresh leukemic cells, *Blood* **61:**596–599.

59. Trinchieri, G., and Santoli, D., 1978, Antiviral activity induced by culturing lymphocytes with tumor-derived or virus-transformed cells. Enhancement of human natural killer cell activity by interferon and antagonistic inhibition of susceptibility of target cells to lysis, *J. Exp. Med.* **147:**1314–1333.

60. Trinchieri, G., Granato, D., and Perussia, B., 1981, Interferon-induced resistance of fibroblasts to cytolysis mediated by natural killer cells: Specificity and mechanism, *J. Immunol.* **126:**335–340.
61. Mangan, K. F., Chikkappa, G., Bieler, L. F., Scharfman, W. B., and Parkinson, D. R., 1982, Regulation of human blood erythroid burst-forming unit (BFU-E) proliferation by T-lymphocyte subpopulations defined by Fc receptors and monoclonal antibodies, *Blood* **59:**990–996.
62. Degliantoni, G., Perussia, B., Mangoni, L., and Trinchieri, G., 1985, Inhibition of bone marrow colony formation by human natural killer cells and by natural killer cell–derived colony-inhibiting activity, *J. Exp. Med.* **161:**1152–1168.
63. Degliantoni, G., Murphy, M. Kobayashi, M., Francis, M. K., Perussia, B., and Trinchieri, G., 1985, Natural killer (NK) cell–derived hematopoietic colony-inhibiting activity and NK cytotoxic factor. Relationship with tumor necrosis factor and synergism with immune interferon, *J. Exp. Med.* **162:**1512–1530.
64. Herrmann, F., Schmidt, R. E., Ritz, J., and Griffin, J. D., 1987, *In vitro* regulation of human hematopoiesis by natural killer cells: Analysis at a clonal level, *Blood* **69:**246–254.
65. Zoumbos, N., Raefsky, E., and Young, N., 1986, Lymphokines and hematopoiesis, *Prog. Hematol.* **14:**201–227.
66. Nathan, D. G., Chess, L., Hillman, D. G., Clark, B., Breard, J., Merler, E., and Housman, D. E., 1978, Human erythroid burst forming unit (BFU-E): T cell requirement for proliferation *in vitro*, *J. Exp. Med.* **147:**324–339.
67. Herrmann, F., Cannistra, S. A., and Griffin, J. D., 1986, T cell–monocyte interactions in the production of humoral factors regulating human granulopoiesis *in vitro*, *J. Immunol.* **136:**2856–2861.
68. Munker, R., Gasson, J., Ogawa, M., and Koeffler, H. P., 1986, Recombinant human TNF induces production of granulocyte-monocyte colony-stimulating factor, *Nature* **323:**79–82.
69. Trinchieri, G., and Perussia, B., 1984, Human natural killer cells: Biologic and pathologic aspects, *Lab. Invest.* **50:**489–513.
70. Scala, G., Allavena, P., Djeu, J. Y., Kasahara, T., Ortaldo, J. R., Herberman, R. B., and Oppenheim, J. J., 1984, Human large granular lymphocytes are potent producers of interleukin-1, *Nature* **309:**56–59.
71. Trinchieri, G., Matsumoto-Kobayashi, M., Clark, S. C., Sheehra, J., London, L., and Perussia, B., 1984, Response of resting human peripheral blood natural killer cells to interleukin-2, *J. Exp. Med.* **160:**1147–1169.
72. Pistoia, V., Cozzolino, F., Torcia, M., Castigli, E., and Ferrarini, M., 1985, Production of B cell growth factor by a Leu7+, OKM1+ non-T cell with the features of large granular lymphocytes (LGL), *J. Immunol.* **134:**3179–3184.
73. Linch, D. C., Lipton, J. M., and Nathan, D. G., 1985, Identification of three accessory cell populations in human bone marrow with erythroid burst-promoting properties, *J. Clin. Invest.* **75:**1278–1284.
74. Pistoia, V., Ghio, R., Nocera, A., Leprini, A., Perata, A., and Ferrani, M., 1985, Large granular lymphocytes have a promoting activity on human peripheral blood erythroid burst-forming units, *Blood* **65:**464–472.
75. Kasahara, T., Djeu, J. Y., Dougherty, S. F., and Oppenheim, J. S., 1983, Capacity of human large granular lymphocytes (LGL) to produce multiple lymphokines: Interleukin 2, interferon and colony stimulating factor, **131:**2379–2385.
76. Stanley, E. R., Bartocci, A., Patinkin, D., Rosendaal, M., and Bradley, T. R., 1986, Regulation of very primitive multipotent hemopoietic cells by hemopoietin-1, *Cell* **45:**667–674.
77. Sieff, C. A., Tsai, S., and Faller, D. V., 1987, Interleukin 1 induces cultured human endothelial cell production of granulocyte-macrophage colony-stimulating factor, *J. Clin. Invest.* **79:**48–51.

78. Zoumbos, N. C., Gascon, P., Djeu, J. Y., and Young, N. S., 1985, Interferon is a mediator of hematopoietic suppression in aplastic anemia *in vitro* and possibly *in vivo*, *Proc. Natl. Acad. Sci* USA **82**:188–192.

79. Broxmeyer, H. E., Lu, L., Platzer, E., Feit, C., Juliano, L., and Rubin, B. Y., 1983, Comparative analysis of the influences of human gamma, alpha, and beta interferons on human multipotential (CFU-GEMM), erythroid (BFU-E) and granulocyte-macrophage (CFU-GM) progenitor cells, *J. Immunol.* **131**:1300–1305.

80. Djeu, J. Y., Stocks, N., Zoon, K., Stanton, G. J., Timonen, T., and Herberman, R. B., 1982, Positive self regulation of cytotoxicity in human natural killer cells by production of interferon upon exposure to influenza and herpes virus, *J. Exp. Med.* **156**:1222–1234.

81. Perussia, B., Fanning, V., and Trinchieri, G., 1985, A leukocyte subset bearing HLA-DR antigens is responsible for *in vitro* alpha interferon production in response to viruses, *Nat. Immun. Cell Growth Regul.* **4**:120–137.

82. Bandyopadhyay, S., Perussia, B., Trinchieri, G., Miller, D. S., and Starr, S. E., 1986, Requirement for HLA-DR positive accessory cells in natural killing of cytomegalovirus-infected fibroblasts, *J. Exp. Med.* **164**:180–195.

83. Klimpel, G. R., Fleischmann, R., and Klimpel, K. D., 1982, Gamma interferon (IFNγ) and IFNα/β suppress murine myeloid colony formation (CFU-C): Magnitude of suppression is dependent upon level of colony-stimulating factor (CSF), *J. Immunol.* **129**:76–80.

84. Murphy, M., Loudon, R., Kobayashi, M., and Trinchieri, G., 1986, Gamma interferon and lymphotoxin, released by activated T cells, synergize to inhibit granulocyte-monocyte colony formation, *J. Exp. Med.* **164**:263–279.

85. Stone-Wolfe, D. S., Yip, Y. K., Kelker, H. C., Le, J., Henriksen-Destafano, D., Rubin, B. Y., Rinderknecht, E., Aggarwal, B. B., and Vilcek, J., 1984, Interrelationship of human interferon-gamma with lymphotoxin and monocyte cytotoxin, *J. Exp. Med.* **159**:828–843.

86. Wright, S. C., and Bonavida, B., 1982, Studies on the mechanism of natural killer (NK) cell-mediated cytotoxicity (CMC). I. Release of cytotoxic factors specific for NK-sensitive target cells (NKCF) during coculture of NK effector cells with NK target cells, *J. Immunol.* **129**:433–439.

87. Cuturi, M. C., Murphy, M., Costa-Giomi, M. P., Weinmann, R., Perussia, B., and Trinchieri, G., 1987, Independent regulation of tumor necrosis factor and lymphotoxin production by human peripheral blood lymphocytes, *J. Exp. Med.* **165**:1581–1594.

88. Broxmeyer, H. E., Williams, D. E., Lu, L., Cooper, S., Anderson, S. L., Beyer, G. S., Hoffman, R., and Rubin, B. Y., 1986, The suppressive influences of human tumor necrosis factors on bone marrow hematopoietic progenitor cells from normal donors and patients with leukemia: Synergism of tumor necrosis factor and interferon-γ, *J. Immunol.* **136**:4487–4495.

89. Murphy, M., Perussia, B., and Trinchieri, B., 1988, Effects of recombinant tumor necrosis factor, lymphotoxin and immune interferon on proliferation and differentiation of enriched hematopoietic precursor cells. *Exp. Hematol.* **16**:131–138.

90. Ortaldo, J. R., Ransom, J. R., Sayers, T. J., and Herberman, R. B., 1987, Analysis of cytostatic/cytotoxic lymphokines: Relationship of natural killer cytotoxic factor to recombinant lymphotoxin, recombinant tumor necrosis factor, and leukoregulin, *J. Immunol.* **137**:2857–2863.

91. Wright, S. C., and Bonavida, B., 1987, Studies on the mechanism of natural killer cell–mediated cytotoxicity. VII. Functional comparison of human natural killer cytotoxic factors with recombinant lymphotoxin and tumor necrosis factor, *J. Immunol.* **138**:1791–1798.

92. Trinchieri, G., Kobayashi, M., Rosen, M., Loudon, R., Murphy, M., and Perussia, B., 1986, Tumor necrosis factor and lymphotoxin induce differentiation of human myeloid cell lines in synergy with immune interferon. *J. Exp. Med.* **164**:1206–1225.

93. Spies, T., Morton, C. C., Nedospasou, S. A., Fiers, W., Pious, D., and Strominger, J. L.,

1986, Genes for the tumor necrosis factors α and β are linked to the major histocompatibility complex, *Proc. Natl. Acad. Sci. USA* **83:**8699–8702.

94. Pennica, D., Nedwin, G. E., Hayflick, J. S., Seeburg, P. H., Derynk, R., Palladino, M. A., Kohr, W. J., Aggarwal, B. B., and Goeddel, D. V., 1984, Human tumor necrosis factor: Precursor structure, expression and homology to lymphotoxin, *Nature* **312:**724–729.

12

Characterization and Functions of the Natural Suppressor Cell Systems

Tom Maier, James H. Holda, Ken Lee Choi, and Henry N. Claman

Natural suppressor (NS) cells are potent, nonspecific, major histocompatibility complex (MHC)-unrestricted inhibitors of immune responses. They have been characterized as having the "null" phenotype; i.e., they do not have the usual markers or characteristics of mature T cells, B Cells, or macrophages and are, for example, Thy^-, Ig^-, and nonadherent. They differ from natural killer (NK) cells in that they show no ability to kill NK targets such as YAC-1.

1. NS Cell Systems: Where They Are Found

Cells with the functional and phenotypic characteristics of NS cells have been found in a number of locations and after a variety of procedures (Table I). The effector cells in all these sites bear the "null" phenotype, are not MHC-restricted, and are nonspecific in their suppressive activity. They are not cytolytic and do not kill natural killer (NK) targets.

Tom Maier, James H. Holda, Ken Lee Choi, and Henry N. Claman • Departments of Medicine and Microbiology/Immunology, University of Colorado School of Medicine, Denver, Colorado 80262.

TABLE I
Natural Suppressor Systems

Systems	References
Definite	
Fetal and newborn tissues	19, 24–26, 31, 33, 108–110, 124
Adult bone marrow	37–42, 46, 125
After total lymphoid irradiation	47, 51, 52, 54, 92, 109, 131, 140
After cyclophosphamide	60–67
During graft vs. host disease	69–71, 114, 122
Possible	
Pregnancy	73, 77, 78, 80, 128
After strontium-89	81, 82

1.1. Fetal and Newborn Tissues

The foundation for self, nonself reactivity is normally "learned" by the immune system during the fetal and neonatal period.[1,2] If foreign antigen is introduced into the animal during this period, the animal will become tolerant to it and subsequently will react to it as it would to self.[1–3]

During the fetal/newborn period, mammals have an extremely reduced capacity to mount a productive immune response to foreign antigens.[4–9] One reason for this may be the potent suppressive activity that is found in fetal and newborn lymphoid tissues at this time.[10–20] This suppression is antigen-nonspecific and MHC-unrestricted. It is first apparent in the fetal liver and spleen at about 16 days' gestation.[11,13,16,21,22] After birth this suppressive activity has been mainly found in the spleen, where it rapidly declines over the next 2–3 weeks.[20,23–25]

The cells responsible for this suppressive activity were generally characterized as being Thy$^+$ T cells in early work on this subject.[10–12,17,18,20] However, many later studies, which generally employed monoclonal antibodies instead of polyclonal antiserum, have characterized the cells responsible for fetal/newborn suppression as being Thy$^-$ and negative for most other T cell markers and therefore probably not a T cell.[19,24–33] The cells have also been found to be Ig$^-$ and thus are not mature B cells.[24,30–33] The suppressor cells have generally been found to be nonadherent to nylon wool (NW), Sephadex, G-10, and plastic or glass plates, and they do not phagocytize carbonyl iron.[17–19,22,24,25,27,29,32,33] They are also Ia$^-$ and Mac-1$^-$ and therefore are not mature macrophages.[24,26,30,32] Finally, the fetal/newborn suppressor cells are not NK cells, because they have no cytolytic activity toward NK targets such as YAC-1.[24–26] In fact, NK activity does not appear until after the fetal/

newborn suppressor activity is gone.[34–36] Thus, although there is no consensus, it appears that at least a major part of the suppressive activity found in fetal/newborn liver and spleen is caused by a nonadherent, Thy⁻, Ig⁻, non-NK cell population—i.e., NS cells.

1.2. Adult Bone Marrow (BM)

Another location where NS activity appears to be normally present is adult BM. BM cells contain a population of cells that have very potent suppressor activity.[37–46] This suppression is not MHC-restricted and is antigen-nonspecific. The cells responsible for this suppression are nonadherent Thy⁻ cells which do not lyse NK targets.[37,38,40–42,46] The suppressor cells are also usually found to lack surface Ig, although they may express Fc receptors and therefore might be counted as surface Ig⁺ under some experimental conditions.[37,38,41,42,44,45] Although NS activity is present in adult BM, there is no evidence that it does not also exist in fetal/newborn BM. However, too few experiments have been done with fetal/newborn BM to know with certainty.

1.3. After Total Lymphoid Irradiation (TLI)

NS activity is not prominent in normal spleen but can be found there after a variety of manipulations. One of these is TLI. TLI is a procedure consisting of high-dose, fractionated irradiation administered to the major lymphoid tissues while protecting the long bones of the limbs.[47,48] After TLI treatment the animal is very immunosuppressed, with suppressor cells appearing in the spleens and peripheral blood.[24,47–54] In mice these suppressor cells disappear during the 3–4 weeks following the cessation of treatment.[24,47,49,51,52] The suppressor cells that appear in the spleens after TLI have been extensively characterized and found to be NS cells.[24,47,51,52,54]

1.4. After Cyclophosphamide (CY)

Injecting a large dose of CY into an animal induces a transient state of immunological unresponsiveness.[55–58] Recovery from CY treatment is accompanied by potent suppressor activity in the spleen.[59–67] The suppressor activity is MHC-nonrestricted and antigen-nonspecific. The suppressor cell has generally been characterized as nonadherent, Thy⁻, and Ig⁻.[61,64,65,67] Therefore the cell responsible for CY-induced suppressor activity appears to also be an NS cell.

1.5. During Chronic Graft versus Host Disease (GVHD)

When chronic GVHD is induced across minor histocompatibility barriers in a murine system, a state of immunosuppression results.[68,69] This is associated with the appearance of a very potent nonspecific, nonrestricted suppressor cell in the spleens of the recipients.[69] These suppressor cells remain in the spleen for many months.[70] The cells responsible for this suppression have been characterized as Thy⁻, L3T4⁻, and Lyt-2⁻ and therefore not T cells. They are also not B cells, as they are Ig⁻. The suppressor cells are nonadherent to NW, Sephadex G-10, and plastic, and they are esterase-negative, so they are not macrophages. The GVHD spleen cells are also not cytolytic or cytostatic toward YAC-1. The cells responsible for the suppressive activity in GVHD spleen are therefore NS cells.[70,71]

1.6. During Pregnancy

All the above locations have suppressor activity that has been definitely shown to be caused at least in part by NS cells. There are several situations where profound nonspecific and nonrestricted suppression may also be associated with NS activity. These include during pregnancy and after treatment with radioactive strontium-89 (^{89}Sr).

During pregnancy the mother is carrying what is essentially an allograft. Why the fetus is not attacked and aborted by the immune system has been the subject of considerable research. One finding that is thought to bear on the mother's tolerance of the fetus is the presence of suppressor activity in both fetal[10-33,72,73] and maternal tissue.[73-80] This suppression is noted locally in the decidua and draining lymph nodes of the uterus[75-80] but can be seen less prominently systemically.[29,73,74] The suppression is nonspecific and MHC-nonrestricted.[77,78] Similar to the fetal/newborn suppressor activity, there is still no real consensus about the phenotype of the cells responsible. However, a number of studies have characterized the suppressor cells as being nonadherent, Thy⁻, and Ig⁻.[73,76-78] Thus NS cells may be responsible.

1.7. After Strontium-89

When the BM-seeking, radioactive compound ^{89}Sr is injected into an animal, the spleen soon develops potent, nonspecific, MHC-nonrestricted suppressor activity.[81,82] The cell responsible for this splenic suppressor activity is Thy⁻, Ig⁻, and nonadherent to NW, and it does not phagocytize carbonyl iron.[81,82] Therefore the cells responsible again may be NS.

1.8. General Characteristics of NS Cell Locations

There are two important characteristics of all the locations where NS cell activity is found. The first is that they are all environments of considerable hematopoiesis. Fetal/newborn liver and spleen and adult BM all contain active hematopoietic tissue. After TLI treatment, the shielded bone marrow quickly autotransfuses the spleen, which becomes a site of rapid stem cell turnover.[47,48] During GVHD in irradiated recipients, there is considerable proliferation of donor hematopoietic cells in the spleen.[68,70] After ^{89}Sr is injected into an animal, it quickly destroys the BM, and the spleen subsequently takes over as the primary hematopoietic organs.[81,82] High-dose CY treatment causes generalized lymphohematopoietic destruction, followed by cell proliferation in the spleen and elsewhere.[83] Finally, during pregnancy there is rapid cell proliferation in the uterus and to a lesser degree in the maternal BM.[84]

The second important characteristic of all the NS-containing locations is that they are all environments where tolerance can occur. The fetal/newborn period, as mentioned above, is not only one where self-tolerance is established, but is also a period when the induction of acquired tolerance can occur.[1-3] It has been postulated that BM may be the site of some tolerance induction throughout life.[85-87] TLI[47,54,88-92] and CY[93-100] treatments are both practical means of producing an immunosuppressive environment where tolerance induction is possible. In BM transplantation, host immunosuppression is essential to the induction of tolerance to the graft and the host.[101-104] During pregnancy, although true long-term tolerance is never really induced, there is nevertheless immunological tolerance between the fetus (graft) and the mother (host).[105,106] The tolerogenic potential of ^{89}Sr treatment is less clear.

2. NS Cells: Development and Mode of Action

Figure 1 is a diagram of a theoretical developmental pathway of NS cells. From the section above it is apparent that NS cells have so far been found only in sites of intense hematopoiesis. We can therefore assume that this local environment is conducive for NS cell development. Dissecting just what cells and factors make up this environment and are required for NS cell development is very difficult because of the complexity involved. However, we can make a few general comments about this area.

2.1. What Stimulates NS Development?

There are a number of known growth factors that could be involved in the development of NS cells from pluripotent stem cells through pNS

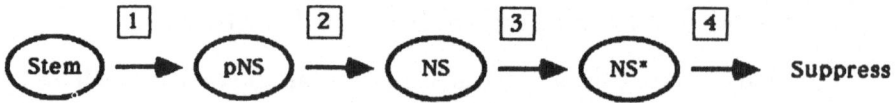

FIGURE 1. Schema of the theoretical developmental pathway of NS cells. NS cells originally come from a pluripotent stem cell through a pre-NS (pNS) to a NS cell. NS cells have the potential to become suppressive (NS*) if the proper "signals" are present. Numbers 1 and 2 represent various growth and competency factors, which are required. Number 3 represents the various "signals" to which NS cells can respond to become fully suppressive NS*. Number 4 represents the actual suppressive mechanism(s) of NS cells.

cells (pathways 1 and 2 in Fig. 1). Among these is IL-3, which can support the development of several cell lineages.[107] A suppressor monocyte/mast cell from the spleens of newborn mice has been grown in IL-3.[108] Therefore, IL-3 may be involved normally in NS cell development in hematopoietic locations. Con A supernatant (CAS) may also contain factors that are required for NS development.[109,110] Other factors probably involved in NS cell development include the many colony-stimulating factors (CSF) that have been shown to be required for the *in vitro* propagation of various hematopoietic cell lineages.[107]

There may also be hormonal influences on NS cell development. For example, during pregnancy many hormones are released by both the mother and the fetus which are suspected of influencing immune reactivity, including suppressor cell development during pregnancy.[111–113] There are no doubt major hormonal shifts that occur after all the treatment procedures listed in Table I. These could play a major role in NS cell development in these situations.

2.2. What Stimulates NS Effector Function?

In Fig. 1 we have made the distinction between potentially active NS and activated NS cells (NS*) which express suppressive activity (pathway 3). This is based on findings from the GVHD-induced NS cell system. First, it was found that T cells were required in the B10.D2 inoculum in order for NS activity to develop in BALB/c (600R) recipients (Table II).[114] This showed that T cells are required for the *in vivo* expression of NS activity. Second, it was found that GVHD-induced NS activity declines slowly over a period of several months.[70,114] However, the ability to suppress a T cell mitogen response such as Con A wanes much more slowly than a B cell mitogen response such as LPS (Fig. 2).[114]

These two findings suggested that NS cells were responding to activated T cell "signals" to express their suppressive potential. *In vivo* B10.D2 T cells are required because they probably respond to and become activated by the host BALB/c antigens. These activated T cells "signal" NS cells which then become activated to express their suppressive ability. Thus,

TABLE II
*T and Non-T Cells Are Required for in Vivo
Induction of Maximum Suppression*

Cell type and no. injected[a]			Suppression of Con A[b] CPM (% Supp)	Suppression of LPS[b] CPM (% Supp)
Untreated spleen	Anti-Thy + C' spleen	NW-NA LNC		
Control: normal BALB/c			356,000	93,000
Experimental: B10.D2				
4×10^7	—	—	35,000 (90)	5,000 (95)
—	4×10^7	—	320,000 (10)	69,000 (25)
—	—	2×10^7	264,000 (26)	52,000 (43)
—	2×10^7	2×10^7	85,000 (76)	5,000 (95)
—	3.3×10^7	6.7×10^7	67,000 (81)	9,000 (90)
—	3.8×10^7	2.2×10^6	83,000 (77)	26,000 (72)

[a] B10.D2 cells of the indicated type were injected into BALB/c (600 r) on day 0, and their spleens were assayed 10 days later. Normal BALB/c = mice neither irradiated nor injected. The cell types injected were either untreated spleen cells or anti-Thy 1.2 + C'–treated spleen cells (i.e., T cell-depleted) or nylon wool nonadherent (NW-NA) lymph node cells (LNC) (i.e., T cell-enriched) or a mixture of the latter two.
[b] 1.0×10^5 GVHD spleen cells were added to 2.5×10^5 normal BALB/c + Con A (4 μg/ml) or LPS (10 μg/ml). Percent suppression was calculated versus 2.5×10^5 normal BALB/c spleen cells + the appropriate mitogen.

at early times in the GVHD system when T cell "signals" are vigorously generated *in vivo*, the NS cells suppress both a Con A and an LPS response equally well because they have been activated fully *in vivo*. However, with time, the activity of T cells declines (perhaps because NS cells feed back and down-regulate them), and the NS cells are no longer activated *in vivo*. They must then come into contact with activated T cell "signals" *in vitro* to become fully reactivated to express their suppressive abil-

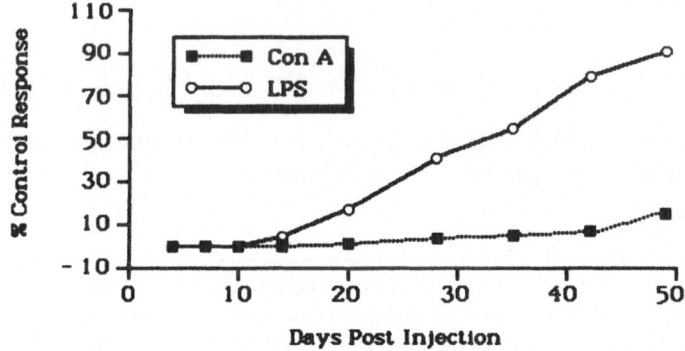

FIGURE 2. GVHD spleen cells lose the ability to suppress the LPS response of B10.D2 cells with time. BALB/c (600 r) recipients were injected on day 0 with 5×10^7 B10.D2 spleen cells. At various times subsequently, 2.5×10^5 GVHD spleen cells were cocultured with 2.5×10^5 normal B10.D2 spleen cells plus either Con A (4 μg/ml) or LPS (10 μg/ml). Present control response is vs. 2.5×10^5 normal B10.D2 spleen cells plus the appropriate mitogen.

ity. Thus spleen cells from "late" GVHD (i.e., >40–50 days) will suppress a Con A response (where the mitogen directly activates T cells to generate lots of "signals") but not an LPS response well (because there are few T cell "signals"). They will, however, suppress an allogeneic LPS response better than a syngeneic LPS response,[70,114] presumably because the allogeneic indicator cells respond to the suppressor population as in a mixed leukocyte reaction, thereby activating T cells to release lymphokines which, in turn, activate the suppressors.

Therefore the NS cells would only *appear* to be less able to suppress a non-T cell response such as an LPS response. If an exogenous source of activated T cell "signals" is added to these LPS responses, then the late GVHD NS cells should be just as capable of suppressing it as a Con A response which internally generates its own "signal." Figure 3A shows that if CAS (source of activated T cell "signals") is added to an LPS suppression assay, the late GVHD spleen cells are able to suppress the response.[46,114]

It was found that two lymphokines were able to substitute for the CAS in activating NS suppressive ability.[46] These were IFY-γ and IL-2 (Fig. 3B,C). Using anti-IFN-γ, it has been shown that IFN-γ is responsible for NS cell activation. Anti-IFN-γ is able to remove the ability of NS cells to suppress a Con A response (Fig. 4). Moreover, anti-IFN-g removes the ability of not only IFN-γ but also IL-2 and CAS to enhance NS suppression of an LPS response (Fig. 5). By these criteria, IFN-γ is the molecule responsible for NS cell activation, and IL-2 acts indirectly by its ability to stimulate IFN-γ synthesis. These findings have been extended to two other NS systems: adult BM[46] and newborn spleen.[25]

Work with cloned NS cells from both TLI and newborn spleen has shown that these suppressor cells respond to histamine to become more suppressive.[115–117] Thus histamine may be another factor besides IFN-γ which activates NS cells to become functional.

2.3. Effector Mechanism of NS Activity (Fig. 1, Pathway 4)

By definition, NS cells are suppressive. The commonest assays for NS activity measure suppression of proliferation. These include suppression

FIGURE 3. Late GVHD spleen cells significantly suppress a B10.D2 LPS response only with added lymphokines. The control LPS response of 2.5×10^5 B10.D2 spleen cells is 74,000 cpm (100%). When lymphokines are added, this response is 67,000 (for CAS, 15 μl), 63,000 (for IFN-γ, 300 U), and 90,000 (for IL-2, 3,000 U). When 2.5×10^5 day 70 GVHD spleen cells (without lymphokines) are added to the LPS response of normal spleen cells, the response is 87% of control. Addition of CAS, IFN-γ, or IL-2 to cultures containing responders, mitogen, and GVHD NS cells enhances their suppressive ability of the LPS proliferative response. The lymphokines themselves had little effect on the LPS response of the indicator B10.D2 LPS response control.

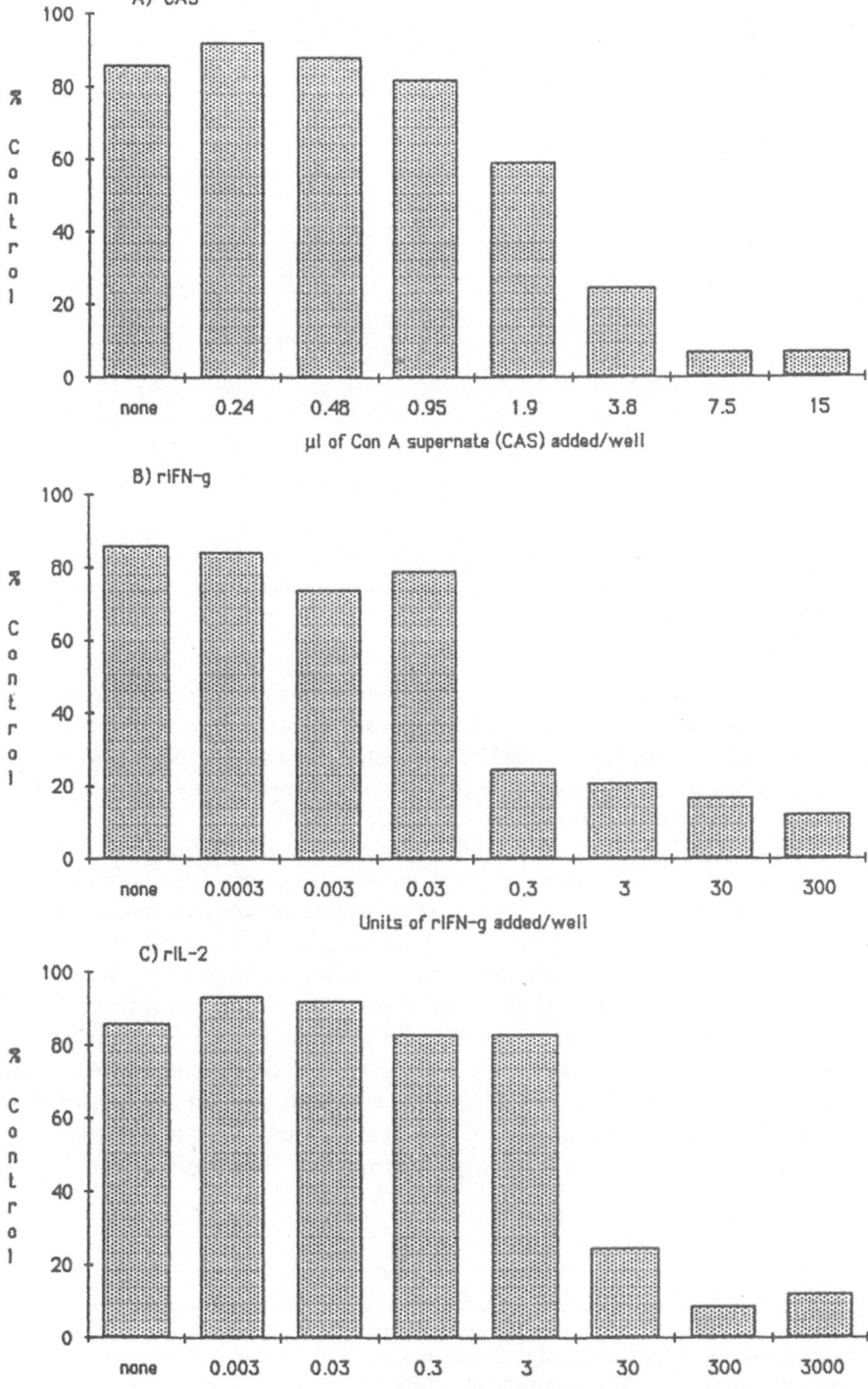

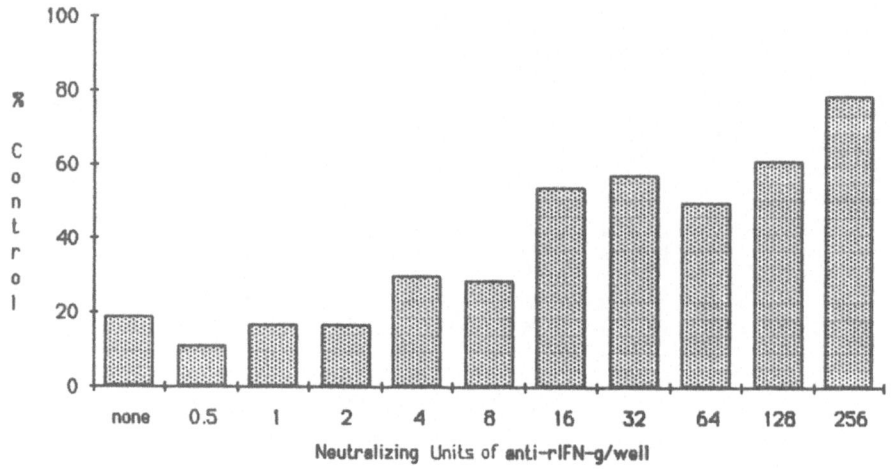

FIGURE 4. Anti-IFN-γ antibody removes the ability of late (48 day) GVHD spleen cells to suppress a B10.D2 spleen cell Con A response. 2.5×10^5 B10.D2 spleen cells were mixed with 2×10^5 GVHD spleen cells plus Con A. Increasing amounts of anti-IFN-γ antibody were added to these cultures.

of MLR and mitogen cultures. NS activity is also commonly assayed by inhibition of Ig production. The suppression carried out by NS cells is neither antigen-specific nor MHC-restricted. However, NS cells are not just merely antiproliferative in activity. For example, they will not inhibit the proliferation of a number of tumors.[108–110]

Because both NS and NK cells have the "null" phenotype and have nonspecific, non-MHC-restricted activity, many groups have looked to see if NS cells are also cytotoxic. There is no cytotoxic activity toward YAC-1 in the murine newborn spleen.[24–26] In fact, such NK activity does not develop in the mouse until after the newborn NS activity has disappeared.[34–36,118] Adult BM is also a very poor source of NK activity. [118,119] The spleens in all the locations listed in Table I, except during pregnancy, have very little if any NK activity associated with them.[24,54,67,71,82,108,109,120–122] Also, it is very difficult, in cell populations rich in NS activity, to demonstrate the presence of cytotoxic activity either *in vitro* or *in vivo*.[54,122] It thus appears than NS cells, unlike NK and NC

FIGURE 5. Anti-IFN-γ antibody removes the enhanced ability of late (48 day) GVHD spleen cells to suppress B10.D2 spleen cell LPS response. 2.5×10^5 normal B10.D2 spleen cells were mixed with 2×10^5 GVHD spleen cells plus LPS. This gave 51% of control response. To these cultures were added CAS (15 μl), rIFN-γ (6.7 μl), or rIL-2 (67 μl). These lymphokines all enhance suppression of the LPS response to near 100%. Increasing amounts of anti-IFN-γ antibody were added to these cultures.

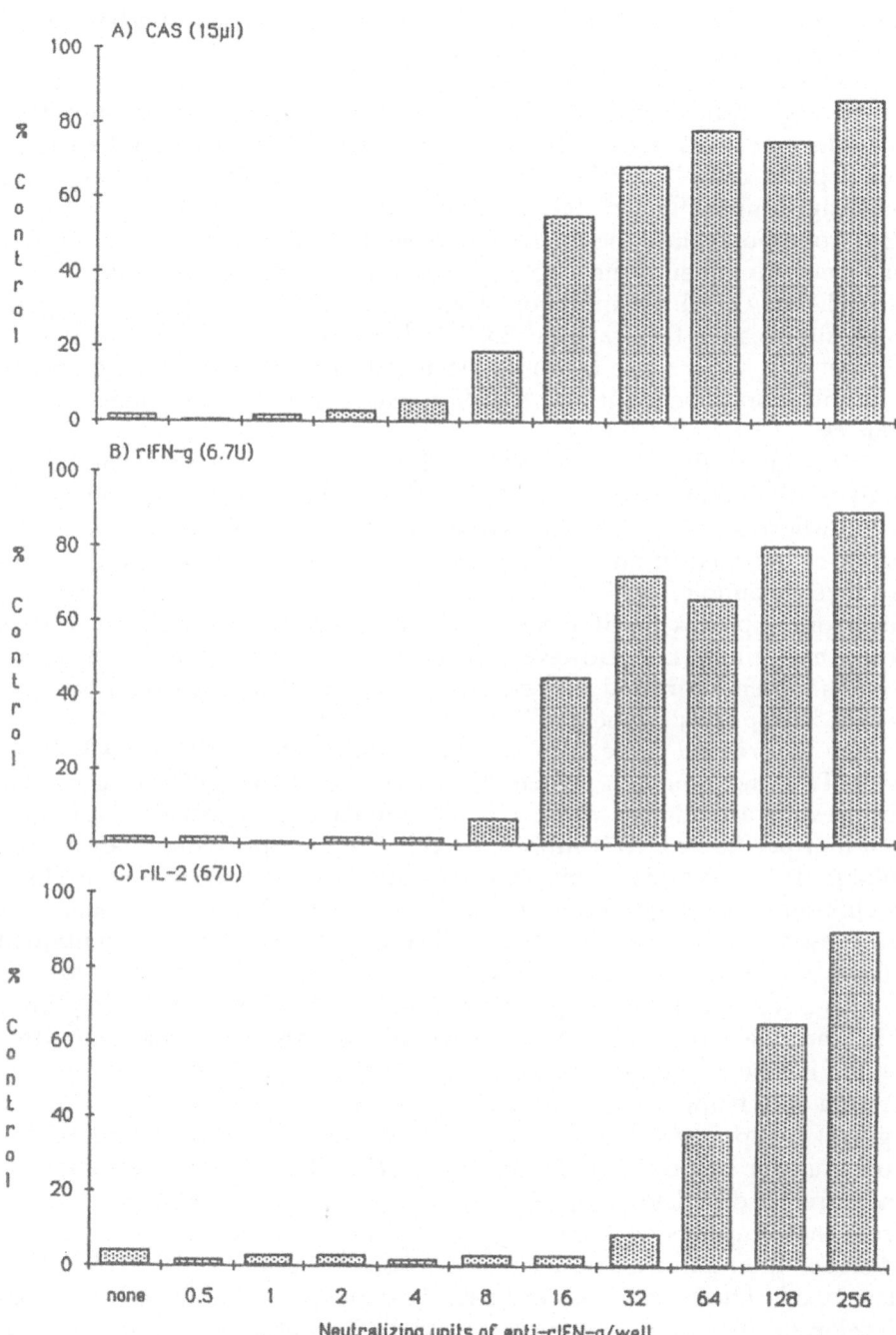

cells, do not carry out their immunosuppressive activity via a cytolytic mechanism.[108,122,123] This does not preclude a cytostatic mechanism of action, however.

Because NS cells can suppress an immune response at low suppressor-to-target ratios (often well below 1:1), there has been speculation that NS cells may work via the release of an inhibitory factor(s). When newborn spleen cells,[14,20,32,124] adult BM,[125] or tissues (e.g., decidua) from pregnant animals[76,79,126–130] are cultured they have been found to release compounds into the media that have the ability to inhibit the targets in a similar manner to the whole cell populations. These results imply that NS cells may work via inhibitory factor release. Similar attempts to generate suppressive factors from TLI[109] have been less successful. It may be that if a suppressive factor is produced, it is somewhat labile and/or concentration-dependent and therefore difficult to find in some circumstances.

It is interesting that NS cells can inhibit only if added at the beginning of an immune response.[39,42,62,108,124] This may tell us something about where and how NS must work. This indicates that after some point has been reached in an immune response, NS cells are no longer effective. For example, NS cells may inhibit IL-2 receptor expression and therefore suppress proliferation in an MLR because the primed T cells could not utilize IL-2. However, in this same example, once IL-2 receptors had been expressed on reactive T cells, the NS cells would be ineffective in suppressing the proliferation of these cells.

In this regard there is some recent work with cloned NS cells from both TLI and newborn spleens.[109,110] These cloned cells have similar suppressive activities to whole NS cell populations in that they are antigen-nonspecific and MHC-unrestricted in their suppression. They do not inhibit IL-1 secretions by macrophages, nor do they interfere with HT-2 proliferation in response to IL-2, so it is tempting to speculate that suppression is at an early stage of T cell activation (i.e., as mentioned above, maybe IL-2 secretion or IL-2 receptor expression is inhibited).

In a series of interesting experiments[31,49,50] with NS cells from both TLI and newborn spleen, it was proposed that NS cells had a dual effect on the immune response. It was proposed that NS cells inhibit the proliferation of T helper (T_H) cells, while at the same time allowing (or encouraging?) T suppressor (T_S) cells to develop. This was found by adding NS-containing spleen cell populations to an MLR. The MLR was inhibited as measured 5 days later by proliferation. However, if the cells were taken from these suppressed MLR cultures at 5 days and then assayed in a new MLR for their suppressive activity, it was found that specific T_S cells had developed. These findings were used to explain the specific T_S cells that develop in mice after TLI treatment and which also develop in acquired tolerance in newborn mice.

The question arises whether NS cells have any activity *in vivo*. It has been shown that newborn spleen cells will inhibit GVHR in adult F_1 recipients of parental spleen cells.[11,13,16] They have also been shown to inhibit Ia induction *in vivo*.[28] This also indicates another possible mechanism by which NS cells may inhibit immune responses. If NS cells are capable of inhibiting Ia expression on accessory cells, which require Ia on their surface to activate T_H cells, for example, then this may be one way they can down-regulate a response.

Finally, cloned NS cells if injected *in vivo* are also capable of inhibiting $P \rightarrow F_1$ GVHRs.[131] This was measured by the ability of these cloned NS cells to suppress both GVHR-induced splenomegally and death.

3. *In Vivo* Relevance of NS Cell Activity

3.1. Tolerance Induction and Maintenance

As mentioned before, one of the important features of the locations where NS activity is found (Table I) is that they are all environments where tolerance induction may occur.[1–3,47,54,85–106,122] Thus, considering the activity of NS cells, their locations, and their responsiveness to activated T cell lymphokines, it seems entirely possible that NS cells may have a role in the development and maintenance of self-tolerance or of acquired immunological unresponsiveness.[54,122]

NS cells could act in the following way. When T cells are stimulated by self or foreign antigens in an environment rich in NS cells, the lymphokines they release stimulate the suppressive activity of NS cells. This increased NS activity in turn acts as a negative feedback loop to inhibit the development of fully functional T cell clones. The work of Okada *et al.*[31,50] outlined above indicates that the T_S cells may be refractory to such NS-mediated down-regulation. If so, this would mean that while NS cells were down-regulating specific T_H clones, specific T_S clones to the same antigen may survive and thus supply the specific down-regulation usually seen in the mature animal.

3.2. Hematopoietic Regulation

The second feature of NS locations is that they are all environments of considerable hematopoiesis. Thus NS activity, which shows such potent nonspecific inhibition, may also control cell growth in these sites, perhaps by suppression of growth factor production.[44,122] Therefore, one normal function of NS cells may be to regulate the proliferation of hematopoietic tissue.

3.3. Disease States

As shown in Table I, NS activity occurs after immunosuppressive/cytoreductive treatments including TLI,[24,47–54] CY,[55–67] and [89]Sr.[81,82] NS activity also develops in chronic GVHD that occurs after BM transplants.[46,69–71,114,122] The NS activity is probably at least partly responsible for the generalized immunosuppression seen in all these situations. This may be especially true of BM transplantation where the BM inoculum,[37–46] the immunosuppressive/cytoreductive pretransplant regimen,[37–46,54–67] and the frequent complication of GVHD[46,69–71,114,122] all have NS cell activity associated with them.

Both TLI and CY are used as treatments for several autoimmune diseases such as systemic lupus erythrematosus (SLE) and rheumatoid arthritis.[47,53,58,66,132–141] The efficacy of these treatments may be at least partially related to the induced NS activity, which may down-regulate self-reactive clones.

IFN-γ is mainly produced by T cells.[142,143] IFN-γ has been shown to inhibit BM hematopoiesis,[142–148] and patients with idiopathic aplastic anemia have increased levels of IFN-γ in their serum and BM.[147,148] This is consistent with the finding that patients treated to lower their T cell numbers show improved BM function.[149] The success of this treatment may occur because it removes the T cells that produce the excess IFN-γ, which directly inhibits the proliferating cells in the BM.[142–149] Alternatively, IFN-γ may also inhibit hematopoiesis indirectly by stimulating NS activity in the BM.[46]

It has been proposed that local nonspecific suppression of the maternal immune system may play an important role in protection of the fetus.[73–80] Thus, NS cell activity, which may be present during pregnancy, may be involved in down-regulating maternal antifetal reactivity.

Finally, as mentioned above, there seems to be an inverse relationship between NS and NK activities. Thus NK activity does not appear to develop in the mouse until after the NS activity has declined.[34–36,118] Also, it is difficult to find NK activity in locations where NS activity is present.[25,26,31,54,67,71,82,109,118–121] Therefore it is not unreasonable to speculate that the reason for this is that NS cells may in some way down-regulate NK activity. This would be important because NK cells have been shown to be involved with limiting cancer metastasis *in vivo*.[150–152] NK cell development and activity have already been shown to be under the regulatory influence of macrophages and their products.[119,153–156] NS cells may suppress NK cells in a similar fashion.

4. Potential Therapeutic Use of NS Cell Activity

4.1. TLI

TLI has been used for a number of years as an experimental procedure to induce tolerance with relatively few harmful side effects.[47,140] In murine systems it has been found that after TLI treatment a mouse will accept allogeneic skin grafts, with the subsequent development of donor skin specific T_S cells.[88,91,92] This work has been successfully extended to other species and other tissues (e.g., kidney).[140] The reason that TlI is thought to work is because of the large numbers of NS cells generated by this procedure.[47–54,87–92,140] As mentioned above, the NS cells may down-regulate reactive T_H cells while allowing specific T_S cells to develop.[31,49,50]

TLI has also been proposed as a treatment for several autoimmune diseases, including SLE and arthritis.[47,53,132–141] Treatment with TLI of several animal models of these two diseases has been encouraging.[132–134,140] Pilot studies with severely involved patients have also been promising.[134–141] The theoretical basis for treating autoimmune diseases with TLI is that the NS cells generated by the treatment would down-regulate the self-reactive clones causing the disease, while again allowing the development of specific T_S cells which would reduce the chances of relapse.

Of course other immunosuppressive/cytoreductive treatments are used for similar reasons. CY treatment is used to treat arthritis[141] and as a pretransplant procedure for BM transplantation.[101,102,104] The success of CY in these procedures may be partially due to the generation of large numbers of NS cells,[61,64–67] which could work as outlined above: suppression of reactive clones while allowing specific T_S cell development.

4.2. BM Transplantation

NS cell activity could be of great importance in BM transplantation, where recipients often develop chronic GVHD, which is associated with abnormal immune responses and a high incidence of suppressor cell activity.[157–163] Because NS cells are found in mice with chronic GVHD[68–71] and also in normal murine and human BM,[37–46] the ability to understand and regulate NS cell activity could be of great clinical importance in recipient survival. For example, it has been found in experiments designed to decrease the incidence of GVHD in BM recipients that the complete removal of T cells from BM grafts resulted in much lower graft acceptance.[164,168] However, if a small number of T cells are added back to the T cell-depleted BM, then the graft acceptance rate increases greatly.[165,166] One explanation for this is that T cells within the BM inoculum are required to generate lymphokines which activate NS cells, which help prevent graft rejection.

There is a growing awareness of the potential tolerance-inducing capability present in BM itself. Several groups have been studying the ability of BM to induce tolerance allogeneic barriers. In one system, if irradiated recipients are injected with both syngeneic and allogeneic BM, the recipient develops specific tolerance toward the allogeneic BM donor as measured by skin graft acceptance.[169,170] In the second system, injection of allogeneic BM into an antilymphocyte serum (ALS)-treated recipient renders that animal tolerant of a subsequent skin graft from the BM donor.[43,45,171–173] This tolerogenic producing activity in the BM has been associated with a cell that closely resembles NS cells.

In autologus systems where animals are irradiated and then injected with their own BM, followed by allogeneic organ transplants, the transplanted organs survive substantially longer than controls.[174–177] Again it appears that the BM was capable of causing immune unresponsiveness to occur in these recipients. Although the cells responsible were not phenotyped, this may have been caused by NS cell activity present in the BM.

Finally, there is growing interest in the fetus as a source of transplantable tissue. This is especially true of fetal liver as a source of hematapoietic tissue for BM transplantation,[178–181] the rationale being that fetal liver is a good source of stem cells while being relatively free of mature reactive cells, which could cause GVHD.[178–181] Thus donor–recipient matching is thought to be less critical. In this regard, there is considerable evidence that fetal liver contains a large amount of endogenous NS cell activity.[16,21,22] This is another reason why fetal liver would be very good as a donor tissue for BM transplants. The high NS activity in fetal liver may help down-regulate any harmful reactions toward either the graft or the host. However, the low number of mature T cells, as a source of lymphokines required to activate these NS cells, may be a problem for graft acceptance.

4.3. Consequences of Any Cytoreductive Treatment

A common treatment protocol for cancer is to use a cytotoxic regimen specific for proliferating cells in the hope of eliminating the rapidly dividing tumor cells. However, if used systemically, this also has the undesired side effect of being cytoreductive for BM and many BM-derived cells. At some time after ceasing the treatment, there will be a rapid repopulation of these cells. As mentioned above, NS cell activity has been found in association with rapid stem cell proliferation. Therefore, it is likely that there will be a strong NS cell component to any of these treatments. This NS cell activity may be counterproductive here, because if the tumor is not eliminated, the NS cells may down-regulate any reactive cells that may have removed the remaining tumor cells. Thus we need to know just what role NS cells may play here to make these treatments more efficacious.

4.4. Using or Manipulating NS Cell Activity to Induce Tolerance

Finally, if NS cells are involved, or can be manipulated *in vivo* or *in vitro* to become involved, in tolerance induction and maintenance, then understanding how this occurs is very important. For example, if NS cell activity in BM is helpful in suppressing BM graft-rejecting immune mechanisms *in vivo*, then understanding this fact could be very helpful. As pointed out before, it appears that removal of all the T cells from the BM graft prior to transplant has often had the deleterious effect of greater graft rejection.[164–168] It therefore appears that a small number of T cells may be required for good graft take. However, this brings back the problem of increased probability of GVHD development in the recipient.[167] This situation poses a dilemma of the "Catch-22" variety. However, if NS cell activity is one of the reasons why T cells are required in the BM inoculum (to supply the lymphokines to activate NS cells, which then inhibit the immune responses which may reject the graft), it may be possible to activate the NS cells in the BM inoculum *in vitro* prior to transplantation and *in vivo* after transplantation if the proper lymphokines are known. This may allow the elimination, or further reduction, of T cells from the BM inoculum without the deleterious side effect of increased graft rejection. Also, it may be possible in the future to grow large numbers of highly enriched NS cells *in vitro*. They could then be injected with the BM inoculum to reduce the chances of graft rejection and GVHD development.

5. Directions of Future Research

5.1. Growth and Developmental Factors

In Fig. 1, pathways 1 and 2 probably have many growth and differentiation factors feeding into them. At present we know little about what these factors might be or how they may work. Some of the cloning work might be helpful in this regard. It has been found that an NS-like suppressor cell can be grown from newborn spleen with WEHI-3B supernatant.[108] WEHI-3B supernatant is a very potent source of IL-3, which is known to be a pluripotent growth factor for a number of cell types, most prominently monocytes/mast cells.[107]

5.2. Induction and Mechanism of Suppression

Pathway 3 in Fig. 1 represents what we believe is the induction of NS cell suppression. It has been shown that NS cells can respond to IFN-γ to actually express their suppressive potential.[25,46] Histamine may be another activation signal for NS cells.[115–117] There probably are a number of "signals" that NS cells respond to.

The actual suppressor mechanism (pathway 4, Fig. 1) is also poorly understood. Do NS cells require cell contact to exert their inhibition, or is a suppressor factor produced? A suppressor factor could work at a distance from the NS cell itself. There is conflicting evidence as to whether NS cells produce a suppressor factor. Many studies have shown that NS cell-containing populations do produce an inhibitory factor,[14,20,32,76,79,124–130] but others have not seen factor production.[109] In this regard, we have some preliminary data to indicate that indomethacin can remove some of the suppressive ability of NS cells from a number of locations (unpublished observations). Are prostaglandins and/or leukotrienes produced by the NS cells themselves, and do these arachidonate metabolites then suppress immune responses directly or indirectly?

5.3. Lineage of NS Cells

One of the major unsolved questions facing NS cell research is, What is the linage of these cells? Almost every hematopoietic cell has at one time or another been suspected of causing the suppression we term natural suppression. In this regard it is important to realize that all the locations of NS activity listed in Table I contain cells of all hematopoietic lineages, in various stages of development. Very little is known of the phenotypic characteristics of these cells.[6,168,182,183] This has undoubtedly contributed to the confusion concerning the cells, which are responsible for suppression. Therefore, it is probably best to take a cautious approach in interpreting phenotypic data concerning these cells.

5.4. Relationship(s) with Other Suppressor Cells

A question related to the possible lineage of NS cells is how NS cells relate to other cells with known suppressor activities.

5.4.1. T_S Cells

T_S cells have been studied for almost 20 years now. The major difference between T_S cells and NS cells is that T cells in general are antigen-specific and MHC-restricted in their activities.[184] It is therefore hard to envisage that NS cells are T cells unless they are not using the T cell receptor (at least as it is now known) in their functions.

A more intriguing idea is that nonspecific NS cells may in fact induce antigen-specific T_S cells to develop in environments rich in NS activity. This idea was first proposed to explain how nonspecific "null," non-T cell, suppressor cells may work in a newborn system.[26] It was extended by other work on the NS cells found in both TLI and newborn spleen.[31,49,50] In this later work it was shown that TLI and newborn spleen allowed for the development of specific T_S cells and also inhibited the proliferative response in an MLR culture.[49,50] In this regard, if NS cells do produce prostaglandins, this may be involved in T_S cell induction, as prostaglandins are known inducers of T_S cell activity.[185–188] These ideas need to be studied further, as they could show how NS cells are involved in the development of specific tolerance even though they themselves are nonspecific.

5.4.2. Macrophages

Macrophages have suppressive capabilities.[189–192] This suppression, like that caused by NS cells, is antigen-nonspecific and MHC-unrestricted. However, NS cells and macrophages differ in a number of phenotypic and functional characteristics. NS cells are esterase-negative and non-adherent to such things as plastic, Sephadex G-10, and nylon wool. NS cells are also generally MAC-1$^-$ and Ia$^-$. Thus by a number of criteria, NS cells are not mature macrophages. However, NS cells may be somehow related to macrophages.[27,28,30,192] Perhaps NS cells are an early, less differentiated stage of the macrophage lineage.

There have recently been interesting reports of two separate macrophage populations.[189] One has high levels of surface Ia. These macrophages are thought to induce good immune responses. The second macrophage population has low Ia expression. These macrophages are thought to inhibit immune responses. Is there some relationship between these later macrophages and NS cells?

5.4.3. NK Cells and Natural Cytotoxic (NC) Cells

The relationship of NS cells and NK and NC cells has been reviewed recently[122] but will be summarized here. Both NK and NC cells are cells with the "null" phenotype, which have been found in certain circumstances to have the ability to suppress immune responses.[193–197] The suppression carried out by these cells is antigen-nonspecific and MHC-unrestricted, like NS cells. However, the major difference between these cells and NS cells is that both NK and NC cells are by definition cytotoxic, and the suppressive action of these two cells is thought to be via a cytolytic mechanism.[123] This is in contrast to NS cells, which are not cytotoxic,[122] and therefore their suppressive action cannot be via a cytolytic mechanism. Also as mentioned before, NK activity and NS activity have an inverse relationship,[34–36,118] which may involve NS cells' regulating NK development. Of course an alternative explanation could be that NS cells are just an early cell type in the NK development pathway.

5.5. Phenotypic Markers

One of the most pressing needs to further the studies of NS cells is for the development of antibodies to NS cell markers. By definition, the NS cell has the "null" phenotype. This of course does not mean that NS cells have no unique antigenic markers. What it means is that at the present time there is no unique marker that is known to be on NS cells. The development of antibodies that would detect NS cells and their precursors would go a long way in solving many of the lineage questions of NS cells.

5.6. Genetics

The genetics of NS cell activity have not been studied well. Is NS activity found in all strains of mice in equivalent quantities? We have begun to look into these questions. We find that there may indeed be genetic differences in the expression of NS cell activity. Table III shows in summary form some of the strains we have looked at and the locations of NS activity. Even though the number of strains is not great, it can be seen that there may be high and low NS strains and that NS activity is not MHC-linked.

These results could have important implications. First, it is apparent that experiments on NS cell activity performed on different strains, especially a low strain such as BALB/c and a high strain such as B10.D2, may not be comparable. Second, it raises the question of whether there are several different cells that can carry out the suppressive activities seen in the locations listed in Table I. In other words, are the low-NS strains just low in one of several cells capable of inhibiting immune responses? It is not at all unreasonable to think that several different cell types may be responsible for the suppression seen in newborn spleen, for example. If this is indeed the case, it would explain a lot of the confusion that has developed about the cells responsible for this suppression. For example, the spleen cells from both BALB/c and C57BL/6 newborn mice can suppress an immune response, but the suppression may be caused predominantly by NS cells in C57BL/6 and pre-T cells in BALB/c. Another complication in this regard is that within any one strain, the suppression by a given cell type may rise and fall relative to the other suppressive cell types with time. Thus, until this area is resolved, it may be better to speak of natural suppressor *activity* instead of natural suppressor cells.

TABLE III
Strain Difference in Natural Suppressor (NS) Activity

NS system[a]	Strain	NS activity[b]
Newborn spleen	B10.D2	High
	BALB/c	Low
	(BALB/c × B10.D2)F$_1$	Intermediate
	C57BL/6	High
Adult bone marrow	B10.D2	High
	BALB/c	Low
	(BALB × B10.D2)F$_1$	Intermediate
	C57BL/6	High
Cyclophosphamide-treated	B10.D2	High
	BALB/c	Low

[a] Newborn spleens were from animals ≤3 days old. Adult bone marrow (BM) was from animals 6–12 weeks old. Cyclophosphamide (CY) spleens were from adult (6–12 weeks) animals that had received 200 mg/kg of CY IP 9–10 days before.
[b] NS activity is defined as high if the cells suppress a syngeneic Con A response ≥80% at a 1:1 suppressor:target ratio, and the LPS suppression can be enhanced with IFN-γ. NS activity is defined as low if the cells suppress a syngeneic Con A response ≤30% at a 1:1 suppressor:target ratio, and the LPS suppression cannot be enhanced with IFN-γ.

5.7. "Pure" NS Cell Populations

To answer definitively the questions that have been posed above, there will have to be "pure" NS cell clones available. There have been several different clones produced from TLI and newborn spleens.[108–110] These cloned suppressor cells have many of the characteristics of the whole NS cell populations. However, there are also some disconcerting differences. When cells are cloned with growth factors, there is always the question of how similar the resulting clone is to the original cell. Clones by necessity are developed under very artificial conditions and with tremendous selective pressures that the original cell never faced. However, if the investigator is lucky, the clone may be similar enough in both functional and phenotypic characteristics that much can be learned from it. Therefore, even with the obvious constraints imposed by cloning, there is still much to be gained by producing more NS clones.

ACKNOWLEDGMENTS. The work from this laboratory was supported by NIH grants AI-12685 and AI-07035. T.M. is supported by a fellowship from the Arthritis Foundation and a grant from the American Cancer Society, J.H.H. is supported by a fellowship from the Leukemia Society, and K.L.C. is supported by a fellowship from the Province of Ontario Ministry of Health. We thank Dr. M. Shepard and Genentech for the rIFN-γ and the anti-IFN-γ, Roberta Dustin for technical assistance, and Kathy Utschinski for preparation of this manuscript.

References

1. Billingham, R. E., Brent, L., and Medawar P. B., 1953, "Actively acquired tolerance" of foreign cells, *Nature* **172**:603–606.
2. Streilein, J. W., 1979, Neonatal tolerance: Towards an immunogenetic definition of self, *Immunol. Rev.* **46**:125–146.
3. Billingham, R. E., Brent, L., and Medawar, P., 1956, Quantitative studies on tissue transplantation immunity. III. Actively acquired tolerance, *Proc. R. Soc. Lond. (Biol.)* **238**:357–415.
4. Adler, W. H., Takiguchi, T., Marsh, B., and Smith, R. T., 1970, Cellular recognition by mouse lymphocytes *in vitro*. II. Specific stimulation by histocompatibility antigens in mixed cell culture, *J. Immunol.* **103**:984–1000.
5. Fidler, J. M., Chiscon, M. O., and Golub, E. S., 1972, Functional development of the interacting cells in the immune response. II. Development of immunocompetence to heterologous erythrocytes *in vitro*, *J. Immunol.* **109**:136–140.
6. Spear, P. G., Wang, A.-L., Rutishauser, U., and Edelman, G. M., 1973, Characterization of splenic lymphoid cells in fetal and newborn mice, *J. Exp. Med.* **138**:557–573.
7. Stites, D. P., Carr, M. C., and Fudenberg, H. H., 1974, Ontogeny of cellular immunity in the human fetus: Development of responses to phytohemagglutinin and to allogeneic cells, *Cell. Immunol.* **11**:257–271.
8. Spear, P. G., and Edelman, G. M., 1974, Maturation of the humoral immune response in mice, *J. Exp. Med.* **139**:249–263.
9. Rabinowitz, S. G., 1976, Measurement and comparison of the proliferative and antibody responses of neonatal, immature and adult murine spleen cells to T-dependent and T-independent antigens, *Cell. Immunol.* **21**:201–216.
10. Mosier, D. E., and Johnson, B. M., 1975, Ontogeny of mouse lymphocyte function. II. Development of the ability to produce antibody is modulated by T lymphocytes, *J. Exp. Med.* **141**:216–226.
11. Skowron-Cendrzak, A., and Ptak, W., 1976, Suppression of local graft-versus-host reactions by mouse fetal and newborn spleen cells. *Eur. J. Immunol.* **6**:451–452.
12. Morse, H. C., Prescott, B., Cross, S. S., Stashak, P. W., and Baker, P. J., 1976, Regulation of the antibody response to type III pneumococcal polysaccharide. V. Ontogeny of factors influencing the magnitude of the plaque-forming cell response. *J. Immunol.* **116**:279–287.
13. Ptak, W., and Skowron-Cendrzak, A., 1977, Fetal suppressor cells: Their influence on the cell-mediated immune responses, *Transplantation* **24**:45–51.
14. Bassett, M., Coons, T. A., Wallis, W., Goldberg, E. H., and Williams, R. C., 1977, Suppression of stimulation in mixed leukocyte culture by newborn splenic lymphocytes in the mouse, *J. Immunol.* **119**:1855–1857.
15. Mosier, D. E., Mathieson, B. J., and Campbell, P. S., 1977, Ly phenotype and mechanism of action of mouse neonatal suppressor T cells, *J. Exp. Med.* **146**:59–73.
16. Globerson, A., and Umiel, T., 1978, Ontogeny of suppressor cells. II. Suppression of graft-versus-host and mixed leukocyte culture responses by embryonic cells, *Transplantation* **26**:438–443.
17. Argyris, B. F., 1978, Suppressor activity in the spleen of neonatal mice, *Cell. Immunol.* **36**:354–362.
18. Hardy, B., and Mozes, E., 1978, Expression of T cell suppressor activity in the immune response of newborn mice to a T-independent synthetic peptide, *J. Immunol.* **35**:757–762.
19. Calkins, C. E., and Stutman, O., 1978, Changes in suppressor mechanisms during postnatal development in mice, *J. Exp. Med.* **147**:87–97.
20. Pavia, C. S., and Stites, D. P., 1979, Immunosuppressive activity of murine newborn spleen cells. I. Selective inhibition of *in vitro* lymphocyte activation, *Cell. Immunol.* **42**:48–60.

21. Rabinovich, H., Umiel, T., Reisner, Y., Sharon, N., and Globerson, A., 1979, Characterization of embryonic liver suppressor cells by peanut agglutinin, *Cell. Immunol.* **47**:347–355.
22. Monden, M., Staruch, A. J., and Fortner, J. G., 1982, A partial characterization of suppressor cells in rat fetal liver cells, *Cell. Immunol.* **68**:16–24.
23. Argyris, B. F., 1981, Role of suppressor cells in immunological maturation, *Transplantation* **31**:334–338.
24. Oseroff, A., Okada, S., and Strober, S., 1984, Natural suppressor (NS) cells found in the spleen of neonatal mice and adult mice given total lymphoid irradiation (TLI) express the null surface phenotype, *J. Immunol.* **132**:101–110.
25. Maier, T., and Holda, J. H., 1987, Natural suppressor (NS) activity from neonatal spleen is responsive to IFN-γ, *J. Immunol.* **138**:4075–4084.
26. Rodriguez, G., Andersson, G., Wigzell, H., and Peck, A. B., 1979, Non–T cell nature of the naturally occurring spleen-associated suppressor cell present in the newborn mouse, *Eur. J. Immunol.* **9**:737–746.
27. Piguet, P. F., Irle, C., and Vassalli, P., 1981, Immunosuppressor cells from newborn mouse spleen are macrophages differentiating *in vitro* from monoblastic precursors, *Eur. J. Immunol.* **11**:56–61.
28. Snyder, D. S., Lu, C. Y., and Unanue, E. R., 1982, Control of macrophage Ia expression in neonatal mice: Role of splenic suppressor cell, *J. Immunol.* **128**:1458–1465.
29. Hoskin, D., Hooper, D. C., and Murgita, R. A., 1982, Naturally occurring non-T suppressor in pregnant and neonatal mice: Some functional and phenotypic characteristics, *Am. J. Reprod. Immunol.* **3**:72–77.
30. Peeler, K., Wigzell, H., and Peck, A. B., 1983, Isolation and identification of the naturally occurring newborn spleen-associated suppressor cells: A mixed monocyte/mast cell population with separable suppressor activities, *Scand. J. Immunol.* **17**:443–453.
31. Okada, S., and Strober, S., 1982, Spleen cells from adult mice given total lymphoid irradiation (TLI) or from newborn mice have similar regulatory effects in mixed leukocyte reaction (MLR). II. Generation of antigen-specific suppressor cells in the MLR after the addition of spleen cells from newborn mice, *J. Immunol.* **129**:1892–1897.
32. Jadus, M. R., and Peck, A. B., 1986, Naturally occurring, spleen associated suppressor activity of the newborn mouse: Biochemical and functional identification of three monokines secreted by newborn suppressor-inducer monocytes, *Scand. J. Immunol.* **23**:35–44.
33. Hooper, D. C., Hoskin, D. W., Gronvik, K. D., and Murgita, R. A., 1986, Murine neonatal spleen contains natural T and non-T suppressor cells capable of inhibiting adult alloreactive and newborn autoreactive T-cell proliferation, *Cell. Immunol.* **99**:461–475.
34. Kiessling, R., Klein, E., Pross, H., and Wigzell, H., 1975, "Natural" killer cells in the mouse. II. Cytotoxic cells with specificity for mouse Moloney leukemia cells: Characteristics of the killer cell, *Eur. J. Immunol.* **5**:117–121.
35. Stutman, O., Paige, C. J., and Figarella, E. F., 1978, Natural cytotoxic cells against solid tumors in mice. I. Strain and age distribution and target cell susceptibility. *J. Immunol.* **121**:1819–1826.
36. Lanza, E., and Djeu, J. Y., 1982, Age-independent natural killer cell activity in murine peripheral blood, in. *NK Cells and Other Natural Effector Cells* (R. B. Herberman, ed.), Academic Press, New York, pp. 335–340.
37. Duwe, A. K., and Singhal, S. K., 1979, The Immunoregulatory role of bone marrow. II. Characterization of suppressor cell inhibiting the *in vitro* antibody response, *Cell. Immunol.* **43**:372–381.
38. Corvese, J. S., Levy, E. M., Bennett, M., and Cooperband, S. R., 1980, Inhibition of an *in vitro* antibody response by a suppressor cell in normal bone marrow, *Cell. Immunol.* **49**:293–306.

39. Dorshkind, K., Klimpel, G. R., and Rosse, C., 1980, Natural regulatory cells in murine bone marrow: Inhibition of *in vitro* proliferative and cytotoxic responses to alloantigens, *J. Immunol.* **124:**2584–2588.
40. Dorshkind, K., and Rosse, C., 1981, Nonspecific inhibition of alloantigen-induced proliferation by bone marrow natural regulatory cells, *Transplant. Proc.* **13:**1182–1186.
41. Dorshkind, K., and Rosse, C., 1982, Physical, biologic, and phenotypic properties of natural regulatory cells in murine bone marrow, *Am. J. Anat.* **164:**1–17.
42. Bains, M. A., McGarry, R. C., and Singhal, S. K., 1982, Regulatory cells in human bone marrow: Suppression of an *in vitro* primary antibody response, *Cell. Immunol.* **74:**150–161.
43. Gozzo, J. J., Crowley, M., Maki, T., and Monaco, A. P., 1982, Functional characteristics of a Ficoll-separated mouse bone marrow cell population involved in skin allograft prolongation, *J. Immunol.* **129:**1584–1588.
44. Soderberg, L. S. F., 1985, Rabbit bone marrow suppressor cells block the production or release of a soluble bone marrow growth factor, *Cell. Immunol.* **92:**313–320.
45. De Fazio, S., Hartner, W. C., Monaco, A. P., and Gozzo, J. J., 1985, Mouse skin graft prolongation with donor strain bone marrow and antilymphocyte serum: Surface markers of the active bone marrow cells, *J. Immunol.* **135:**3034–3038.
46. Holda, J. H., Maier, T., and Claman, H. N., 1986, Natural suppressor activity in graft-vs-host spleen and normal bone marrow is augmented by IL 2 and interferon-γ, *J. Immunol.* **137:**3538–3543.
47. Kotzin, B. L., and Strober, S., 1984, Total lymphoid irradiation, *Clin. Immunol. Allergy* **4:**331–358.
48. White, D. J. G., 1984, Total lymphoid irradiation, in: *Transplantation Immunology: Clinical and Experimental* (R. Y. Calne, ed.), Oxford University Press, Oxford, U.K., pp. 339–346.
49. King, D. P., Strober, S., Kaplan, H. S., 1981, Suppression of the mixed leukocyte response and of graft-vs-host disease by spleen cells following total lymphoid irradiation (TLI), *J. Immunol.* **126:**1140–1145.
50. Okada, S., and Strober, S., 1982, Spleen cells from adult mice given total lymphoid irradiation or from newborn mice have similar regulatory effects in the mixed leukocyte reaction. I. Generation of antigen-specific suppressor cells in the mixed leukocyte reaction after the addition of spleen cells from adult mice given total lymphoid irradiation, *J. Exp. Med.* **156:**522–538.
51. May, R. D., Slavin, S., and Vitetta, E. S., 1983, A partial characterization of suppressor cells in the spleens of mice conditioned with fractionated total lymphoid irradiation (TLI), *J. Immunol.* **131:**1108–1114.
52. Weigensberg, M., Morecki, S., Weiss, L., Fuks, Z., and Slavin, S., 1984, Suppression of cell-mediated immune responses after total lymphoid irradiation (TLI). I. Characterization of suppressor cells of mixed lymphocyte reaction. *J. Immunol.* **132:**971–978.
53. Kotzin, B. L., Strober, S., Kansas, G. S., Terrell, C. P., and Engleman, E. G., 1984, Suppression of pokeweed mitogen-stimulated immunoglobulin production in patients with rheumatoid arthritis after treatment with total lymphoid irradiation, *J. Immunol.* **132:**1049–1055.
54. Strober, S., 1984, Natural suppressor (NS) cells, neonatal tolerance, and total lymphoid irradiation: Exploring obscure relationships, *Annu. Rev. Immunol.* **2:**219–237.
55. Kerckhaert, J. A., Hofhuis, F. M., and Willers, J. M., 1977, Effects of variation in time and dose of cyclophosphamide injection on delayed hypersensitivity and antibody formation, *Cell. Immunol.* **29:**232–237.
56. Goidl, E. A., Cusano, A., Redner, R., Innes, J. B., Weksler, M. E., and Siskind, G. W., 1979, Studies on the control of antibody synthesis. XV. Effect of nonspecific immunodepression on antibody affinity, *Cell. Immunol.* **47:**293–303.

57. Shand, F. L., and Howard, J. G., 1979, Induction *in vitro* of reversible immunosuppression and inhibition of B cell receptor regeneration by defined metabolites of cyclophosphamide, *Eur. J. Immunol.* **9:**17–21.
58. Riera, C. M., Galmarini, M., and Serra, H. M., 1983, Specific suppression of humoral and delayed hypersensitivity responses by cyclophosphamide in an experimental model of autoimmunity, *Am. J. Reprod. Immunol.* **4:**71–75.
59. L'age-Stehr, J., and Diamantstein, T., 1978, Studies on induction and control of cell-mediated autoimmunity. II. Prevention of induction and activity of autoreactive T cells by suppressor cells and by a suppressive serum factor, *Eur. J. Immunol.* **8:**624–628.
60. L'age-Stehr, J., and Diamantstein, T., 1978, Induction of autoreactive T lymphocytes and their suppressor cells by cyclophosphamide, *Nature* **271:**663–665.
61. McIntosh, K. R., Segre, M., and Segre, D., 1979, Inhibition of the humoral response by spleen cells from cyclophosphamide-treated mice, *Immunopharmacology* **1:**165–173.
62. Braciale, V., and Parish, C. R., 1980, Inhibition of *in vitro* antibody synthesis by cyclophosphamide-induced suppressor cells, *Cell. Immunol.* **51:**1–12.
63. Wander, R. H., and Hilgard, H. R., 1981, Activation and suppression of graft-vs-host reactions by cyclophosphamide, *J. Immunol.* **126:**901–904.
64. McIntosh, K. R., Segre, M., and Segre, D., 1982, Characterization of cyclophosphamide-induced suppressor cells, *Immunopharmacology* **4:**279–289.
65. Greeley, E. H., Segre, M., and Segre, D., 1982, Supressor cells in cyclophosphamide-treated autoimmune mice, *Immunopharmacology* **4:**355–363.
66. Greeley, E. H., Segre, M., and Segre, D., 1985, Modulation of autoimmunity in NZB mice by cyclophosphamide-induced, nonspecific suppressor cells, *J. Immunol.* **134:**847–851.
67. Segre, M., Tomei, E., and Segre, D., 1985, Cyclophosphamide-induced suppressor cells in mice: Suppression of the antibody response *in vitro* and characterization of the effector cells, *Cell. Immunol.* **91:**443–454.
68. Hamilton, B. L., and Parkman, R., 1983, Acute and chronic graft-versus-host disease induced by minor histocompatibility antigens in mice, *Transplantation* **36:**150–155.
69. Holda, J. H., Maier, T., and Claman, H. N., 1985, Graft-vs-host reactions (GVHR) across minor murine histocompatibility barriers. I. Impairment of mitogen responses and suppressor phenomena, *J. Immunol.* **134:**1397–1402.
70. Maier, T., Holda, J. H., and Claman, H. N., 1985, Graft-vs-host reactions (GVHR) across minor murine histocompatibility barriers. II. Development of natural suppressor cell activity, *J. Immunol.* **135:**1644–1651.
71. Holda, J. H., Maier, T., and Claman, H. N., 1985, Murine graft-versus-host disease across minor barriers: Immunosuppressive aspects of natural suppressor cells, *Immunol. Rev.* **88:**87–105.
72. Daya, S., and Clark, D. A., 1986, Production of immunosuppressor factor(s) by preimplantation human embryos, *Am. J. Reprod. Immunol.* **11:**98–101.
73. Hoskin, D., Hooper, D. C., and Murgita, R. A., 1983, Naturally occurring non-T suppressor cells in pregnant and neonatal mice: Some functional and phenotypic characteristics, *Am. J. Reprod. Immunol.* **3:** 72–77.
74. Vanderbeeken, Y., Vlieghe, M. P., and Delespesse, G., 1981, Regulation of the lymphocyte response to phytohemagglutinin during pregnancy: Role of adherent cells and prostaglandins, *Am. J. Reprod. Immunol.* **1:**233–235.
75. Kearns, M., and Lala, P. K., 1983, Life history of decidual cells: A review, *Am. J. Reprod. Immunol.* **3:**78–82.
76. Slapsys, R., and Clark, D. A., 1983, Active suppression of host-versus-graft reaction in pregnant mice. V. Kinetics, specificity, and *in vivo* activity of non-T suppressor cells localized to the genital tract of mice during first pregnancy, *Am. J. Reprod. Immunol.* **3:**65–71.
77. Clark, D. A., Slapsys, R., Croy, B. A., Krcek, J., and Rossant, J., 1984, Local active

suppression by suppressor cells in the decidua: A review, *Am. J. Reprod. Immunol.* **5**:78–83.

78. Mohammad, M., Saadi, M., el Balhaa, G. R., Mayer, G., and Chateaureynaud, P., 1984, Rat pregnancy induces two suppressor cell activities, *Am. J. Reprod. Immunol.* **6**:152–158.

79. Chaouat, G., and Kolb, J.-P., 1985, Immunoactive products of placenta. IV. Impairment by placental cells and their products of CTL function at effector stage, *J. Immunol.* **135**:215–222.

80. Clark, D. A., Mowbray, J., Underwood, J., and Lidell, H., 1987, Histopathologic alterations in the decidua in human spontaneous abortion: Loss of cells with large cytoplasmic granules, *Am. J. Reprod. Immunol.* **13**:19–22.

81. Merluzzi, V. J., Levy, E. M., Kumar, V., Bennett, M., and Cooperband, S. R., 1978, *In vitro* activation of suppressor cells from spleens of mice treated with radioactive strontium, *J. Immunol.* **121**:505–512.

82. Levy, E. M., Bennett, M., Kumar, V., Fitzgerald, P., and Cooperband, S. R., 1980, Adoptive transfer of spleen cells from mice treated with radioactive strontium: Suppressor cells, natural killer cells, and "hybrid resistance" in recipient mice, *J. Immunol.* **124**:611–618.

83. Raveche, E. S., Laskin, C. A., Rubin, C., Tjio, J. H., and Steinberg, A. D., 1983, Comparison of stem-cell recovery in autoimmune and normal strains, *Cell. Immunol.* **79**:56–67.

84. Gleicher, N. (ed.), 1985, *Principles of Medical Therapy in Pregnancy,* Plenum, New York, pp. 49–51.

85. Chervenak, R., Cohen, J. J., and Miller, S. D., 1983, Clonal abortion of bone marrow T cell precursors: T cells acquire specific antigen reactivity prethymically, *J. Immunol.* **131**:1688–1692.

86. Kast, W. M., DeWaal, L. P., Melief, C. J. M., 1984, Thymus dictates major histocompatibility complex (MHC) specificity and immune response gene phenotype of class II MHC-restricted T cells but not of class I MHC-restricted T cells, *J. Exp. Med.* **160**:1752–1766.

87. Hurme, M., and Sihvola, M., 1985, *In vivo* activation of autoreactive cytotoxic T lymphocytes after bone marrow transplantation: Evidence for the prethymic specificity of T cells? *J. Immunol.* **135**:1108–1112.

88. Slavin, S., Strober, S., Fuks, Z., and Kaplan, H. S., 1977, Induction of specific tissue transplantation tolerance using fractionated total lymphoid irradiation in adult mice: Long-term survival of allogeneic bone marrow and skin grafts, *J. Exp. Med.* **146**:34–48.

89. Zan-Bar, I., Slavin, S., and Strober, S., 1978, Induction and mechanism of tolerance to bovine serum albumin in mice given total lymphoid irradiation (TLI), *J. Immunol.* **121**:1400–1404.

90. Slavin, S., and Strober, S., 1979, Induction of allograft tolerance after total lymphoid irradiation (TLI): Development of suppressor cells of the mixed leukocyte reaction (MLR), *J. Immunol.* **123**:942–946.

91. Strober, S., 1984, Strategies promoting allograft acceptance: An overview, *Fed. Proc.* **43**:261–262.

92. Strober, S., Okada, S., and Oseroff, A., 1984, Role of natural suppressor cells in allograft tolerance, *Fed. Proc.* **43**:263–265.

93. Berenbaum, M. C., 1963, Prolongation of homograft survival in mice with single doses of cyclophosphamide, *Nature* **200**:84.

94. Aisenberg, A. C., 1967, Studies on cyclophosphamide-induced tolerance to sheep erythrocytes, *J. Exp. Med.* **125**:833–845.

95. Stockman, G. D., and Trentin, J. J., 1972, Cyclophosphamide-induced tolerance to equine α globulin and to equine-anti-mouse-thymocyte globulin in adult mice. I. Studies on antigen and drug requirements, *J. Immunol.* **108**:112–118.

96. Starzl, T. E., Groth, C. G., Putnam, C. W., Corman, J., Halgrimson, C. G., Penn, I., Husberg, B., Gustafsson, A., Cascardo, S., Geis, P., and Iwatsuki, S., 1973, Cyclophosphamide for clinical renal and hepatic transplantation, *Transplant. Proc.* **5**:511–516.

97. Caves, P. K., Dong, E. Jr., and Shumway, N. E., 1973, Immunosuppression with cyclophosphamide in dogs following cardiac transplantation, *Transplant. Proc.* **5**:517–521

98. Nirmul, G., Severin, C., and Taub, R. N., 1973, Mechanisms and kinetics of cyclophosphamide-induced specific tolerance to skin allografts in mice, *Transplant. Proc.* **5**:675–678.

99. Shin, T., Mayumi, H., Himeno, K., Sanui, H., and Nomoto, K., 1984, Drug-induced tolerance to allografts in mice. I. Difference between tumor and skin grafts, *Transplantation* **37**:580–584.

100. Mayumi, H., Himeno, K., Shin, T., and Nomoto, K., 1985, Drug-induced tolerance to allografts in mice. IV. Mechanisms and kinetics of cyclophosphamide-induced tolerance, *Transplantation* **39**:209–215.

101. Gale, R. P., 1982, Progress in bone marrow transplantation in man, *Surv. Immunol. Res.* **1**:40–66.

102. Storb, R., 1983, Human bone marrow transplantation, *Transplantat. Proc.* **15**:1379–1384.

103. Sado, T., Kamisaku, H., and Kubo, E., 1985, Strain difference in the radiosensitivity of immunocompetent cells and its influence on the residual host-vs-graft reaction in lethally irradiated mice grafted with semiallogeneic bone marrow, *J. Immunol.* **134**:704–710.

104. Soderling, C. C. B., Song, C. W., Blazar, B. R., and Vallera, D. A., 1985, A correlation between conditioning and engraftment in recipients of MHC-mismatched T cell–depleted murine bone marrow transplants, *J. Immunol.* **135**:941–946.

105. Lancet, 1975, Editorial: The fetus as a homograft, *Lancet* **2**:535–536.

106. Stein-Werblowsky, R., 1981, On the retention of the fetus as a homograft, *Am. J. Reprod. Immunol.* **1**:180–181.

107. Watson, J. D., and Prestidge, R. L., 1983, Interleukin 3 and colony-stimulating factors, *Immunol. Today* **4**:278–280.

108. Jadus, M. R., and Parkman, R., 1986, The selective growth of murine newborn-derived suppressor cells and their probable mode of action, *J. Immunol.* **136**:783–792.

109. Hertel-Wulff, B., Okada, S., Oseroff, A., Strober, S., 1984, *In vitro* propagation and cloning of murine natural suppressor (NS) cells, *J. Immunol.* **133**:2791–2796.

110. Schwadron, R. B., Gandour, D. M., and Strober, S., 1985, Cloned natural suppressor cell lines derived from the spleens of neonatal mice, *J. Exp. Med.* **162**:297–310.

111. Szekeres-Bartho, J., Csernus, V., Hadnagy, J., and Pacsa, A. S., 1983, Progesterone-prostaglandin balance influences lymphocyte function in relation to pregnancy, *Am. J. Reprod. Immunol.* **4**:139–141.

112. Mohammad, M., Saadi, M., el Balhaa, G. R., Chateaureynaud, P., and Nayer, G., 1984, Rat pregnancy immunoregulatory circuits are progestation hormonal status, decidual tissue, embryo-trophoblast and late pregnancy changes dependent. *Am. J. Reprod. Immunol.* **6**:159–166.

113. Hansen, P. J., Bazer, F. W., and Segerson, E. C. Jr., 1986, Skin graft survival in the uterine lumen of ewes treated with progesterone, *Am. J. Reprod. Immunol.* **12**:48–54.

114. Maier, T., Holda, J. H., and Claman, H. N., 1985, Synergism between T and non-T cells in the *in vivo* and *in vitro* expression of graft-vs.-host disease–induced natural suppressor cells, *J. Exp. Med.* **162**:979–992.

115. Khan, M. M., Melmon, K. L., Fathman, C. G., Hertel-Wulff, B., and Strober, S., 1985, The effects of autacoids on cloned murine lymphoid cells: Modulation of IL 2 secretion and the activity of natural suppressor cells, *J. Immunol.* **134**:4100–4106.

116. Khan, M. M., and Melmon, K. L., 1985, Are autacoids more than theoretic modulators of immunity? *Clin. Immunol. Rev.* **4**:1:30.

117. Khan, M. M., Marr-Leisy, D., Verlander, M. S., Bristow, M. R., Strober, S., Goodman,

M., and Melmon, K. L., 1986, The effects of derivatives of histamine on natural suppressor cells, *J. Immunol.* **137**:308–314.

118. Herberman, R. B., Nunn, M. E., and Lavrin D. H., 1975, Natural cytotoxic reactivity of mouse lymphoid cells against syngeneic and allogeneic tumors. I. Distribution of reactivity and specificity, *Int. J. Cancer* **16**:216–229.

119. Uchida, A., 1983/84, Lack of spontaneous and inducible natural killer cell activity in human bone marrow: Presence of adherent suppressor cells, *Nat. Immun. Cell Growth Regul.* **3**:181–192.

120. Baley, J. E., and Schacter, B. Z., 1985, Mechanisms of diminished natural killer cell activity in pregnant women and neonates, *J. Immunol.* **134**:3042–3048.

121. Haller, O., and Wigzell, H., 1977, Suppression of natural killer cell activity with radioactive strontium: Effector cells are marrow dependent, *J. Immunol.* **118**:1503–1506.

122. Maier, T., Holda, J. H., and Claman, H. N., 1986, Natural suppressor (NS) cells: Members of the LGL regulatory family, *Immunol. Today* **7**:312–315.

123. Targan, S., Brieva, J., Newman, W., and Stevens, R., 1985, Is the NK lytic process involved in the mechanism of NK suppression of antibody-producing cells? *J. Immunol.* **134**:666–669.

124. Argyris, B. F., 1981, Suppressor factor produced by neonatal mouse spleen cells, *Cell. Immunol.* **62**:412–424.

125. Atkinson, M. J., Mortari, F., Saffran, D. C., McCain, G. A., and Singhal, S. K., 1986, Regulation of normal and autoimmune responses by bone marrow derived mediators, in: *Mediators of Immune Regulation and Immunotherapy* (S. K. Singhal and T. L. Delovitch, eds.), Elsevier, New York, pp. 51–57.

126. Rubinstein, A., Koren, Z., and Murphy, R. A., 1982, Suppression of maternal lymphocyte mitogenic responses by supernatants from short-term placental cell cultures, *Am. J. Reprod. Immunol.* **2**:260–264.

127. Chaouat, G., and Chaffaux, S., 1984, Placental products induce suppressor cells of graft versus host reaction, *Am. J. Reprod. Immunol.* **6**:107–111.

128. Clark, D. A., Chaput, A., Walker, C., and Rosenthal, K. L., 1985, Active suppression of host-vs-graft reaction in pregnant mice. VI. Soluble suppressor activity obtained from decidua of allopregnant mice blocks the response to IL 2, *J. Immunol.* **134**:1659–1664.

129. Clark, D. A., Slapsys, R., Chaput, A., Walker, C., Brierley, J., Daya, S., and Rosenthal, K. L., 1986, Immunoregulatory molecules of trophoblast and decidual suppressor cell origin at the maternofetal interface, *Am. J. Reprod. Immunol.* **10**:100–104.

130. Chaouat, G., Kolb, J. P., Riviere, M., and Lankar, D., 1986, Immunoactive products of placenta. V. Soluble factors fom murine placenta can block effector stages of maternal antipaternal cell-mediated immunity, *Am. J. Reprod. Immunol.* **12**:70–77.

131. Strober, S., Palathumpat, V., Schwadron, R., and Hertel-Wulff, B., 1987, Cloned natural suppressor cells prevent lethal graft-vs-host disease, *J. Immunol.* **138**:699–703.

132. Kotzin, B. L., and Strober, S., 1979, Reversal of NZB/NZW disease with total lymphoid irradiation, *J. Exp. Med.* **150**:371–378.

133. Theofilopoulos, A. N., Balderas, R., Shawler, D. L., Izui, S., Kotzin, B. L., Strober, S., and Dixon, F. J., 1980, Inhibition of T cell proliferation and SLE-like syndrome of MRL/1 mice by whole body or total lymphoid irradiation, *J. Immunol.* **125**:2137–2142.

134. McCune, W. J., Buckley, J. A., Belli, J. A., and Trentham, D. E., 1982, Immunosuppression by fractionated total lymphoid irradiation in collagen arthritis, *Arth. Rheum.* **25**:532–539.

135. Field, E. H., Strober, S., Hoppe, R. T., Calin, A., Engleman, E. G., Kotzin, B. L., Tanay, A. S., Calin, H. J., Terrell, C. P., and Kaplan, H. S., 1983, Sustained improvement of intractable rheumatoid arthritis after total lymphoid irradiation, *Arth. Rheum.* **26**:937–946.

136. Brahn, E., Helfgott, S. M., Belli, J. A., Anderson, R. J., Reinherz, E. L., Schlossman, S. F., Austen, K. F., and Trentham, D. E., 1984, Total lymphoid irradiation therapy in

refractory rheumatoid arthritis: Fifteen- to forty-month followup, *Arth. Rheum.* **27**:481–488.

137. Strober, S., Tanay, A., Field, E., Hoppe, R. T., Calin, A., Engleman, E. G., Kotzin, B., Brown, B. W., and Kaplan, H. S., 1985, Efficacy of total lymphoid irradiation in intractable rheumatoid arthritis: A double-blind, randomized trial, *Ann. Intern. Med.* **102**:441–449.

138. Strober, S., Field, E., Hoppe, R. T., Kotzin, B. L., Shemesh, O., Engleman, E., Ross, J. C., and Myers, B. D., 1985, Treatment of intractable lupus nephritis with total lymphoid irradiation, *Ann. Intern. Med.* **102**:450–458.

139. Ben-Chetrit, E., Gross, D. J., Braverman, A., Weshler, Z., Fuks, Z., Slavin, S., and Eliakim, M., 1986, Total lymphoid irradiation in refractory systemic lupus erythematosus, *Ann. Intern. Med.* **105**:58–60.

140. Strober, S., 1986, Use of total lymphoid irradiation in autoimmunity and transplantation: Cellular mechanisms, in: *Mediators of Immune Regulation and Immunotherapy* (S. K. Singhal and T. L. Delovitch, eds.), Elsevier, New York, pp. 204–212.

141. Decker, J. L., Malone, D. G., Haroui, B., Wahl, S. M., Schrieber, L., Klippel, J. H., Steinberg, A. D., and Wilder, R. L., 1984, Rheumatoid arthritis: Evolving concepts of pathogenesis and treatment, *Ann. Intern. Med.* **101**:810–824.

142. Kirchner, H., 1984, Interferons, a group of multiple lymphokines, *Springer Semin. Immunopathol.* **7**:347–374.

143. Stiehn, E. R., Kronenberg, L. H., Rosenblatt, H. M., Bryson, Y., and Merigan, T. C., 1982, Interferon: Immunobiology and clinical significance, *Ann. Intern. Med.* **96**:80–93.

144. Broxmeyer, H. E., Lu, L., Platzer, E., Feit, C., Juliano, L., and Rubin, B. Y., 1983, Comparative analysis of the influences of human gamma, alpha and beta interferons on human multipotential (CFU-GEMM), erythroid (BFU-E) and granulocyte-macrophage (CFU-GM) progenitor cells, *J. Immunol.* **131**:1300–1305.

145. Zoumbos, N. C., Djeu, J. Y., and Young, N. S., 1984, Interferon is the suppressor of hematopoiesis generated by stimulated lymphocytes *in vitro*, *J. Immunol.* **133**:769–774.

146. Broxmeyer, H. E., Cooper, S., Rubin, B. Y., and Taylor, M. W., 1985, The synergistic influence of human interferon-γ and interferon-α on suppression of hematopoietic progenitor cells is additive with the enhanced sensitivity of these cells to inhibition by interferons at low oxygen tension *in vitro*, *J. Immunol.* **135**:2502–2506.

147. Raefsky, E. L., Platanias, L. C., Zoumbos, N. C., and Young, N. S., 1985, Studies of interferon as a regulator of hematopoietic cell proliferation, *J. Immunol.* **135**:2507–2512.

148. Zoumbos, N. C., Gascon, P., Djeu, J. Y., and Young, N. S., 1985, Interferon is a mediator of hematopoietic suppression in aplastic anemia *in vitro* and possibly *in vivo*, *Proc. Natl. Acad. Sci. USA* **82**:188–192.

149. Champlin, R. E., 1985, Aplastic anemia, in: *Hematopoietic Stem Cells* (D. W. Golde and F. Takaku, eds.), Marcel Dekker, New York, pp. 205–227.

150. Barlozzari, T., Leonhardt, J., Wiltrout, R. H., Herberman, R. B., and Reynolds, C. W., 1985, Direct evidence for the role of LGL in the inhibition of experimental tumor metastases, *J. Immunol.* **134**:2783–2789.

151. Wiltrout, R. H., Herberman, R. B., Zhang, S.-R., Chirigos, M. A., Ortaldo, J. R., Green, K. M. Jr., and Talmadge, J. E., 1985, Role of organ-associated NK cells in decreased formation of experimental metastases in lung and liver, *J. Immunol.* **134**:4267–4275.

152. Lotzova, E., Savary, C. A., and Herberman, R. B., 1986, Antileukemia reactivity of endogenous and interleukin-2-activated natural killer cells, in: *Natural Immunity, Cancer and Biological Response Modification* (E. Lotzova and R. B. Herberman, eds.), Karger, Basel, Switzerland, pp. 177–195.

153. Combe, B., Pope, R., Darnell, B., Kincaid, W., and Talal, N., 1984, Regulation of natural killer cell activity by macrphages in the rheumatoid joint and peripheral blood, *J. Immunol.* **133**:709–713.

154. Minato, N., Amagai, T., Yodoi, J., Diamanstein, T., and Kano, S., 1985, Regulation of the growth and functions of cloned murine large granular lymphocyte lines by resident macrophages, *J. Exp. Med.* **162:**1161–1181.

155. Parhar, R. S., and LaLa, P. K., 1985, Changes in the host natural killer cell population in mice during tumor development. 2. The mechanism of suppression of NK activity, *Cell. Immunol.* **93:**265–279.

156. Lotzova, E., and Savary, C. A., 1986, Regulation of NK cell activity by suppressor cells, in: *Immunobiology of Natural Killer Cells* (E. Lotzova and R. B. Herberman, eds.), Volume II, CRC Press, Boca Raton, FL, pp. 163–177.

157. Noel, D. R., Witherspoon, R. P., Storb, R., Atkinson, K., Doney, K., Mickelson, E. M., Ochs, H. D., Warren, R. P., Weiden, P. L., and Thomas, E. D., 1978, Does graft-versus-host disease influence the tempo of immunologic recovery after allogeneic human marrow transplantation? An observation on 56 long-term survivors, *Blood* **51:**1087–1105.

158. Tsoi, M.-S., Storb, R., Dobbs, S., Kopecky, K. J., Santos, E., Weiden, P. L., and Thomas, E. D., 1979, Nonspecific suppressor cells in patients with chronic graft-vs-host disease after marrow grafting, *J. Immunol.* **123:**1970–1976.

159. Tsoi, M.-S., Storb, R., Dobbs, S., Medill, L., and Thomas E. D., 1980, Cell-mediated immunity to non-HLA antigens of the host by donor lymphocytes in patients with chronic graft-vs-host disease, *J. Immunol.* **125:**2258–2262.

160. Tsoi, M.-S., 1982, Immunological mechanisms of graft-versus-host disease in man, *Transplantation* **33:**459–464.

161. Tsoi, M.-S., Dobbs, S., Brkic, S., Ramberg, R., Thomas, E. D., and Storb, R., 1984, Cellular interactions in marrow-grafted patients. II. Normal monocyte antigen-presenting and defective T-cell-proliferative functions early after grafting and during chronic graft-versus-host disease, *Transplantation* **37:**556–561.

162. Witherspoon, R. P., Matthews, D., Storb, R., Atkinson, K., Cheever, M., Deeg, H. J., Doney, K., Kalbfleisch, J., Noel, D., Prentice, R., Sullivan, K. M., and Thomas, E. D., 1984, Recovery of *in vivo* cellular immunity after human marrow grafting: Influence of time postgrafting and acute graft-versus-host disease, *Transplantation* **37:**145–150.

163. Brkic, S., Tsoi, M.-S., Mori, T., Lachman, L., Gillis, S., Thomas, E. D., and Storb, R., 1985, Cellular interactions in marrow-grafted patients. III. Normal interleukin 1 and defective interleukin 2 production in short-term patients and in those with chronic graft-versus-host disease, *Transplantation* **39:**30–35.

164. Vallera, D. A., Soderling, C. C. B., Carlson, G. J., and Kersey, J. H., 1982, Bone marrow transplantation across major histocompatibility barriers in mice. II. T cell requirement for engraftment in total lymphoid irradiation-conditioned recipients, *Transplantation* **33:**243–248.

165. Storb, R., Doney, K. C., Thomas, E. D., Appelbaum, F., Buckner, C. D., Clift, R. A., Deeg, H. J., Goodell, B. W., Hackman, R., Hansen, J. A., Sanders, J., Sullivan, K., Weiden, P. L., and Witherspoon, R. P., 1982, Marrow transplantation with or without donor buffy coat cells for 65 transfused aplastic anemia patients, *Blood* **59:**236–246.

166. Storb, R., Prentice, R. L., Thomas, E. D., Appelbaum, F. R., Deeg, H. J., Doney, K., Fefer, A., Goodell, B. W., Mickelson, E., Stewart, P., Sullivan, K. M., and Witherspoon, R. P., 1983, Factors associated with graft rejection after HLA-identical marrow transplantation for aplastic anaemia, *Br. J. Haematol.* **55:**573–585.

167. Mitsuyasu, R. T., Champlin, R. E., Gale, R. P., Ho, W. G., Lenarsky, C., Winston, D., Selch, M., Elashoff, R., Giorgi, J. V., Wells, J., Terasaki, P., Billing, R., and Feig, S., 1986, Treatment of donor bone marrow with monoclonal anti-T-cell antibody and complement for the prevention of graft-versus-host disease: A prospective, randomized, double-blind trial, *Ann. Intern. Med.* **105:**20–26.

168. Prendergast, M. M., Bradstock, K. F., Broomhead, A. F., Hughes, W. G., Kabral, A., Berndt, M. C., and Tiver, K., 1986, Monoclonal antibody analysis of canine hemopoietic cells: Role of IA-like and THY-1 antigens in bone marrow engraftment, *Transplantation* **41:**565–571.

169. Ildstad, S. T., and Sachs, D. H., 1984, Reconstitution with syngeneic plus allogeneic or xenogeneic bone marrow leads to specific acceptance of allografts or xenografts, *Nature* **307:**168–170.

170. Ildstad, S. T., Wren, S. M., Sharrow, S. O., Stephany, D., and Sachs, D. H., 1984, *In vivo* and *in vitro* characterization of specific hyporeactivity to skin xenografts in mixed xenogeneically reconstituted mice (B10 + F344 Rat→B10), *J. Exp. Med.* **160:**1820–1835.

171. Wood, M. L., and Monaco, A. P., 1984, Induction of unresponsiveness to skin allografts in adult mice disparate at defined regions of the H-2 complex. I. Effect of donor-specific bone marrow in ALS-treated mice, *Transplantation* **37:**35–39.

172. Hartner, W. C., De Fazio, S. R., Make, T., Markees, T. G., Monaco, A. P., and Gozzo, J. J., 1986, Prolongation of renal allograft survival in antilymphocyte-serum-treated dogs by postoperative injection of density-gradient-fractionated donor bone marrow, *Transplantation* **42:**593–597.

173. Simpson, M. A., Maki, T., Gozzo, J. J., and Monaco, A. P., 1983, Characterization of antilymphocyte-induced suppressor cells, *Transplant. Proc.* **15:**740–743.

174. Main, J. M., and Prehn, R. T., 1957, Fate of skin homografts in X-irradiated mice treated with homologous marrow, *JNCI* **19:**1053–1064.

175. Bridges, J. B., Loutit, J. F., and Micklem, H. S., 1960, Transplantation immunity in the isologous mouse radiation chimaera, *Immunology* **3:**195–213.

176. Rapaport, F. T., Bachvaroff, R. J., Akiyama, N., Sato, T., and Ferrebee, J. W., 1980, Specific allogeneic unresponsiveness in irradiated dogs reconstituted with autologous bone marrow, *Transplantation* **30:**23–30.

177. Hartnett, L. C., Dittmer, J. E., and Schwadron, R. B., 1984, Acceptance of cardiac allografts by lethally irradiated rats repopulated with syngeneic bone marrow: Role of class I and class II alloantigens, *Transplantation* **37:**378–382.

178. Gale, R. P., 1980, Concepts of fetal liver transplantation, in: *Fetal Liver Transplantation: Current Concepts and Future Directions* (G. Lucarelli, T. M. Fliedner, and R. P., Gale, eds.), Excerpta Medica, Amsterdam, pp. 247–256.

179. O'Reilly, R. J., Pollack, M. S., Kapoor, N., Kirkpatrick, D., and Dupont, B., 1983, Fetal liver transplantation in man and animals, in: *Recent Advances in Bone Marrow Transplantation* (R. P. Gale, ed.), Alan R. Liss, New York, pp. 799–830.

180. Stitzel, K. A., Champlin, R., and Gale, R. P., 1983, Fetal liver cell transplantation in dogs: A possible alternative source of hematopoietic cells for transplantation, in: *Recent Advances in Bone Marrow Transplantation* (R. P. Gale, ed.), Alan R. Liss, New York, pp. 831–840.

181. Prummer, O., Raghavachar, A., Werner, C., Calvo, W., Carbonell, F., Steinbach, I., and Fliedner, T. M., 1985, Fetal liver transplantation in the dog. I. Restoration of hemopoiesis with cryopreserved fetal liver cells from DLA-identical siblings, *Transplantation*, **39:**349–355.

182. Basch, R. S., and Berman, J. W., 1982, Thy-1 determinants are present on many murine hematopoietic cells other than T cells, *Eur. J. Immunol.* **12:**359–364.

183. Muller-Sieburg, C. E., Whitlock, C. A., and Weissman, I. L., 1986, Isolation of two early B lymphocyte progenitors from mouse marrow: A committed pre-pre-B cell and a clonogenic Thy-1 hematopoietic stem cell, *Cell* **44:**653–662.

184. Dorf, M. E., and Benacerraf, B., 1984, Suppressor cells and immunoregulation, *Annu. Rev. Immunol.* **2:**127–158.

185. Fulton, A. M., and Levy J. G., 1981, The induction of nonspecific T suppressor lymphocytes by prostaglandin E_1, *Cell. Immunol.* **59:**54–60.

186. Fischer, A., Durandy, A., LeDeist, F., and Griscelli, C., 1985, The role of PGE_2 in the induction of suppressor cells in humans, in: *Prostaglandins and Immunity* (J. S. Goodwin, ed.), Martinus Nihjoff, Boston, pp. 55–97.

187. Fischer, A., Le Deist, F., Durandy, A., and Griscelli, C., 1985, Separation of a population of human T lymphocytes that bind prostaglandin E2 and exert a suppressor activity, *J. Immunol.* **134:**815–819.

188. Almawi, W. Y., and Pope, B. L., 1986, Induction of suppression by a murine nonspecific suppressor-inducer cell line (M1-A5). II. The role of prostaglandins, *J. Immunol.* **136**:1982–1987.

189. Ju, S.-T., and Dorf, M. E., 1985, Functional analysis of cloned macrophage hybridomas. IV. Induction and inhibition of mixed lymphocyte responses, *J. Immunol.* **134**:3722–3730.

190. Schaefer, A. E., Scafuri, A. R., Fredericksen, T. L., and Gilmour, D. G., 1985, Strong suppression by monocytes of T cell mitogenesis in chicken peripheral blood leukocytes, *J. Immunol.* **135**:1652–1660.

191. Shibata, Y., and Volkman, A., 1985, The effect of hemopoietic microenvironment of splenic suppressor macrophages in congenitally anemic mice of genotype S1/S1^d, *J. Immunol.* **135**:3905–3910.

192. Kato, K., Yamamoto, K.-I., and Kimura, T., 1985, Migration of natural suppressor cells from bone marrow to peritoneal cavity by live BCG, *J. Immunol.* **135**:3661–3668.

193. Abruzzo, L. V., and Rowley, D. A., 1983, Homeostasis of the antibody response: Immunoregulation by NK cells, *Science* **222**:581–585.

194. Tilden, A. B., Abo, T., and Balch, C. M., 1983, Suppressor cell function of human granular lymphocytes identified by the HNK-1 (Leu 7) monoclonal antibody, *J. Immunol.* **130**:1171–1175.

195. Brieva, J. A., Targan, S., and Stevens, R. H., 1984, NK and T cell subsets regulate antibody production by human *in vivo* antigen-induced lymphoblastoid B cells, *J. Immunol.* **132**:611–615.

196. Brieva, J. A., and Stevens, R. H., 1984, Involvement of the transferrin receptor in the production and NK-induced suppression of human antibody synthesis, *J. Immunol.* **133**:1288–1292.

197. Abruzzo, L. V., Mullen, C. A., Rowley, D. A., 1986, Immunoregulation by natural killer cells, *Cell. Immunol.* **98**:266–278.

13

Cytokine Production by CD3⁻ Large Granular Lymphocytes

John R. Ortaldo

1. Introduction

The original concept of immune surveillance, as first postulated by Ehrlich[1] and later modified by Burnet and Thomas,[2] proposed that "the immune system plays a very central role in resistance against the development of detectable tumors." An updated version of the hypothesis[3] would include the following concepts: (1) Transformed cells express surface antigens or other structures that are recognized by one or more components of the immune system, which would encompass not only classical T and B cell immunity but also various components of the natural resistance system; (2) one or more components of the natural or induced immunological effector mechanisms can eliminate transformed cells or impede their progression or growth. This updated hypothesis would also apply to the immunological regulation of infection by other pathogens such as viruses, bacteria, protozoans, and other parasites.

The earliest studies by immunologists emphasized the products and regulation of specifically sensitized lymphocytes. T cells are a major subpopulation of small lymphocytes that depend on the thymus, where they mature and develop functional activity. Studies conducted over the past decade have indicated that T cells are a highly heterogeneous collection of cells with a variety of clones, each restricted in reactivity to a particular antigen or set of cross-reactive antigens plus self major histocompatibility complex molecules. However, T cells have virtually no detectable sponta-

John R. Ortaldo • Laboratory of Experimental Immunology, Biological Response Modifiers Program, Division of Cancer Treatment, National Cancer Institute–FCRF, Frederick, Maryland 21701-1013.

neous cytotoxic activity; rather, they must be activated by exposure to specific antigen in the presence of accessory cells such as macrophages, and they require a cascade of specific regulatory cytokines. In addition, there is a considerable latent period (usually 7–10 days) required for T cells to develop their initial reactivity.

In addition to specific T cells, recent studies have also proposed natural resistance mechanisms as antitumor effectors.[5,6] During the past decade these effector mechanisms have been studied in great detail, and evidence clearly indicates that they play a significant role *in vivo* in preventing the development and metastatic spread of tumor cells (see the chapter by Herberman, this volume). Natural resistance mechanisms can be related to several categories of effector cells. Macrophages can be induced to become tumoricidal as well as antiparasitic and antibacterial. Evidence now exists that macrophages function both by phagocytosis of microbes and by exerting direct lytic and cytostatic effects against many tumor cells. In addition, macrophages produce a variety of cytokines that may have antitumor effects or that may produce or augment cytotoxic activity in other cells.

Natural killer (NK) cells are another component of the natural resistance system. NK cells were determined in the early 1970s to be spontaneously cytotoxic effector cells. Detailed studies regarding the nature and regulation of these cells revealed them to be a small subset of nonadherent, nonphagocytic, Fc-receptor-positive leukocytes. [5–8] The majority of human NK activity (approximately $\geq 95\%$) is mediated by $CD3^-$, $Leu19^+$ (NKH1), $CD16^+/$ large granular lymphocytes (LGL). These LGL also demonstrate high levels of antibody-dependent cellular cytotoxicity that is directly related to the expression of the Fc-receptor for IgG ($Fc\gamma R$) or the CD16 antigen.

Recently, a $CD3^+$ T cell was also shown to mediate the killing of K562 cells (NK activity), apparently via the CD3-associated T cell receptor complex. However, this population exists at a low frequency, and its contribution to the lytic activity in fresh leukocyte preparations has been difficult to assess accurately. The present use of "NK cells" as a collective term to identify all cells with this function has led to a great deal of confusion regarding which of the cell populations is being addressed. In this chapter we will attempt to avoid this confusion by using the term "NK" only in references to the distinct function associated with non-T (or $CD3^-$) LGL.

Further studies [8–11] have revealed that most NK activity is associated with a population of LGL, distinct from macrophages, that have unique patterns of cytotoxic activity. In the mouse, there exists a population, termed natural cytotoxic (NC) cells, that seems to differ somewhat from cells mediating NK activity. The equivalent of NC cells has not been reported in other species.[8,9] However, NC cells may be a subpopulation of LGL. An effector cell termed the K cell, which is responsible for antibody-dependent cellular cytotoxicity, has been examined in mice, rats, and humans. The data indicate that the majority of K cells also exhibit NK activity.[10] K cells mediate their cytolytic function via antibodies, which may occur naturally or be produced by antigen-specific immune B cells.

Granulocytes are also an important component of natural resistance systems. Most of the evidence regarding natural resistance by granulocytes refers to their activity against microbial diseases. Although *in vitro* cytolytic activity of granulocytes has been studied, conclusive *in vivo* data regarding their role against tumor spread has not been reported.[5,8]

Most natural resistance mechanisms are regulated by soluble factors and, in some instances, by cellular interactions.[5–7,11,12] However, the mechanism by which natural effector cells are regulated has been most thoroughly examined with the soluble cytokines. Interleukin 1 (IL-1), IL-2, and the interferons (IFN) have been the most intensively studied with regard to regulation of natural effector mechanisms.[11–13] IFN(α,β,γ) are very potent activators of NK activity and macrophage function. The strong NK-activating potential of IL-2 has also been documented. A variety of other agents, including prostaglandins, leukotrienes, and tumor promoters, can inhibit NK activity. Thus, as with immune lymphocytes, natural resistance is subject to a complex system of regulation by a variety of agents.

2. Interferon Production

Interferon (IFN) was the first cytokine demonstrated to be produced by cells with NK activity by Trinchieri et al.[14] as the result of the interaction of effector and target cells. This initial observation was extended to account for IFN induction by virus-infected allogeneic fibroblasts. The cells in these studies were shown to be non-T cells, Fc receptor-positive lymphocytes and were classified as NK cells. In accordance with these observations, LGL were shown to produce IFNα during contact with NK-susceptible target cells.[15] In highly purified human LGL, IFNα was produced in response to various polyclonal stimulants, including *Staphylococcus* enterotoxin A, concanavalin A (Con A), phytohemagglutinin (PHA), bacterial preparations (*Corynebacterium parvum*), and *Mycoplasma*-infected target cell lines.[16]

In the course of these studies, Djeu and co-workers[16] also collected evidence indicating that CD3⁻ LGL are able to secrete IFNγ (previously termed immune IFN). Until that time, only CD3⁺ cells had been thought to produce IFNγ. However, in both the murine[17] and human[18] systems, highly enriched CD16⁺ LGL were shown to secrete IFNγ. When we studied the mechanism by which IL-2 activates LGL for cellular lysis,[18] we found that high levels of IFNγ could be stimulated in NK-active cells by recombinant IL-2. CD3⁺ T lymphocytes can also produce significant amounts of IFNγ, but they must either be activated with lectins (PHA, Con A) or receive multiple signals (IL-2 and antigen).[19] These results collectively indicate that the CD3⁻ LGL is capable of not only lysing tumor and virus-infected targets, but also of secreting IFNα and γ in response to these target cells.

3. Interleukin Production

3.1. Interleukin 2

That human LGL secrete IL-2 was first reported by Domzig and Stadler,[20] who demonstrated that highly enriched, T cell-depleted LGL produce appreciable levels of IL-2 in response to lectins (PHA, Con A). This study was extended by Kasahara et al.,[21] who used highly enriched CD3⁻ LGL obtained from discontinuous Percoll density gradients to examine both the quantity of IL-2 produced and kinetics of lectin stimulation of CD3⁻ LGL and CD3⁺ T cells. These studies indicated that CD3⁻ LGL made a significant proportion of the "early" (0–48 h) IL-2, whereas the later IL-2 (>96 h) was produced by CD3⁺ T cells and was highly regulated by macrophages and their products. In other experiments, it was demonstrated that CD3⁻ LGL could secrete IL-2 in response to NK-susceptible but not NK-resistant targets.[22] Moreover, CD3⁺ T cells could not secrete IL-2 after coculture with either NK-susceptible or NK-resistant targets. These results indicate that LGL can be stimulated by target cells or lectins resulting in the secretion of IL-2 that positively regulates their NK activity as well as other immune effector cells.

3.2. Interleukin 1

Like many cell types, purified CD3⁻ LGL have also been shown to secrete IL-2.[23] It is interesting, however, that LGL are the only nonmyelomonocytic leukocytes that have been shown to produce amounts of IL-1α/β. Studies have shown that neoplastic T and B cell lines secrete significant quantities of IL-1, but their counterparts have not been found in normal leukocytes. More recent studies have examined the quantities of IL-1α and IL-1β produced by CD3⁻ LGL (Galli et al., in preparation). These studies have indicated that NK-active CD3⁻ LGL primarily secrete IL-1β and small but appreciable amounts of IL-1α. In regard to function, IL-1 from CD3⁻ LGL has been shown to stimulate proliferation of human fibroblasts and to sustain T lymphocyte responses to lectins,[21] both of which are well-established and accepted characteristics of macrophage-derived IL-1. In addition, recent studies have confirmed these findings with genetic probes for IL-1α/β, coupled with enumeration of soluble IL-1 levels in supernatants (Galli *et al.*, in preparation).

3.3. Interleukin 4

Since CD3⁻ LGL can produce IL-1 and IL-2, it was of interest to examine their ability to secrete B cell regulatory factors. LGL have been shown to have both positive and negative effects on the regulation of B cell growth and Ig secretion (see chapter by Kumagai *et al.*, this volume). Studies have demonstrated that highly purified CD3⁻ LGL with cytolytic activity can secrete B cell growth factor (IL-4),[24,25] which is a major proliferative signal for B cells. It is not clear whether LGL also produce factors like B cell differentiation factor (BCDF) that regulate B cell Ig secretion.

3.4. Colony-Stimulating Factor

CD3⁻ LGL have also been reported to secrete colony-stimulating factor (CSF-I).[21] Consistent levels of CSF as determined by the promotion of colony and cluster formation in adherent cell-depleted human bone marrow were reported after PHA and Con A stimulation. In contrast to the effect on IL-2 production, PHA inhibited the secretion of CSF by LGL. These results indicate that either CSF production is regulated differently or different subsets are responsible for IL-2 and CSF secretion.

4. Cytotoxic and Cytostatic Factors

In addition to numerous immunoregulatory factors, recent studies have demonstrated the release of soluble cytolytic factors from human and murine LGL.[26,27] One of the factors, produced during the interaction of mouse spleen cells and NK-susceptible target cells, has been termed natural killer cytotoxic factor (NKCF).[28] As with the immunoregulatory factors, human NKCF has been found after incubation of human peripheral blood lymphocytes or highly purified LGL with NK-susceptible target cells or mitogens.[27,29] The general biochemical characteristics of human and mouse NKCF have been previously reviewed in some detail.[30–32] NKCF appears to be a relatively unstable secreted protein. Using molecular sieving columns, human and mouse NKCF has been found to have molecular weights in the range of 20,000–40,000 daltons, with a peak of activity at 20,000 and another at 40,000 daltons.[30–32] Several observations support the hypothesis that NKCF is involved in the lytic mechanism of CD3$^-$ LGL cells: (1) NKCF is produced by highly purified populations of LGL; (2) the ability of various target cells to stimulate NKCF release correlates well with their susceptibility to lysis by NK active cells; (3) NKCF has selective lytic activity against NK-susceptible target cells; and (4) the amount of NKCF activity released from effector cells can be substantially augmented by pretreatment with IFN, paralleling the ability of this cytokine to boost NK activity.

In addition to NKCF, lymphotoxin and tumor necrosis factor (TNF) have also been shown to be produced by CD3$^-$ LGL.[33] However, both molecules demonstrate high levels of activity against a much narrower range of target cells than are affected by NK active cells, with L929 and WEHI-164 being perhaps the most susceptible targets available to both recombinant cytokines. We have found that most NK-susceptible human and mouse targets are not susceptible to growth inhibition or cytolysis by either of these recombinant molecules.[33,34] However, TNF is released after activation of CD3$^-$ LGL with either NK-susceptible target cells or recombinant IL-2. Studies with neutralizing antibodies are currently examining the role of these cytostatis/cytolytic factors in cellular lysis. In addition, several recent studies have suggested that TNF may augment specific T cell responses in mixed lymphocyte cultures.[35]

In addition to the cytotoxic/cytostatic factors discussed above, other cytokines have been shown to be either selectively toxic or to affect target cell susceptibility to lysis. IFNγ and IL-1 have been reported to lyse or inhibit the growth of some selected targets.[32] The extent to which these cytokines possess antitumor effects *in vivo* remains unclear; however, their ability to mediate such direct effects warrants further consideration.

5. Production of Factors by Cloned LGL

The finding that IL-2 can propagate CD3$^-$ LGL clones *in vitro* has provided a means for determining, at the clonal level, the association between subpopulations of LGL and cytokine production.[36] After examining numerous clones for function, we found that both cytotoxic and noncytotoxic LGL clones produce IFNα/γ, IL-1, Il-2, and IL-4 (Table I).[37] IFNγ was the most frequently produced lymphokine (>85%), further emphasizing the potential physiological relevance of this lymphokine. The other cytokines were produced by approximately one-third of the clones. No clear association was seen between the expression of different surface markers on the LGL clones and cytokine production. This finding is in contrast to the results from fresh CD3$^-$ LGL (see below). These results indicate the capacity of LGL to secrete multiple cytokines, many of which were previously associated with more defined subsets of leukocytes.

TABLE I
Summary of Cloned CD3⁻ LGL Cytokine Production[a]

Cytokine	Cytotoxic	No. positive/no. tested (%)	
IFNγ	No	32/37	(86)
	Yes	3/3	(100)
IL-1	No	7/48	(14)
	Yes	2/6	(33)
IL-2	No	14/41	(34)
	Yes	3/8	(37)
IL-4	No	5/24	(21)
	Yes	1/1	(100)

[a]Summarized from ref. 37.

6. Cell Phenotype and Stimuli

One of the interesting characteristics of freshly isolated CD3⁻ LGL is the diversity of phenotypes and stimuli that result in high levels of cytokine production. Table II summarizes the data regarding the phenotypes that result in the production of ILs, IFNs, and cytotoxic/cytostatic factors.[16,21–23,27] Although the CD16⁺ (Fc receptor) phenotype is consistently present, the remaining markers are partially segregated. Most of the ILs are made of CD2⁺, CD16⁺, CD11⁻ LGL, whereas IFNs (IFNα/γ) are made by CD2⁺, CD16⁺, CD11⁺ LGL. Segregation of cytokines on

TABLE II
Relationship of Phenotype to Cytokine Production

Phenotype[a]	Stimulus	Function
CD16⁺, CD2⁺/⁻, CD11⁻, HLA-DR⁺	Lipopolysaccharide, silica, IL-2	IL-1
CD16⁺, CD2⁺, CD11⁻, HLA-DR⁺/⁻	Lectin, K562 cells[b]	IL-2
CD16⁺, CD2⁺, CD11⁻, HLA-DR⁻	Lectin, K562 cells	IL-4
CD16⁺	Lectin	CSF
CD16⁺, CD2⁺, CD11⁺, HLA-DR⁻	Lectin, K562 cells	IFNα
CD16⁺, CD2⁺, CD11⁺, HLA-DR⁺	Lectin, K562 cells	IFNγ
CD16⁺, CD2⁺, CD11⁺, HLA-DR⁻	IFNα, IL-2	NKCF, cytotoxic
CD16⁺	Lectin, K562 cells	TNFα

[a]All cells were ≤2% CD3⁺.
[b]Either K562 cells or K562-soluble membranes can be used.

the basis of HLA-DR expression was seen only with IL-1 and IFNγ, where both were produced by this minor subset. Further studies with freshly isolated CD3⁻ LGL are needed to determine the exact subset responsible for producing CSF and TNFα. In addition, the subset of CD3⁻ LGL active in factor production might vary with the stimulus, although no evidence exists to either support or dispute this hypothesis.

In addition to phenotypic heterogeneity, different stimuli can significantly alter the levels and type of cytokines produced. Table III summarizes the stimuli that most effectively induce production of ILs, IFNs, and cytotoxic/cytostatic factors. With the exception of CSF, the abilities of NK target cells, lectins (PHA, Con A), IL-2, and bacterial products have been compared. Lectins are the best stimuli for the production of IL-2, IL-4, and CSF. The best stimulus of IFNα is virus-infected target cells, although NK target cells are quite effective. IFNγ is most effectively induced by IL-2. The cytotoxic/cytostatic factors are released in large quantities after most effector-target cell interactions. Such interactions are a requirement for cytokine release; however, often the target cell–triggered release can be augmented by coincubation with agents such as IFNα and IL-2.

One of the most interesting events seen in CD3⁻ LGL is the activation of these cells by recombinant IL-2. Recombinant IL-2 activates CD3⁻ LGL to increased cytotoxicity, to produce IFNγ, and to proliferate in the absence of detectable IL-2 receptor (defined by TAC).[18,38] Recent evidence indicates that CD3⁻ LGL express high levels of the β chain of the IL-2 receptor. It seems that this molecule regulates the functions of LGL and causes expression of the IL-2 receptor chair α (defined by anti-TAC monoclonal antibody).

TABLE III
Most Effective Stimuli for Induction of LGL Cytokine Production

Function	Stimulus
Interleukin	
IL-1	Bacterial products
IL-2	Lectin
IL-4	Lectin
CSF	Lectin
Interferon	
IFNα	Virus-infected target cells
IFNγ	IL-2
Cytotoxic/cytostatic	
NKCF	NK target cells
TNFα	NK target cells

Table IV summarizes our recent studies regarding the abilities of recombinant IL-2 to induce cytokine production in CD3$^-$ LGL or CD3$^+$ T lymphocytes. By analyzing cytokine secretion and Northern blots, we determined that recombinant and natural IL-2 can induce significant levels of transcription, translation, and secretion of IFNα/γ and IL-1α/β as well as low levels of IL-2 and TNFα. Although the levels of many of these factors are not quantitatively as high as can be induced with lectins or NK target cells, they represent an interesting and unique activation event. This phenomenon is not observed in normal CD3$^+$ T cells.

7. Conclusion and Remarks

7.1. Potential Role for NK Cells in Immunity

Immune surveillance against neoplasia by natural and specific (adaptive) effector cells has been examined in great detail during the past decade. However, the activities of natural resistance mechanisms in other surveillance systems have begun to be recognized. The postulates and concepts formulated by studying anti-tumor mechanisms will surely apply elsewhere. Natural effector cells have been identified as responsible, at least in part, for mediating marrow allograft rejection. The NK activity against virus-infected cells has been studied extensively, and the data suggest that these natural effector cells play an important role in resistance against many viral pathogens *in vivo*. The defense of the host against not only tumor and viral pathogens but also bacterial, parasitic, and fungal pathogens requires the participation and interaction of leukocytes with

TABLE IV
Recombinant IL-2-Induced Factor Production by CD3$^{-/+}$
Peripheral Lymphocytes

	CD3$^-$		CD3$^+$	
Cytokine	Secreted[a]	mRNA[b]	Secreted	mRNA
IFNα	+[c]	NT[d]	−	−
IFNγ	+ +	+ +	−	−
IL-1α	+	+	−	−
IL-1β	+ +	+ +	−	−
IL-2	+	NT	−	−
TNFα	+	NT	−	−

[a] Determined by bioassay and/or specific antibody.
[b] Determined by Northern blot analysis on total cellular RNA.
[c] −, Not detected; +, low level of activity detected; + +, high level of activity detected.
[d] NT, not tested.

many humoral factors. The ensuing discussion will emphasize the role of natural effectors in tumor immunity, but it is relevant to immune surveillance in general.

In a sense, the predictions regarding immune surveillance establish a set of criteria for determining immune effector cells that are analogous in every way to Koch's postulates on the etiology of infectious diseases. Although many agents augment or suppress leukocyte function *in vivo* and *in vitro*, most of them are not entirely selective. In addition, even if a suppressive treatment completely depletes one effector mechanism, the possibility exists that an alternative effector mechanism will replace the suppressed function. Alternatively, resistance may result from cooperation between two effector mechanisms, and elimination of one would be sufficient to increase disease or tumor incidence. Since soluble factors (e.g., IFN, prostaglandins) are known to augment natural resistance mechanisms, these same agents may modulate a variety of other effector mechanisms *in vivo*. This nonselectivity makes it difficult to dissect out the relative *in vivo* roles of various effector mechanisms. In determining which effector mechanism mediates a specific event, one must consider these pitfalls and difficulties.

Another level of complexity arises from the fact that many lymphoid and myelomonocytic cells are capable of producing a variety of cytokines. T cells can produce many cytokines, IFN (α,γ) and IL-2, chemotactic factors, which affect macrophages and NK active cells. Macrophages can produce IFNα, which affects T cell and NK activity. CD3$^-$ LGL produce IFN(α,γ), IL-1, IL-2, IL-4, and CSF, which affects other immune system components. These findings emphasize the importance of the noncytolytic capacities of natural effector cells.

Our first line of defense against tumor cells is the natural resistance system, composed of NK active cells, macrophages, and NC cells, with the specific immune T cell being the second line of defense.[3] For tumor models, we postulate a similar scheme of primary and secondary lines of defense, with cooperation between natural and specific immune systems. These hypotheses are based on (1) the presence of cytotoxic T lymphocytes in primary and secondary antitumor responses; (2) cytotoxic T cells which exhibit lymphoproliferative responses against tumor cells and develop specific cytotoxic T cells; (3) and release of a variety of lymphokines, including macrophage migration and inhibitory factor, leukocyte migration factor, and a leukocyte adherence inhibitory factor, from T cells following contact with tumor cells; and (4) the antitumor effects mediated by adoptive transfer of immune T cells into tumor-bearing mice.

Briefly, the evidence for a possible role of CD3$^-$ LGL in surveillance against tumors includes the following findings (see chapter by Herberman for more details): (1) NK active cells can accumulate at the sites of inflammation in primary and transplantable tumors; (2) NK active cells have a natural and rapidly activated ability to lyse tumors; (3) NK active cells can inhibit the development of metastases, especially during the blood-borne phase; (4) an increased tumor incidence has been shown in animals and individuals with depressed NK activity; and (5) an early and profound depression of NK activity is related to tumor incidence and suggests an interference with natural resistance mechanisms.

Such observations confirm the role of CD3⁻ LGL as antimetastatic effectors. Most of the present data indicate that NK active cells play an important role against tumor metastases in the vasculature of the lungs and the liver,[38–40] although few reports support the role of NK active cells at the primary tumor site. However, these studies have been performed with the bias that NK lytic activity is the major antitumor mechanism. Because these cells secrete a variety of factors that modulate tumor growth either directly, by acting on tumor cells, or indirectly, by regulating other immune system components, the potential of CD3⁻ LGL to mediate antitumor effects through soluble factors must be considered further. Collectively, these data suggest that CD3⁻ LGL may function *in vivo* as an important first-line defense against tumor growth and microbial infection. It is now necessary to determine more directly the role of NK active cells in immune surveillance.

7.2. Potential Interactions of NK Cell Cytokines

Figure 1 diagrams a hypothetical role CD3⁻ LGL cytokine production and its regulation of specific immunity. Since the cytotoxic activity of CD3⁻ LGL represents a first line of defense, it is reasonable to propose that this extends to their secretory abilities. It has been established that no sensitization is needed for CD3⁻ LGL to produce these factors, since cytokines are triggered in unsensitized individuals by virus-infected target cells, tumor target cells, bacterial products, etc. The activation and release of CD3⁻ LGL cytokines could initiate and enhance numerous specific and nonspecific components of the immune system. T cell immunity, which generally requires macrophage products, could be enhanced by the IL-1,

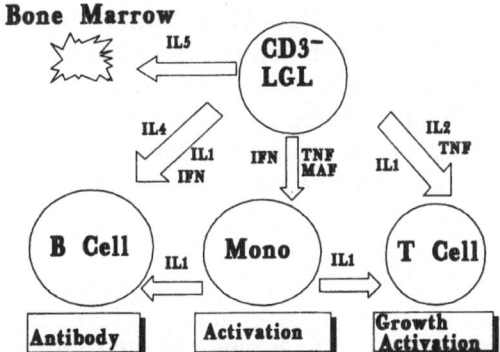

FIGURE 1. Proposed immunoregulatory role of cytokines from CD3⁻ LGL.

IL-2, and TNF that CD3⁻ LGL produce. In addition, there have been reports that CD3⁻ LGL present antigen to T cells.[41,42] Thus, preliminary activation or sensitization might occur in the absence of macrophages. Alternatively, the production of IFNγ and TNF might activate and recruit macrophages to sites of antigen or tumors and lead to enhanced T cell-mediated immunity.

The same concepts might be postulated regarding B cell stimulation, since IL-1, IL-4, and IFNγ are produced by CD3⁻ LGL, and these cytokines have been shown to be important in B cell growth and differentiation. Alternatively, the activation and recruitment of macrophages by IFNγ and TNF production might enhance antibody production.

In addition to specific immunity, the cytokine products of CD3⁻ LGL, IFNγ, and TNFα have been shown to activate numerous macrophage functions including their cytolytic activity. Therefore, it is quite conceivable that the direct activation of other components of the natural immune system, including neutrophils and macrophages, might be another function of the cytokines produced by CD3⁻ LGL.

Finally, the ability of CD3⁻ LGL to produce CSF has broad implications. Since all of the components of the immune system are derived from bone marrow progenitors, the present information is consistent with the ability of CD3⁻ LGL not only to activate local effector cells but to recruit new waves of immunological precursors. This concept is basic to immune surveillance, since the renewal of immune components is necessary to maintain an intact immunologically competent host. Therefore, it is quite conceivable that when CD3⁻ LGL contact microbes or tumors, one of their most important functions as a first-line defense mechanism is to activate and recruit new immune cells from both the natural and the specific immune systems.

Previous studies that have demonstrated the role of NK cells have utilized antibodies (asialo GM_1, NK1.1, or OX-8) that were directed against these cells to eliminate their function. With the increasing availability of human and murine lymphokines and monoclonal antibodies against them, experiments could be performed that would determine whether the secretory activity of CD3⁻ LGL are involved in functions of these cells. For example, the antiviral role of LGL in HSV-1 infections could be examined by employing anti-IFNs or combinations of anti-asialo GM_1 and reconstitution with IFNα or IFNγ. The knowledge of the mechanism of action, whether by direct cellular cytotoxicity or by factor release the interaction with other components of the immune system, will be useful in using recombinant materials in treatments of neoplasia and infectious diseases.

7.3. Summary

Our data demonstrate that, in addition to their cytotoxic activity, CD3⁻ LGL produce a number of cytokines. Clearly, IFNα and IL-2 have a positive self-regulatory function on the growth and cytotoxic activity of CD3⁻ LGL. This self-regulatory role of cytokines was supported by both subset analysis of fresh CD3⁻ LGL and their *in vitro* clones. However, the data also indicate that the noncytolytic activity of LGL may serve an accessory function for other leukocytes. Furthermore, the production of cytokines by noncytotoxic LGL indicate that these cells may have broader immunological functions than merely mediating NK activity.

References

1. Ehrlich, P., 1957, Immune surveillance, in: *The Collected Papers of Paul Ehrlich*, Volume 2 (F. Himmelweit, ed.), Pergamon, Oxford, U.K., pp. 550–562.
2. Burnet, F. M., 1970, The concept of immunological surveillance, *Prog. Exp. Tumor Res.* **13:**1–27.
3. Herberman, R. B., 1983, Immune surveillance hypothesis: Updated formulation and possible effector mechanisms, in: *Progress in Immunology*, Volume 5, Fifth International Congress of Immunology (Y. Yamamura and T. Tada, eds.), Academic Press, Orlando, FL, pp. 1157–68.
4. Lanier, L. L., Le. A. M., Cwirla, S., Federspiel, N., Phillips, J. H., 1986, Antigenic, functional, and molecular genetic studies of human natural killer cells and cytotoxic T lymphocytes not restricted by the major histocompatibility complex, *Fed. Proc.* **45:**2823–2828.
5. Herberman, R. B., 1980, *Natural Cell Mediated Immunity Against Tumors*, Academic Press, New York.
6. Herberman, R. B., and Ortaldo, J. R., 1981, Natural killer cells: Their role in defenses against disease, *Science* **214:**24–30.
7. Wiltrout, R. H., and Varesio, L., 1988, Macrophage-mediated cytotoxicity and supression, in: *Textbook of Immunophsiology: Role of Cells and Cytokines in Immunity and Inflammation*, (J. J. Oppenheim and E. Shevach, eds.), Oxford University Press, New York (in press).
8. Ortaldo, J. R., and Herberman, R. B., 1984, Heterogeneity of natural killer cells, in: *Annual Reviews of Immunology*, Volume 2 (W. E. Paul, G. G. Fathman, and H. Metzger, eds.), Annual Reviews, Palo Alto, CA, pp. 359–394.
9. Lattime, E. C., Pecoraro, G. A., and Stutman, O., Natural cytotoxic cells against solid tumors in mice. III. Comparison of effector cell antigenic phenotype and target cell recognition structures with those of NK cells, *J. Immunol.* **126:**2011–2019.
10. Timonen, T., Ortaldo, J. R., and Herberman, R. B., 1981, Characteristics of human large granular lymphocytes and relationship to natural killer and K cells, *J. Exp. Med.* **153:**569–582.
11. Welsh, R. M., 1984, Natural killer cells and interferon, *CRC Crit. Rev. Immunol.* **5:**55–93.
12. Stutman, O., and Lattime, E., 1985, Lymphokines and natural cell-mediated cytotoxicity, in: *Lymphokines* (T. Landy, ed.), Academic Press, New York, pp. 107–125.
13. Ortaldo, J. R., and Herberman, R. B., 1987, Augmentation of natural killer cells, in: *Immunobiology of Neutral Killer Cells* (E. Lotzova and R. B. Herberman, eds.), CRC Press, Boca Raton, FL, pp. 145–162.
14. Trinchieri, G., Santoli, D., Dee, R. R., and Knowles, B. B., 1978, Antiviral activity by culturing lymphocytes with tumer-derived or virus transformed cells: Identification of the antiviral activity as interferon and characterization of the human effector lymphocyte subpopulation, *J. Exp. Med.* **147:**1229–1231.
15. Saksela, E., Timonen, T., and Cantell, K., 1980, Cellular interactions in the augmentation of human NK activity by interferon, *Ann. N.Y. Acad. Sci.* **350:**102–111.

16. Djeu, J. Y., Timonen, T., and Herberman, R. B., 1982, Production of interferon by human natural killer cells in response to mitogens, viruses, and bacteria, in: *NK Cells and Other Natural Effector Cells* (R. B. Herberman, ed.), Academic Press, New York, pp. 669–675.

17. Suzuki, R., Handa, K., Itoh, K., and Kumagai, K., 1983, Natural killer (NK) cells as a responder to interleukin 2 (IL-2). I. Proliferative response and establishment of cloned cells, *J. Immunol.* **130**(2):981–987.

18. Ortaldo, J. R., Mason, A. T., Gerald, J. P., Henderson, L. E., Farrar, W., Hopkins, R. F. III, Herberman, R. B., and Rabin, H., 1984. Effects of natural and recombinant IL-2 on regulation of IFNγ production and natural killer activity: Lack of involvement of the Tac antigen for these immunoregulatory effects, *J. Immunol.* **133**:779–792.

19. Young, H. A., and Ortaldo, J. R., 1987, One signal requirement for gamma interferon production from large granular lymphocytes, *J. Immunol.* **139**:724–727.

20. Domzig, W., and Stadler, B. M., 1982, The relationship between human natural killer cells and interleukin 2, in: *NK Cells and Other Natural Effector Cells* (R. B. Heberman, ed.), Academic Press, New York, pp. 409–420.

21. Kasahara, T., Djeu, J. Y., Dougherty, S. F., and Oppenheim, J. J. 1983, Capacity of human large granular lymphocytes (LGL) to produce multiple lymphokines: Interleukin 2, interferon, and colony stimulating factor, *J. Immunol.* **131**(5):2379–2385.

22. Scala, G., Djeu, J. Y., Allavena, P., Kasahara, T., Ortaldo, J. R., Herberman, R. B., and Oppenheim, J. J., 1986, Cytokine secretion and noncytotoxic functions of human large granular lymphocytes, in: *Immunobiology of Naturlal Killer Cells*, Volume 2 (E. Lotzova and R. B. Herberman, eds.), CRC Press, Boca Raton, FL, pp. 133–144.

23. Scala, G., Allavena, P., Djeu, J. Y., Kasahara, T., Ortaldo, J. R., Herberman, R. B., and Oppenheim, J. J., 1984, Human large granular lymphocytes are potent producers of interleukin-1, *Nature* **309**:56–59.

24. Pistoia, V., Cozzolino, Fl, Torcia M., Castigli, E., Ferrarini, M. J., 1985, Production of B cell growth factor by a Leu-7$^+$, OKM1$^+$ non-T cell with the features of large granular lymphocytes (LGL), *J. Immunol.* **134**:3179–3184.

25. Procopio, A. D., Allavena, P., and Ortaldo, J. R., 1985, Noncytotoxic functions of natural killer (NK) cells: Large granular lymphocytes (LGL) produce a B cell growth factor (BCGF), *J. Immunol.* **135**:3264–3271.

26. Wright, S. C., and Bonavida, B., 1981, Selective lysis of NK-sensitive target cells by soluble mediator released from murine spleen cells and human peripheral blood lymphocytes, *J. Immunol.* **125**:1561–1571.

27. Blanca, I., Herberman, R. B., and Ortaldo, J. R., 1985, Human natural killer cytotoxic factor. Studies on its production, specificity, and mechanism of interaction with target cells, *Nat. Immun. Cell. Growth Regul.* **4**:48–59.

28. Wright, S. C., and Bonavida, B., 1982, Studies on the mechanism of natural killer (NK) cell mediated cytotoxicity (CMC). I. Release of cytotoxic factors specific for NK-sensitive target cells (NKCF) during co-culture of NK effector cells with NK target cells, *J. Immunol.* **129**:433–439.

29. Farram, E., and Targan, S. R., 1983, Identification of human natural killer soluble cytotoxic factors (NKCF) derived from NK-enriched lymphocyte populations: Specificity of generation and killing, *J. Immunol.* **130**:1252–1256.

30. Wright, S. C., Wilber, S. M., and Bonavida, B., 1985, Biochemical characterization of natural killer cytotoxic factors, in: *Mechanisms of Cell-mediated Cytotoxicity*, Volume 2 (P. Henkart and E. Martz, eds.), Plenum, New York, pp. 179–192.

31. Ortaldo, J. R., Blanca, I., and Herberman, R. B., 1985, Studies of human natural killer cytotoxic factor (NKCF): Characterization and analysis of its mode of action, in: *Mechanisms of Cell-mediated Cytotoxicity*, Volume 2 (P. Henkart and E. Martz, eds), Plenum, New York, pp. 203–220.

32. Herberman, R. B., Reynolds, C. W., and Ortaldo, J. R., 1985, Mechanism of cytotoxicity by natural killer (NK) cells, *Annu. Rev. Immunol.* **4**:651–680.

33. Sayers, T. J., Ransom, J. R., Denn, A. C. III, Herberman, R. B., and Ortaldo, J. R., 1986, analysis of a cytostatic lymphokine produced by incubation of lymphocytes with tumor cells: Relationahip to leukoregulin and distinction from recombinant lymphotoxin, recombinant tumor necrosis factor, and natural killer cytotoxic factor, *J. Immuol.* **137**:385–390.

34. Ortaldo, J. R., Ransom, J. R., Sayers, T. J., and Herberman, R. B., 1986, Analysis of cytostatic/cytotoxic lymphokines: Relationship of natural killer cytotoxic facctor to recombinant lymphotoxin, recombinant tumor necrosis factor, and leukoregulin, *J. Immunol.* **137**:1–7.

35. Yamamoto, R. S., Ware, C. F., Granger G. A., 1986, The human LT system. XI. Identification of LT and "TNF-like" forms from stimulated natural killers, specific and nonspecific cytotoxic human T cells in vitro. *J. Immunol.* **137**:1878–1884.

36. Allavena, P., and Ortaldo, J. R., 1984, Characteristics of human NK clones: Target specificity and phenotype, *J. Immunol.* **132**:2363–2367.

37. Allavena, P., Scala, G., Djeu, J. Y., Procopio, A. D., Oppenheim, J. J., Herberman, R. B., and Ortaldo, J. R., 1985, Production of multiple cytokines by clones of human large granular lymphocytes, *Cancer Immunol. Immunother.* **19**:121–126.

38. Kawase, I., Brooks, C. G., Kuribayashi, K., Olabuenaga, S., Newman, W., Gillis, S., and Henney, C. S., 1983, Interleukin 2 induces gamma-interferon production: Participation of macrophages and NK-like cells, *J. Immunol.* **131**:288–292.

39. Wiltrout, R. H., Herberman, R. B., Zhang, S. R., Chirigos, M. A., Ortaldo, J. R., Green, K. M., and Talmadge, J. E., 1985, Role of organ-associated NK cells in decreased formation of experiemtnal metastases in lung and liver, *J. Immunol.* **134**:4267–4275.

40. Barlozzari, T., Leonhardt, J., Wiltrout, R. H., Herberman, R. B., and Reynolds, C. W., 1985, Direct evidence for a role of LGL in the inhibition of experimental tumor metastases, *J. Immunol.* **134**:2783–2789.

41. Burlington, D. B., Djeu, J. Y., Wells, M. A., Kiley, S. C., and Quinnan, G. V. Jr., 1984, Large granular lymphocytes provide an accessory function in the *in vitro* development of influenza A virus–specific cytotoxic T. cells, *J. Immunol.* **132**:3154–3158.

42. Scala, G., Allavena, P., Ortaldo, J. R., Herberman, R. B., and Oppenheim, J. J., 1985, Subsets of human large granular lymphocytes (LGL) exhibit accesory cell functions, *J. Immunol.* **134**:3049–3055.

Summary of Part III

For some time the importance of both specific antibody (Ab) and cell-mediated immune responses in a variety of disease settings has been generally appreciated. However, it has only recently become clear that these antigen-specific responses are both positively and negatively regulated by a variety of cells with broader specificities. These latter cells include a number of effector cell populations found within the NIS, including macrophages, monocytes, and LGL with NK or natural suppressor (NS) activities. Considering the ability of these natural effector cells both to suppress and to up-regulate almost all types of immune responses, it now seems probable that the immunoregulatory functions of the NIS may be among the most important nontumoricidal activities of this system. The reviews in this section have discussed the role of natural effector cells in (1) regulation of the Ab response, (2) regulation of cell-mediated immunity, (3) control of normal hematopoietic stem cell growth and differentiation, (4) natural suppressor activity, and (5) in the production of a wide variety of cytokines.

The review by Kumagai *et al.* (Chapter 9) discusses the regulatory effects of cells with NK activity on Ab responses. It seems clear that these cells can inhibit immunoglobulin (Ig) secretion and the differentiation of B cells. Cells with NK activity appear to suppress the Ab response early in the activation process, possibly by interfering with antigen presentation or in the inhibition of B cell differentiation into Ig-producing cells. Recent studies further suggest that the secretion of INF-γ by cells with NK activity may play a significant role in enhancing Ab responses to T cell-dependent antigens but may also result in the suppression of many T cell-independent responses. Since several autoimmune diseases such as systemic lupus erythematosus (SLE) and rheumatoid arthritis show enhanced Ab production as well as reduced NK levels, this decrease in NK activity may be causally related to such immune disorders (see chapter by Merrill, this volume). Additional studies in chronically NK-depleted animals and in patients with a variety of immune disorders with abnormal Ig production should help to further clarify the role of natural effector cells in the overall Ab response.

The chapter by Tilden and Clement (Chapter 10) reviews the data suggesting that natural immunoregulatory cells are involved in the control of a variety of cell-mediated immune responses. These data indicate that one major function of natural effector cells may be to suppress the proliferation of a variety of different cell types, including T cells, which actively participate in many cell-mediated immune responses. In contrast, these immunoregulatory cells can also enhance certain cellular responses by secreting a variety of immunostimulatory lymphokines. The authors suggest that an important area of future research will be to define the activation requirements, cellular interactions, and immunoregulatory mechanisms of natural effector cells. In addition, future studies need to clarify the extent to which the NIS controls cell-mediated immune responses, in both a normal setting and in a wider variety of pathological conditions involving abnormal cellular responses. These latter conditions might include a number of animal models as well as patients with either immunodeficiency or autoimmune disorders. At present, it is not clear whether abnormal regulation of the cellular immune response by the NIS plays any role in the development of these conditions. However, such information may be critical in defining the exact mechanism of pathogenesis in these disorders.

A large body of direct and indirect evidence now suggests that the NIS, and in particular cells with NK activity, may play a very important role in the control of hematopoietic progenitor cells. The chapter by Trinchieri et al. (Chapter 11) reviews these data in detail. In brief, cells with NK activity can produce a variety of factors that both enhance (CSF, BSF-1, IL-1, IL-2) and suppress (TNF, IFN) hematopoiesis. There are now compelling experimental and clinical data that these cells are an important cellular mediator of certain types of pathological disregulation of hematopoiesis. This evidence is further discussed by Chan and Winton (Chapter 17) on the role of the NIS in the development of aplastic anemia and neutropenia. Although it is generally believed that these pathological conditions reflect abnormalities in normal immunoregulatory systems, evidence that cells with NK activity play a role in maintaining physiological hematopoietic homeostasis is much less compelling. However, a variety of in vivo and in vitro systems both in experimental animals and in man are being studied to address this issue further for normal erythropoiesis, granulopoiesis, and megakaryocytopoiesis.[1] It is hoped that these studies will increase our understanding of how important the NIS may be in maintaining normal hematopoietic homeostasis.

The previous chapters have all discussed how the NIS might enhance as well as suppress a variety of immunological responses. In general, the data are strongest that natural effector cells are potent suppressors of these responses. In the chapter by Maier *et al.* (Chapter 12), the authors discuss the characterization and function of a population of effector cells known as natural suppressor (NS) cells. It is likely that these cells account for much of the suppressor activity previously discussed in relationship to Ab production and cell-mediated immunity and in hematopoietic stem cell regulation. NS cells are LGL in morphology but, unlike NK cells, are present in bone marrow, thoracic duct lymph, and other lymphoid organs. They appear early after birth, decrease when NK activity begins to appear, and are observed during pregnancy and in some periods of graft versus host disease (GVHD). At present it is not clear whether NS cells are immature cells that will acquire other functions as they mature (i.e., NK, NC, or LAK) or whether they represent an entirely separate population of cells. The direct evidence that is available regarding NS function suggests that these cells may suppress the development of specific immune responses by decreasing Ia expression *in vivo*. However, these cells have also been hypothesized to play a role in suppressing GVH responses, in controlling the development of self-tolerance or acquired unresponsiveness, in regulating normal hematopoiesis, and in preventing the development of antifetal responses during pregnancy. Although the mechanism(s) of NS activity are not clear, NS cells may function directly on target lymphoid cells or indirectly via the production of factors, stimulation of suppressor cells, or inhibition of helper activity. Further studies both on the functional capabilities of these cells and in further characterization of these cells in humans are a prerequisite to understanding their clinical significance. In addition, a great deal of work needs to be done to characterize these cells with NS activity in animal models. Are NS cells a distinct population of natural effector cells, or is NS activity a function found in a variety of cell types that also demonstrate NK or NC activity or the generation of immunomodulatory lymphokines? The review by Maier *et al.* discusses the need for such characterization as well as a variety of experimental approaches to do this.

It is clear from all the previous chapters that the production of cytokines by natural effector cells is an important function of these cells and could account for many of the physiological and pathological effects observed. The review of cytokine production in $CD3^-$ LGL by Ortaldo (Chapter 13) summarizes the factors produced by these cells as well as their potential role in the regulation of a variety of immune responses. The cytokines produced by $CD3^-$ LGL include IFN, IL-1, IL-2, IL-4, IL-5, and cytotoxic factors like TNF and NKCF as well as factors that increase macrophage migration (NK-LCF) and activation (NK-MAF). With this number of factors and their variety of effects, it is not difficult to envision how the NIS may interact both positively and negatively with other cells within the immune system.

In summary, the present data clearly demonstrate a strong potential for natural effector cells to regulate the development of a number of immune responses. A fair amount of direct and indirect evidence has also been obtained that is consistent with this hypothesis. Future experiments need to further investigate this possibility by further examining (1) the extent of immunoregulation by the NIS in both physiological and pathological settings, (2) factors controlling these immunoregulatory functions of the NIS, (3) the cellular heterogeneity associated with the various immunoregulatory functions of natural effector cells, and (4) the mechanisms by which natural effector cells control the development of Ab responses, specific cell-mediated immunity, or normal hematopoietic homeostasis.

References

1. Gewirtz, A. M., Xu, W. Y., and Mangan, K. F., 1987, Role of natural killer cells in comparison with T lymphocytes and monocytes, in the regulation of normal human megakaryocytopoiesis *in vitro*, J. Immunol. **139:**2915–2924.

PART IV
INVOLVEMENT OF THE NIS IN
BONE MARROW AND
ALLOGRAFT REJECTION

14

Involvement of Natural Effector Cells in Bone Marrow Transplantation and Hybrid Resistance

Ichiro Nakamura

1. Introduction

Natural resistance to hemopoietic allograft can be defined as the nonadaptive host reaction that causes graft failure in a nonimmunized host. Such restrictions notwithstanding, the phenomenon is heterogenous and can encompass a variety of effector mechanisms directed against distinct cell surface entities. The key word is "nonadaptive." However, since the nonadaptive nature of a reaction cannot always be tested, misinterpretation can result. Exposure of the prospective allograft recipients to a lethal dose of ionizing radiation does not completely eliminate the adaptive components of the host immune system.[1,2] Although conventional effector mechanisms by T and B lymphocytes are excluded by the above definition, these cells may play a role as a secondary component or as regulators of nonadaptive responses. Therefore, the study of natural resistance to hemopoietic cells must take into account the possible multiplicity of the effector mechanisms and the complexity of regulatory influences. Our inability to define the mechanisms precisely and to understand the factors influencing the ultimate fate of the grafted cells has led to conflicting interpretations of the unique phenomenon discovered nearly 30 years ago

Ichiro Nakamura • Department of Pathology, School of Medicine and Biomedical Sciences, State University of New York at Buffalo, Buffalo, New York 14214.

by Snell, Boyse, Cudkowicz, and other pioneers.[3–8] For a comprehensive treatment of this subject, the reader is referred to the excellent review recently written by Bennett.[9]

In this article, we will attempt to place pertinent observations in perspective, though subjective it certainly will be. The critical issues, as we see them, include the nature of target structures recognized and the mechanism of effector functions involved. At first, we will discuss the genetic control of natural resistance as it relates to the nature of target structures recognized. The emphasis will be on "hybrid resistance," in which hemopoietic grafts of parental inbred strain donors are resisted by F_1 hybrid animals. This type of resistance is, in theory, simpler than the "allogeneic" resistance associated with graft exchanges between a pair of inbred strains, because the host reactivity against codominantly expressed alloantigens of the graft can be largely excluded. Although the primary effectors involved appear to be the same in the two types of resistance, the mechanism of resistance may not always be identical.

2. Genetic Control of the Target Structure

Snell[3] and Snell and Stevens[4] reported that certain lymphomas of parental origin grew less well in F_1 hybrids than in syngeneic mice. Moreover, the mice preimmunized with parental thymocytes or the survivors of the lymphoma grafts had no alteration in resistance to the parental lymphomas. These were the first descriptions of the nonadaptive nature of hybrid resistance. Although preimmunization had no effect in these examples, neoplastic cells could have growth requirements distinct from normal cells. Therefore, the discovery[5–7] that normal parental bone marrow or lymphoid cells are apparently subject to the same type of resistance as the lymphomas was highly significant. Shortly afterward, it was demonstrated that the resistance is controlled by a single major locus linked to the H-2 and, more specifically, to the D end of the H-2.[10–12] Resistance of the graft recipient was associated with heterozygosity at this locus, the hybrid or hemopoietic histocompatibility 1 (Hh-1). Interpretation of this phenomenon has been controversial, however. On the one hand, since the resistance occurs in a lethally irradiated host in which immunological competence has presumably been eliminated, the depressed survival and growth of the grafted cells in an allogeneic environment were thought to be nonimmunological—i.e., not dependent on host reaction to the graft.[7,8] Cudkowicz, however, argued that hybrid resistance is an acute histoincompatibility reaction mounted by the host against "parental" determinants not present in the resistant hosts.[13–15] Several characteristics of hybrid resistance can be listed in support of this contention, as discussed elsewhere in this article. The difficulties were nevertheless quite evident. Here we will focus our attention on two salient questions: noncodominant expression of the putative target determinants for resistance, and relationship with class I genes.

The assumption that hybrid resistance is a histocompatibility reaction implies that the relevant target structure of resistance is not codominantly expressed in the resistant F_1 hybrid. In certain cases, simple recessive inheritance may provide a satisfactory explanation—for example, the case of C57BL/10 (B10, H-2^b), C3H (H-2^k) and F_1 hybrid between them. It is well established that the resistance occurs as shown in Table I. In this situation, the resistance of F_1 mice to parental B10 bone marrow cells is under control by the Hh-1 locus, and the target structures are controlled by the Hh-1^b allele. The allele Hh-1^k of C3H strain is assumed to control certain structures that are not recognized as a target for resistance; i.e., the allele is "silent." We then assume that this allele is dominant over the Hh-1^b. The last assumption needed is that C3H mice are nonresponders for the structures controlled by the Hh-1^b.

Cudkowicz[16] showed, in fact, that this unresponsiveness is recessive and is controlled by at least two independent autosomal genes that are present in B10.BR (H-2^k). According to this explanation, the resistance of F_1 mice to B10 grafts is the result of acquisition of the responsiveness from B10 parent, and the reaction is directed against a set of recessive determinants expressed in B10 but not in F_1 mice. That B10 mice fail to reject C3H or F_1 graft is attributed to the silent nature of the determinants controlled by the Hh-1^k allele. It is noteworthy that the assumed dominant gene in C3H does not have to be an allele of the Hh-1 locus, but its close linkage to the H-2 complex is evident from the strain distribution.[15]

Another strain combination involving B10, DBA/2 (H-2^d), and F_1 mice shows a different pattern (Table II). With the apparent exception of subline DBA/2Cum,[13] DBA/2 mice are only weakly resistant to B10 bone marrow allograft.[17] Therefore, the same assumption of a recessive Hh-1 allele in B10 and contribution of responsiveness to F_1 mice by the B10

TABLE I
*Resistance and Susceptibility to Parental
or Allogeneic Bone Marrow Graft in Mice
of B10 and C3H Strains and Their
F_1 Hybrid*[a]

	Recipient		
Donor	B10	F_1	C3H
B10	S	R	S
F_1	S	S	S
C3H	S	S	S

[a]Letters S and R refer to susceptible and resistant, respectively.

TABLE II
Resistance and Susceptibility to Parental or
Allogeneic Bone Marrow Graft in Mice of
B10 and DBA/2 Strains and Their
F_1 Hybrids[a]

	Recipient		
Donor	B10	F_1	DBA/2
B10	S	R	Rw
F_1	S	S	S
DBA/2	R	S	S

[a] Letters R, Rw, and S refer to resistant, weakly resistant, and susceptible, respectively.

parent could be made, since the resistance of F_1 mice to B10 bone marrow allograft in this strain combination is directed to the Hh-1[b]-controlled determinants. Weak resistance exhibited by some F_1 hybrid mice to parental DBA/2 has been attributed to determinants not controlled by H-2-linked genes[18] and hence would not prevent us from assigning a "silent," dominant Hh-1 allele for DBA/2.

Again, a dominant gene not controlled by the Hh-1 locus but which prevents the expression of Hh-1[b] allele can suffice, although it would have to be closely linked to H-2 because all H-2[d] strains, regardless of non-H-2 background, appear to possess a gene with the same function. Unlike the case of C3H and B10 discussed above, however, B10 mice clearly reject DBA/2 bone marrow grafts, but not F_1 grafts. Thus, the situation emerges which Snell[18] referred to as "corecessive," i.e., F_1 hybrids appear to express neither of the structures borne by the parental strains.

Implicit in this terminology is that the genes responsible for the structures in question are allelic. However, the structures recognized by B10 mice on DBA/2 graft need not be controlled by the same gene that prevented the expression of the Hh-1[b] allele. The relevant structures are either not expressed on F_1 hybrid cells or expressed only at a density below the functional threshold. It is not even necessary to assume identical mechanisms of resistance; for instance, F_1 hybrid may reject B10 cells by effectors that directly recognized Hh-1[b]-controlled determinants, whereas B10 may reject DBA/2 cells by another mechanism—e.g., antibody-dependent cellular cytotoxicity (ADCC) by way of anti-H-2[d] natural antibody. It seems obvious from the above examination of some typical cases that the genetic pattern of resistance is complex. The question of allelic relationship becomes somewhat moot, as it is now clear that the genomic structure of the H-2 complex may differ substantially from one haplotype to another.[20]

Another important issue is the non-H-2-linked genetic control of the responsiveness (or the resistance status) of the graft recipient. Thus, the expression of target structures is only one of the genetic factors that affect responsiveness. The influence of these genetic factors, presumably not pertaining to the target expression, has been referred to as "*Ir* gene" effect,[16] in analogy with the terminology used for adaptive immunity. Whether an *Ir* gene effect in adaptive immunity is partly or wholly due to self-tolerance is irrelevant here. We will discuss this issue further in relation to NK cell activity under a separate heading. Here we will continue our discussion on the genetic control of target structures and especially the relationship, or the absence of it, with class I H-2 determinants. It should be evident from the above that the traditional analysis of the genetic control of target structures by studying the responsiveness of F_1 hybrids between different strains must be interpreted with caution, because the responsiveness is not determined solely by the composition of self determinants. Although B10 background appears to confer responsiveness against *Hh* incompatibilities, this has been shown not to apply to H-2^k-associated determinants.[21,22] Moreover, hybrid resistance is not limited to strain combinations involving C57BL background.[23] The potential involvement of multiple *Hh* loci[24] may be substantially reduced by using *H-2* congenic strains, such as the B10 series. Some of the latter, however, are known to contain sizable segments of donor-derived genetic material outside the *H-2*.[25] If the expression of certain target determinants in a strain of mice were to be tested by using the mice as graft donors rather than recipients, similar problems would be evident. Possible involvement of multiple determinants, the detection of which would depend on the recipient used, can be difficult to sort out.

To alleviate some of these difficulties, we employed an *in vivo* equivalent of cold-target inhibition assay of hybrid resistance to detect the expression of a particular set of target determinants specified by the *Hh-1*^{b} allele of C57BL/6 (B6) or B10 strain.[26,27] In this analysis, potential inhibitor cells were rendered incapable of DNA synthesis by a supralethal dose of irradiation and injected into irradiated host animals along with or prior to the bone marrow graft. The proliferation of grafted bone marrow cells is then measured by incorporation of a radiolabeled precursor into DNA. If resistance is mediated through specific recognition event, the presence of excess targets would favor the survival of the graft. In principle, this assay should be applicable to any specific resistance as long as the inhibitor cells employed reach the spleen or another target organ at a sufficiently high frequency.

Initial studies[26] were performed using lymphomas as potential inhibitors and verified that these lymphomas share the target determinants with normal bone marrow cells, as suggested previously for some of the lymphomas.[28] This approach as a means of testing specificity of resistance was introduced by Iorio *et al.*[29] using lymphoma cells. As pointed out by Trentin *et al.*,[30] the same mechanism explains why the resistance is readily overridden by increasing the number of grafted bone marrow cells.

We studied the expression of H-2^b-controlled target structures in various strains of mice by using normal hemopoietic cells from these mice as inhibitors to block hybrid resistance of B6D2F$_1$ (H-$2^{b/d}$) and B10C3F$_1$ (H-$2^{b/k}$) mice to parental B6 and B10 bone marrow cells, respectively. Our findings were as follows.[27]

1. Regardless of the background, all H-2^b strains express the Hh-1^b-controlled determinants.
2. Homozygosity of the H-$2D^b$ allele is associated with the Hh-1^b phenotype with the exceptions noted below under (5). The above two points confirmed the conclusion of Daley and Nakamura,[26] in which tumor cells were used as inhibitors of bone marrow graft rejection.
3. Strains bearing the H-$2D^s$ allele are indistinguishable from H-$2D^b$ strains, indicating that the same allele of the Hh-1 locus occurs in the H-2^b and H-2^s haplotypes. It should be noted that the results are not entirely concordant with the conclusion reached by Clark and Harmon.[31] These authors analyzed the genetic control of resistance to EL-4 lymphoma of B6 origin in nonirradiated F$_1$ hybrid mice and found that the Hh-1^s allele is distinct from the Hh-1^b, since (B10 × B10.S)F$_1$ mice were resistant to EL-4. We feel that the resistance to this tumor, as detected in the long-term assay, is not solely dictated by natural resistance controlled by the Hh-1 genes.
4. Strains bearing a recombination between the H-$2D$ region and Qa/Tla region indicate that the Hh-1 locus is located to the left of the most proximal class I locus $Q1$ in the Qa region.
5. Two haplotypes in which recombination occurred between a *non*-H-2^b chromosome and H-2^b chromosome at a position centromeric to the class I H-$2D^b$ locus are negative for Hh-1^b. The first haplotype is H-2^j or H-2^{ja}, representing strains I/St, WB, and B10.WB. The second is a recombinant containing the D^b gene instead of the L^d gene, in the recently described strain B10.RQDB.[32] Thus, the Hh-1 locus in H-2^b chromosome is located to the left of the H-$2D^b$ locus.
6. Consistent with the above and in agreement with a previous study,[33] the two mutations H-2^{bm13} and H-2^{bm14} of the H-$2D^b$ gene have no effect on Hh-1^b. These data are summarized in Table III. An alternative possibility is that the recombination in H-2^j or H-2^{ja} and B10.RQDB chromosome resulted in a transfer of a dominant gene to a *cis* position with regard to the recessive Hh-1^b gene and prevented the expression of the latter, as in heterozygotes where the same dominant gene acts from a *trans* position. Thus, geneti-

TABLE III

Summary of the H-2 Haplotypes and Strains Expressing the Hh-1[b] Phenotype as Tested by Competitive Inhibition in Vivo[a]

Haplotype	H-2 K	I A	I E	S	D	L	Inhibition	Strain
a	k	k	k	d	d	d	−	B10.A, A/J, A.Tla[b]
as1	k	k	k	s	s	—	+	B10.S(8R)
b	b	b	b	b	—	b	+	B6, B10, BALB.B10, 129/J, D1.LP, B6.K1, B6.K2, B6.Tla[a], C3H.B10
bm13	b	b	b	b	—	bm13	+	B6.C-H-2[bm13]
bm14	b	b	b	b	—	bm14	+	B6-H-2[bm14]
d	d	d	d	d	d	d	−	DBA/2, BALB/cKh, B10.D2/n, B6-H-2[d]-Thy-1[a]
f	f	f	f	f	f	—	−	RFM
g	d	d	d	d	—	b	+	HTG, B10.HTG
h2	k	k	k	d	—	b	+	B10.A(2R)
h4	k	k	b	b	—	b	+	B10.A(4R)
i	b	b	b	b	d	d	−	HTI
i5	b	b	k	d	d	d	−	B10.A(5R)
j/ja	j	j	j	j	—	b	−	I/St, WB, B10.WB(69NS)
k	k	k	k	k	k	—	−	C3H, B10.BR
oz1	b	b	b	b	k	—	−	B6.AK1
p	p	p	p	p	p	—	−	C3H.NB
q	q	q	q	q	q	q	−	DBA/1
s	s	s	s	s	s	—	+	SJL, A.SW, B10.S, B10.S(12R)
t2	s	s	s	s	d	d	−	A.TH, B10.S(7R)
y2	q	q	q	q	d	d	−	B10.T(6R)
yb1	q	q	q	q	d	b	−	B10.RQDB
bxd or dxb							−	(B6 × DBA/2)F$_1$, (B10.D2 × B10)F$_1$
bxk							−	(B10 × C3H)F$_1$
h2xb							+	[B10.A(2R) × B10]F$_1$
i5xb							−	[B10.A(5R) × B10]F$_1$
sxb							+	(SJL × B6)F$_1$
sxd							−	(SJL × DBA/2)F$_1$

[a] Lethally irradiated (900 rads) B6D2F$_1$ or B10C3F$_1$ mice received IV parental B6 (5×10^6) or B10 (1.5×10^6) bone marrow cells. Two inocula of 10^8 irradiated (2000 rads) inhibitor cells each were given IV to the F$_1$ mice 24 and 3 h before bone marrow transplantation. Inhibitor cells were splenocytes from mice irradiated (750 rads) and bone marrow (7×10^6) reconstituted 10–20 days earlier. Splenic [^{125}I]UdR uptake was measured on day 5. Except for the haplotypes in which the existence of both D and L loci is confirmed, the presumed single D region locus is assigned here to the D locus for the sake of convenience. For the H-2[b] haplotype, the single D region locus is assigned to the L following the suggestion by Reyes et al.[34] and Maloy and Colligan.[35] The crossing over that yielded the haplotype of B10.RQDB strain could have been unequal and is not evidence by itself for the L locus assignment of the D region b allele.

cally the *Hh-1* locus can be separated from the class I *H-2D* locus in the *H-2*^{b} chromosome. These mapping data, therefore, support the view that it is not the class I gene products *per se* that are recognized. However, this still does not rule out the possibility that the putative structures controlled by the *Hh-1* genes are recognized only in association with appropriate class I gene products. Thus, the H-2D^b-positive recombinants lacking Hh-1^b expression may retain the restricting element but have lost the restricted structure specified by the *Hh-1*^{b} gene. Unless a reciprocal recombinant is found in which *Hh-1*^{b} is expressed in combination with a *H-2D* allele distinct from *b* or *s*, the above possibility is not completely excluded. For example, the mouse genome is known to harbor a number of endogenous viral genes,[36] at least one of them not very far from the *H-2D* region.[37] The products of these genes may be recognized by themselves or in the context of class I gene products. Indeed, Snell[38] and Clark and Harmon[31] suggested that a viral antigen might be the presumed target controlled by the *Hh-1* locus.

To test the possible involvement of class I product another way, we recently selected and cloned several variants of a H-2^b lymphoma in which class I H-2D^b is not expressed. These cells express the K^b molecule, indicating that the expression defect is not generalized to all class I genes. When tested for the expression of the Hh-1^b phenotype by the *in vivo* competition assay, these cells proved to be Hh-1^b-positive, thus demonstrating that this phenotype does not depend on the expression of the class I H-2D^b molecules.[39] The case made for the *Hh-1* genes by no means excludes the possibility that some of the other *Hh* loci known to be *H-2*-linked are class I or that they require recognition in the context of appropriate class I products. At this point, however, we believe that *Hh-1* locus is a previously unidentified locus or cluster of genes located in the *D* region or between the *D* and *S* regions.

Structural similarities between the MHC of higher vertebrates suggest that analogous genes could exist in other animals. One locus known to be between the *S* and the *D* region of the *H-2* is *Neu-1*, coding for neuraminidase (sialidase) isozymes.[40] In the rat[41] and man,[42] the counterpart of *Neu-1* locus is linked to the MHC. In the rat, indeed, a similar phenomenon known as allogeneic lymphocyte cytotoxicity (ALC) is controlled by MHC-linked genes.[43] What the genes of the *Hh-1* locus code for is entirely unknown. It is possible that the products of these genes are themselves recognized on the cell surface as the target for resistance, assuming a *trans*-acting control of expression. Conversely, the *Hh-1* genes may be coding for enzymes with limited polymorphism, and these enzymes may be involved in the synthesis or modification of the cell surface structures recognized by the effector cells.

For example, if natural resistance to hemopoietic cells is a remnant of relatively primitive recognition systems, it would not be surprising if these enzymes participate in the synthesis of cell surface glycoconjugates, such as glycoproteins and glycolipids. It might be the oligosaccharide structures that are recognized as the actual target for resistance by lectin-like receptors. Indeed, a possibility exists that the *Hh-1* locus codes for a limited number of glycosyltransferases (we thank Barbara Knowles for suggesting this possibility). The attractiveness of this hypothesis is also evident in view of the accumulating evidence that the cell surface structures recognized by NK cells on NK-reactive tumors are certain oligosaccharides.[44–48] The possible mechanism(s) by which the effector cells of resistance recognize these cell surface structures are discussed in the next section.

3. NK Cell Involvement in Hemopoietic Resistance

The view that hybrid resistance and related phenomena reflect the ability of host animals to react against the graft assumes the existence of an effector mechanism that recognizes the histoincompatibility. When we confine our discussion to the situations in which adaptive effector mechanisms can be largely excluded (discussed later), the evidence for a major role of NK cells in hemopoietic resistance is substantial, although mostly correlative.[31,49–53] The correlations include maturation of both functions during the fourth week of postnatal development, relative insensitivity to radiation, similar dependence on anatomical site, thymus independence, and dependence on bone marrow integrity as assessed by the effect of ^{89}Sr. Hybrid resistance is augmented by interferons and interferon inducers, as is NK activity.[26,54,55] Although sensitivity to cortisone appeared to differentiate NK activity from hybrid resistance,[56] a later study indicated that the discrepancy was observed only when resistance to bone marrow was evaluated on day 5 rather than on day 7 or 8.[57] Subsequently, attempts were made to obtain more direct evidence.

In vivo administration of an antiserum raised against asialo GM_1 abrogates NK cell activity as well as resistance to hemopoietic allograft.[58] Lotzová, et al.[59] showed that anti–NK 1.1 alloantiserum can abolish NK cell activity and bone marrow allograft rejection by the recipient. These data clearly establish that NK cells are an essential component of the resistance mechanism. Furthermore, Warner and Dennert[60] showed that a small number of cells from a cloned cell line with NK-like activity could restore the ability of NK-defective mice to reject allogeneic, but not semi-syngeneic, bone marrow graft in a specific manner. It was not clear, however, whether the injected cells directly mediated the resistance or recruited endogenous effectors. Moreover, the relationship between the cloned cells and the fresh splenic NK cells was not entirely evident, since

CTLs are capable of acquiring NK-like activity during *in vitro* cultivation under similar conditions.[61]

Experiments designed to test the possibility that NK cells derived from F_1 mice preferentially exert a cytotoxic effect *in vitro* on parental target cells have met only with limited success. Harmon *et al.*[62] found that lymphoma EL-4, which is susceptible to resistance *in vivo* by $H-2D^b$ heterozygous F_1 hybrid mice, is more susceptible to natural killing *in vitro* by spleen cells of resistant F_1 mice than cells from nonresistant mice. However, the difference was slight. Kumar *et al.*[63] found that the B6D2F$_1$ cells mediating lysis of EL-4 cells *in vitro* differ from NK cells cytotoxic for YAC-1 cells. However, the properties of NK cells cytotoxic for YAC-1 seemed closer to the characteristics of hybrid resistance. Bordignon *et al.*[64] showed that clonogenic hemopoietic progenitors from parental B6 mice, as compared to syngeneic F_1 hybrid, are preferentially inactivated by NK-enriched fraction of F_1 hybrid spleen cells during *in vitro* incubation. The specificity of reaction could be shown by competitive inhibition. Although this model seems to reflect *in vivo* resistance more closely than previous models, the reaction is weak, and the specificity is only relative, since syngeneic progenitors are also inactivated to a lesser degree. Several other laboratories found that NK cells are cytotoxic for bone marrow cells, but did not observe genetic specificity.[65–70] In humans this is difficult to observe or predict, since no definitive information is available for the existence of natural resistance for hemopoietic grafts and since homozygosity at a relevant locus, if any, would be infrequent.

Genetic control of NK activity provides circumstantial evidence for the role of NK cells in hemopoietic resistance, but only in the following context. It is not the level of NK activity *per se*, as evaluated by *in vitro* cytotoxicity, that determines the presence or absence of resistance. When there is resistance due to the *Hh*-controlled incompatibility, NK levels correlate with the strength of resistance. The only apparent exception to this rule is the mouse of SJL strain. The SJL mice are defective in both natural killing of susceptible target and in ADCC, even after stimulation with interferon or interferon inducers.[71] These mice are also susceptible to all alllo- and xenografts so far tested.[72] Of the two H-2^k strains, AKR and CBA, the former is generally low in NK activity but is a good responder to bone marrow allografts of different *H-2* types.[21] CBA mice, on the other hand, are known for their high NK activity but are susceptible to allografts from several donor strains.

Mutant alleles at the *beige* locus cause reduced NK activity in homozygotes or heterozygotes with two mutant alleles and at the same time reduce hemopoietic resistance.[54,73] However, NK activity as measured in long-term assays is normalized after interferon-mediated activation, and so is the ability to reject bone marrow allograft. We interpret these data to mean that the specificity of resistance is determined by the *Hh* genes, but among the recipients sharing the same *Hh* genotype, the level of resistance is affected by the strength of NK cell activity. Therefore, at least a part of the *"Ir* gene effect" described by Cudkowicz[15] is related to the level of NK activity. NK cell activity is controlled by multiple loci, including a few that are linked to the MHC.[31,62,74,75]

Although NK cells are of primary importance in natural resistance to hemopoietic grafts, there is evidence that other cell types are involved either as effectors or regulators. As may be expected, the apparent contribution of various cell types seems to depend heavily on the assay method and duration, the anatomical site, the target cells employed, and the genetic combination. The lack of increased resistance by deliberate immunization of the host by parental tumor cells is not tested in many studies in which nonirradiated F_1 mice are inoculated with parental tumors. In most cases the tumor-bearing hosts survive long enough to mount adaptive immune responses. In some of these studies, the involvement of non-NK mechanisms is quite evident.[76,77] This type of resistance to parental tumors has also been referred to as hybrid resistance, and some of the early literature on this phenomenon, the "hybrid effect," is discussed by Sanford.[77]

More difficult to interpret are the situations in which NK cells apparently function as the initial and, presumably, the most important resistance mechanism, followed perhaps by adaptive immune responses. Participation of T cells, either as effectors or regulators of other effectors, in lethally irradiated animals is evident in some systems,[79] as might be expected from the elegant study of Aizawa et al.[2] These authors showed that alloreactive T cells survive even supralethal doses of ionizing radiation and contribute to the rejection of bone marrow allograft. The possibility that alloreactive T cells act to augment or enhance resistance is not limited to the graft recipient. If the graft contains such T cells, the reaction with host alloantigens could conceivably trigger local activation, or suppression, of NK or other effector functions within a relatively short time.

Although intraperitoneal inoculation of allogeneic cells required 2 days to activate splenic NK activity measured as a whole,[80] an intravenous injection of allogeneic cells is likely to trigger a much earlier local activation. Such a mechanism may explain the interesting observation on the involvement of class I and class II loci in hybrid resistance where a sensitive assay was employed to detect resistance to low doses of parental bone marrow allografts.[81] An assay requiring more than a few days after grafting is potentially subject to this type of influence. It should be noted that the T cell content of bone marrow is itself a property influenced by genetic and other factors, C57BL/6 being one of the strains with a relatively high bone marrow T cell content.[82] Moreover, it has been well documented that the outcome of graft versus host reaction depends not only on the extent of MHC mismatch but also on the haplotype combination.

When the proliferation of grafted hemopoietic stem and progenitor cells is the parameter to quantitate resistance, the production of colony-stimulating factors clearly plays a role. A consequence of interferon induction, for example, by poly I:C is an enhancement in hemopoiesis which, to some extent, counteracts the augmented resistance caused by increased effector activity (unpublished data). The modulation of resistance by interferons requires macrophage participation.[55,83] Moreover, there is indication that certain levels of interferon production may be needed to sustain resistance, since antiinterferon antibodies administered to otherwise untreated recipients result in reduced resistance, as shown by Afifi *et al.*[55] and Cudkowicz and Gresser (unpublished study). Presumably this explains, at least in part, the dependence of hemopoietic resistance on macrophages.[84,85] Evidence has been presented that elimination of labeled tumor cells or bone marrow cells from the lung may be governed by a distinct mechanism not necessarily associated with the recipient's NK activity or resistance to bone narrow allograft.[86,87] The data obtained earlier by Riccardi *et al.*,[88] indeed, suggest that elimination of parental bone marrow cells from the lung of F_1 hybrid recipients does not agree with the specificity of resistance predicted for the particular strain combination.

Hybrid resistance is often measured by the growth of grafted cells several days posttransplantation in heavily irradiated recipients or by the mortality of the recipients of leukemic grafts. These assays measure the medium- to long-term consequences produced by the progeny of the original graft. The direct effect of resistance on the original graft is not measured. When periodic sampling of the grafted cells during the first 24 h in the resistant host was done, the rejection was found to be a rapid process. The first of this type of study was conducted by retransplantation of the rescued graft to a secondary host syngeneic to the donor of the first graft.[17,89] It was found that the rejection process begins within several hours and virtually ends 48 h posttransplantation. Once this initial phase of elimination is over, the surviving cells start to proliferate. Thus, one needs to focus on this initial period of elimination to study the direct interaction between the effector and its target. With this in mind, Carlson and colleagues studied the elimination of radiolabeled target cells from the recipient using as target either tumor cells[86] or normal or cultured bone marrow cells.[87] These studies suggested that the transplanted cells are in fact eliminated from the host spleen within 24 h, rather than prevented from proliferation through a cytostatic effect. The temporal requirement for optimum inhibition of hybrid resistance by inhibitor cells also indicates that the inhibitor cells must be present during this time span.[26]

We recently examined the kinetics of elimination, from the host spleen, of clonogenic hemopoietic stem/progenitor cells which form visible colonies in semisolid culture media *in vitro*. Both normal hemopoietic cells that are dependent on specific growth factors and leukemic cells with autonomous growth potential were employed (unpublished). Depending on the number of cells injected, more than 50% of the clonogenic myeloid progenitors can be eliminated from the spleen of a resistant host within 4–6 h after these cells reach the host spleen. Thus, the elimination process *in vivo* is efficient, essentially comparable in kinetics to the cytolytic *in vitro* activity of NK cells against highly sensitive lymphoma targets. This is clearly in contrast to what we observed in the attempts to reproduce the process *in vitro*. *In vitro*, the reaction of effector cells to the specific target of hybrid resistance is very inefficient, and in most cases the specificity itself is not easily demonstrated.[62–64]

The reason for this incongruity between *in vivo* and *in vitro* reactions is not known, and it is likely to be the clue to the mechanism of this resistance. An obvious possibility is that a small number of specific effectors capable of directly recognizing the *Hh*-controlled target structures are positioned in the critical microenvironment where the hemopoietic target cells preferentially settle. These cells may constitute only a fraction of the entire NK cell pool isolated from the spleen. Presumably, the size and the repertoire of this population are subject to modification by the genetic and other factors. The second possibility, that the resistance is mediated by nonspecific effectors that are activated by a specific recognition event, seems untenable, because, first of all, the presumed specific recognition event itself would have to be unique and controlled by *Hh* genes. Moreover, a bystander effect would be expected if the hypothesis were true, but this was not observed when irradiated parental tumor cells were injected into F_1 mice a few hours before transplantation of syngeneic F_1 bone marrow graft.[26] Finally, the resistance *in vivo* may be mediated by a mechanism different from natural killing mediated *in vitro* by NK cells.

Conceivably, a key element that exists *in vivo* is missing from *in vitro* assays: unstable target structures or receptors that are lost or readily modified, and involvement of soluble factors that confer target structures, receptor specificities, or both. The possibility that the target structures for resistance are either unstable or readily modified *in vitro* is somewhat unlikely, since cultured tumor cells or bone marrow cells are as susceptible to rapid killing *in vivo*, or as potent as competitive inhibitors, as cells freshly obtained from intact mice.[26,87] However, it is conceivable that the putative target structures are rapidly modulated within minutes of exposure to the *in vivo* environment. Loss of specific reactivity by NK cells is a possibility not easily excluded. Indeed, the issue was raised soon after the discovery of natural killing *in vitro* to explain the reactivity of NK cells with various tumors.[90–93]

Recently, however, Warner and Dennert[94] presented evidence that in at least some allogeneic combinations, natural antibodies may provide the specific link between NK cells and bone marrow allograft. Characterization of the antibody isotype and specificity is still needed. A critical test would be whether the avidity of the antibody is such that *H-2* heterozygous cells not susceptible to resistance escape killing.[95] Although there is no evidence that hybrid resistance to parental graft is mediated by the same mechanism,[94] the possibility clearly exists. Not only can an active immune response against self H-2 molecules be induced in F_1 mice by parental tumors,[96,97] but also such antibodies have been found in unimmunized mice.[98] Existence of natural autoantibodies with specificities for other self cell surface antigens is well established.[99–102]

Natural antibodies specific for MHC-controlled alloantigens have also been described.[103,104] However, the results obtained by others are not always consistent with the data of Warner and Dennert.[92] For example, possible involvement of natural alloreactive antibodies was not demonstrated in ALC in rats.[105] These authors tested nude rats rendered B lymphocyte–defective by anti-μ chain antibody treatment from birth onward and found an increase rather than decrease in resistance. Serum transfer experiments also suggested that normal serum from responsive rats does not appreciably contribute to resistance. A similar study on the effect of anti-μ on hybrid resistance in mice was performed earlier by Brodt *et al.*[106] An increase rather than a decrease in resistance was observed. Since NK activity was increased in these mice as the result of anti-μ treatment, this may have caused an overall increase of resistance in spite of presumably decreased natural antibody. To explain hybrid resistance directed against the *Hh-1* locus–controlled structures by ADCC mechanism, natural antibodies with the same specificity would have to exist in the resistant F_1 mice. Finally, an unexplored possibility is a specific recognition event that requires soluble lectins.

ACKNOWLEDGMENTS. Preparation of the manuscript by Julieann Kostyo is gratefully acknowledged. I would like to thank Steven G. Kaminsky and Vita Milisauskas for critically reading the manuscript. Recent work cited has been supported in part by the National Institutes of Health grants DK-13969 and CA-12844 and American Cancer Society grant IM-434.

References

1. Kataoka, Y., and Sado, T., 1975, The radiosensitivity of T and B lymphocytes in mice. *Immunology* **29**:121–130.
2. Aizawa, S., Sado, T., Kamisaku, H., and Kubo, E., 1980, Cellular basis of the immunohematologic defects observed in short-term semiallogeneic B6C3F$_1$→C3H chimeras: Evidence for host-versus-graft reaction initiated by radioresistant T cells, *Cell. Immunol.* **56**:47–57.
3. Snell, G. C., 1958, Histocompatibility genes of the mouse. II. Production and analysis of isogenic resistant lines, *JNCI* **21**:843–877.
4. Snell, G. D., and Stevens, L. C., 1961, Histocompatibility genes of mice. III. *H-1* and *H-4*, two histocompatibility loci in the first linkage group, *Immunology* **4**:366–379.

5. Boyse, E. A., 1959, The fate of mouse spleen cells transplanted into homologous and F₁ hybrid hosts, *Immunology* **2**:170–181.

6. Cudkowicz, G., and Cosgrove, G. E., 1961, Modified homologous disease following transplantation of parental bone marrow and recipient liver into irradiated F₁ mice, *Transplant. Bull.* **27**:90–94.

7. McCulloch, E. A., and Till, J. E., 1963, Repression of colony-forming ability of C57BL hematopoietic cells transplanted into nonisologous hosts, *J. Cell. Comp. Physiol.* **61**:301–308.

8. Hellström, K. E., 1963, Differential behaviour of transplanted mouse lymphoma lines in genetically compatible homozygous and F₁ hybrid mice, *Nature* **199**:614–615.

9. Bennet, M., 1987, Biology and genetics of hybrid resistance, *Adv. Immunol.* **41**:333–445.

10. Holmberg, L. A., Miller, B. A., and Ault, K. A., 1984, The effect of natural killer cells on the development of syngeneic hematopoietic progenitors, *J. Immunol.* **133**:2933–2939.

11. Cudkowicz, G., and Stimpfling, J. H., 1964, Hybrid resistance to parental marrow grafts: Association with the K region of the *H-2* locus, *Science* **144**:1339–1340.

12. Cudkowicz, G., and Stimpfling, J. H., 1965, Hybrid resistance controlled by *H-2* region: Correction of data, *Science* **147**:1056.

13. Cudkowicz, G., 1965, Hybrid resistance to parental hematopoietic cell grafts: Implications for bone marrow chimeras, in: *La Greffe des Cellules Hématopoiétiques Allogéniques* (G. Methé, J. L. Amiel, and L. Schwarzenberg, eds.), Centre National de la Recherche Scientifique, Paris, pp. 207–219.

14. Cudkowicz, G., 1965, The immunogenetic basis of hybrid resistance to parental marrow grafts, in: *Isoantigens and Cell Interactions* (J. Palm, ed.), Wistar Press, Philadelphia, pp. 37–52.

15. Cudkowicz, G., 1968, Hybrid resistance to parental grafts of normal and neoplastic hemopoietic cells, in: *The Proliferation and Spread of Neoplastic Cells* (E. Frei III, ed.), Williams and Wilkins, Baltimore, pp. 661–691.

16. Cudkowicz, G., 1971, Genetic control of bone marrow graft rejection. I. Determinant-specific difference of reactivity in two pairs of inbred mouse strains, *J. Exp. Med.* **134**:281–293.

17. Cudkowicz, G., and Bennett, M., 1971, Peculiar immunobiology of bone marrow allografts. I. Graft rejection by irradiated "responder" mice, *J. Exp. Med.* **134**:83–102.

18. Cudkowicz, G., and Rossi, G. B., 1972, Hybrid resistance to parental DBA/2 grafts: Independence from the *H-2* locus. I. Studies with normal hemopoietic cells, *JNCI* **48**:131–139.

19. Snell, G. D., 1976, Recognition structures determined by the H-2 complex. *Transplant. Proc.* **8**:147–156.

20. Stephan, D., Sun, H., Fischer Lindahl, K., Meyer, E., Hämmerling, G., Hood, L., and Steinmetz, M., 1986, Organization and evolution of D region class I genes in the mouse major histocompatibility complex, *J. Exp. Med.* **163**:1227–1244.

21. Cudkowicz, G., and Lotzová, E., 1973, Hemopoietic cell-defined components of the major histocompatibility complex of mice, *Transplant. Proc.* **5**:1399–1405.

22. Cudkowicz, G., and Warner, J. F., 1979, Natural resistance of irradiated 129-strain mice to bone marrow allografts: Genetic control by the H-2K region, *Immunogenetics* **8**:13–26.

23. Lotzová, E., 1977, Resistance to parental, allogeneic and xenogeneic hemopoietic grafts in irradiated mice, *Exp. Hematol.* **15**:215–235.

24. Cudkowicz, G., and Nakamura, I., 1983, Genetics of the murine hemopoietic-histocompatibility system: An overview, *Transplant. Proc.* **15**:2058–2063.

25. Klein, D., Tewarson, S., Figueroa, F., and Klein, J., 1982, The minimal length of the differential segment in *H-2* congenic lines, *Immunogenetics* **16**:319–328.

26. Daley, J. P., and Nakamura, I., 1984, Natural resistance of lethally irradiated F₁ hybrid

mice to parental marrow grafts is a function of H-2/Hh-restricted effectors, *J. Exp. Med.* **159**:1132–1148.

27. Daley, J. P., Wroblewski, J. M., Kaminsky, S. G., and Nakamura, I., 1987, Genetic control of the target structures recognized in hybrid resistance, *Immunogenetics* **26**:21–30.

28. Bonmassar, E., and Cudkowicz, G., 1976, Suppression of allogeneic lymphomas in spleens of irradiated mice: Importance of the D end of the H-2 complex, *J. Immunol.* **117**:697–700.

29. Iorio, A. M., Neri, M., Enrico, P., and Bonmassar, E., 1981, Inhibition of hybrid resistance to lymphomas by inactivated tumor cells in lethally irradiated mice, *Transplantation* **32**:355–362.

30. Trentin, J. J., Gallagher, M. T., and Lotzová, E., 1976, Xenogeneic and genetic resistance to bone marrow transplantation: Relationship to leukemia surveillance, *Transplant. Proc.* **8**:463–468.

31. Clark, E. A., and Harmon, R. C., 1980, Genetic control of natural cytotoxicity and hybrid resistance, *Adv. Cancer Res.* **31**:227–285.

32. Savarirayan, S., Lafuse, W. P., and David, C. S., 1985, Recombination between H-2D and H-2L genes: Identification, characterization, and gene order, *Transplant. Proc.* **17**:702–706.

33. Morgan, G. M., and McKenzie, I. F. C., 1981, Implication of the *H-2L* locus in hybrid histocompatibility (Hh-1), *Transplantation* **31**:417–422.

34. Jenkins, N. A., Copeland, N. G., Taylor, B. A., and Lee, B. K., 1982, Organization, distribution, and stability of endogenous ecotropic murine leukemia virus DNA sequences in chromosomes of *Mus musculus, J. Virol.* **43**:26–36.

35. Reyes, A. A., Schöld, M., and Wallace, R. B, 1982, The complete amio acid sequence of the murine transplantation antigen H-2D^b as deduced by molecular cloning, *Immunogenetics* **16**:1–9.

36. Maloy, W. L., and Coligan, J. E., 1982, Primary structure of the H-2D^b alloantigen. II. Additional amino acid sequence information, localization of a third site of glycosylation and evidence for *K* and *D* region specific sequences, *Immunogenetics* **16**:11–22.

37. Meruelo, D., Kornreich, R., Rossomando, A., Pampeno, C., Mellor, A. L., Weiss, E. H., Flavell, R. A., and Pellicer, A., 1984, Murine leukemia virus sequences are encoded in the murine major histocompatibility complex. *Proc. Natl. Acad. Sci. USA* **81**:1804–1808.

38. Snell, G. D., 1979, Recent advances in histocompatibility immunogenetics, *Adv. Genet.* **20**:291–355.

39. Milisauskas, V. K., Kaminsky, S. G., and Nakamura, I., 1987, Class I H-2D^b determinants are not involved in hybrid resistance to parental H-2^b/Hh-1^b bone marrow allograft, *Eur. J. Immunol.* **17**:1043–1049.

40. Figueroa, F., Klein, D., Tewarson, S., and Klein, J., 1982, Evidence for placing the *Neu-1* locus within the mouse *H-2* complex, *J. Immunol.* **129**:2089–2093.

41. VandeBerg, J. L., Bittner, G. N., Meyer, G. S., Kunz, H. W., and Gill, T. J. III, 1981, Linkage of neuraminidase and α-mannosidase to the major histocompatibility complex in the rat. *J. Immunogenet.* **8**:239–242.

42. Oohira, T., Nagata, N., Akaboshi, I., Matsuda, I., and Naito, S., 1985, The infantile form of sialidosis type II associated with congenital adrenal hyperplasia: Possible linkage between HLA and the neuraminidase deficiency gene, *Hum. Genet.* **70**:341–343.

43. McNeilage, L. J., Heslop, B. F., Heyworth, M. R., and Gutman, G. A., 1982, Natural cytotoxicity in rats: Strain distribution and genetics, *Cell. Immunol.* **72**:340–350.

44. Stutman, O., Dien, P., Wisun, R. E., and Lattime, E. C., 1980, Natural cytotoxic cells against solid tumors in mice: Blocking of cytotoxicity by D-mannose, *Proc. Natl. Acad. Sci. USA* **77**:2895–2898.

45. MacDermott, R. P., Kienker, L. J., Bertovich, M. J., and Muchmore, A. V., 1981, Inhibition of spontaneous but not antibody-dependent cell-mediated cytotoxicity by simple sugars: Evidence that endogenous lectins may mediate spontaneous cell-mediated cytotoxicity, *Immunology* **44**:143–152.

46. Young, W. W. Jr., Durdik, J. M., Urdal, D., Hakomori, S.-I., and Henny, C. S., 1981, Glycolipid expression in lymphoma cell variants: Chemical quantity, immunologic reactivity, and correlations with susceptibility to NK cells, *J. Immunol.* **126**:1–6.

47. Pohajdak, B., Wright, J. A., and Greenberg, A. H., 1984, An oligosaccharide biosynthetic defect in concanavalin A–resistant Chinese hamster ovary (CHO) cells that enhances NK reactivity *in vitro* and *in vivo*, *J. Immunol.* **133**:2423–2429.

48. Dennis, J. W., and Laferté, S., 1985, Recognition of asparagine-linked oligosaccharides on murine tumor cells by natural killer cells, *Cancer Res.* **45**:6034–6040.

49. Kiessling, R., Hochman, P. S., Haller, O., Shearer, G. M., Wigzell, H., and Cudkowicz, G., 1977, Evidence for a common or similar mechanism for natural killer cell activity and resistance to hemopoietic grafts, *Eur. J. Immunol.* **7**:655–663.

50. Trentin, J. J., Kiessling, R., Wigzell, H., Gallagher, M. T., Datta, S. K., and Kulkarni, S. S., 1977, Bone marrow transplantation immunology, in: *Experimental Hematology Today* (S. J. Baum and G. D. Ledney, eds.), Springer-Verlag, New York, pp. 179–183.

51. Lotzová, E., and Savary, C. A., 1977, Possible involvement of natural killer cells in bone marrow graft rejection, *Biomed. Exp.* **27**:341–344.

52. Cudkowicz, G., 1978, Natural resistance to foreign hemopoietic and leukemia grafts, in: *Natural Resistance Systems Against Foreign Cells, Tumors, and Microbes* (G. Cudkowicz, M. Landy, and G. M. Shearer, eds.), Academic Press, New York, pp. 3–30.

53. Cudkowicz, G., and Hochman, P. S., 1979, Do natural killer cells engage in regulated reactions against self to ensure homeostasis? *Immunol. Rev.* **44**:13–41.

54. Kaminsky, S. G.., and Cudkowicz, G., 1980, Natural killing and resistance to marrow grafts: Correlation in four beige mutant mouse lines, *Fed. Proc.* **39**:466 (abstract).

55. Afifi, M. S., Kumar, V., and Bennett, M., 1985, Stimulation of genetic resistance to marrow grafts in mice by interferon-α/β, *J. Immunol.* **134**:3739–3745.

56. Hochman, P. S., and Cudkowicz, G., 1977, Different sensitivities to hydrocortisone of natural killer cell activity and hybrid resistance to parental grafts, *J. Immunol.* **119**:2013–2015.

57. Lotzová, E., and Savary, C. A., 1981, Parallelism between the effect of cortisone acetate on hybrid resistance and natural killing, *Exp. Hematol.* **9**:766–774.

58. Okumura, K., Habu, S., and Shimamura, K., 1982, The role of asialo GM1[+] (GA1[+]) cells in the resistance to transplants of bone marrow or other tissues, in: *NK Cells and Other Natural Effector Cells* (R. B. Herberman, ed.), Academic Press, New York, pp. 1527–1533.

59. Lotzová, E., Savary, C. A., and Pollack, S. B., 1983, Prevention of rejection of allogeneic bone marrow transplants by NK 1.1 antiserum, *Transplantation* **35**:490–494.

60. Warner, J. F., and Dennert, G., 1982, Effects of a cloned cell line with NK activity on bone marrow transplants, tumour development and metastasis *in vivo*, *Nature* **300**:31–34.

61. Brook, C. G., and Henney, C. S., 1985, Interleukin-2 and the regulation of natural killer cell activity in cultured cell populations, *Contemp. Top. Mol. Immunol.* **10**:63–92.

62. Harmon, R. C., Clark, E. A., O'Toole, C., and Wicker, L. S., 1977, Resistance of *H-2* heterozygous mice to parental tumors. I. Hybrid resistance and natural cytotoxicity to EL-4 are controlled by the *H-2D-Hh-1* region, *Immunogenetics* **4**:601–607.

63. Kumar, V., Leuvano, E., and Bennett, M., 1979, Hybrid resistance to EL-4 lymphoma cells. I. Characterization of natural killer cells that lyse EL-4 cells and their distinction from marrow-dependent natural killer cells, *J. Exp. Med.* **150**:531–547.

64. Bordignon, C., Daley, J. P., and Nakamura, I., 1985, Hematopoietic histoincompatibility reactions by NK cells *in vitro:* Model for genetic resistance to marrow grafts, *Science* **230**:1398–1401.

65. Hansson, M., Kiessling, R., and Andersson, B., 1981, Human fetal thymus and bone marrow contain target cells for natural killer cells, *Eur. J. Immunol.* **11**:8–12.

66. Hansson, M., Beran, M., Andersson, B., and Kiessling, R., 1982, Inhibition of *in vitro* granulopoiesis by autologous and allogeneic human NK cells, *J. Immunol.* **129**:126–132.

67. O'Brien, T., Kendra, J., Stephens, H., Knight, R., and Barrett, A. J., 1983, Recognition

and regulation of progenitor marrow elements by NK cells in the mouse, *Immunology* **49:**717–725.

68. Holmberg, L. A., Miller, B. A., and Ault, K. A., 1984, The effect of natural killer cells on the development of syngeneic hematopoietic progenitors, *J. Immunol.* **133:**2933–2939.

69. Mangan, K. F., Hartnett, M. E., Matis, S. A., Winkelstein, A., and Abo, T., 1984, Natural killer cells suppress human erythroid stem cell proliferation in vitro, *Blood* **63:**260–269.

70. Degliantoni, G., Perussia, B., Manogni, L., and Trinchieri, G., 1985, Inhibition of bone marrow colony formation by human natural killer cells and by natural killer cell–derived colony-inhibiting activity, *J. Exp. Med.* **161:**1152–1168.

71. Kaminsky S. G., Nakamura, I., and Cudkowicz, G., 1983, Selective defect of natural killer and killer cell activity against lymphomas in SJL mice: Low responsiveness to interferon inducers, *J. Immunol.* **130:**1980–1984.

72. Cudkowicz, G., 1975, Genetic control of resistance to allogeneic and xenogeneic bone marrow grafts in mice, *Transplant. Proc.* **7:**155–159.

73. Harrison, D. E., and Carlson, G. A., 1983, Effects of the beige mutation and irradiation on natural resistance to marrow grafts, *J. Immunol.* **130:**484–489.

74. Petranyi, G., Kiessling, R., and Klein, G., 1975, Genetic control of "natural killer" lymphocytes in the mouse, *Immunogenetics* **2:**53–61.

75. Klein, G. O., Klein, G., Kiessling, R., and Kärre, K., 1978, *H-2*-associated control of natural cytotoxicity and hybrid resistance against RBL-5, *Immunogenetics* **6:**561–569.

76. Phillips-Quagliata, J. M., Walker, M. C., and Marsili, M. A., 1985, Towards a mechanism for hybrid resistance to the BALB/c plasmacytoma, MCP-11, in: *Genetic Control of Host Resistance to Infection and Malignancy* (E. Skamene, ed.), Alan R. Liss, New York, pp. 723–728.

77. Williams, R. M., Eig, B. M., and Singer, D. E., 1980, Preliminary analysis of hybrid resistance to histocompatible P815 utilizing bone marrow and thymus epithelium radiation chimeras, in: *Genetic Control of Natural Resistance to Infection and Malignancy* (E. Skamene, P. A. Kongshavn, and M. Landy, eds.), Academic Press, New York, pp. 477–483.

78. Sanford, B. H., 1967, Evidence for immunological resistance to a parental line tumor by F_1 hybrid hosts, *Transplantation* **5:**557–559.

79. Von Melchner, H., and Bartlett, P. F., 1983, Mechanisms of early allogeneic marrow graft rejection. *Immunol. Rev.* **71:**31–56.

80. Clark, E. A., and Holly, R. D., 1981, Activation of natural killer (NK) cells in vivo with H-2 and non-H-2 alloantigens, *Immunogenetics* **12:**221–235.

81. Drizlikh, G., Schmidt-Sole, J., and Yankelevich, B., 1984, Involvement of the K and I regions of the H-2 complex in resistance to hemopoietic allografts, *J. Exp. Med.*, **159:**1070–1082.

82. Cerottini, J.-C., Nordin, A. A., and Brunner, K. T., 1970, In vitro cytotoxic activity of thymus cells sensitized to alloantigens, *Nature* **227:**72–73.

83. Djeu, J. Y., Heinbaugh, J. A., Holden, H. T., and Herberman, R. B., 1979, Role of macrophages in the augmentation of mouse natural killer cell activity by poly I:C and interferon, *J. Immunol.* **122:**182–188.

84. Lotzová, E., and Cudkowicz, G., 1974, Abrogation of resistance to bone marrow grafts by silica particles. Prevention of the silica effect by the macrophage stabilizer poly-2-vinylpyridine *N*-oxide, *J. Immunol.* **113:**798–803.

85. Cudkowicz, G., and Yung, Y. P., 1977, Abrogation of resistance to foreign bone marrow grafts by carrageenans. I. Studies with the antimacrophage agent Seakem carrageenan, *J. Immunol.* **119:**483–487.

86. Carlson, G. A., Taylor, B. A., Marshall, S. T., and Greenberg, A. H., 1984, A genetic analysis of natural resistance to nonsyngeneic cells: The role of *H-2*, *Immunogenetics* **20:**287–300.

87. Carlson, G. A., Marshall, S. T., and Kiesche, A., 1986, Early events in natural resistance to bone marrow transplantation. Use of radiolabeled bone marrow cells, *Transplantation* **41**:688–694.
88. Riccardi, C., Santoni, A., Barlozzari, T., and Herberman, R. B., 1981, *In vivo* reactivity of mouse natural killer (NK) cells against normal bone marrow cells, *Cell. Immunol.* **60**:136–143.
89. Cudkowicz, G., and Bennett, M., 1971, Peculiar immunobiology of bone marrow allografts. II. Rejection of parental grafts by resistance F_1 hybrid mice, *J. Exp. Med.* **134**:1513–1528.
90. Koide, Y., and Takasugi, M., 1977, Determination of specificity in natural cell-mediated cytotoxicity by natural antibodies, *JNCI* **59**:1099–1105.
91. Kay, H. D., Bonnard, G. D., and Herberman, R. B., 1979, Evaluation of the role of IgG antibodies in human natural cell-mediated cytotoxicity against the myeloid cell line K562, *J. Immunol.* **122**:675–685.
92. Kall, M. A., and Koren, H. S., 1979, Human natural killing: Evidence for both serum-independent and serum-dependent cytotoxic mechanisms, *Cell. Immunol.* **47**:57–68.
93. Chow, D. A., Wolosin, L. B., and Greenberg, A. H., 1981, Genetics, regulation, and specificity of murine natural antitumor antibodies and natural killer cells, *JNCI* **67**:445–453.
94. Warner, J. F., and Dennert, G., 1985, Bone marrow graft rejection as a function of antibody-directed natural killer cells, *J. Exp. Med.* **181**:563–576.
95. Bennett, M., 1972, Rejection of marrow allografts. Importance of H-2 homozygosity of donor cells, *Transplantation* **14**:289–298.
96. Risser, R., and Grunwald, D. J., 1981, Production of anti-self-H-2 antibodies by hybrid mice immune to a viral tumor, *Nature* **289**:563–568.
97. Kawashima, K., Watanabe, E., Isobe, K.-I., Ogura, M., Nagura, E., Yamada, K., Sobue, I., Mizoguchi, K., Ito, Y., Nagai, Y., and Nagashima, I., 1982, Production of anti-self-H-2 antibodies by C3D2F₁ mice hyperimmune to L cell/L1210 hybrids and L1210 leukemia cells. *Cell. Immunol.* **67**:279–286.
98. Černý-Provaznik, R., Van Mourik, P., Limpens, J., Leupers, T., and Iványi, P., 1985, Anti-MHC immunity detected prior to intentional alloimmunization. III. Natural autoreactive H-2-specific antibodies. *Immunogenetics* **21**:491–504.
99. Aoki, T., Boyse, E. A., and Old, L. J., 1966, Occurrence of natural antibody to the G (Gross) leukemia antigen in mice, *Cancer Res.* **26**:1415–1419.
100. Nowinski, R. C., and Kaehler, S. L., 1974, Antibody to leukemia virus: Widespread occurrence in inbred mice, *Science* **185**:869–871.
101. Martin, S. E., and Martin, W. J., 1975, Anti-tumor antibodies in normal mouse sera, *Int. J. Cancer* **15**:658–664.
102. Wolosin, L. B., and Greenberg, A. H., 1979, Murine natural anti-tumor antibodies. I. Rapid *in vivo* binding of natural antibody by tumor cells in syngeneic mice, *Int. J. Cancer* **23**:519–529.
103. Wolosin, L. B., and Greenberg, A. H., 1981, Genetics of natural antitumor antibody production: Antibodies to MHC-linked determinants detected in the serum of unstimulated mice, *J. Immunol.* **126**:1456–1459.
104. Iványi, P., Van Mourik, P., Breuning, M., Kruisbeek, A. M., and Krose, C. J., 1982, Natural H-2-specific antibodies in sera of aged mice, *Immunogenetics* **15**:95–102.
105. Rolstad, B., Fossum, S., Bazin, H., Kimber, I., Marshall, J., Sparshott, S. M., and Ford, W. L., 1985, The rapid rejection of allogeneic lymphocytes by a non-adaptive, cell-mediated mechanism (NK activity), *Immunology* **54**:127–138.
106. Brodt, P., Kongshavn, P., Vargas, F., and Gordon, J., 1981, Increased natural host resistance mechanisms in B lymphocyte-deprived mice, *J. Reticuloendothel. Soc.* **30**:283–289.

15
Natural Cytotoxicity and Allograft Rejection

Pekka Häyry, Arto Nemlander, Jussi Tarkkanen,
Bernadette Ferry, Marita Jaakkola,
Yrjänä Nietosvaara, and Jarkko Ustinov

This book describes the role of natural cytotoxicity in health and disease. As virtually every aspect of natural cytotoxicity is covered by other contributors, we will limit ourselves strictly to the role of natural cytotoxicity and natural killer cells in the rejection of parenchymal organ allografts.

1. Identification of the Natural Killer Cells

Approximately 10 years ago it was observed[1-4] that leukocytes (lymphocytes) of deliberately nonsensitized donors display spontaneous cytotoxicity against several cultured cell line targets *in vitro*. This function, defined as natural killer (NK) activity, differs in many ways from immunologically specific lytic activity mediated by cytotoxic T cells: it is not restricted to the products of the major histocompatibility complex, and it does not follow the rules of immunology—i.e., selectivity, specificity, and memory. Instead, it has a (broad) target "specificity," with the erythroid human cell line K562 and myeloid mouse cell line YAC being among the most sensitive ones. However, the cells with NK function appear to constitute a discrete white cell subset. This subset has been defined by a variety of morphological and surface markers, which also form the basis for

Pekka Häyry, Arto Nemlander, Jussi Tarkkanen, Bernadette Ferry, Marita Jaakkola, Yrjänä Nietosvaara, and Jarkko Ustinov • Transplantation Laboratory, University of Helsinki, Helsinki, Finland.

the isolation and identification of these cells both in circulation and in transplant tissue.

A very important finding was the morphological identification of the NK cells as "large granular lymphocytes" (LGL) by our group.[5,6] After adsorption-elution of human mononuclear leukocytes to appropriate target cells, a relatively homogeneous cell population was obtained with a high cytoplasm-to-nucleus ratio and distinct azurophilic granulation in the cytoplasm. Such cells in human blood were described by Pappenheim and Ferrata in 1911.[7] In electron microscopy the LGL appear as medium-size lymphocytes with round or indented nuclei, condensed chromatin, prominent nucleoli, and extended Golgi apparatus. Coated vesicles and cytoplasmic inclusions in cell cytoplasm are equally seen[8,9] that contain several hydrolyses, including alpha-naphthyl-acetate esterase.[8] At least 70% of human peripheral blood LGL have been shown to have NK activity.

The morphological identification of the LGL and their enrichment by Percoll gradient centrifugation, as described by Timonen and Saksela,[10] made it possible to investigate the subcellular and surface markers of the NK cells. It appeared that the NK cells carried markers to both lymphoid and monocytoid cell lineages.

The alpha-naphthyl acetate activity of the LGL was natrium-fluoride inhibitable,[8] characteristic to monocytic but not lymphocytic cells.[11] Virtually all human NK cells and LGL carry IgG Fc receptors, although the receptor is of lower affinity and binds only aggregated or complex Ig.[12] In addition, many of the LGL and NK cells seem to carry receptors for the complement component C3bi.[13]

More than 50% of all human NK cells and LGL form low-affinity rosettes with sheep erythrocytes at $+4°$ C, but, unlike mature T cells, only a small proportion of them form high-affinity rosettes at $+29°$ C.[12,14] Consequently, and as expected, most of them are reactive with anti-CD2 monoclonal antibodies, such as OKT11. Although the LGL express neither CD3 nor CD4, many of them seem to express CD8 defined, e.g., by OKT8. Finally, most of the NK cells and LGL seem to be reactive with monoclonal antibody HNK-1 (anti-Leu 7), originally described as NK cell-specific.[15] Various markers of the NK cells and the LGL have been reviewed by Trinchieri and Perussia.[16]

In the following we will first discuss whether any characteristic changes are observed in the NK activity and/or the number of LGL in circulating leukocytes after transplantation and in context of graft rejection. Secondly, we will review the studies involved in the isolation and characterization of the NK cell *in situ* and in the transplant parenchyma, and, finally, we will speculate about the possible functions of the NK cells in graft rejection.

2. Changes in Systemic NK Activity after Transplantation and during Rejection

A recent review by Soulillou and Moreau[17] and papers published since then demonstrate one uniform finding: after transplantation (and initiation of intensive immunosuppression), the peripheral NK activity of human renal transplant recipients rapidly declines and, compared to uremic patients or normal healthy controls, remains depressed over long periods of time. Most investigators[18–23] report no change in the peripheral NK activity during rejection. However, controversial opinions exist demonstrating a rise in blood NK activity during episodes of rejection.[24–27]

The observed decline in systemic NK activity is most likely due to the administration of immunosuppressive drugs. Soulillou et al.[28] and Uhteg et al.[29] have demonstrated in the rat that if the recipients are not immunosuppressed, NK activity of leukocytes in the spleen, blood, and (irrelevant) lymph nodes significantly increases during rejection. This observation might also explain the results of investigators who have found occasional increases in the peripheral NK activity in context of acute rejection in man.

It is also conceivable that gradually, with time, this defect is corrected and that many of the recipients with long-surviving grafts display virtually normal NK activity in the blood.[25,30] Interestingly enough, the suppressive effect of immunosuppression on NK activity may be disconcordant in regard to circulating ADCC activity: regardless of a profound depression of the NK activity in renal transplant recipients, the ADCC activity may sometimes remain normal.[18,31] Analysis of peripheral leukocytes with various "NK-specific" markers and by flow cytometry has not added very much knowledge on this.

In a group of long-surviving renal transplant recipients (a follow-up of more than 5 years) with well-functioning grafts, the percentage of Leu-7 (HNK-1) lymphocytes has returned to normal and/or may be even increased compared to normal controls.[32] When this population was dissected further,[33] using double-fluorescent lymphocytes with fluorescein isothiocyanate and phycoerythrin and a pattern of monoclonal antibodies, it was found that the elevation was due particularly to a Leu-7$^+$/Leu-3a$^+$ (HNK-1/OKT4) subpopulation and that a Leu-7$^+$/Leu-15$^+$ subpopulation was proportionally decreased. The possible functions of the Leu-7$^+$/Leu-3a$^+$ subpopulation on renal transplant recipients is unknown.

Hammer's group suggests that the presence of LGL in peripheral circulation might differentiate between acute renal allograft rejection and viral infection.[34] Although during rejection there was an increase in the number of LGL in the graft, as quantitated by fine-needle aspiration biopsy (see later), in viral infection the number of LGL is increased both in the graft and in the peripheral circulation.

Finally, it would not be inconceivable that such conditions as acute rejection and in particular viral infection would reflect in the peripheral NK activity: various lymphokines that are known to activate NK cells *in vitro,* such as IL-2 and gamma-interferon,[16] are presumably secreted during rejection in large quantities at the site of inflammation and even more during a systemic viral infection. On the other hand, iatrogenic administration of alpha-interferon does not seem to alter the NK activity in peripheral circulation.[35,36]

3. Changes in Local NK Activity inside the Rejecting Allograft

Nemlander *et al.*[37] were the first to analyze the frequency of NK cells and LGL *in situ,* at the site of allograft rejection. After mechanical and enzymatical disaggregation of rat renal allografts with collagenase and DNAse, in conditions not affecting surface markers and/or functions of the inflammatory leukocytes,[38] they demonstrated that very high NK activity appeared in the allograft shortly after transplantation, peaking on day 3 or 4 and declining thereafter. Concomitantly, the graft was infiltrated by large numbers of inflammatory lymphocytes with LGL morphology. The peak of NK activity and LGL influx occurred approximately 3 days earlier than the peak of CTL activity (Figs. 1, 2). Concomitantly with the appearance of NK activity and LGL in the graft, the NK activity and the LGL were depleted from the recipient spleen, and concomitantly with the disappearance of the NK activity from the graft, this was restored in the spleen. This suggested to the authors that the NK cells and the LGL might be involved in acute allograft rejection, although their precise function remained nonspecified, and, moreover, that during rejection this cohort is shuttling first from the central lymphoid system to the graft and thereafter to the opposite direction.

These findings were later confirmed by several groups of investigators. Ultrastructurally immature, agranulated lymphocytes capable of killing K562 target cells have been reported at the site of intraperitoneal inflammation after injection of allogeneic L929 cells.[39] Marboe and coworkers[40] demonstrated high numbers of OKM1+-positive leukocytes with structural features of LGL in human endomyocardial biopsies during rejection. Hancock with his group[41] found, as did Nemlander, a high frequency of HNK-1-reactive cells in rejecting human renal allografts very shortly after the onset of rejection, on days 2–3. This population decreased thereafter and was taken over by IL-2 receptor-expressing activated lymphoid (blast) cells. Finally, Weber *et al.*[42] have demonstrated an influx of LGL to rejecting human renal allografts in context of acute episodes of rejection by using fine-needle aspiration biopsy cytology.

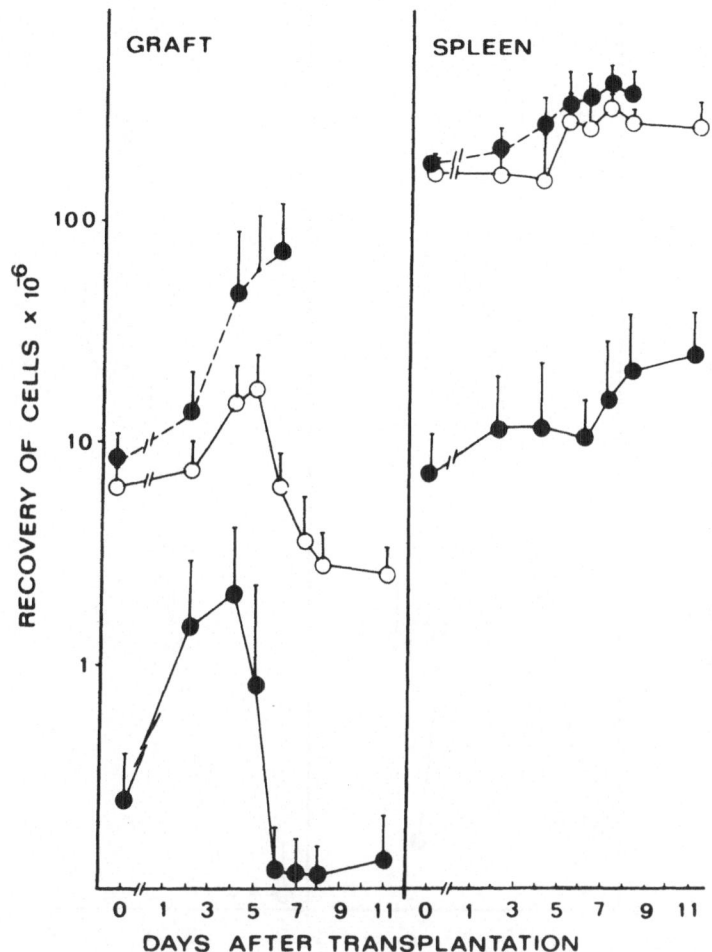

FIGURE 1. Recovery of different cell types from the allograft and in the spleen after renal transplantation in the rat. Each point indicates the mean of 3–5 separate transplants ± SD. Given are the total number of white cells (●----●), the number of lymphocytes (○——○), and the number of LGL recovered from each organ (●——●). Reprinted from *Eur. J. Immunol.* 1983:13:348–350.

As the NK activity in graft inflammatory infiltrate distributed together with the LGL in discontinuous Percoll density gradient fractionation,[43] it may be considered firmly established that LGL with NK activity are a significant part of the graft-infiltrating inflammatory population. These experiments do not demonstrate, however, whether they are of any relevance or necessary for the rejection process or only nonfunctional bystanders.

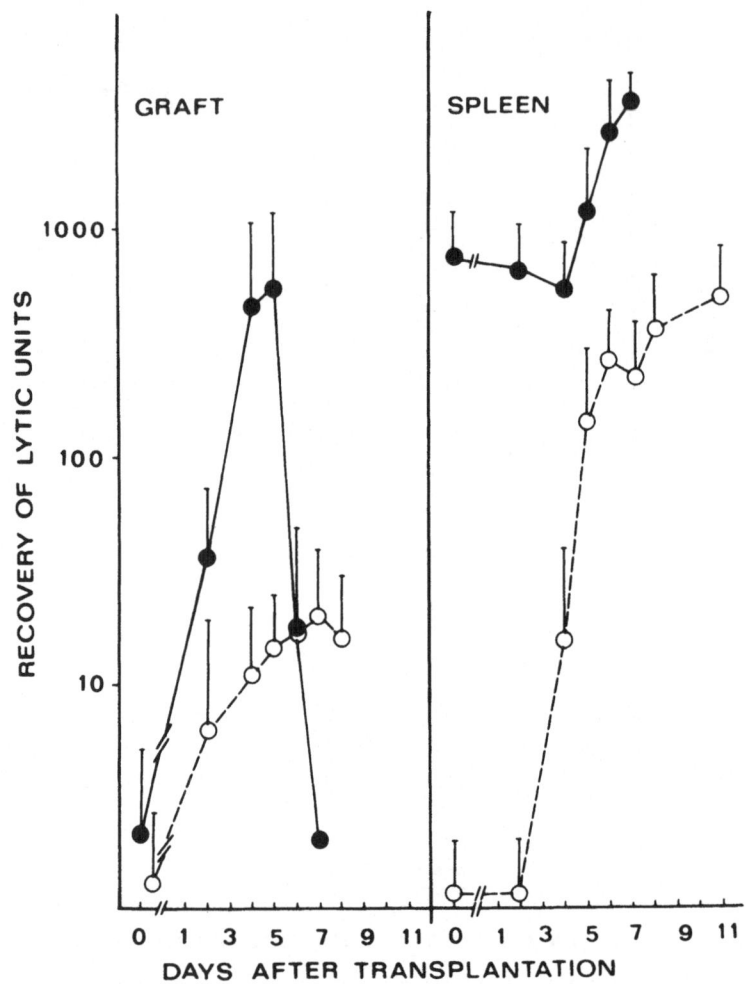

FIGURE 2. Cytotoxic activity in the graft and in the recipient spleen. Each point indicates the mean of 3–5 separate transplants ± SD. (●————●), NK activity to YAC-1 target line; (○————○), CTL activity to donor strain peritoneal exudate target cells (PEC). No cytotoxic activity was demonstrable in the spleen or in the graft to recipient strain PEC (not shown). Reprinted from *Eur. J. Immunol.* 1983:13:348–350.

4. Possible Functions of NK Cells and LGL in Acute Allograft Rejection

The first possibility is that the LGL participate in the rejection process via cytotoxic activity to allograft target cells. This possibility is not entirely unlikely, especially as the LGL and the NK cells have previously

TABLE I
Characteristics of Parenchymal Target Components Using Different
Marker Antibodies

Marker antibody	Endothelial cells	Glomerular epithelial cells	Glomerular mesangial cells	Tubular cells
a-FVIIIrag	100%	NT[a]	NT	NT
a-vimentin	100%	76%	85%	10%
a-cytokeratin	0%	8%	0%	38%
a-desmin	NT	10%	87%	2%
Control antibody	0%	0%	0%	0%

[a] Not tested.

been reported to be involved in the "hybrid resistance" of bone marrow allografts[44] and because they are most likely responsible for a rapid elimination of allogeneic lymphocytes after IV injection to circulation.[45] Unfortunately, no systematic analysis exists on the efficacy of NK cells to damage the cellular components of, e.g., a renal allograft. Some of our unpublished observations suggest, however, that they are not particularly cytotoxic to these target components *in vitro.*

We first isolated and cultured various cellular target components from rat organs—for instance, from vascular endothelial cells of rat heart, glomerular epithelial cells, glomerular mesangial cells, and proximal plus distal tubular cells of rat kidney. The purity of these cell populations was assessed using monoclonal antibodies to structures characteristic for these cell types. As seen in Table I, we have by now obtained essentially pure vascular endothelial cells, and relatively pure (70–90%) glomerular epithelial and mesangial cells, whereas the purity of our tubular target component is still less satisfactory.

When BN strain target cells were exposed to normal resting Lewis spleen cells, the spleen cells exhibiting a strong NK activity toward YAC target cells (positive control) were unable to kill allogeneic epithelial, glomerular mesangial, or tubular cells *in vitro* (Table II). Neither, as expected, were normal spleen cells cytotoxic to Con A-activated BN blast cells (negative control).

In a second experiment we isolated allograft infiltrating inflammatory cells from DA to WF allografts on day 4, when these cells displayed a maximal NK activity toward conventional NK target cells but no or only incipient CTL activity toward relevant allogeneic target cells (Figs. 1, 2). When the graft-infiltrating inflammatory leukocytes (Fig. 3) were tested for cytotoxicity toward DA vascular epithelial cells and DA peritoneal exudate cells (negative control) and toward YAC (positive control), a high cytotoxic activity was recorded against YAC, but none was recorded against the first two targets.

TABLE II
Cytotoxicity of Normal Lewis Spleen Cells to YAC Cells and to Isolated Kidney Components of BN Strain

Effector/target ratio	Percent of specific ^{51}CR release				
	YAC	Endothelial cells	Glomerular mesangial cells	Tubular cells	BN spleen Con A blast cells
100:1	92	12	11	NT[a]	0
50:1	77	9	8	0	0
25:1	59	5	3	0	0
12:1	52	6	5	0	0
Spontaneous release (% of maximal)	34	18	33	53	29

[a] Not tested.

Finally, we cultured nylon-wool-passed DA spleen cells with IL-2 (100 U/ml media) on plastic Petri dishes. After three days, loose cells were washed away and the remaining plastic adherent cells were used for cytotoxicity assays. This method (J. C. Hiserodt, personal communication) yielded 80% pure LGL, which lysed effectively both YAC and P815 target cells, the latter being NK-resistant and lymphokine-activated killer cell (LAK) sus-

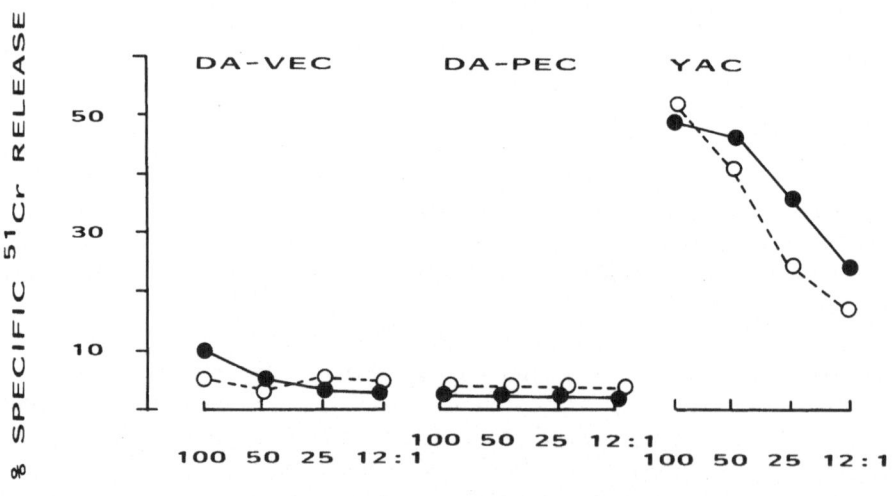

FIGURE 3. Cytotoxicity of DA to WF renal allograft-infiltrating inflammatory leukocytes (●——●) and recipient spleen cells (○----○) to DA vascular endothelial cells (DA-VEC), DA peritoneal exudate cells (DA-PEC), and YAC target cells in a 6-h cytotoxicity assay *in vitro*.

ceptible target line. These IL-2-stimulated LGL expressed also a low level of cytotoxicity to all tested DA-strain parenchymal cell components (Table III). No significant differences in our parenchymal cell components were detected in their susceptibility to lysis by the IL-2-treated splenocytes. Gamma-interferon pretreatment (100 U/ml media for 24 h) of parenchymal cell populations protected the target cells and reduced the cytotoxicity to nearly nondetectable levels. Neither normal spleen cells nor spleen-enriched LGL by Percoll gradient centrifugation killed these parenchymal cell components *in vitro* (Table II).

Although these observations may suggest that the LGL do not carry cytotoxic functions toward graft target components in context of rejection, the results do not demonstrate whether they are of significance for some other reasons. It seems, however, that recent observations speak against their crucial role in the rejection process.

The studies of Nemlander et al.[43] demonstrated that in addition to rejecting allografts, significant NK activity is frequently encountered also in nonrejecting autografts, which always undergo a milder inflammatory episode after the transplantation. Secondly, Bradley et al.,[46] of Morris's group, have shown that rat renal allografts are protected from rejection by cyclosporine and that grafts passively enhanced with a relevant alloantibody contain significant nonspecific cytotoxicity against YAC myeloma cells but no alloantigen-specific cytotoxicity due to cytotoxic T cells. Forbes and co-workers[47] have found that other inflammatory processes, such as experimental infarction of rat heart, generate inflammation with significant accumulation of OX8+ (T-cytotoxic suppressor; NK) inflammatory leukocytes to the affected area. Finally, Heidecke et al.[48] attempted to prolong rat heart allograft survival by an NK cell-specific antibody, rabbit anti-asialo GM_1. Administration of this antibody on days 0, 3, and 6 after

TABLE III

Cytotoxicity of IL-2-Treated[a] DA Spleen Cells to DA Kidney Components

Target population	Percent specific ^{51}Cr release at 100 : 1 effector–target ratio		
	Native	Gamma-interferon[b]	Spontaneous release[c]
YAC	50	NT[d]	10%
P815	65	NT	10%
DA-VEC	21	5	<10%
DA-glom. EPC	21	11	12%
DA-glom. MEC	17	11	15%
DA-tubular cells	23	2	20%

[a] Nylon-wool-passed 3-days IL-2 (100 U/ml media)-treated plastic adherent cells.
[b] Gamma-interferon pretreated (100 U/ml media) for 24 hours.
[c] As percent of maximal release (Triton X–100).
[d] Abbreviations: NT, not tested; VEC, vascular endothelial cells; EPC, glomerular epithelial cells; MEC, glomerular mesangial cells.

transplantation increased the survival of semiallogeneic allografts only marginally, from 8.4 ± 1.2 to 10.6 ± 0.6 days, but failed to generate long-term acceptance. Unfortunately, the treatment dose of anti-asialo GM_1 was so low that the NK cells were not entirely eliminated from the systemic circulation or from the graft inflammatory infiltrate. Thus, taken together, these results speak against the NK cells and the LGL being of any major importance in the inflammatory response of rejection.

Needless to say, additional possibilities remain. The LGL and the NK cells might, e.g., serve as precursor cells and/or function as regulatory cells in the generation of the inflammatory response. The latter possibility has recently been tackled by Hoffman et al.[49] of Simmons's group. They first confirmed, by using sponge matrix allografts of Roberts and Häyry,[50] that NK activity is generated in an allogeneic sponge matrix allograft several days before specific cytolytic T cell activity is observed. The peak of NK activity was observed on day 5 and the peak of CTL activity on day 14 after transplantation. The NK activity on day 14 was due to a $Thy1.2^+$, $Lyt2.2^-$, $L3T4^-$, and asialo GM_1^+ population. Intrasponge injection of anti–asialo GM_1 on days 3, 6, and 9 after grafting resulted in a complete abrogation of both NK and alloimmune CTL activities at the graft site on day 10. Intravenous injection of anti-asialo GM_1 into sponge-bearing mice resulted in severe inhibition of both NK and CTL activities in the sponge. Taken together, these data may suggest either that NK cells develop into autoimmune CTL and/or that NK cells may play another, possibly regulatory role.

5. Summary and Conclusions

After parenchymal organ transplantation, the NK activity in peripheral circulation rapidly decreases and remains depressed over long periods of time. This decrease is most likely due to intensive immunosuppression, as under experimental conditions in nonimmunosuppressed rats, peripheral NK activity after transplantation increases. NK cells codistributing with LGL in discontinuous Percoll density centrifugation infiltrate various types of experimental allografts as well as human allografts, at early stages of rejection, prior to the generation of maximal cytotoxic T lymphocyte activity. The LGL seem not to exhibit cytotoxic functions to allograft cellular components. Systemic administration of anti-asialo GM_1 antibody prolongs only marginally the allograft rejection and does not generate permanent survival in the rat. A possibility exists, however, that the LGL at the graft site perform some regulatory functions and/or act as precursor cells to some inflammatory cell lineages, in particular for CTL.

ACKNOWLEDGMENTS. Financial support for these studies was received in part from the National Institutes of Health, Bethesda, Maryland (grants 1RO1AM26882 and 31730), and in part from the Academy of Finland, Sigrid Juselius Foundation, and the Kidney Foundation, Helsinki, Finland.

References

1. Kiessling, R., Klein, E., and Wigzell. H., 1975, Natural killer cells in the mouse. I. Cytotoxic cells with specificity for mouse Moloney leukemia cells. Specificity and distribution according to genotype, *Eur. J. Immunol.* **5**:112.
2. Jondal, M., and Pross, H., 1975, Surface markers on human B and T lymphocytes. IV. Cytotoxicity against cell lines as a functional marker for lymphocyte subpopulations, *Int. J. Cancer* **15**:596.
3. Matthews, N., MacLaurin, B. P., and Clarge, G. N., 1975, Characterization of the normal lymphocyte population cytolytic to Burkitt's lymphoma cells of the EB1 cell line, *Aust. J. Exp. Biol. Med. Sci.* **53**:389.
4. Herberman, R. B., Nunn, M. E., Holde, H. T., and Lavrin, D. H., 1975, Natural cytotoxic reactivity of mouse lymphoid cells against syngeneic and allogeneic tumors. II. Characterization of effector cells, *Int. J. Cancer* **16**:230.
5. Timonen, T., Ranki, A., and Häyry, P., 1979, Human natural cell-mediated cytotoxicity against fetal fibroblasts. III. Morphological and functional characterization of the effector cells, *Cell. Immunol.* **48**:121.
6. Saksela, E., Timonen, T., Ranki, A., and Häyry, P., 1979, Fractionation, morphological and functional characterization of effector cell responsible for human natural killer activity to fetal fibroblasts and cell line targets, *Immunol. Rev.* **41**:71.
7. Pappenheim, A., and Ferrata, A., 1911, Uber die verschiendenen lymphoiden Zellformen des normalen und pathologischen Blutes, *Folia Haematol.* **10**:78.
8. Grossi, C. E., Cadoni, A., Zicca, A., Leprinik A., and Ferrarini, M., 1982, Large granular lymphocytes in human peripheral blood: Ultrastructural and cytochemical characterization of the granules, *Blood* **59**:277.
9. Payne, C. M., and Glasser, L., 1981, Evaluation of surface markers on normal human lymphocytes containing parallel tubular arrays: A quantitative ultrastructural study, *Blood* **57**:567.
10. Timonen, T., and Saksela, E., 1980, Isolation of human NK cells by density gradient centrifugation, *J. Immunol. Methods* **36**:285.
11. Ranki, A., and Häyry, P., 1979, Histochemical distinction between lymphocytic and monocytic alpha-naphthyl acetate (ANAE) esterases, *J. Clin. Lab. Immunol.* **1**:333.
12. Santoli, D., Trinchieri, G., Moretta, L., Zmijewski, C. M., and Koprowski, H., 1978, Spontaneous cell-mediated cytotoxicity in humans: Distribution and characterization of the effector cells, *Clin. Exp. Immunol.* **33**:309.
13. Perussia, B., Trinchieri, G., Lebman, D., Jankiewicz, J., Lange, B., and Rovera, G., 1982, Monoclonal antibodies that detect differentiation surface antigens on human myelomonocytic cells, *Blood* **59**:382.
14. West, W. H., Cannon, G. B., Kay, H. D., Bonnard, G. D., and Herberman, R. B., 1977, Natural cytotoxic reactivity of human lymphocytes against a myeloid cell line: Characterization of effector cells, *J. Immunol.* **118**:355.
15. Abo, T., and Balch, C. M., 1981, A differentiation antigen of human NK and K cells identified by a monoclonal antibody (NKH-1), *J. Immunol.* **127**:1024.

16. Trinchieri, G., and Perussia, B., 1984, Biology of disease. Human natural killer cells: Biologic and pathologic aspects. *Lab. Invest.* **50:**489.
17. Soulillou, J. P., and Moreau, J. F., 1982, Natural cytotoxicity in organ allotransplantation, *Heart Transplant.* **2:**52.
18. Lipinski, M., Tursz, T., Kreis, H., Finale, Y., and Amiel, J. L., 1980, Dissociation of natural killer cell activity and antibody-dependent cell-mediated cytotoxicity in kidney allograft recipients receiving high-dose immunosuppressive therapy, *Transplantation* **29:**214.
19. Moreau, J. F., Ythier, A., and Soulillou, J. P., 1981, Natural killer activity in kidney allograft recipients, *Transplant. Proc.* **13:**1610.
20. Charpentier, B., Lang, P., Martin, B., and Fries, D., 1981, Unspecific cytotoxic capacities of peripheral blood lymphocytes (PBL) from transplant patients, *Transplant. Proc.* **13:**1614.
21. Kovitavongs, T., cited in: Cosimi A. B., Jeannet, M., 1981, Immunosuppression and immunodeficiency (workshop reports, p. 1684), *Transplant. Proc.* **13:**1682.
22. Ramsey, K. M., Djeu, J. Y., and Rook, A. H., 1984, Decreased circulating large granular lymphocytes associated with depressed natural-killer cell activity in renal transplant recipients, *Transplantation* **38:**351.
23. Ellis, T. M., Berry, C. R., Mendez-Picon, G., Goldman, M. H., Lee, H. M., and Mohanakumar, T., 1982, Lack of association of human renal allograft rejection and circulating K-cell, NK-cell, or total T-cell levels, *Clin. Immunol. Immunopathol.* **25:**335.
24. Ono, Y., Hirabayashi, S., Ohshima, S., and Koide, Y., 1981, Natural cell-mediated cytotoxicity in long-term kidney transplant recipients, *Transplant. Proc.* **13:**1508.
25. Guillou, P. J., Hegarty, J., Ramsden, C., Davison, A. M., Will, E. j., and Giles, G. R., 1982, Changes in human natural killer activity early and later after renal transplantation using conventional immunosuppression, *Transplantation* **33:**414.
26. Hegarty, J. H., Ramsden, C. W., Dowd, P. S., Giles, G. R., and Guillou, P. J., 1983, Sequential changes in NK and ADCC effector cell function after organ transplantation in man, *Biomed. Pharmacother.* **37:**266.
27. Cooksey, G., Robins, R. A., and Blamey, R. W., 1984, Natural killer cells in renal allograft rejection, *Br. J. Surg.* **71:**874.
28. Soulillou, J. P., Vie, H., Moreau, J. F., Peyrat, M. A., and Blandin, F., 1983, Increased NK activity in rats rejecting heart allografts, *Transplantation* **36:**726.
29. Uhteg, L. C., Kupiecweglinski, J. W., Rocher, L. L., Salomon, D. R., Tilney, N. L., and Carpenter, C. B., 1986, Systemic natural killer activity following cardiac engraftment in the rat—lack of correlation with graft survival, *Cell. Immunol.* **100:**274.
30. Waltzer, W. C., Bachvaroff, R. J., Arnold, A., Anaise, D., and Rapaport, F. T., 1985, Immunological consequence of renal transplantation and immunosuppression. I. Alterations in human natural killer cell activity, *J. Clin. Immunol.* **5:**78.
31. Prince, H. E., Ettenger, R. B., Dorey, F. J., Fine, R. N., and Fahey, J. L., 1984, Azathioprine suppression of natural killer activity and antibody-dependent cellular cytotoxicity in renal transplant recipients, *J. Clin. Immunol.* **4:**312.
32. Fregona, L., Guttmann, R. D., and Jean, R., 1985, HNK-1⁺ (Leu-7) and other lymphocyte subsets in long term survivors with renal allotransplants, *Transplantation* **39:**25.
33. Legendre, C. M., Guttmann, R. D., Hou, S. K., and Jean, R., 1985, 2-color immunofluorescence and flow-cytometry analysis of lymphocytes in long-term renal allotransplant recipients—identification of a major Leu-7⁺/Leu-3⁺ subpopulation, *J. Immunol.* **135:**1061.
34. Nguyen, L., Hammer, C., Dendorfer, U., Castro, L., Schleibner, C., and Land, W., 1985, Changes in large granular lymphocyte size and number in kidney transplant patients during rejection and viral infection, *Transplant. Proc.* **17:**2110.
35. Nguyen, L. H., Hammer, C., Dendorfer, U., Castro, L. A., Schleibner, S., and Land, W., 1985, Impact of viral infection, interferon-alpha2, and acute rejection on size and number of large granular lymphocytes in renal transplant patients, *Transplant. Proc.* **17:**2530–2531.

36. Kelly, A. P., Schooley, R. T., Rubin, R. H., and Hirsch, M. S., 1984, Effect of interferon alpha on natural killer cell cytotoxicity in kidney transplant recipients, *Clin. Immunol. Immunopathol.* **32:**20.
37. Nemlander, A., Saksela, E., and Häyry, P., 1983, Are "natural killer" cells involved in allograft rejection? *Eur. J. Immunol.* **13:**348.
38. Von Willebrand, E., Soots, A., and Häyry, P., 1979, *In situ* effector mechanisms in rat kidney allograft rejection. I. Characterization of the host cellular infiltrate in rejecting allograft parenchyma, *Cell. Immunol.* **46:**309.
39. Gregory, C. D., and Atkinson, M. E., 1984, Large agranular lymphocytes—early non-specific effector cells in allograft rejection in the mouse, *Immunology* **53:**257.
40. Marboe, C. C., Knowles, D. M., Chess, L., Reemtsma, K., and Fenoglio, J. J. Jr., 1983, The immunologic and ultrastructural characterization of the cellular infiltrate in acute cardiac allograft rejection: Prevalence of cells with the natural killer (NK) phenotype, *Clin. Immunol. Immunopathol.* **27:**141.
41. Hancock, W. W., Gee. D., De Moerloose, P., Rickles, F. R., Ewan, V. A., and Atkins, R. C., 1985, Immunohistological analysis of serial biopsies taken during human renal allograft rejection. Changing profile of infiltrating cells and activation of the coagulation system, *Transplantation* **39:**430..
42. Weber, B., Welte, M., Hammer, C., Stadler, J., Koller, C., Caspo, C., Land, W., Hillebrand, G., and Castro, M., 1984, Increase of natural killer cells in rejecting kidney grafts, *Transplant. Proc.* **16:**1177.
43. Nemlander, A., Soots, A., and Häyry, P., 1984, *In situ* effector pathways of allograft destruction. 1. Generation of the "cellular" effector response in the graft and the graft recipient, *Cell. Immunol.* **89:**409.
44. Kiessling, R., Hochman, P. S., Haller, O., Shearer, G. M., Wigzell, H., and Cudkowicz, G., 1977, Evidence for a similar or common mechanism of natural killer cell activity and resistance to hemopoietic grafts, *Eur. J. Immunol.* **6:**655.
45. Rolstad, B., Fossum, S., Bazin, H., Kimber, I., Marshall, J., Sparshott, S. M., and Ford, W. L., 1985, The rapid rejection of allogeneic lymphocytes by a non-adaptive, cell-mediated mechanism (NK activity), *Immunology* **54:**127.
46. Bradley, J. A., Mason, D. W., and Morris, P. J., 1985, Evidence that rat renal allografts are rejected by cytotoxic T cells and not by nonspecific effectors, *Transplantation* **39:**169.
47. Forbes, R. D., Lowry, R. P., Gomersall, M., and Blackburn, J., 1985, Comparative immunohistologic studies in an adoptive transfer model of acute rat cardiac allograft rejection, *Transplantation* **40:**77.
48. Heidecke, C. D., Araujo, J. L., Kupiec-Weglinski, J. W., Abbud-Filho, M., Araneda, D., Stadler, J., Siewert, J., Strom, T. B., and Tilney, N. L., 1985, Lack of evidence for an active role for natural killer cells in acute rejection of organ allografts, *Transplantation* **40:**441.
49. Hoffman, R. A., Jordan, M. L., Ascher, N. L., and Simmons, R. L., 1987, The role of natural killer cells is the allograft response, *Transplant. Proc.* **19:**342–344.
50. Roberts, P. J., and Häyry, P., 1976, Effector mechanisms in allograft rejection. I. Assembly of "sponge matrix" allografts, *Cell. Immunol.* **26:**160.

Summary of Part IV

From results presented in previous chapters of this book, it has become evident that NK cells can recognize and directly lyse a variety of cell types, including tumor cells and virus-infected cells (Chapters 1–5). In addition, Trinchieri *et al.* (Chapter 11) have summarized the evidence that suggests that under some conditions, natural effector cells are able to regulate the growth or differentiation of some bone marrow stem cells. Also, Greenberg (Chapter 8) has summarized the evidence that natural effector cells can infiltrate into tissue during the process of inflammation. Collectively, these observations have induced a number of investigators to study the role of natural effector cells in resistance to bone marrow and parenchymal organ allografts. Several possible mechanisms exist by which these may participate in graft rejection. The mechanism of graft rejection may be direct, whereby natural effector cells directly bind to and lyse target allograft cells, or release toxic mediators which act directly on grafted cells in the microenvironment. Alternatively, natural effector cells may act indirectly through the elaboration of lymphokines which may stimulate T and B lymphocytes or macrophages to mediate graft rejection. Thus natural effector cells may participate in a cascade of events, which includes participation of other leukocytes, and which ultimately results in graft rejection.

The studies that provide evidence on the role of the NIS in bone marrow graft rejection have been summarized by Nakamura (Chapter 14). His review of the available information emphasizes the complexity of the problem and the difficulties in assigning a definitive role for NK cells in bone marrow graft resistance. There appear to be at least two components to the rejection of bone marrow grafts. These are the nonadaptive (natural resistance) and the adaptive (largely T lymphocyte-mediated) components of bone marrow rejection. It seems clear that the predominant mechanism of hemopoietic resistance is the adaptive mechanism. However, in situations where specific immune responsiveness is absent, a role can be observed for the NIS. Although most of the evidence is of a correlative nature, more definitive proof for a role of NIS has been reported by Murphy et al.,[1] who have shown that lymphocytes from SCID mice that have depressed T and B lymphocyte responses but normal levels of NK activity mediate bone marrow rejection quite effectively. The degree to which the NIS participates in resistance to bone marrow grafts is strain-dependent and therefore may suggest heterogeneity in the natural effector population, perhaps at the level of recognition. This point seems particularly relevant, since most evidence suggests that natural resistance to hemopoietic grafts can be a direct effect, because lysis of donor bone marrow cells can be easily detected. It would seem critical to better understand the nature of the antigen receptor(s), which allows recognition of bone marrow cells by the NIS. The issue of specificity becomes even more important when one considers the possibility that NK cells might preferentially recognize and lyse relatively small populations of clonogenic stem cells.[2] To test this hypothesis directly, it will be necessary to purify populations of various bone marrow stem cells and determine their sensitivity to lysis by NK cells.

In addition to the potential direct effects of NK cells, there is considerable circumstantial and correlative evidence that the NIS may act indirectly by enhancing the adaptive component of bone marrow rejection. Circumstantial evidence is provided based on the ability of LGL to produce numerous immunoregulatory cytokines (see Ortaldo, Chapter 13). Such cytokines could serve to indirectly enhance bone marrow rejection episodes mediated by T lymphocytes or by antibody. In fact, it has been reported that CD3⁻ LGL act as accessory cells in T cell–depleted mice for immunoglobulin production by B cells.[3] Thus, one area for future study should be to determine the effect of LGL-produced cytokines on various subsets of bone marrow cells and the degree to which such factors could restore bone marrow rejection in NK-depleted mice.

If LGL-mediated effects contribute to bone marrow allograft rejection in humans, it may be that depression of NK activity would allow successful engraftment of transplants through the use of fewer transplanted cells. Such an approach might also decrease the frequency of graft versus host disease encountered in the clinic. Since NK cells appear to be more radioresistant than other lymphocytes, it may be that such cells could only be eliminated by concurrent treatment of the patient with antibodies specific for NK cells. Overall, it seems apparent that the NIS plays a role in hemopoietic resistance, but the relative importance of that role in comparison to adaptive immunity and its actual biologic significance with regard to bone marrow graft survival remain unclear.

Because LGL with NK activity have been shown to infiltrate into tissue, accumulate in sites of inflammation, and produce a variety of cytokines, they have also been considered as candidates for participation in organ allograft rejection. Häyry et al. (Chapter 15) have summarized the relevant literature regarding the potential role of natural effector cells in the rejection of organ allografts. Over the past 5–10 years, a variety of investigators have made positive correlations between levels of NK activity and episodes of allograft rejection. Most interesting of these was the observation of Häyry and colleagues that high levels of NK activity could be detected in rat renal allografts within several days of transplantation. Similar findings were subsequently reported for human cardiac[4] and renal[5] transplants. Most evidence for a role for natural effector cells in rejection of human parenchymal organ grafts has been observed in renal allograft patients.[5,6] Lefkowitz et al.[6] noted a significant increase in peripheral blood NK activity during episodes of rejection in renal transplant patients maintained on cyclosporin A immunosuppression. Thus NK activity was stimulated during the rejection period, but it was unclear whether this augmented NK activity contributed to the rejection process or was merely coincident to it.

Several approaches need to be considered to clarify the role of natural effector cells in the rejection of human allografts. First, putative natural effector cells isolated from transplant patients need to be better characterized in terms of their phenotype. Most studies have used the Leu7 marker as an indicator of LGL. More definitive evidence could be gained if cells from peripheral blood and from within the transplant could be typed using the Leu11 and Leu19 (NKH-1) markers. Second, since the best evidence of a role for LGL in animal models comes from cells isolated from the grafts at early time points, it would seem important to be able to obtain biopsies from patients early in the rejection process. Since the infiltration of LGL precedes the appearance of cytotoxic T cells during renal graft rejection in rats, animal experiments should be designed to determine whether these early-infiltrating LGL are providing immunoregulatory or recruitment stimuli for T cells. Such experiments might be done by conclusively depleting LGL by repeated administration of anti-asialo GM_1 or anti-NK1.1 sera during the early stages of transplantation. Similarly, the development of transplant models in SCID mice could determine the degree to which graft rejection could occur in the absence of both T and B lymphocyte–mediated responses, when NK activity is reasonably normal. It would also seem important to systematically determine the degree to which various normal tissues are susceptible to NK activity or the cytokines are produced by LGL. Overall, the results summarized by Nakamura and Häyry et al. support an association between the NIS and graft rejection. Results obtained from studies of bone marrow graft rejection suggest that natural effectors may function both directly to lyse or growth-inhibit various stem cell populations and indirectly through production of inhibitory cytokine(s). The available data from allograft rejection studies suggest that such a role is most likely to be largely indirect, perhaps through amplification of activities mediated by non-NK cells.

References

1. Murphy, W. J., Kumar, V., and Berrett, M., 1987, Rejection of bone marrow allografts by mice with severe combined immune deficiency (SCID). Evidence that natural killer cells can mediate the specificity of marrow graft rejection, *J. Exp. Med.* **165**:1212–1217.
2. Lotzova, E., 1986, NK cell role in regulation of the growth and functions of hemopoietic and lymphoid cells, in: *Immunobiology of Natural Killer Cells*, Vol. 2 (E. Lotzova and R. B. Herberman, eds.), CRC Press, Inc., Boca Raton, Fl. pp. 89–105.
3. Brenner, M. K., Vyakarnam, A., Reittie, J., Wimperis, J. Z., Grob, J. P., Hoffbrand, A. V., and Prentice, H. G., 1987, Human large granular lymphocytes induce immunoglobulin synthesis after bone marrow transplantation, *Eur. J. Immunol.* **17**:43–47.
4. Marboe, C. C., Knowles, D. M., Chess, L., Reemtsma, K., and Fenoglio, J. J. Jr., 1983, The immunologic and ultrastructural characterization of the cellular infiltrate in acute cardiac allograft rejection: Prevalence of cells with the natural killer (NK) phenotype, *Clin. Immunol. Immunopathol.* **27**:141–151.
5. Weber, B., Welte, M., Hammer, C., Stadler, J., Koller, C., Caspo, C., Land, W., Hillebrand, G., and Castro, M., 1984, Increase of natural killer cells in rejecting kidney grafts, *Transplant. Proc.* **16**:1177–1178.
6. Lefkowitz, M., Jorkasky, D., and Kornbluth, J., 1987, Increase in natural killer activity in cyclosporin-treated renal allograft recipients during rejection, *Hum. Immunol.* **19**:139–149.

Part V
Involvement of the NIS in Various Pathological Conditions

16

Involvement of Natural Effector Cells in Graft versus Host Disease

JOHN CLANCY JR.

1. Introduction

1.1. Definition and Significance of GVHD

Graft versus host disease (GVHD) is the consequence of transplanting a graft of immunocompetent allogeneic lymphoid cells into a host unable to reject them.[1] Many of the clinical studies on GVHD have resulted after bone marrow transplantation to patients with leukemia, aplastic anemia, or immunodeficiency.[2,3]

Numerous studies have documented that for GVHD to occur, a critical number of donor alloreactive T lymphocytes must be present in the graft.[4] It is hypothesized that such T cells, in responding to either disparate host class I or II major and even multiple minor histocompatibility antigens (HA), secrete lymphokines that recruit other host as well as donor reactive cells to various target tissues.[4,5] Those target tissues affected include all lymphoid tissues as well as skin, liver, lungs, gastrointestinal tract, and salivary glands.[1–9]

Depending on the dose of cells injected, the degree of histoincompatibility between donor and host, the immune status of the host, and a number of unknown factors, two types of GVHD have been described: acute and chronic.[2,3] Acute GVHD occurs within 3 months after transplantation and is characterized by skin rash, hepatic dysfunction, and in

JOHN CLANCY JR. • Department of Anatomy, Loyola University Medical Center, Maywood, Illinois 60153.

some species diarrhea and cachexia.[2,3,10,11] Chronic GVHD usually occurs beyond 3 months posttransplantation in humans and is characterized as a general autoimmune reaction manifested by skin, conjunctiva, salivary gland, and oral lesions as well as chronic liver disease and weight loss.[2,3,12] Autoantibodies, donor lymphocytes reactive to recipient cells, and suppressor cells have also been reported.[13,14]

1.2. Why Is a Role for Natural Effector Cells Suspected?

The pathological lesions associated with GVHD are extensive. Although evidence exists for the generation and presumed involvement of specific HA-reactive cytotoxic T cells in GVHD-associated tissue lesions, the breadth and extent of these lesions implicate a role for nonspecific or nonselective natural effector (NE) mechanisms as well.[15] For the purpose of this review, a NE cell will be defined as a cell cytotoxic to various nucleated targets without known prior antigenic stimulation.

Studies by Lopez et al.[16] demonstrated that all seven patients receiving allogeneic bone marrow or fetal liver transplants whose peripheral blood lymphocytes (PBL) exhibited normal cytotoxicity (27–73% ^{51}Cr release) of herpes simplex virus type 1-infected fibroblasts (HSV-1) pretransplantation developed GVHD posttransplantation. However, all six patients with low pretransplant cytotoxicity (2–19% ^{51}Cr release) to HSV-1-infected cells showed no evidence of GVHD. The authors correlated the presence of host natural HSV-1 effectors, before aggressive cytoreduction, with "stimulating" GVHD. However, the phenotype of HSV-1 natural effector cells was not examined.

One of the problems with assessing NE cell activity after allogeneic cellular transplantation is to separate those cells of the host responsible for natural resistance and possible graft failure[17] as well as those cells in the graft with potential antileukemic effects[18,19] from donor and host NE cells either responsible for or contributing to the pathophysiology of GVHD. The former cell type will be dealt with by another chapter in this book and is presumably not a factor in GVHD, because by definition the graft must take for GVHD to occur. Although this review will deal with NE cytotoxicity against erythroblastoid, lymphoma, or viral-infected cell lines by cells developing during GVHD as a consequence of or in concert with alloactivation, it is entirely possible that the same effector cell could be lytic to residual or reappearing leukemic cells.[19]

2. Evidence *against* a Role of NE Cells in Initiating GVHD

2.1. Human

When PBL from patients with aplastic anemia or leukemia were examined 30–100 days post-HLA identical bone marrow transplantation, there was no significant association between levels of cells able to kill the erythroblastoid line K562 after 18–21 h of culture and the presence of clinically manifested acute or chronic GVHD.[20] It is unfortunate that this study was not performed with the more conventional 4-h assay so that the control K562 killing may not have been 60–70%. A recent study examined T cell-enriched PBL populations from 19 mostly leukemic patients having received largely HLA-matched bone marrow. The study found no differences between patients with and without GVHD and the levels of OKT8, 4, and Leu 7 expression by PBL.[21] Also, when 96–99% T cells were removed from the HLA-matched bone marrow inoculum given to seven leukemia patients, GVHD did not develop. In addition, cells able to kill K562 targets in 4 h were present in the peripheral blood by 2–4 weeks and functioned at a higher level of efficiency than normal, nontransplanted controls through 12–14 weeks after transplantation.[22] In addition, cells able to kill the T cell line HSB2 and EBV transformed lymphoblastoid cell lines (B-LCL) were present from 4–14 weeks after transplantation even without the presence of clinically manifested GVHD or cytomegalovirus (CMV) infection. The authors concluded that the development of normal NE and presumably induced NE activity was not dependent on either donor T cells or detectable host GVHD and CMV infection.

There was an early appearance of B73.1 receptor-positive, T3-, K562 lytic cells over T3$^+$ cells in 16 severe combined immunodeficient (SCID) patients receiving T-depleted marrow and who had little or no GVHD.[23] Another study examined immunological reconstitution following Campath-1-treated bone marrow transplantation from an HLA-identical sibling.[24] Although Campath-1 depleted the donor bone marrow of T as well as B cells, no information was given on its effect on K562 killers. Nevertheless, this report delineated the emergence of higher levels of K562 killers within 1 month and Leu 7, class II MHC, and OKT10-positive cells by 3 months after transplantation. However, high standard deviations prevented the phenotypes from being significantly greater than controls, and K562 killing was presented as 2.7 times that of controls with no mention of significance. Nonetheless, the authors state that there was no correlation between K562 killing and presence of GVHD.

Finally, when 23 patients with non-Hodgkin's lymphoma were transplanted with their own marrow after it was purged *in vitro* with anti-B1 and complement, which removed not only very immature B cells but also tumor cells, NKH1-positive and functional K562 killer cells predominated early posttransplantation followed by B and then T cell subsets.[25] The authors demonstrated some cytotoxicity against cryopreserved autologous tumor cells 1 month posttransplantation and concluded that functional NK cells predominate early after autologous bone marrow transplantation.

It thus appears that soon after allogeneic HLA-matched and T cell-depleted autologous bone marrow transplantation, there is an elevation in peripheral blood of the Leu 7^+ K562 killer cell. Whether this cell is only relatively elevated because it simply survived all pretransplant ablation procedures or is truly the first cell type to appear posttransplant has not been evaluated. Also, the fact that the appearance of a K562 killer cell is independent of clinical GVHD does not rule out the presence of subclinical GVHD contributing to its appearance. Since clinical GVHD is not manifest unless a critical number of donor alloreactive T cells are transfused, the existence of subclinical and transient GVHD is possible.

2.2. Mouse

When acute GVHD was induced in unirradiated (C57BL/6 × DBA/2)F1 mice with two injections of 5×10^7 C57BL/6 (B6) spleen cells IV over a 2-week period, their spleen and PBL cells were deficient in NE cells against YAC-1 lymphoma targets even after 7 days.[26] However, such spleen cells exhibited concomitant increased killing of antibody-coated chicken RBCs and allo-P815 targets. If the dose of cells was reduced by half, a more chronic GVHD developed along with an elevation of YAC-1 killing. Such studies suggest that levels of YAC-1 effector cells may depend on the extent or severity of GVHD, and this in turn may depend on the dose of cells injected. In this study, no information was given on the genotype and phenotype of YAC-1 killers in chronic GVHD.

When acute GVHD was induced in unirradiated (B6 × CBA)F1 mice with a single IV injection of B6 spleen cells, there was an elevation in YAC-1 killing by GVHD spleen cells on day 12.[27] However, when B6 donors or F1 recipients were treated with anti–asialo GM_1 (aGM_1) antibody, GVH-induced splenomegaly was only slightly impaired even though the antibody abolished YAC-1 killers for 2 days. Multiple injections of aGM_1 did not affect splenomegaly. In addition, potentiation of YAC-1 killing by polyinosinic-polycytidylic acid (pIC) had no significant effect on splenomegaly. These studies indicate that elevated YAC-1 killing during murine GVHD does not cause or necessarily result from GVHD-induced splenomegaly.

2.3. Rat

The local popliteal GVH (PGVH) reaction was extensively described by Ford et al.[28] and represents a useful model to delineate the GVH inducing capacity of transferred lymphoid cell populations in rats. This subclinical reaction is induced more readily by footpad inoculation of the CD4 (W3/25) than CD8 (Ox8) subset to T cells[29] and results in the preferential accumulation of host B cells over host T cells in the local draining popliteal lymph node producing very significant enlargement.[30]

Rat peripheral blood lymphocytes (PBL) and spleen cells (SC) can be

TABLE I
Ability of Percoll-Fractionated Nonadherent Lewis
Spleen Cells to Cause a Popliteal GVH Reaction in
(DA × LEW)F1 Hybrids

Percoll fraction	Percent LGL[a]	Weight (mg)[a]
2[b]	38	10.1
3[b]	18	16.4
4[b]	2	45.6[d]
5[b]	1	14.0
2 + 4[c]	19	41.8[d]

[a] Four animals injected per group, and only the means reported.
[b] 5×10^6 cell injected.
[c] 5×10^6 from each population injected.
[d] Significantly larger than any of the other weights at $p < 0.05$.

enriched for the putative YAC-1 killer cell or large granular lymphocyte (LGL) by discontinuous Percoll gradients.[31] When such gradients were performed on Lewis (LEW:RT1^1) SC and single or combined fractions used to induce PGVH reactions in (DA × Lew) F1 recipients, a pattern similar to that shown in Table I was uniformly obtained. The results demonstrate that the less dense fractions that were enriched for LGL and YAC-1 effector cells[31] were associated with less lymph node enlargement. In addition, when fractions enriched for LGL were combined with those without LGL, there was no modulation of the PGVH reaction produced. The results of this study indicate that low-density LGL-enriched fractions are not as efficient as high-density LGL-deficient fractions in producing the PGVH reaction. However, this is not surprising, as LGL are largely CD8$^+$, and inocula predominating in this phenotype have proved to be markedly inferior to CD4$^+$ cells in producing the PGVH reaction.[30] This study should perhaps be extended by adding enriched LGL to pure CD4$^+$ populations.

3. Evidence *for* a Role of NE Cells in Initiating GVHD

Because GVHD is a systemic disease, one should examine various target tissues for the involvement of NE cells. Particular attention will be paid to whether NE cells induced the tissue lesion or were recruited into it.

3.1. Human

3.1.1. Blood

When a group of 23 patients with either severe aplastic anemia (SAA) or leukemia were transplanted with largely HLA-identical or identical-twin bone marrow, four patients developed acute lethal GVHD and had a higher level of K562 killer cells 10–30 days posttransplantation than those without GVHD.[32] None of the patients developed chronic GVHD. It is unfortunate that K562 killing by donor marrow and recipient blood was performed on only one of these GVHD patients, because that patient had a much higher PBL K562 killing level pretransplant than any other patient and also received marrow with almost no lytic activity. Nevertheless, this report clearly demonstrates that cells lytic for K562, but not antibody-coated L1210, were activated or present in increased numbers during acute lethal GVHD.

In another study, 45 SAA or leukemia patients were transplanted with HLA-identical marrow, and pre- as well as posttransplant K562 and IgG-coated P-815 killing followed with reference to whether the patient developed grade 0–IV acute GVHD. There was no correlation between the lytic activities of donors or recipients and the subsequent development of grades II–IV GVHD. However, when 13 patients were tested at days 5, 12, 19, and 26 posttransplantation, 8 of the 10 patients that developed grades I–IV GVHD showed high K562 killing but not P815 killing.[33] Thus both of the above reports show a strong correlation between K562 killer cells in blood and GVHD.

Phenotypic analysis performed on the peripheral blood and bone marrow of leukemic patients with GVHD via double immunofluorescence revealed that there was a very significant increase in CD8$^+$, CD10$^+$, Leu7$^+$, DR$^+$ PBL in patients with acute GVHD, but CD8$^+$, Leu7$^-$, DR$^-$ cells predominated in epidermal infiltrates.[34] Treatment of donor marrow with anti–pan T (CD6) and anti-CD8 resulted in little GVHD, with cells representing the Leu7$^+$ CD8$^-$ population being the first to appear.[35]

It appears that a population of LGL may play a significant role in maintaining B cell function *in vivo* after T cell-depleted non-GVHD bone marrow transplantation. A state of immunodeficiency has been reported to occur frequently during chronic GVHD which can affect both T and B cell functions.[36] Although it is possible that an innate failure at cooperation ensues during GVHD,[37] the generation of suppressor cells has been documented.[14] However, recent studies indicate that the B cell defect in some of these patients may occur because of an inability of their B cells to undergo a second round of ontogeny and respond to either mitogens or T cell helper factors.[38] In contrast, when 11 patients with leukemia were transplanted with T cell-depleted marrow, immunoglobulin and specific antibody levels were well maintained.[39] Examination of preparations containing 50–70% Leu 11b$^+$ LGL from these patients revealed that they were found to secrete interleukin 2, inferferon gamma, and B cell differentiation factors. These cells were not lytic LGL, as incubation with antibody to the B chain of LFA-1, which reduced K562 killing by up to 58%, did not influence this B cell helper function.[40] Since a higher number of LGL circulate 3–8 weeks posttransplantation,[22,39] they may indeed have a regulatory role in B cell differentiation.

Leukemia patients receiving allogeneic marrow in whom significant GVHD develops have a reduced risk of leukemia relapse compared to patients in whom GVHD does not develop.[41] Accordingly, leukemia blasts were cryopreserved pretransplantation, and the leukemic patient was transfused with anti-T12-treated bone marrow. Although the patient did not develop GVHD, at 12 weeks posttransplantation cells were found in PBL of donor genotype which expressed the NKH1$_A$ phenotype and were lytic for not only K562 but pretransplant leukemic targets.[42] Thus the development of NE cells with GVHD may be beneficial, since they may have an antileukemic effect.

3.1.2. Liver

An inflammatory periportal infiltrate has been well described in both human and rodent GVHD[43] and is evident ultrastructurally in bile duct epithelial and adjacent hepatocyte lesions.[44] When specimens from seven normal autopsy livers were compared to nine marrow recipients without and 10 with GVHD, the most striking difference was a marked increase in the concentration of Leu7$^+$ cells, CD8$^+$ (suppressor/cytotoxic) lymphocytes, and macrophages with few CD4$^+$ leukocytes in the periportal infiltrates of those patients with GVHD.[45] Although the GVHD patients exhibited bile duct lesions and increased HLA-DR expression by the lining epithelium, there was no leukocyte infiltration of the bile duct lining. Thus, the role for Leu7$^+$ and CD8$^+$ cells within the infiltrates in GVHD bile duct lesions remains unclear.

3.1.3. Skin

Skin biopsies were examined immunohistologically from a group of 17 matched and 11 unmatched leukemic patients receiving bone marrow transplants. There was an increase in epidermal infiltrating CD4$^+$ and CD8$^+$ cells, some of which also expressed CD25 (IL-2 receptor) and HLA-DR antigens within lichenoid lesions of 14 patients with GVHD.[46] No B lymphocytes, HNK1$^+$ or CD10$^+$ cells were detected.

3.1.4. Rectum

When rectal biopsies were examined from eight leukemic patients with GVHD 2–28 weeks posttransplantation, there was an increase in CD8$^+$ cells both in the lamina propria and within the epithelium.[47] There was also no significant change in the number of Leu7$^+$ or UCHM1$^+$ (monocytes) leukocytes within the lamina propria and epithelial lining.

Thus, whereas two reports documented an increase in K562 killers in the blood of GVHD patients, an increase in Leu7$^+$ cells was present only in liver periportal infiltrates. Although Leu7$^+$ cells did not appear to contribute directly to bile duct lesions or to skin lichenoid and rectal lesions, they could secrete lymphokines and other cytokines could either recruit other cells or cause damage directly to epithelial or endothelial cells.

3.2. Mouse

When unirradiated 7- to 8-week-old $(CBA \times A)F1$ mice were injected with $5-7.5 \times 10^7$ A strain spleen and lymph node cells, there was a significant increase in the ability of GVHD F1 compared to control cells to kill YAC-1 lymphoma cells, peaking at day 3 after cell injection and falling below control values by day 15.[48] Lymph nodes exhibited elevated YAC-1 killers by day 5 that peaked on day 15. The thymus also had elevated YAC-1 killers from days 5–15 and returned to control levels by day 30 only to become markedly elevated on days 80–140 after GVH induction. Subsequent studies have implicated high YAC-1 killers with GVH-associated tissue damage, particularly in the thymus.[49] Depletion of mature YAC-1 killers with anti-aGM$_1$ prior to transplantation did not prevent GVH-associated splenomegaly and early partial immunosuppression of PFC response to SRBC but did prevent tissue damage to the thymus and severe persistent immunosuppression.[50] In addition, bg/bg lymphoid cells are deficient in YAC-1 killer cells but can readily induce GVH splenomegaly and partial immunosuppression without moderate-severe thymic lesions and persistent immunosuppression.[51] These studies imply a distinction between the donor cell type causing splenomegaly from that killing YAC-1 and causing tissue lesions.

GVHD is frequently associated with severe intestinal lesions.[5] When 5- to 7-week-old unirradiated $(CBA \times BALB/c)F1$ mice were injected with 6×10^7 CBA spleen cells, increased spleen cell YAC-1 killing by donor and host cells was present by day 7, and significant splenomegaly was present on day 21, when YAC-1 killing was still elevated.[52] Normal intestinal intraepithelial lymphocytes (IEL) morphologically resemble LGL but are thymus-dependent.[53] NE activity for YAC-1 by donor-derived IEL was markedly augmented in parallel with splenomegaly in these same $(CBA \times BALB/c)F1$ mice during GVHD.[54] The authors speculated that donor NE cells are recruited into the epithelial lining of the intestine during GVHD as a result of lymphokines secreted locally in the gut and may contribute to intestinal lesions.

Further evidence for the role of YAC-1 killer cells in GVHD was the ability of anti–asialo GM$_1$ (aGM$_1$) antibody *in vivo* to prevent lethal minor-determinant GVHD in irradiated LP/J mice.[55] The treatment was ineffective if applied to the donor or to the donor cells *in vitro* before GVHD induction. Even though there may be some effect on antigen-presenting cells by aGM$_1$ antibody,[56] the results of this study must be interpreted with caution, as aGM$_1$ has been shown to react not only with YAC-1 killers but also with alloreactive cytotoxic T lymphocytes,[57] which may develop an aGM_1 receptor during GVHD and become the putative effector cell.

The L-leucyl-L-leucine methyl ester (Leu-Leu-OMe) is selectively toxic for YAC-1 killers and cytotoxic T lymphocytes.[58] However, it spares alloantigen-induced proliferation and IL-2 production by $CD4^+$ T helper cells. Leu-Leu-OMe completely prevented GVHD across the major C57BL/6J(B6) into $(B6 \times DBA/2)F1$ combination. This finding supports the fact that a Leu-Leu-OMe-sensitive cell is necessary for GVHD to occur. However, further studies indicate that the GVHD cell is not the YAC-1 killer, because treatment of donor cells at a lower concentration of $125\mu M$ Leu-Leu-OME eliminated NK activity but did not prevent lethal GVHD.[58]

When 1×10^7 T cell-enriched B10.BR spleen and 1×10^7 T cell-depleted bone marrow cells were given to lethally irradiated and H-2k identical CBA mice, classical lethal GVHD ensued with 100% lethality by 21 days. When an ultrastructural, phenotypic, and genotypic analysis was performed of the cells found in association with necrotic epidermal cells, they were found to be of donor origin and to contain a small number of membrane-bound dense paranuclear granules and parallel tubular arrays classically found in NK cells.[59] These cells were also $aGM1^+$, $Thy1^+$ and appeared to be the same cells recently described by Romani et al.[60] The cells were not isolated from the skin, so no functional assays were performed.

Thus an aGM_1^+, Leu-Leu-OMe-sensitive cell has been implied by numerous studies to be involved in the effector phase of GVHD. Whether the same cell is involved in all the murine studies mentioned above will have to await future studies.

3.3. Rat

3.3.1. Lymphoid Tissue

We have been inducing acute GVHD in the rat by injecting $1-1.5 \times 10^6$ DA or Lewis (L) spleen and lymph node cells per gram body weight into unirradiated 6- to 13-week-old $(DA \times L)F1$ animals. The disease produced is very reproducible at this dose, with all of the animals dying by days 21–28.[6,8,9] When the level of ^{51}Cr release from a canine lymphoma and sarcoma as well as human rhabdomyosarcoma cell lines was analyzed sequentially during GVHD, cells lytic for these targets were present at 7 days in the spleen, lymph nodes, thymus, and PBL.[61] However, by 14 days, the spleen lytic activity was significantly below control levels and remained so until death. Subsequent studies using the conventional YAC-1 target cell essentially confirmed these results but found the time course to be accelerated in that elevated YAC-1 killing peaked on day 4 but was significantly below control levels by day 10 (Table II). When spleen cells from day 18 were mixed with those from control or day 4 GVHD rats, there

TABLE II
Level of YAC-1, DAUDI Killing, and Relative LGL in
Nonadherent Spleen Cells from GVHD and Control
(DA × LEW)F1 Rats

Animal group	YAC-1 (Percent ^{51}Cr)	DAUDI (Release[a])	Percent LGL
Control	32 ± 2	5 ± 1	3.1 ± 0.6
GVHD (days)			
4	51 ± 4[b]	9 ± 1[b]	6.4 ± 0.5[b]
7	36 ± 4	4 ± 0	5.7 ± 0.7[b]
10	17 ± 2[b]	5 ± 1	4.0 ± 0.3[b]
14	9 ± 2[b]	3 ± 1	2.0 ± 0.6
18–21	10 ± 1[b]	1 + 0	0.8 ± 0.1[b]

[a] The effector:target ratio was 50:1.
[b] Significantly different from controls at $p < 0.05$.

was no alteration in the level of cytotoxicity observed by control or day 4 cells alone. Thus there did not appear to be a suppressor cell for YAC-1 killing emerging during late GVHD. Differential cell counts revealed that there was a relative increase in LGL from GVH versus control spleens from days 2–10 (Table II).

Phenotypic and genotypic analysis of nonadherent cells harvested from PBL, spleen, and lymph nodes of GVHD and control rats demonstrates an increase in donor and host T cells expressing the rat equivalent of CD4 and CD8 phenotypes during the first week of acute GVHD. In addition, some donor non-B cells also express class II MHC antigens. After the first week there was a marked shift to host cells of the CD8, B, and null phenotype. Thus the cause of death was attributed to acute GVHD with elevated numbers of host B, null, and CD8$^+$ cells, with very few donor cells within their lymphoid tissue. Because the LGL number and activity were decreased, it is presumed that the host CD8$^+$ cells remaining were not morphologic LGL.

3.3.2. Liver

The liver in rat GVHD is significantly enlarged, and the periportal area and sinusoids are infiltrated with proliferating cells.[8] During the progression of acute GVHD induced in irradiated BN (RT1^n) rats with LEW (RT1^1) bone marrow, an increased concentration of LGL has been harvested via mechanical disaggregation from whole liver pieces.[62] In addition, the liver infiltrate has been further characterized in the rat with monoclonal antibodies to several leukocyte subtypes and demonstrates an increase in class II antigen-expressing phagocytes and T cell subtypes.[63]

Ultrastructurally, most of the cells in the infiltrate are large, nongranular lymphocytes (Fig. 1), but the sinusoids contained a significant number of granulated cells (Fig. 2).

A granular cell analogous to LGLs has been described ultrastructurally to be present in rat liver sinusoids and has been called the pit cell.[64] Because of the increased level of such granulated cells present in GVHD liver sinusoids, livers from exsanguinated GVHD and control rats were perfused through the portal vein with a balanced salt solution, and the perfusate was collected for morphologic, phenotypic, and functional analysis. Such a liver perfusion technique largely spared the periportal infiltrate and thus differs from the mechanically harvested cells. The results demonstrated an absolute 1.5–3 times increase in CD3 (Ox19), CD8, CD4, B, and aGM1$^+$ and LGL during the first week of GVHD. CD3, CD8, B, and aGM1$^+$ cells remained elevated on day 14, but only CD8, B, and aGM1$^+$ cells remained elevated on days 18–21 (Beilman, Stechschutte, Clancy, submitted). YAC-1 killing was increased on days 4–7 but decreased on days 14–21. Granule counts were also performed on LGL at intervals and found to be elevated 30% on day 7 and decreased to 50% of control levels by day 21. The data thus indicated that increased YAC-1 killing on day 7, and decreased killing on days 18–21 was a function of LGL granularity and not an expression of CD8 and aGM$_1$ phenotypes by cells within the infiltrate. Furthermore, whereas it appears that granulated cells do not contribute to the periportal infiltrate, they do contribute to the sinusoidal infiltrate. The function of the granulated cells in the liver sinusoidal cells are not known, but they may be important in removing donor allogeneic cells, as has recently been proposed by Rolstad *et al.*[65]

4. Role of IL-2 and IL-2 Receptor-Bearing Cells on NE Cells and GVHD Progression

Interleukin 2 is a powerful lymphokine secreted by CD4$^+$ cells and LGL.[66] Besides being critical for *in vitro* antigen-specific cytotoxic T (CTL) cell differentiation as well as maintenance of NK activity, IL-2 is necessary for the generation of lymphokine-activated killer cells (LAK), which are very efficient at killing a variety of NK-sensitive and -insensitive tumors.[67] Because IL-2 is produced in mixed lymphocyte cultures, it seems reasonable to examine the role of IL-2 administration and IL-2 receptor-bearing cells on NK cells during GVHD.

Studies with aplastic and leukemic patients after HLA-matched marrow transplantation demonstrated an impairment of CTL generation in patients with acute and chronic GVHD.[68] Addition of IL-2 to cultures from these patients restored CTL activity to normal levels in the acute GVHD but not the chronic GVHD group, probably because of excess suppressor cells in chronic GVHD patients.[14] Unfortunately, the levels of

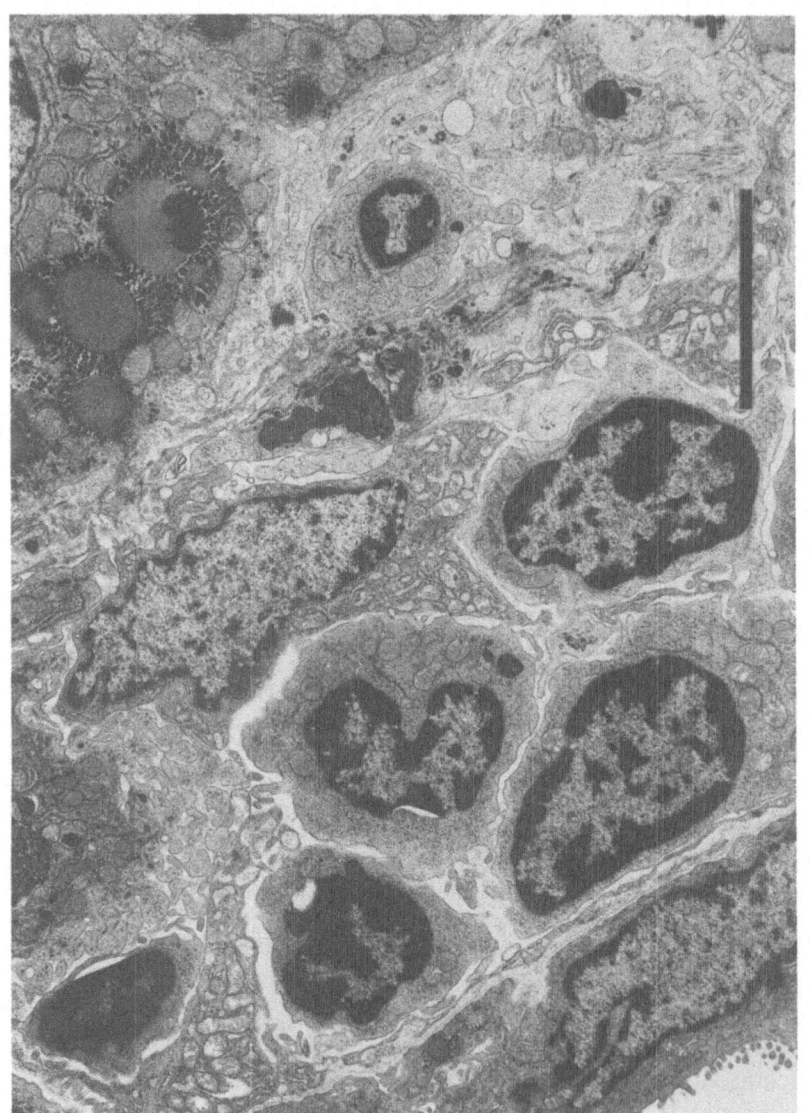

FIGURE 1. Periportal area from a day 7 GVHD rat. The bile duct epithelium is at lower left, and edge of hepatocyte at upper right. Most of the cells in the infiltrate are nongranular large lymphocytes. Scale bar: 5 μm.

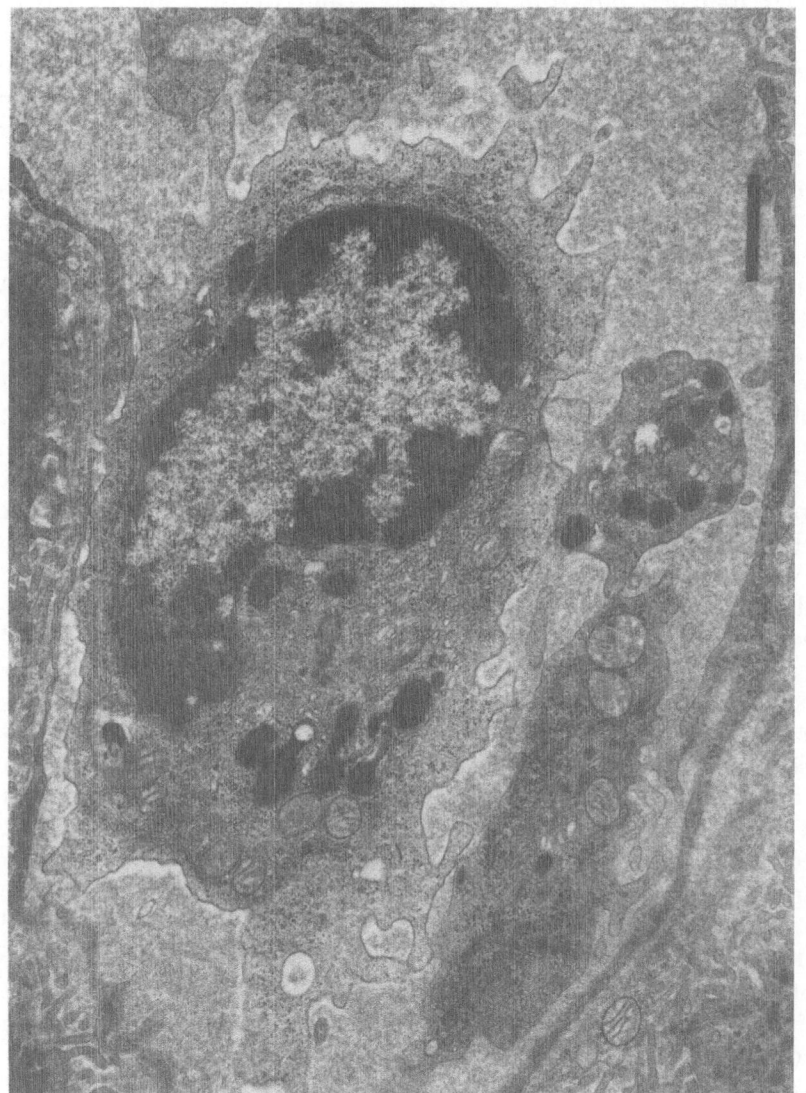

FIGURE 2. Liver sinusoid from a day 4 GVHD rat. The number and granule concentration of such a pit cell are increased at this stage. Such cells are very rare in the periportal areas. Scale bar: 1 μm.

IL-2 receptor-bearing cells, LAK cells, or NK cells were not evaluated in either GVHD group.

When C3H/Hej mice were lethally irradiated and transplanted with 5×10^6 H-2 matched but H-3, 7, 13, MLs, and Ly-10 disparate CBA/J bone marrow, few developed GVHD.[69] IL-2 production by recipient spleen cells after stimulation with Con A did not reach control levels until 2 months posttransplantation. When marrow recipients were given 1000 units of recombinant IL-2 (rIL-2) three times a week for four weeks, allospecific CTL could be readily generated *in vivo*, but there was no increase in GVHD within the IL-2-treated mismatched group. LAK and NK cell activities were not evaluated. The studies should be repeated on a group of animals where there is a higher incidence of GVHD.

Thus when CBA mice were lethally irradiated and given 10^7 H-2k identical but H1, 3, 7–9, 12, Tla, Mls, Ly1, and Ly2 disparate B10.BR marrow and spleen cells, 40% of these animals died of GVHD by 39 days.[70] When mice so injected received 2000 units of rIL-2 three times a week after cell injection, lethality was 100% by 39 days. Depletion of Thy 1.2 cells from donor B10.BR marrow blocked development of GVHD in either the rIL-2-treated or the nontreated group. Even though rIL-2 had no effect on recovery of marrow cellularity, the results clearly demonstrated that an IL-2 responsive Thy 1.2^+ population of cells was responsible for lethal GVHD in this model. The authors reasoned that IL-2 can be given to T cell–depleted allogeneic marrow graft recipients without exacerbating GVHD.

There have not been studies in which LAK and NK cell activity and IL-2 receptor (CD25) expression were monitored during GVHD ensuing in either the presence of absence of exogenous IL-2. Our studies in progress indicate that there is a 5-15% enhanced expression of Ox39 (CD25) by lymphoid cells during GVHD in the rat. This increase correlates with the level of autoradiographically assessed [^{3}H]thymidine-incorporating cells harvested from these tissues. In contrast, whereas early GVHD spleen cells express enhanced ability to kill YAC-1 targets, there was little alteration in their ability to kill LAK-sensitive DAUDI targets (Table II).[71] When animals with GVHD were injected daily with rIL-2, splenomegaly was approximately 1.5–2 times that of other GVHD rats, and the IL-2-injected animals died about 20–30% sooner. GVHD rIL-2 rats also demonstrated a 50–70% elevation in $CD25^+$ cells over the GVHD only group. Also noteworthy in this regard is the ability of an antibody to $CD25^+$ cells (ART18) to prevent a popliteal GVH reaction[72] and, when used in conjunction with cyclosporin A, to prevent GVHD.[73] In addition, antibody to IL-2 alone has been reported to prevent GVHD against minor HA antigens in mice.[74]

Thus rIL-2 and $CD25^+$ cells appear to be able to exacerbate GVHD in mice and rats. The genotype and phenotype of the cell being activated as well as the role of NK and LAK cells in GVHD are thus being explored.

5. Conclusions

Although there is no doubt that a foreign major or minor MHC-reactive T cell initiates GVHD, the role of non-allo-MHC NE cells in GVHD is still somewhat controversial. Whereas it is clear that some elevation of cells able to kill K562 or YAC-1 targets occurs in many human and rodent GVHD systems and may contribute to tissue lesions, such cells may also be the first to appear after non-GVHD-inducing marrow transplants. Some reports have presented morphological evidence that cells of the Leu7 phenotype occur in GVHD skin and gut lesions and that YAC-1 killers are responsible for murine thymic lesions. Nevertheless, their presence could be coincidental and result from factors secreted by alloactivated T cells.

The description of Leu11b⁺, non-K562 lytic, granulated cells after T cell-depleted marrow transplantation, which may help in B cell differentiation suggests a possible beneficial role for some Leu11b⁺ granulated cells. The apparent increase in severity of GVHD after rIL-2 injection and decrease in GVH after anti-CD25 injection supports the role for IL-2 and CD25⁺ cells in GVHD progression. Although it is highly likely that IL-2 would activate T cells and thereby accelerate the pathophysiologic cascade of GVHD effectors, it could also activate NE cells. It would therefore seem rational in any future studies of the pathogenesis of GVHD to examine the involvement of IL-2-responsive NE cells, their genotype, and phenotype. Because such cells may play an important role in preventing the reoccurrence of leukemia in marrow transplant recipients, their control may also be beneficial to the patient.

ACKNOWLEDGMENTS. This work was supported by grants AI-20677 and AI-23718 from the NIH. The author thanks Linda Fox for her help with the electron micrographs and Judy Maples for her careful typing of the manuscript.

References

1. Billingham, R. E., Defendi, V., Silvers, W. K., and Steinmuller, D., 1962, Quantitative studies on the induction of tolerance of skin homografts and on runt disease in neonatal rats, *JNCI* **28**:365–435.
2. Santos, G. W., Hess, A. D., and Vogelsang, G. B., 1985, Graft-versus-host reactions and disease, *Immunol. Rev.* **88**:169–192.
3. Storb, R., and Thomas, E. D., 1985, Graft-versus-host disease in dog and man: The Seattle experience, *Immunol. Rev.* **88**:215–238.
4. Korngold, R., and Sprent, J., 1985, Surface markers of T cells causing lethal graft-vs-host disease to class I vs class II H-2 differences, *J Immunol.* **135**:3004–3010.
5. Piquet, P. F., 1985 GVHR elicited by products of class I or class II loci of the MHC: Analysis of the response of mouse T lymphocytes to products of class I and class II loci of the MHC in correlation with GVHR-induced mortality, medullary aplasia, and enteropathy, *J. Immunol.* **135**:1637–1643.

6. Clancy, J. Jr., and Mauser, L., 1981, The absolute level of IgG Fi and C3 receptor-positive T, B, and null leukocytes within various lymphoid compartments during acute graft-versus-host disease in the adult rat, *Transplantation* **32**:401–408.

7. Stuart, S. P., Klein, R. M., and Clancy, J. Jr., 1987, Kinetics of mast cell, fibroblast, and epidermal cell poliferation during acute graft-versus-host disease in the neonatal rat, *J. Invest. Dermatol.* **88**:369–374.

8. Klein, R. M., Clancy, J., and Stuart, S., 1982, Acute lethal graft-versus-host disease stimulates cellular proliferation in the adult rat liver, *Cell Tissue Kinet.* **15**:651–660.

9. Clancy, J. Jr., Klein, R. M., and Weddle, S. L., 1981, Stimulation of cellular proliferation in the adult rat submandibular gland during acute graft-versus-host disease, *Transplantation* **31**:296–299.

10. Champlin, R. E., and Gale, R. P., 1984, Role of bone marrow transplantation in the treatment of hematologic malignancies and solid tumors: Critical review of syngeneic, antologous, and allogeneic transplants, *Cancer Treat. Rep.* **68**:145–161.

11. O'Reilly, R. J., 1983, Allogeneic bone marrow transplantation: Current status and future directions, *Blood* **62**:941–964.

12. Sullivan, K. M., Shulman, H. M., Storb, R., Weiden, P. L., Witherspoon, R. P., Schubert, M. M., Atkinson, K., and Thomas, E. D., 1981, Chronic graft-versus-host disease in 52 patients: Adverse natural course and successful treatment with combination immunosuppression, *Blood* **57**:267–276.

13. Tsoi, M.-S., Storb, R., Dobbs, S., Medill, L., and Thomas, E. D., 1980, Cell mediated immunity to non-HLA antigens of the host by donor lymphocytes in patients with chronic graft-versus-host disease, *J. Immunol.* **125**:2258–2262.

14. Maier, T., Holda, J. H., and Claman, H. N., 1985, Graft-vs-host reactions (GVHR) across minor murine histocompatibility barriers. II. Development of natural suppressor cell activity, *J. Immunol.* **135**:1644–1651.

15. Streilein, J. W., 1972, Pathologic lesions of GVH disease in hamsters: Antigenic target versus "innocent bystanders," *Prog. Exp. Tumor Res.* **16**:396–408.

16. Lopez, C., Kirkpatrick, D., Sorell, M., and O'Reilly, R. J., 1979, Association between pretransplant natural killer and graft-versus-host disease after stem cell transplantation, *Lancet* **2**:1103–1107.

17. Lotzova, E., and Cudkowicz, G., 1973, Assistance of irradiated F1 hybrid and allogeneic mice to bone marrow grafts of NZB donors, *J. Immunol.* **110**:791–800.

18. Bortin, M. M., Truitt, R. L., Rimm, A. A., and Bach, F. H., 1979, Graft versus leukemia reactivity induced by alloimmunization without augmentation of graft versus host reactivity, *Nature* **281**:490–491.

19. Lotzova, E., Savary, C. A., and Herberman, R. B., 1987, Induction of NK cell activity against fresh human leukemia in culture with interleukin 2, *J. Immunol.* **138**:2718–2727.

20. Livnat, S., Seigneuret, M., Storb, R., and Prentice, R. L., 1980, Analysis of cytotoxic effector cell function in patients with leukemia or aplastic anemia before and after marrow transplantation, *J. Immunol.* **124**:481–490.

21. Mizuno, S., Morishima, Y., Kodera, Y., Ohno, R., Yokomaku, S., Sao, H., and Yoshikawa, S., 1986, Gamma-interferon production capacity and T lymphocyte subpopulation after allogeneic bone marrow transplantation. *Transplantation* **41**:311–315.

22. Rooney, C. M., Wimperis, J. Z., Brenner, M. K., Patterson, B. J., Hoffbrand, A. V., and Prentice, H. G., 1986, Natural killer cell activity following T-cell depleted allogeneic bone marrow transplantation, *Br. J. Haematol.* **62**:413–420.

23. DeVillartay, J. P., LeDeist, F., Griscelli, C., and Fischer, A., 1987, Dissociation between onset of natural killer E-rosette forming cells and of T3-positive cells following HLA-mismatched T cell depleted bone marrow transplantation, *Clin. Exp. Immunol.* **67**:406–414.

24. Parreira, A., Smith, J., Hows, J. M., Smithers, S. A., Apperley, J., Rombos, Y., Gordon-Smith, E. C. and Catovsky, D., 1987, Immunological reconstitution after bone marrow transplant with Campath-1 treated bone marrow, *Clin. Exp. Immunol.* **67**:142–150.

25. Anderson, K. C., Ritz, J., Takvorian, T., Coral, F., Daley, H., Canellos, G. P., Schlossman, S. F., and Nadler, L. M., 1987, Hematologic engraftment and immune reconstitution posttransplantation with anti-B1 purged autologous bone marrow, *Blood* **69**:597–604.

26. Pattengale, P. K., Ramstedt, U., Gidlund, M., Orn, A., Axberg, I., and Wigzell, H., 1983, Natural killer activity in (C57BL/6×DAB/2)F1 hybrids undergoing acute and chronic graft-vs.-host reaction, *Eur. J. Immunol.* **13**:912–919.

27. Varkila, K., and Hurme, M., 1985, Natural killer (NK) cells and graft-versus-host disease (GVHD): No correlation between the NK cell levels and GVHD in the murine P-F1 model, *Immunology* **54**:121–126.

28. Ford, W. L., Burr, W., and Simonsen, M., 1970, A lymph node weight assay for the graft-versus-host activity of rat lymphoid cells. *Transplantation* **10**:258–266.

29. Brideau, R. J., Carter, P. B., McMaster, W. R., Mason, D. W., and Williams, A. F., 1980, Two subsets of rat T lymphocytes defined with monoclonal antibodies, *Eur. J. Immunol.* **10**:609–615.

30. Rolstad, B., Fossum, S., Hunt, S. V., and Ford, W. L., 1986, The host component of the popliteal lymph node graft-versus-host reaction. Selective representation of lymphocyte subsets and the requirement of alloantigenic incompatibility between donor cells and activated host B cells, *Scand. J. Immunol.* **23**:589–598.

31. Reynolds, C. W., Timonen, T., and Herberman, R. B., 1981, Natural killer (NK) cell activity in the rat. I. Isolation and characterization of the effector cells, *J. Immunol.* **127**:282–287.

32. Dokhelar, M.-C., Wiels, J., Lipinski, M., Tetaud, C., Devergie, A., Gluckman, E., and Tursz, T., 1981, Natural killer cell activity in human bone marrow recipients. Early reappearance of peripheral natural killer activity in graft-versus-host disease, *Transplantation* **31**:61–65.

33. Gratama, J. W., Lipovich-Oosterveer, M. A., Ronteltap, C., Sinnige, L. G. F., Van der Griend, R. J., and Bolhuis, R. L. H., 1985, Natural immunity and graft-versus-host disease, *Transplantation* **40**:256–260.

34. Favrot, M., Janossy, G., Tidman, N., Blacklock, H., Lopez, E., Bofill, M., Lambert, I., and Morgenstein, G., 1983, T cell regeneration after allogeneic bone marrow transplantation, *Clin. Exp. Immunol.* **54**:59–72.

35. Janossy, G., Prentice, H. G., Grob, J.-P., Ivory, K., Tidman, N., Grundy, J., Griffiths, P. D., and Hoffbrand, A. V., 1986, T lymphocyte regeneration after transplantation of T cell depleted allogeneic bone marrow, *Clin. Exp. Immunol.* **63**:577–586.

36. Lum, L. G., Seigneuret, M. C., Storb, R. F., Witherspoon, R. P., and Thomas, E. D., 1981, *In vitro* regulation of immunoglobulin synthesis after marrow transplantation. I. T-cell and B-cell deficiencies in patient with and without graft-versus-host disease, *Blood* **58**:431–438.

37. Dosch, H. M., and Gelfand, E. W., 1981, Failure of T and B cell cooperation during graft-versus-host disease, *Transplantation* **31**:48–50.

38. Matsue, K., Lum, L. G., Witherspoon, R. P., and Storb, R., 1987, Proliferative and differentiative response of B cells from human marrow graft recipients to T cell-derived factors, *Blood* **69**:308–315.

39. Brenner, M. K., Reittie, J. E., Grob, J.-P., Wimperis, J. Z., Stephens, S., Patterson, J., Hoffbrand, A. V., and Prentice, H. G., 1986, The contribution of large granular lymphocytes to B cell activation and differentation after T-cell-depleted allogeneic bone marrow transplantation, *Transplantation* **42**:257–261.

40. Brenner, M. K., Vyakarnam, A., Reittie, J. E., Wimperis, J. Z., Grob, J. P., Hoffbrand, A. V., and Prentice, H. G., 1987, Human large granular lymphocytes induce immunoglobulin synthesis after bone marrow transplantation, *Eur. J. Immunol.* **17**:43–47.

41. Weiden, P. L., Flournoy, N., Thomas, E. D., Prentice, R., Fefer, A., Buckner, C. D., and Storb, R., 1979, Antileukemic effect of graft-versus-host disease in human recipients of allogeneic-marrow grafts, *N. Engl. J. Med.* **300**:1068–1073.

42. Hercend, T., Takvorian, T., Nowill, A., Tantravahi, R., Moingeon, P., Anderson, K. C., Ytheir, A., and Ritz, J., 1986, Characterization of natural killer cells with antileukemia activity following allogeneic bone marrow transplantation, *Blood* **67**:722–728.

43. Slavin, R., and Santos, G., 1973, The graft versus host reaction in man after bone marrow transplantation: Pathology, pathogenesis, clinical features, and implication, *Clin. Immunol. Immunopathol.* **1**:472–498.

44. Bernnan, D., Gisselbrecht, C., Devergie, A., Feldman, G., Gluckman, E., Martz, M., and Bviron, M., 1980, Histological and ultrastructural appearance of the liver during graft-versus-host disease complicating bone marrow transplantation, *Transplantation* **29**:236–244.

45. Dilly, S. A., and Sloane, J. R., 1985, An immunohistological study of human hepatic graft-versus-host disease, *Clin. Exp. Immunol.* **62**:545–553.

46. Sloane, J. P., Thomas, J. A., Imrie, S. F., Easton, D. F., and Powles, R. L., 1984, Morphological and immunohistological changes in the skin in allogeneic bone marrow recipients, *J. Clin. Pathol.* **37**:919–930.

47. Dilly, S. A., and Sloane, J. P., 1987, Changes in rectal leukocytes after allogeneic bone marrow transplantation, *Clin. Exp. Immunol.* **67**:151–158.

48. Roy, C., Ghayur, T., Kongshavn, P. A. L., and Lapp, W. S., 1982, Natural killer activity by spleen, lymph node, and thymus cells during graft-versus-host reaction, *Transplantation* **34**:144–146.

49. Ghayur, T., Seemayer, T. A., and Lapp, W. S., 1987, Kinetics of natural killer (NK) cell cytotoxicity during the graft-versus-host reaction. Relationship between NK cell activity, T and B activity, and development of histopathological alterations, *Transplantation* **44**:254–260.

50. Ghayur, T., Seemayer, T. A., and Lapp, W. S., 1988, Prevention of marine graft-versus-host disease by inducing and eliminating ASGM1+ cells of donor origin, *Transplantation* **45**:586–590.

51. Ghayur, T., Seemayer, T. A., Kongshavn, P. A. L., Gartner, J. G., and Lapp, W. S., 1987, Graft-versus-host (GVH) reactions in the beige mouse: An investigation of the role of host and donor natural killer (NK) cells in the pathogenesis of GVH disease, *Transplantation* **44**:261–267.

52. Borland, A., Mowat, A. I., and Parrott, D. M. V., 1983, Augmentation of intestinal and peripheral natural killer cell activity during graft-versus-host reaction in mice, *Transplantation* **36**:513–519.

53. Guy-Grand, D., Griscelli, C., and Vassalli, P., 1978, The moust gut T lymphocyte, a novel type of T cell, *J. Exp. Med.* **148**:1661–1677.

54. Mowat, A., 1986, Evidence that Ia+ bone-marrow-derived cells are stimulus for the intestinal phase of the murine graft-versus-host reaction, *Transplantation* **42**:141–144.

55. Charley, M. R., Mikhael, A., Bennett, M., Gilliam, J. N., and Sontheimer, R. D., 1983, Prevention of lethal, minor-determinate graft-versus-host disease in mice by the *in vivo* administration of anti–asialo GMI, *J. Immunol.* **131**:2101–2103.

56. Charley, M. R., Mikhael, A., Hoot, G., Hackett, J., and Bennett, M., 1985, Studies addressing the mechanism of anti–asialo GM₁ prevention of graft-versus-host disease due to minor histocompatibility antigenic differences, *J. Invest. Dermatol.* **85**:121S–123S.

57. Ting, C.-C., Bluestone, J. A., Hargrove, M. E., and Loh, N.-N., 1986, Expression and function of asialo GM₁ in alloreactive cytotoxic T lymphocytes, *J. Immunol.* **137**:2100–2106.

58. Thiele, D. L., Charley, M. R., Calomeni, J. A., and Lipsky, P. E., 1987, Lethal graft-versus-host disease across major histocompatibility barriers: requirement for leucyl-leucine methyl ester sensitive cytotoxic T cells. *J. Immunol.* **138**:51–57.

59. Guillen, R. J., Ferrara, J., Hancock, W. W., Messadi, D., Fonferko, E., Burakoff, S. J., and Murphy, G. F., 1986, Acute cutaneous graft-versus-host disease to minor histocompatibility antigens in murine model, *Lab. Invest.* **55**:35–42.

60. Romani, N., Stingl, G., Tschachler, E., Witmer, M. D., Steinman, R. M., Shevch, E. M.,

and Schuler, G., 1985, The Thy-1-bearing cell of murine epidermis. A distinctive leukocyte perhaps related to natural killer cells, *J. Exp. Med.* **161**:1368–1383.

61. Clancy, J. Jr., Mauser, L., and Chapman, A. L., 1983, Level and temporal pattern of naturally cytolytic cells during acute graft-versus-host disease (GVHD) in the rat, *Cell. Immunol.* **79**:1–10.
62. Renkonen, R., and Hayry, P., 1984, Bone marrow transplantation in the rat. I. Histologic correlations and quantitation of cellular infiltrates in acute graft-versus-host disease, *Am. J. Pathol.* **117**:462–470.
63. Stet, R. J. M., Thomas, C., Koudstaal, J., Hardonk, M. J., Hulstaert, C. E., and Nieuwenhuis, P., 1986, Graft-versus-host disease in the rat: Cellular changes and major histocompatibility complex antigen expression in the liver, *Scand. J. Immunol.* **23**:81–89.
64. Wisse, E., Van't Noordende, J. M., Van der Menlen, J., and Deams, W. T., 1976, The pit cell: Description of a new cell occurring in rat liver sinusoids and peripheral blood, *Cell Tissue Res.* **173**:423–435.
65. Rolstad, B., Fossum, S., Bazin, H., Kimber, I., Marshall, J., Sparshott, S. M., and Ford, W. L., 1985, The rapid rejection of allogeneic lymphocytes by a non-adaptive, cell-mediated mechanism (NK activity), *Immunology* **54**:127–138.
66. Kasahara, T., Djeu, J. Y., Dougherty, S. F., and Oppenheim, J. J., 1983, Capacity of human large granular lymphocytes (LGL) to produce multiple lymphokines: Interleukin 2, interferon, and colony stimulating factor, *J. Immunol.* **131**:2379–2385.
67. Grimm, E. A., Mazumder, A., Zhang, H. Z., and Rosenberg, S. A., 1982, Lymphokine-activated killer cell phenomenon. Lysis of natural killer–resistant fresh solid tumor cells by interleukin 2–activated antologous human peripheral blood lymphocytes, *J. Exp. Med.* **155**:1823–1841.
68. Mori, T., Tsoi, M.-S., Gillis, S., Santos, E., Thomas, E. D., and Storb, R., 1983, Cellular interactions in marrow-grafted patients. I. Impairment of cell-mediated lympholysis associated with graft-vs-host disease and the effect of interleukin 2, *J. Immunol.* **130**:712–716.
69. Merluzzi, V. J., Welte, K., Last-Barney, K., Mertelsmann, R., Souza, L., Savage, D. M., Quinn, D., and O'Reilly, R. J., 1985, Production and response to interleukin 2 *in vitro* and *in vivo* after bone marrow transplantation in mice, *J. Immunol.* **134**:2426–2430.
70. Malkovsky, M., Brenner, M. K., Hunt, R., Rastan, S., Dore, C., Asherson, G. L., Prentice, H. G., and Medawar, P. B., 1986, T-cell depletion of allogeneic bone marrow prevents acceleration of graft-versus-host disease induced by exogenous interleukin 2, *Cell Immunol.* **103**:476–480.
71. Clancy, J. Jr., Goral, J., Triner, J., and Ellis, T., 1987, Increased activation of YAC-1 killer (natural killers: NK) versus Daudi killers (lymphokine activated killers: LAK) by spleen cells from control and acute graft versus host disease (AGVHD) rats by recombinant IL-2 (rIL-2), *Anat. Rec.* **218**:25A.
72. Volk, H.-D., Brocke, S., Osawa, H., and Diamantstein, T., 1986, Suppression of the local graft-vs.-host reaction in rats by treatment with a monoclonal antibody specific for the interleukin 2 receptor, *Eur. J. Immunol.* **16**:1309–1312.
73. Diamanstein, T., Volk, H.-D., Tilney, N. L., and Kupiec-Weglinski, J., 1986, Specific immunosuppressive therapy by monoclonal anti-IL-2 receptor antibody and its synergistic action with cyclosporin, *Immunobiology* **172**:391–399.
74. Ferrara, J. L. M., Marion, A., McIntyre, J. F., Murphy, G. F., and Burakoff, S. J., 1986, Amelioration of acute graft vs host disease due to minor histocompatibility antigens by *in vivo* administration of anti-interleukin 2 receptor antibody, *J. Immunol.* **137**:1874–1877.

17

The Natural Immune System and Hematocytopenias

WING C. CHAN and ELLIOTT F. WINTON

1. Introduction

Major advances have been made in the past two decades in our understanding of cellular differentiation and proliferation in the normal immune and hematopoietic systems. These advances have resulted from the application of continually improving technologies that include clonogenic culture of hematopoietic cells, cell separation techniques, the production of monoclonal antibodies that recognize differentiation-associated cell surface antigens, assays of immunocyte function, chromosomal and DNA analysis methods, and the recent purification and molecular cloning of major proteins that regulate the immunohematopoietic system. To complement and challenge our evolving understanding of normal immune and blood cell formation a variety of significant clinical problems with quantitative deficiencies of red cells, platelets, or granulocytes exists (collectively termed hematocytopenias). Although empiric therapy has been successful for some of these clinical problems, there remains a need for rational therapy which in turn requires an understanding of the mechanisms of pathogenesis in these disorders. In the introductory sections that follow, we will briefly review the current understanding of the structure, function, and humoral and cellular interactions involved in normal hematopoiesis—information that suggests a regulatory role for immunocytes in blood cell formation—and present an overview of mechanisms of path-

WING C. CHAN • Department of Pathology and Laboratory Medicine, Emory University School of Medicine, Atlanta, Georgia 30322.   ELLIOTT F. WINTON • Division of Hematology and Oncology, Department of Medicine, Emory University School of Medicine, Atlanta, Georgia 30322.

ogenesis of the hematocytopenias. Following the introductory section we will present clinical and laboratory data that support the concept that cellular immunity is important in the development of severe aplastic anemia, pure red cell aplasia, and chronic neutropenia, with emphasis on the role of large granular lymphocytes (LGL). In the final sections we will review the limitations of current methodologies for elucidating cellular mechanisms in normal and abnormal hematopoiesis and provide a perspective for the direction of future research.

1.1. Normal Hematopoiesis: Structure, Function, and Regulation

The cells of the hematopoietic system are arranged in a hierarchy such that a great many differentiated blood cells are derived from a lesser number of progenitor cells, which in turn are derived from an even lesser number of stem cells. The functional features that distinguish the stem cell include an unlimited capacity for self-renewal (i.e., proliferation without differentiation) in addition to multipotent differentiative capabilities that include all three myeloid lineages (granulocytic, erythrocytic, megakaryocytic) as well as the T and B immunocyte lineages.

The direct progeny of stem cells, the progenitor cells, are functionally defined by their ability to proliferate and differentiate *in vitro*, forming discrete colonies or clones of cells in semisolid media in the presence of various colony-stimulating factors (CSF) with or without erythropoietin (EPO). The progenitor cell that originates the *in vitro* colony is defined by the differentiated cells present in the colony after clonogenic culture. The types of colony-forming cells or units (CFU) present in normal marrow include: multilineage CFU (CFU $_{GEMM}$) resulting in colonies composed of erythrocytic and nonerythrocytic cells such as neutrophilic granulocytes, macrophages, and megakaryocytes; macrophage CFU (CFU_M) or neutrophilic granulocyte CFU (CFU_G) with colonies composed of only macrophages or neutrophilic granulocytes, respectively; CFU resulting in mixed granulocyte macrophage colonies (CFU_{GM}); and erythroid colony-forming cells that result in bursts of multiple closely aggregated erythroid colonies (BFU_E) or single erythroid colonies (CFU_E). The major murine and human hematopoietic CSF have been purified and recently cloned (see ref. 1 for review). These factors include granulocyte macrophage colony-stimulating factor (GM-CSF),[2] granulocyte CSF (G-CSF),[3,4] macrophage CSF (M-CSF),[5] and multilineage CSF (multi-CSF or IL-3).[6] Both multi-CSF and GM-CSF enhance erythroid burst formation and are the probable source of previously described burst-promoting activities (BPA).

As cells differentiate from stem cells to progenitor cells to mature blood cells, there is a decrease in proliferative potential associated with an ordered synthesis of a variety of proteins that characterize the more differentiated cell. The proteins include those that functionally distinguish the cell such as hemoglobin in erythrocytes, myeloperoxidase or lactoferrin in granulocytes, and a number of cell surface proteins. The cell surface proteins, for example, include class II HLA antigens and MY9 antigen, which appear on all hematopoietic progenitor cells, and various lineage-specific proteins such as MY8 that appear only on cells of granulocyte macrophage lineage (see ref. 7 for review).

Multiple cell types appear involved in the regulation of hematopoiesis. Current data indicate that endothelial cells and fibroblasts produce M-, G-, and GM-CSF[8,9]; macrophages produce G- and M-CSF[10] and activated T cells produce GM-CSF and multi-CSF.[6,11,12] With the possible exception of M-CSF, these factors are not constitutively produced by these cells but are produced after stimulation. For example, fibroblasts have recently been shown to produce GM-CSF after exposure to IL-1,[8] and monocytes exposed to rGM-CSF produce M-CSF.[10] Negative feedback loops to down-regulate stimulated blood cell proliferation have been postulated, but at present there is little consensus on this matter.

1.2. T Cells, NK Cells, and Normal Hematopoiesis

T cells have an enhancing effect on growth of all cell lineages in *in vitro* hematopoiesis,[11,13,14] probably relating to their production of GM-CSF and multi-CSF noted above. Production of factors by T cells requires activation, and it is of interest that both activated T cells and progenitor cells express class II major histocompatibility complex antigens. Related to this is the observation that T cell modulation of BFU_E growth may be DR-restricted,[15] suggesting that direct cell–cell cooperation between T cells and hematopoietic cells is required.

Subsets of T cells have been shown to inhibit normal *in vitro* granulopoiesis. E-rosette (ER)$^+$ cells have been separated into those with and those without receptors for the F_c portion of IgG (i.e., Tγ cells). Spitzer and Verma[16] showed that bone marrow Tγ cells can inhibit CFU_{GM}. On the other hand, Bacigalupo *et al.*[17] have shown that Tγ cells could not inhibit CFU_{GM} unless they have been stimulated with PWM. Perhaps, the isolation procedures (E- and EA-rosetting) *per se*, are able to activate Tγ cells under some experimental conditions. It should also be noted that the Tγ cells contain both bona fide T cells and NK cells. The exact contribution of each cannot be determined from the reports.

The effect of NK cells has been studied, but the population of cells examined by different investigators may be different. In some studies, low-density fractions in a Percoll gradient were used; other studies use HNK1$^+$ or B73.1$^+$ cells. In general, a suppressive effect on bone marrow granulocytes and erythroid CFU was observed[18,19] despite the fact that secretion of CSF could be demonstrated from CD3-depleted light-density

Percoll fractions. Herrmann et al.[20] reported an interesting approach in using different clones of NK cells to study the effect on different CFU. Enriched fractions of colony-forming cells were used to minimize the effect of other cells in the culture. The NK clones showed different patterns and potency of inhibition of colony-forming units.

In some of the experiments discussed above, conditioned medium was used instead of the cell population under study. In most cases, the conditioned medium was able to reproduce the effects of the cell population under study. Crude conditioned medium contains numerous factors, and it is difficult to determine which factors were responsible for the observed effect. Gamma interferon (INFγ) has been implicated as a suppressive factor by the presence of INFγ in the conditioned medium and the ability of anti-INFγ to partially reverse the inhibitory effect.[18,20] The addition of INFγ and lymphotoxin has been shown to have potent inhibitory effects not observed with either alone, suggesting that INFγ may act synergistically with other lymphokines secreted by hematopoietic suppressor cells.[19,21] The potential role of NK cells in normal hematopoiesis is presented in detail in Chapter 5.

1.3. Pathogenic Mechanisms in Hematocytopenias

Although there are multiple, well-defined pathogenic mechanisms that result in hematocytopenias, there are many more that wait to be elucidated. Cytopenias may result from an acquired disorder in cellular DNA, defects in regulatory cells or humoral regulatory factors, and either humoral or cellular autoimmunity. The target for the disorder may include any of the cells in the hierarchy from stem cell to the more mature cells or the regulatory cells or factors themselves.

Myelodysplastic syndromes with acquired clonal chromosomal abnormalities are an example of cytopenias that often result from disordered stem cell DNA. Recently, the genes for GM-CSF, IL-3, and the receptor for M-CSF (fms oncogene product) have been localized to the long arm of chromosome 5.[22] With the clinical definition of a myelodysplastic syndrome associated with the deletion of the long arm of chromosome 5 (the 5q-syndrome[23]), a molecular genetic explanation for the cytopenias seems close at hand.

Defective EPO production in chronic renal failure largely accounts for the anemia of that disorder as recently demonstrated by successful clinical application of cloned human EPO to correct the anemia in these patients.[24] One suspects that similar deficiencies in the production of CSFs or IL-3 may be defined as these molecules become available for clinical use. An IgG serum inhibitor of EPO has been described,[25] and similar mechanisms involving the CSFs, although not recognized, seem inevitable.

Autoantibodies directed at specific stages of erythrocyte, granulocyte, and megakaryocyte production are well documented. These include autoantibodies against BFU_E or CFU_E seen in pure red cell aplasia[26] and Coombs-positive hemolytic anemias, similar progenitor cell[27] or mature neutrophil-directed autoantibodies[28] in neutropenic disorders, and megakaryocyte progenitor[29] or platelet-directed antibodies[30] in amegakaryocytic thrombocytopenia or ITP, respectively. Evidence for cell-mediated autoimmunity directed at various cells of hematopoietic lineage is discussed in sections 2 through 6 below.

2. Cellular Immunity and Severe Aplastic Anemia

2.1. Clinical Observations

Severe aplastic anemia (SAA) is a disease characterized by marked failure of hematopoiesis associated with a hypoplastic bone marrow (for review see ref. 31). Often all three hematopoietic lineages are affected with severe neutropenia, thrombocytopenia, and anemia, although in some patients one or more lineages may be relatively spared. Although the pathogenic mechanisms involved in SAA are diverse and poorly understood, there are both clinical and experimental observations that support the concept that cellular immunity plays a role in the cytopenias in some of these patients.

One of the most convincing observations regarding the role of cellular immunity in SAA comes from the syngeneic bone marrow transplant experience. Although some patients with SAA transplanted with marrow from an identical twin accept the graft, others fail to engraft unless they are immunosuppressed with high-dose cyclophosphamide prior to marrow infusion.[32–34] It is presumed that in those cases requiring immunosuppression, the cyclophosphamide eliminated hematopoietic suppressor cells that had prevented engraftment.

Perturbation of the immune system with xeonantibodies to lymphocytes or high-dose corticosteroid therapy is also effective in some patients with SAA. For example, Speck and Gluckman[35] observed that of 29 patients with severe aplastic anemia treated with antilymphocyte/thymocyte globulin (ATG) followed by allogeneic bone marrow transplantation (BMT), 12 showed significant hematopoietic reconstitution without any evidence of engraftment. In the first prospective randomized trial with ATG in SAA, patients were assigned to receive either ATG or supportive treatment alone.[36] Eleven out of 21 patients receiving ATG treatment responded, although the hematologic recovery was often incomplete. Patients who did not receive ATG did not improve. This study and others[37,38] confirm the efficacy of ATG in the treatment of some patients with aplastic anemia. Other investigators have demonstrated responses in approximately half of SAA patients treated with high-dose methylprednisolone with or without ATG.[39-41]

Several investigators have examined possible imbalances in subsets of lymphocytes in patients with SAA. For example, Zoumbos et al.[42] reported a decrease in the CD4:CD8 ratio in patients with SAA secondary to both a decrease in CD4+ cells and an increase in CD8+ cells. Using two-color flow microfluorometric analysis, these investigators found an increased percentage of activated CD8+ cells bearing HLA-DR and IL-2 receptors as detected by anti-Tac in 12 patients with SAA.[43] The cells bearing IL-2 receptors were capable of producing INF and could suppress CFU$_{GM}$ in vitro (vide infra). The authors suggested that activated CD8+ cells may have a role in the pathogenesis of the bone marrow failure. These investigators did not find an increase in NK cells as defined by the presence of CD16 or Leu-7 surface antigens. Other investigators using single-color microfluorometry have not demonstrated significant T cells subset imbalances in patients with SAA.[44]

2.2. Laboratory Observations Regarding Pathogenesis

Although clinical observations had suggested that cellular pathogenic mechanism for SAA was operative in some patients, it was not until the mid 1970s that supporting in vitro data began to appear. In 1976, the group at Sloan-Kettering Institute demonstrated that poor CFU$_{GM}$ growth from the marrow of a patient with SAA could be overcome by the removal of lymphocytes from the marrow.[45,46] Furthermore, the SAA marrow inhibited CFU$_{GM}$ growth when cocultured with normal marrow.

In 1977, Hoffman et al.[47] demonstrated that the peripheral blood lymphocytes from five of seven patients with aplastic anemia markedly inhibited erythroid colony formation of bone marrow from normal individuals. Sera from these patients did not have any inhibitory activity. PBL from normal and multitransfused controls did not inhibit erythroid colony growth.

Studies by Singer et al.[48] suggested the probability that some of the observed inhibitory effects of SAA bone marrow and/or peripheral blood lymphocytes on progenitor cell growth were secondary to alloimmunization related to the multiple prior blood transfusions received by the patients. Using a completely autologous in vitro system to circumvent the concern about alloimmunization, Torok-Storb et al.[49] demonstrated that removal of T cells from peripheral blood permitted BFU$_E$ growth in eight of 32 patients with SAA and that adding back T cells inhibited BFU$_E$ growth in six of these patients.

Bacigalupo et al.[50] studied the effects of T cells (ER+ cells) and adherent cells from the bone marrow of seven patients with severe aplastic anemia and demonstrated suppression of autologous and allogeneic CFU_{GM}. The suppressive activity appeared to reside within the Tγ fraction, and adherent cells did enhance the T cell-mediated CFU_{GM} suppression. Tγ cells from normal bone marrow did not suppress, whereas Tγ cells from SAA patients who had not received any transfusions had suppressive activity. Suda et al.[51] reported that peripheral blood mononuclear cells from nine of 20 patients with SAA showed a mild to moderate degree of suppression of CFU_{GM} when cocultured with normal bone marrow. This suppression was reversed by carbonyl iron treatment of the peripheral blood cells prior to culture, suggesting that adherent cells may be important effector cells or an accessory cell for another effector cell. The difference between this report and the previously cited study[50] may be related to the source of effector cells studied (i.e., bone marrow vs. peripheral blood) or related to the vigor of the monocyte depletion process.

INFs have been shown to inhibit hematopoietic colony growth in vitro[52-55] and Zoumbos et al.[56] have proposed that INFγ may in fact be a mediator of marrow suppression in patients with SAA. In support of this hypothesis, they have observed that INFγ production by lectin-stimulated peripheral blood mononuclear cells from patients was significantly higher than that of control cells, and the spontaneous in vitro production of INF was elevated in over half of the patients.[56] In addition, whereas circulating INF was not detectable in normal individuals, it was detected in 10 of 24 patients with SAA. In a more recent report,[43] these investigators noted that 12 patients with SAA had a striking increase in peripheral blood mononuclear cells bearing the CD8 and DR antigens, and that in five patients tested, these activated cells were also positive for Tac antigen. Furthermore, when Tac$^+$ and Tac$^-$ cells were separated by a cell sorter, only Tac$^+$ cells were shown to be a source of INF and capable of inhibiting hematopoietic colony formation when cultured with normal marrow cells. The pathogenic role of INFγ has recently been questioned by Torok-Strob et al.[57] who failed to find elevated INFγ in the sera of 50 patients with SAA using a recently developed solid-phase radioimmunoassay specific for biologically active INFγ.[58] In addition, although detectable levels of spontaneously produced INFγ were found in the supernatants of peripheral blood mononuclear cells of 18 of 50 patients, none of the supernatants reduced normal hematopoietic colony growth. There was also no correlation between the number of activated T cells or NK cells in the peripheral blood mononuclear cells and the presence of INFγ in the supernatants. More work is clearly needed to resolve the potential role of INFs or other lymphokines in the pathogenesis of the SAA.

In summary, there is well-documented evidence incriminating the cellular immune system in the pathogenesis of at least some cases of aplastic anemia. The cell type implicated in most reports is an E rosette–forming cell (CD2$^+$). In reports where T cell subsets were studied, Tγ cells and CD8$^+$ cells appeared to contain the suppressor population. Although

NK cells have been often implicated to one degree or another as a likely mediator of cellular inhibition of hematopoiesis, there are no definitive data to prove a causal relationship.

3. Cellular Immunity and Pure Red Cell Aplasia

3.1. Clinical Observations

Pure red aplasia (PRCA) is characterized by isolated failure of the marrow's production of red blood cells. There is severe reticulocytopenia in the peripheral blood and a marked but selective decrease in erythroid precursors in the bone marrow with less than 0.5% polychromatophilic or orthochromatophilic normoblasts, with or without 3–4% basophilic normoblasts and pronormoblasts.[59] Thrombocytopoiesis and granulocytopoiesis are unaffected. In the majority of cases, autoimmunity plays a role through either a humoral or a cell-mediated pathogenic mechanism.

PRCA is seen both as an isolated disorder, related to certain drugs,[60] and in combination with a variety of proliferative disorders including thymoma,[59,61–63] B or T cell chronic lymphocytic leukemia,[64–72] myelodysplastic syndrome,[59] and lymphoma.[73,74] The exact incidence of thymoma in PRCA is unknown, but it ranges from less than 10% up to 50% depending on the series.[59,62,63] Although PRCA occurs in only 5% of patients with thymoma, it is highly associated with hypogammaglobulinemia and thymoma. Removal of the thymoma results in remission in approximately 30% of the cases.[75]

The occurrence of PRCA with CLL is rare, and in a recent review only 27 cases were identified in the English language literature.[67] In some cases, the predominant cell type is a typical mature B cell[65–67]; in others it is a T cell.[68–72,76–78] In some of the T cell CLLs, the T cells have been further characterized as Tγ cells or LGL.[69–72,76–78]

The pathogenic mechanism of erythrocytopenia in cases of PRCA appears equally divided between humoral and cellullar mediation. Patients with thymoma, particularly with panhypogammaglobulinemia, are less likely to have serum inhibitors of erythropoiesis but may have cell-mediated suppression.[79]

The majority of cases of PRCA respond to some sort of immunomanipulative therapy. In one large series, 66% of 37 patients responded to a variety of immunosuppressive treatment regimens including corticosteroids, cyclophosphamide, or azathioprine, the last used with or without corticosteroids.[59] More recently, ATG has shown efficacy in treating patients refractory to corticosteroids or cyclophosphamide.[80,81] Treatment of patients with B cell CLL associated with PRCA with alkylating agent and/or corticosteroid therapy usually results in remission of the erythroid aplasia.

3.2. Laboratory Observations Regarding Pathogenesis

Marrow cells from patients with PRCA usually show normal or above-normal erythropoietin-dependent proliferation and differentiation *in vitro*. This observation permitted Krantz *et al.* to demonstrate that plasma from some patients with PRCA inhibited autologous *in vitro* erythropoiesis as quantified by an erythropoietin-dependent heme synthesis assay[82] and to identify the inhibitor as IgG.[83] It was subsequently shown that plasma from some patients with PRCA results in complement-dependent lysis of erythroblasts[83] and in other patients complement-independent inhibition of BFU_E and CFU_E.[26] In one well-studied case, patient IgG was demonstrated to specifically inhibit BFU_E and not CFU_{GM}, and plasma exchange led to long-term remission.[84]

In a number of cases of PRCA, inhibition of *in vitro* erythropoiesis by T cells has been demonstrated and has usually been associated with either B cell or T cell CLL. In 1978, Hoffman *et al.*[68] showed marked inhibition of CFU_E growth by peripheral blood and bone marrow-derived T cells from patients with T cell CLL. Based on the morphologic description, it is not possible to conclude that these were LGL. The inhibiting effect of these T cells was revealed using both autologus and normal marrow cells. Nagasawa *et al.*[70] reported a case of T-CLL with PRCA and hypogammaglobulinemia. The T cells had receptors for the F_c portion of IgG and were LGL by electron microscopic criteria. In coculture experiments, peripheral-blood T cells from this patient inhibited CFU_E formation by marrow cells from a normal donor and immunoglobulin production by normal B cells. The cells were also shown to have ADCC. Other reports of PRCA in association with T cell CLL (or in patients with expanded subpopulations of T cells) have followed and are discussed further below (section 5) and summarized in Table I. In each instance the predominant cell appeared to be an LGL based on one criterion or another including morphology, presence of F_c receptor, or functional assays.

Patients with B cell CLL and PRCA, while not having serum inhibitors to erythropoiesis, appear to have T cell-based pathogenesis for the erythrocytopenia. Mangan *et al.* demonstrated that T cells from four patients with B-CLL and PRCA had reduced BPA production compared to T cells from normals and three patients with idiopathic PRCA.[85] In a subsequent study these investigators reported bone marrow culture studies of two patients with B-CLL PRCA who had excess bone marrow $T\gamma$ cells. Removal of $T\gamma$ cells from the marrow by E-rosetting techniques resulted in a 10-fold increase in CFU_E. In contrast, BFU_E and CFU_{GM} growth was not effected by the $T\gamma$ cells.[65] In a larger study including 30 patients with B cell CLL ranging from early to advanced disease (Rai stages 0 through IV), Mangan and D'Alessandro[86] noted that the number of $T\gamma$ cells infiltrating the marrow increased from 3 to 20 times normal and were highest in the nine patients of the series who had red cell aplasia or

TABLE I
Cases of PRCA with Possible LGL

| Case | Assoc. disease | Phenotype | LGL | NK | ADCC | BFU$_E$/CFU$_E$ Inhibition | | | Ref. |
| | | | | | | Cell-mediated | | SM | |
						Auto.	Allo.		
1	T-CLL	ER$^+$	−	ND	ND	+	+		68
2	T-CLL	ER$^+$, Tγ$^+$	+	UA	+	ND	+	−	70
3		ER$^+$, Tγ$^+$, CD8$^+$	UA	ND	ND	ND	ND	ND	76
4		ER$^+$, Tγ$^+$	+		ND	+	ND	−	77
5	T-CLL	ER$^+$, Tγ$^+$, CD3$^+$, CD8$^+$, CD5$^+$	+	ND	ND	+	−	ND	71
6	T-CLL	CD2$^+$, CD3$^+$, CD8$^+$, CD11$^+$	+	−	ND	+	−^{a}	ND	69
7	T-CLL	ER$^+$, Tγ$^+$, CD2$^+$, CD3$^+$, CD5$^+$, CD8$^+$	UA	−	+	+	−	−	78
8		ER$^+$, CD3$^+$, CD8$^+$, Leu7$^+$, Tγ$^+$	+	−	ND	+^{b}	ND	−	120
9	T-CLL	CD2$^+$, CD3$^+$, CD8$^+$, CD16$^+$	+	ND	ND	+	ND	ND	72
10	EBV with infection	ER$^+$	UA	ND	ND	+	ND	−	121

^{a} Inhibited allogeneic HLA matched but not HLA nonmatched.
^{b} Inhibition confined to CFU$_E$, not affecting BFU$_E$.
Abbreviations: Assoc. disease, associated disease; Auto., autologous; Allo., allogeneic; SM, serum-mediated; ND, not determined; UA, unable to assess.

pancytopenia. In these nine patients, removal of Tγ cells resulted in significantly increase growth of CFU_E and BFU_E. The immunophenotype of the marrow Tγ cells, fully determined in only three patients, was shown to be $CD2^+$, $CD3^+$, $CD8^+$, $CD4^-$, $Leu7^+$. Functional assays of the Tγ cells for ADCC or NK activity were not performed, and it is not clear that these cells were LGL by morphology.

4. Cellular Immunity and Neutropenia

4.1. Clinical Observations

Chronic neutropenia is usually defined by peripheral blood absolute granulocyte counts of less than $1200–1500/\mu l$ (depending on race[87]) without recognized precipitating cause such as drug hypersensitivity or cytotoxicity. Chronic neutropenia is associated with a large number of clinical entities that range from autoimmune disorders such as collagen vascular disease to myelodysplastic syndromes. When the pathogenesis of chronic neutropenia has been elucidated, it has involved acquired genetic abnormalities in stem cells (as suggested by cytogenetic abnormalities in certain myelodysplastic syndromes) and humoral and/or cellular autoimmunity. Within certain clinical entities such as Felty's syndrome, an individual patient's predominant pathogenic mechanism may be humoral, cellular, or a combination of the two.[88] Over the past decade, since the lymphoproliferative disorder involving Tγ cells or LGL has been recognized, the high association of this entity with neutropenia and/or anemia has been noted. The LGL lymphoproliferative disorder has recently been reviewed[89] and is discussed further below (section 6).

4.2. Laboratory Observations Regarding Pathogenesis

In vitro evidence for T cell suppression of granulopoiesis in neutropenia first appeared in 1978 with a report by Abdou *et al.*[90] of suppressor T cells that inhibited CFU_{GM} growth in three patients with Felty's syndrome. These investigators found that T cell-enriched populations from the spleen, bone marrow, and peripheral blood of these patients suppressed CFU_{GM} growth of normal bone marrow, whereas similar cells from patients with rheumatoid arthritis and drug-induced neutropenia failed to demonstate this suppression. Subsequently, Bagby and Gabourel[91] reported that removal of T cells from the marrow of three patients with rheumatoid disorder and neutropenia normalized suppressed CFU_{GM} growth of the patients' marrow cells. The T cells were also shown to be cortisol-sensitive, and prednisone therapy of these patients led to remission of the neutropenia. In another study involving two neutropenia patients, hematosuppressive T cells were identified to have a CD3$^+$, CD8$^+$ phenotype.[92]

In a very large study which included 234 patients with granulopoietic failure, Bagby et al.[93]correlated the results of in vitro studies with a response to prednisone treatment. Lymphocytes suppressive to CFU$_{GM}$ growth, demonstrated by improved colony formation following T cell removal, were found in 39 of the 234 patients tested. In the study, 25 patients were shown to have improved colony growth by the addition of prednisolone to the colony assay. Twenty-four of these 25 had resolution of the neutropenia when treated with prednisone.

5. Cellular Immunity and Thrombocytopenia

Autoimmunity resulting in thrombocytopenia is most often secondary to humoral mechanisms, with the mature platelet as the target for antibody destruction in common immune thrombocytopenic purpura. The megakaryocyte, however, may be the target for humoral immune destruction as was demonstrated in two cases studied by Hoffmann et al.[29] In these thrombocytopenic patients with amegakaryocytosis, complement fixing autoantibodies directed at megakaryocyte CFU (CFU$_{Meg}$) were demonstrated.

The only two cases of probable cell-mediated megakaryocytic thrombocytopenia were reported by Gewirtz et al.[94] In one of the patients, removal of ER$^+$ cells from bone marrow resulted in an increase in CFU$_{Meg}$ and in the second patient removal of macrophages led to improved CFU-$_{Meg}$ growth. Readdition of the T cells and macrophages to the T cell- or macrophage-depleted bone marrow cells in patients 1 and 2, respectively, resulted in a decrease in CFU$_{Meg}$. Macrophage-conditioned medium prepared from the second patient's cells was able to suppress autologous and allogeneic CFU$_{Meg}$. As the CFU$_{Meg}$ assay becomes more widely available, we suspect further examples of cell-mediated thrombocytopenia and amegakaryocytic thrombocytopenia will be described.

6. LGL Proliferative Disorders and Hematocytopenias

6.1. Clinical Observations

The clinical association of an expanded population of LGL in patients with neutropenia was first reported in 1975 by Brouet et al.,[95] but has been recognized as a distinct clinicopathologic entity for only the past several years. The disorder is characterized by moderate lymphocytosis in both the blood and bone marrow, common presence of splenomegaly without lymadenopathy, and polyclonal hypergammaglobulinemia.[96–98]

A portion of these patients have associated rheumatologic disorders such as rheumatoid arthritis, Felty's syndrome, and autoantibodies.[98–102] Although most cases have either mild or moderate associated anemia, the anemia has occasionally been severe.[98,99] The small number of cases of LGL proliferative disorder and PRCA are summarized in Table I. Two of the "PRCA" cases included had associated neutropenia.[69,70] Although the frequency of LGL proliferative disorder in all patients with chronic neutropenia is uncertain, we observed LGL excess in five of 16 consecutive cases of chronic neutropenia at our institution.[99]

Thrombocytopenia is uncommon in this group of patients. It may be associated with aggressive disease[103] where the low level may be explained by reasons other than a specific suppressive effect of the LGL. Sporadic cases with mild to moderate thrombocytopenia (50,000–100,000/mm^3) have been reported. However, it is unclear whether the thrombocytopenia was persistent or transient and whether it could be accounted for by factors other than the lymphocytosis. Of the 11 patients studied by us,[104] three had thrombocytopenia. One of these was mild and corrected spontaneously. Another had chronic liver disease and splenomegaly, and the thrombocytopenia resolved completely on splenectomy, but the neutropenia persisted. There is little clinical evidence indicating that LGL suppress thrombopoiesis or cause a significant level of platelet destruction.

LGL proliferation may be separated into two main types[104]: type A consists of LGL that are CD3$^+$, CD8$^+$ and type B consists of LGL that are CD3$^-$, CD2$^+$, CD8$^{+/-}$. Type A LGL proliferation is almost invariably accompanied by neutropenia and/or anemia.[98,104] Cytopenia appears to be less frequently associated with type B disease but has been reported.[103,105] The degree of cytopenia is often out of proportion to the extent of bone marrow infiltration or splenomegaly present.

An etiological relation between the cytopenia and the lymphoproliferation appears to be likely from their frequent association, especially in type A cases. This is made more plausible by the observation that treatment effective in reducing the lymphocytes is followed by improvement of the neutropenia or anemia.[69,71,98,106] We have observed a case with neutropenia and moderate anemia who had a spontaneous remission of the peripheral blood LGL excess accompanied by normalization of the neutrophil count and hemoglobin level.[107] This spontaneous remission is not unique and has been observed by the authors in another patient (Winton, unpublished observation) and three additional patients reported by others.[108,109] The clinical course of most of these patients is indolent, with little progression over years of observation. Morbidity and mortality are usually confined to recurrent pyogenic infections, and only infrequently progressive lymphoproliferation.[98]

Both cytogenetic and Southern analysis of DNA for T cell receptor gene rearrangements in type A disease provide evidence that the LGL

expansion is clonal in most patients studied. [96,100,104,107,109–111] Loughran
et al. reported trisomy 8 in one patient and trisomy 14 in another along
with a variety of other abnormalities.[100] McKenna et al. reported a variety of cytogenetic abnormalities observed in three patients with LGL lymphoproliferative disorder.[96] In these few cases with cytogenetic abnormalities, no consistent chromosomal pattern is apparent. The availability
of cDNA probes for the constant region of the beta chain of the T cell
antigen receptor has permitted analysis for clonal origin of CD3$^+$ T cell
populations.[112,113] A substantial portion of the type A LGL lymphocytosis patients analyzed by this technique show clonal rearrangement[104,107,110,112] (see Table II).

Retroviral etiology by HTLV-1 has been established for adult T cell
leukemia endemic in southern Japan, the Caribbean, and the southern
United States.[114] Recently, Starkebaum et al.[115] have reported that sera
from six of 12 patients with proliferation of LGL reacted with either or
both the p19/p22 retroviral proteins of HTLV-1 as determined by West-
ern blot. Two of these six patients had rheumatoid arthritis. No reactive
sera were obtained from 59 patients with rheumatoid arthritis without
excess LGL, including 27 patients with Felty's syndrome. The authors have
not found anti-HTLV-1 antibodies in 12 patients with LGL proliferation
and neutropenia (Chan and Winton, unpublished observation).

6.2. Laboratory Observations Regarding Pathogenesis

LGL in type A disease with CD3$^+$, CD8$^+$ phenotype usually have
weak ADCC and very low NK activity. Bone marrow culture studies to
examine the CFU$_{GM}$ in these patients and the effect of addition and depletion of LGL on CFU$_{GM}$ have usually, but not always, failed to demonstrate suppression (see Table II). The granulocytic series in the bone marrow generally show a left shift in maturation with a decrease of the more
mature precursors,[98,104,116] but the granulocytic cellularity is not significantly decreased in many patients.[104] This suggests that if the LGL are
responsible for the neutropenia, it may not act at the level of the CFU$_{GM}$
but on more mature cells. One possible mechanism is the destruction of
the bands and segmented neutrophils by LGL. Pross et al.[117] looked for
evidence of ADCC against neutrophils by patients' PBL. A low level of
cytotoxicity could be demonstrated in the presence of the patients' serum.
We have performed an ADCC assay where the granulocytes and bone
marrow mononuclear cells from an HLA-identical sibling of a patient were
incubated with patient's serum and mononuclear cells. No significant ADCC
against granulocytes and bone marrow cells could be demonstrated (Chan
and Winton, unpublished observation).

Despite the negative findings, further experiments are necessary to
define more clearly the role of LGL in neutropenia. CFU$_{GM}$ inhibitor studies

TABLE II

Cases of Neutropenia with LGL Proliferation

Case	Assoc. disease	Phenotype	LGL	NK	ADCC	CFU$_{GM}$ Inhib.	Anti-NG	Gene rearr.	Ref.
1		ER+, CD8+, Fcγ+	+	ND	+	–	ND	ND	122
2		ER+, CD8+, Fcγ+	+	ND	ND	–	ND	ND	122
3		ER+, CD8+, Fcγ+	+	ND	ND	–	ND	ND	122
4		ER+, Fcγ+	+	ND	+	ND	–	ND	123
5		ER+, Fcγ+	+	ND	+	ND	+	ND	123
6	RF+	ER+, CD3+, CD8+	+	–	+	ND	+[a]	ND	124
7	RF+	ER+, Fcγ+	ND	–	+	ND	+[a]	ND	117
8		CD2+, CD3–, CD8–, CD16+, Leu7–, NKH1–, CD11–	+	+	+	+[b]	ND	ND	103
9		ER–, CD2–, CD3+, CD8+, Fcγ+	+	–	+	+[c]	+[d]	ND	118
10		ER+, CD3+, CD8+, Fcγ+	+	–	+	–	–	ND	118
11		ER+, CD3+, CD8+, Fcγ+	+	–	+	–	+	ND	118
12		ER+, CD3+, CD8+, Fcγ+	+	–	+	–	–	ND	118
13		ER+, CD3+, CD8+, Fcγ+	+	–	+	–	+	ND	118
14	Felty's syndrome	ER+, CD3+, CD8+, CD11–, CD16+, Leu7+	+	–	+	–	+	ND	102
15	Felty's syndrome	CD3+, CD8+, CD11–	+	–	+	–	–	ND	102
16	Felty's syndrome	ER+, CD3+, CD8+, CD11–	+	ND	ND	–	+	ND	102
17	Felty's syndrome	CD3+, CD8+, CD11–, CD16–, Leu7+	+	–	–	–	+	ND	102

(Continued)

Table II (Continued)

Case	Assoc. disease	Phenotype	LGL	NK	ADCC	CFU$_{GM}$ Inhib.	Anti-NG	Gene rearr.	Ref.
18	Felty's syndrome	ER+, CD3+, CD8+, CD11-, CD16+, Leu7+	+	-	-	-	+	ND	102
19	RF+, ANA+	ER-, CD3+, CD8+, Fcγ-, CD7+	+	ND	ND	-	+	ND	125
20	ANA+, APA	ER+, CD3+, CD8+, Leu7+, CD11-, Fcγ-	+	-	-	-	+[d]	ND	100
21	RF+, APA, RBCA	ER+, CD3+, CD8+, CD11+, Fcγ-	+	-	-	+[e]	+	ND	100
22	CAH	CD2+, CD3+, CD8+, CD11+, Leu7-, CD16-	+	-	-	ND	+	R	104
23	ANA+, SAS	CD2+, CD3+, CD8+, CD11-, Leu7+, Fcγ(EA)+, CD16-, NKH1-	+	-	-	ND	-	R	104
24	PG	CD2+, CD3+, CD8+, CD11-, Leu7+, CD16-, NKH1-	+	-	-	+	-	R	104
25	RF+, ANA+	CD2+, CD3+, CD8+, CD11-, Leu7+, CD16-, NKH1-	+	-	+	ND	-	R	104
26	RF+, ANA+	CD2+, CD3+, CD8+, CD11-, Leu7+, CD16-, NKH1-	+	-	+	ND	-	R	104
27		ER+, CD3+, CD8+, LEU7-	+	-	+	-	ND	PC	110
28		ER+, CD3+, CD8+, CD16-, NKH1-	+	+/-	+	+[f]	ND	ND	110

No.	Disease	Phenotype							
29		ER+, CD3+, CD8+, NKH1-	+	+/-	+/-	-	ND	PC	110
30		ER+, CD3+, CD8+, Leu7+/-, CD16+/-, NKH1-	+	-	ND	-	ND	PC	110
31		ER+, CD3+, CD8+, Leu7+/-, CD16+, NKH1-	+	-	+	+[f]	ND	R	110
32		ER+, CD3+, CD8+, Leu7+/-, CD16+/-, NKH1-	+	+/-	ND	+[f]	ND	R	110
33		ER+, CD3-, Leu7+, CD16+, NKH1-	+	+	ND	+[g]	ND	GL	110
34	RA	CD2+, CD3+, CD8+, Leu7+	+	ND	ND	ND	+[d]	*	115
35	RA	CD2+, CD3+, CD8+, Leu7+	+	ND	ND	ND	-	*	115
36	Arthritis	CD2+, CD3+, CD8+, Leu7+	+	ND	ND	ND	-	*	115
37	Arthritis	CD2+, CD3+, CD8+, Leu7+	+	ND	ND	ND	+[d]	*	115
38	RA	CD2+, CD3+, CD8+, Leu7+	+	ND	ND	ND	+[d]	*	115
39	RA	CD2+, CD3+, CD8+, Leu7+	+	ND	ND	ND	+[d]	*	115
40	Arthritis	CD2+, CD3+, CD8+, Leu7+	+	ND	ND	ND	+[d]	*	115
41		CD2+, CD3+, CD8+, Leu7+	+	ND	ND	ND	+[d]	*	115
42		CD2+, CD3+, CD8+, Leu7+	+	ND	ND	ND	+[d]	*	115
43		CD2+, CD3+, CD8+, Leu7+	+	ND	ND	ND	+[d]	*	115

(Continued)

TABLE II (Continued)

Case	Assoc. disease	Phenotype	LGL	NK	ADCC	CFU$_{GM}$ Inhib.	Anti-NG	Gene rearr.	Ref.
44		CD2$^+$, CD3$^+$, CD8$^+$, Leu7$^+$	+	ND	ND	ND	+^{d}	*	115
45		CD2$^+$, CD3$^+$, CD8$^+$, Leu7$^+$	+	ND	ND	ND	−	*	115

[a] Measured with an ADCC assay using patient's serum, cells, and normal neutrophils.
[b] IL-2 activation necessary for suppression.
[c] Suppression obtained with serum and complement.
[d] Increased neutrophil-bound antibody.
[e] Patient multitransfused.
[f] Only demonstrable using cultured LGL.
[g] Inhibition of blood but not bone marrow CFU$_{GM}$.

* Six of the 12 patients were tested and all have Tβ gene rearrangement.

This table included only cases of LGL lymphocytosis with neutropenia where either bone marrow culture study or antineutrophil antibody determination have been performed. Cases that have apparently appeared in multiple publications are listed only once in the table. NK and ADCC activities are designated negative when they were absent or below normal; a +/− designation means variable results were obtained. Phenotyping results listed have been confined to the following markers to maintain uniformity: CD2/(ER), CD3, CD8, CD11, CD16, Leu7, NKH-1, and Fcγ(EA). Results that were equivocal have been excluded.

A large series reported by Newland et al. is not included because of the difficulty of incorporating the cases in the format of the table. Interested readers should consult the original article (Br. J. Haematol. 58:443–446, 1984).

Abbreviations: Assoc. disease, associated disease; CFU$_{GM}$ inhib., CFU$_{GM}$ inhibition; Anti-NG, antineutrophil granulocyte; Gene rearr., gene rearrangement; ND, not determined; RF, rheumatoid factor; APA, antiplatelet antibody; RBCA, red blood cell antibody; ANA, antinuclear antigen; CAH, chronic active hepatitis; R, clonally rearranged; SAS, severe aphous stomatitis; PG, pyoderma gangrenosa; PC, polyclonal; GL, germ line; RA, rheumatoid arthritis.

are especially relevant in patients with a hypocellular granulocytic series. Since cell-to-cell interaction may be necessary, an experiment with a period of coculture of LGL and marrow cells in the presence of the patient's own serum should be included. Furthermore, the effect of an inhibitory population may be obscured by the presence of many cells other than granulocytic precursor cells. Concentrating precursor cells before testing is therefore desirable. It is also possible that only a minor population of LGL have been "activated" to be inhibitory, and the effector:target ratio used *in vitro* is insufficient to demonstrate any effect. Since it is not possible for practical reasons to use a high effector:target ratio, activating the LGL population by IL-2 or other stimuli before culture may reveal the inhibitory property of the cell.

The failure to demonstrate ADCC activity in many instances could be due to a number of factors. LGL may specifically destroy granulocytes at a certain stage of differentiation such as metamyelocytes or band forms. The ^{51}Cr cytotoxicity assay may not be sensitive enough to detect the destruction of a small percentage of such cells present in the target population. Late granulocytic cells also take up ^{51}Cr poorly, compounding the difficulty in the detection of destruction of minor populations. ADCC is generally performed using peripheral blood LGL which may be functionally different from LGL in the bone marrow and other organs. Another possibility is that LGL form conjugates with neutrophils without actually killing them, and such conjugates are sequestered and destroyed in the reticuloendothelial system. An *in vitro* cytotoxicity assay would fail to detect such a pathogenic mechanism. We would like to point out that splenectomy performed in such patients is usually ineffective in improving the neutropenia. If the last postulate is true, the sequestration and destruction of neutrophils have to take place in multiple sites in addition to the spleen.

Type B LGL may also be associated with neutropenia.[103,105] This type of LGL corresponds to the subsets of peripheral blood CD3$^-$, NK cells. Studies on the effect of NK-enriched populations and NK clones on normal hematopoiesis generally show inhibition. The degree of inhibition and the effector population are somewhat variable in different reports. Even NK clones with similar phenotypes have variable activity.[20] It appears likely that the pathogenic expansion of NK cells in the patient would be associated with variable degrees and variable forms of cytopenia depending on the hematopoietic suppressive activity of the expanded population or clone. Very few studies have been performed on the mechanism of cytopenia in type B patients. Koizumi *et al.*[103] reported a decrease in CFU$_{GM}$ in bone marrow cells from a type B patient, but the patient's LGL failed to inhibit allogeneic bone marrow CFU$_{GM}$ unless they were preincubated with IL-2. IL-2 culture supernatant could suppress CFU$_{GM}$ and the effect was reversible with anti-IFNγ.

Because many of the patients with LGL proliferation and neutropenia are seropositive for rheumatoid factor and antinuclear antibodies, the role of autoantibodies in the associated cytopenia needs further investigation. To our knowledge, there have been very few reports of autoimmune hemolytic anemia or antibody-mediated thrombocytopenia in these patients,[100] although a number of investigators have reported the presence of antineutrophil antibodies in LGL-neutropenia patients (see Table II). A variety of techniques have been used for detecting serum- and neutrophil-associated antibodies, and the significance of the presence of serum neutrophil-reactive antibodies is sometimes difficult to interpret. Starkebaum and co-workers[115] have reported that nine of 12 neutropenic patients with LGL proliferation have neutrophil-reactive IgG as determined by a radiolabeled Fab-anti-F(ab)₂ technique. We have examined the sera of five patients for the presence of antineutrophil antibodies, kindly performed by Dr. Gerald Logue (Buffalo, NY), and only one showed a borderline increase.

Van der Veen et al.[118] studied five neutropenic patients with LGL lymphocytosis and could detect neutrophil-bound IgG in only one patient. In two other patients, the sera show neutrophil-reactive antibodies which could be removed by adsorption with normal platelets, indicating that they were probably directed against HLA antigens. Inhibition of CFU_{GM} and BFU_E could be demonstrated by the serum from one patient, and the inhibitory activity could again be removed by adsoprtion with pooled normal donor platelets. The authors concluded from their study that "autoantibodies against neutrophils or neutrophil precursors or circulating immune complexes do not seem to play an important role in K-cell lymphocytosis/neutropenia syndrome."

7. Critique of Investigation Methods

There are several major limitations of the *in vitro* study of the interaction between the immune system and hematopoiesis

1. Heterogeneity and definition of the effector population. Up to the late 1970s, there were few markers available for the definition of different subsets of lymphocytes. T lymphocytes have been equated to cells that rosette with sheep erythrocytes (E), and this property was used for the separation of T cells from non-T cells. E rosette–forming cells are heterogenous. Not only do they contain multiple T cell subpopulations, they may also include NK cells. The availability of monoclonal antibodies has allowed us to dissect the T cell into further subpopulations. However, it is quite clear that the CD8- and CD4-positive populations are not homogeneous. The definition of NK cells is even less satisfactory. A small population of cells with classical T cell markers can mediate NK activity.[119]

The majority of NK cells appear to have few T cell characteristics, and even within this population there is considerable heterogeneity in the expression of a number of "NK" markers such as Leu 7, CD16(Leu11), NKH1, and NKH2. Most of the experiments reviewed above in sections 2–6 designed to study the function of a certain population of cells *in vitro* used rather crude separation methods to enrich and deplete subpopulations of lymphocytes. The results obtained therefore represent the combined effect of enrichment or depletion of multiple populations of cells and not a single unique subset of cells. Furthermore, the isolation technique by itself may have altered the property of the cell under study. Isolation of Tγ cells is usually performed by E-rosetting followed by rosetting with antibody-coated ox erythrocytes. The binding of sheep red blood cells to CD2 and the F_c portion of IgG to F_c receptors may activate a cell and elicit some function not observed in resting cells. The depletion of monocytes and macrophages in various experiments was done with different techniques which had different efficacy. Depletion by plastic adherence would have left significantly more monocyte/macrophages than nylon wool or carbonyl iron depletion.

2. Cellular composition of the bone marrow cells used for culture. The population of cells used for culture study is very heterogeneous, and progenitor cells in fact constitute only a very minor subset. The presence of numerous non-progenitor cells may obscure the effect of a suppressor population. This may be brought about by a number of different mechanisms. The presence of a high proportion of nontarget cells may block effector-progenitor-cell interaction if cell-to-cell contact is necessary. The presence of other lymphocytes or macrophages may enhance or inhibit the action of the effector cell population. If lymphokines are responsible for suppression, the factor may be adsorbed on many of the cells present other than the progenitor cells, reducing the observed effects.

3. Lack of definition of the culture medium. It is important to add a source of growth factors to culture bone marrow progenitor cells and also to have serum that will support the growth of colonies. Both of these contain many factors, many of which may have opposing effects on colony growth. Hematopoietic suppressor cells may act by secreting inhibitory factors whose effect may not be apparent in the presence of strong growth-promoting activities. Furthermore, it is possible that the suppressor cells act *in vivo* by inhibiting growth-promoting properties of the microenvironment which may be provided for by the addition of extraneous factors *in vitro*. On the other hand, some components of the added factors may act synergistically with certain populations of cells (or their

secreted factors) to bring about an inhibitory effect which would not be observed *in vivo*. Some factors may also activate a population of cells to suppress hematopoiesis, which would not otherwise happen. These considerations illustrate the possibility of obtaining false-negative and false-positive results and may also account for some variability and contradictions in reported *in vitro* studies.

4. Definition of the culture condition. Hematopoietic suppressor cells, if present, would act in the microenvironment of the bone marrow. It is not possible to reproduce the exact bone marrow microenvironment *in vitro*, but the standard culture condition used does not allow intimate cellular interaction, and suppression requiring a period of cell contact would not be observed. Factors secreted by suppressor cells that would act on short ranges would also be ineffective. This is addressed in some experiments by a period of preincubation between bone marrow cells and presumed inhibitory cells before culture. The solution is only partial, since this is at best a crude simulation of the conditions *in vivo*, and the precise relationship between different cellular populations could not be reproduced.

8. Perspective on the Investigation of Immune-Mediated Cytopenia

Some measures can be taken to overcome the limitations on *in vitro* studies mentioned in the preceding section. The availability of many monoclonal antibodies can partially alleviate the problem of the uncertainty of the cellular composition of the population under study. To isolate a population of cells for study, negative selection may be used to avoid activation of the cell population. Another approach is to obtain clones of effector cells and determine the effect of the clones and their products on hematopoiesis. This approach has been used by Herrmann *et al.*[20] to study the effect NK cells with a variety of phenotypes on hematopoiesis. It can also be used to study cytopenic states provided there are means of selecting and cloning the cell of interest. One drawback is that the cells used are in an activated state and may express functions not normally associated with the cells *in vivo*.

Regarding the interference of other bone marrow cells in the same culture on progenitor cells/effector interaction, it would be ideal to be able to isolate pure populations of progenitors to study. Although this would increase the sensitivity of the system, interference by other cells has not been eliminated. Furthermore, even if we identify the surface phenotypes of various progenitor populations and isolate them acccordingly, supplemental growth factors may be necessary for their growth in the absence of other cells. There has been a remarkable development in the isolation and cloning of hematopoietic growth factors. It is hoped that in the near future, pure recombinant growth factors and defined medium will be available for culture work. This would eliminate the variables introduced by the use of crude factors and serum.

The cytopenia associated with LGL lymphocytosis is the main cause of morbidity and mortality in this condition, and its pathogenesis warrants further study. In the case of pure red cell aplasia, evidence for suppression of erythropoiesis by the LGL is strong. However, little work has been done to investigate the pathogenesis of the less striking degree of anemia which is also commonly observed. We also know little about the pathogenesis of neutropenia in this disorder. Although suppression of CFU_{GM} has not been demonstrated in the majority of studies, further studies should be performed especially with modifications to overcome some of the limitations imposed by an *in vitro* system as discussed earlier. The role of ADCC should be investigated further as suggested in the preceding sections. Although the inhibition of progenitor cell growth, differentiation, and cytolysis of mature and immature granulocytic cells are likely pathogenic pathways, one should not lose sight of possible novel mechanisms. For example, LGL may immobilize neutrophils in a number of organs causing premature destruction which would not be detected by the culture or cytotoxicity assays.

The hematocytopenias seen in patients with LGL proliferation present both a challenge and an opportunity for the investigator. Fortunately, these disorders are not usually life-threatening, and in individual patients there may be ample time to perform laboratory studies using cells obtained on multiple occasions before and after treatment. As the techniques used to investigate cells and cellular interactions at the molecular and cellular levels are continually refined, we can look forward to a more complete understanding of the etiology and pathogenesis of this and related disorders and the design of rational therapy.

ACKNOWLEDGMENTS. Supported in part by USPHS grant 5-ROI-AI20376.

References

1. Metcalf, D., 1985, The granulocyte-macrophage colony-stimulating factors, *Science* **229**:16–22.
2. Wong, G. G., Witek, J. S., Temple, P. A., Wilkens, K. M., Leary, A. C., Luxenberg, D. P., Jones, S. S., Brown, E. L., Kay, R. M., Orr, E. C., Shoemaker, C., Golde, D. W., Kaufman, R. J., Hewick, R. M., Wang, E. A., and Clark, S. C., 1985, Human GM-CSF: Molecular cloning of the complementary DNA and purification of the natural and recombinant proteins, *Science* **228**:810–815.
3. Souza, L. M., Boone, T. C., Gabrilove, J., Lai, P. H., Zsebo, K. M., Murdock, D. C., and Chazin, V. R., 1986, Recombinant human granulocyte colony-stimulating factor: Effects on normal and leukemic myeloid cells, *Science* **232**:61–64.

4. Nagata, S., Tsuchiya, M., Asano, S., Kaziro, Y., Yamazaki, T., Yamamoto, O., Hirata, Y., Kubota, N., Oheda, M., Nomura, H., and Ono, M., 1986, Molecular cloning and expression of cDNA for human granulocyte colony-stimulating factor, *Nature* **319**:415–418.

5. Kawasaki, E. S., Ladner, M. B., Wang, A. W., Van Arsdell, J., Warren, M. K., Coyne, M. Y., Schweickart, V. L., Lee, M.-T., Wilson, K. J., Boosman, A., Stanley, E. R., Ralph, P., and Mark, D. F., 1985, Molecular cloning of a complementary DNA encoding human macrophage-specific colony-stimulating factor (CSF-1), *Science* **230**:291.

6. Yang, Y.-C., Ciarletta, A. B., Temple, P. A., Chung, M. P., Kovacic, S., Witek-Giannotti, J. S., Leary, A. C., Kriz, R., Donahue, R. E., Wong, G. G., and Clark, S. C., 1986, Human IL-3 (multi-CSF): Identification by expression cloning of a novel hematopoietic growth factor related to murine IL-3, *Cell* **47**:3–10.

7. Foon, K. A., and Todd, R. F., III, 1986, Immunologic classification of leukemia and lymphoma, *Blood* **68**:1–31.

8. Zucali, J. R., Dinarello, C. A., Oblon, D. J., Gross, M. A., Anderson, L., and Weiner, R. S., 1986, Interleukin 1 stimulates fibroblasts to produce granulocyte-macrophage colony-stimulating activity and prostaglandin E_2, *J. Clin. Invest.* **77**:1857–1863.

9. Knudtzon, S., and Mortenson, B. T., 1975, Growth stimulation of human bone marrow cells in agar culture by vascular cells, *Blood* **46**:937–943.

10. Horiguchi, J., Warren, M. K., and Kufe, D., 1987, Expression of the macrophage-specific colony-stimulating factor in human monocytes treated with granulocyte-macrophage colony-stimulating factor, *Blood* **69**:1259–1261.

11. Cline, M. J., and Golde, D. W., 1974, Production of colony-stimulating activity by human lymphoctyes, *Nature* **248**:703–704.

12. Ruscetti, F. W., and Chervenick, P. A., 1975, Release of colony-stimulating activity from thymus-derived lymphocytes, *J. Clin. Invest.* **55**:520–527.

13. Nathan, D. G., Chess, L., Hillman, D. G., Clark, B., Breard, J., Merler, E., and Housman, D. E., 1978, Human erythroid burst-forming unit: T-cell requirement for proliferation *in vitro*, *J. Exp. Med.* **147**:324–336.

14. Geissler, D., Lu, L., Bruno, E., Yang, H. H., Broxmeyer, H. E., and Hoffman, R., 1986, The influence of T-lymphocyte subsets and humoral factors on colony formation by human bone marrow and blood megakaryocyte progenitor cells *in vitro*, *J. Immunol.* **137**:2508–2513.

15. Torok-Storb, B., and Hansen, J. A., 1982, Modulation of *in vitro* BFU-E growth by normal Ia-positive T-cells is restricted by HLA-DR, *Nature* **298**:473.

16. Spitzer, G., and Verma, D. S., 1982, Cells with $F_c\gamma$ receptors from normal donors suppress granulocytic macrophage colony formation, *Blood* **60**:758–766.

17. Bacigalupo, A., Podesta, M., Mingari, M. C., Moretta, L., Piaggio, G., Van Lint, M. T., Durando, A., and Marmont, A. M., 1981, Generation of CFU-C/suppressor T cells *in vitro*: An experimental model for immune-mediated marrow failure, *Blood* **5**:491–496.

18. Mangan, K. F., Hartnett, M. E., Matis, S. A., Winkelstein, A., and Abo, T., 1984, Natural killer cells suppress human erythroid stem cell proliferation *in vitro*, *Blood* **63**:260–269.

19. Degliantoni, G., Perussia, B., Mangoni, L., and Trinchieri, G., 1985, Inhibition of bone marrow colony formation by human natural killer cells and by natural killer cell-derived colony-inhibiting activity, *J. Exp. Med.* **161**:1152–1168.

20. Herrmann, F., Schmidt, R. E., Ritz, J., and Griffin, J. D., 1987, *In vitro* regulation of human hematopoiesis by natural killer cells, *Blood* **69**:246–254.

21. Murphy, M., London, R., Kobayashi, M., and Trinchieri, G., 1986, γ-Interferon and lymphotoxin, released by activated T-cells, synergize to inhibit granulocyte/monocyte colony formation, *J. Exp. Med.* **114**:263–279.

22. LeBeau, M. M., Westbrook, C. A., Diaz, M. O., Larson, R. A., Rowley, J. D., Gasson, J. C., Golde, D. W., and Sherr, C. J., 1986, Evidence for the involvement of GM-CSF and FMS in the deletion (5q) in myeloid disorders, *Science* **231**:984–987.

23. Wisniewski, L. P., and Hirshhorn, K., 1983, Acquired partial deletions of the long arm of chromosome 5 in hematologic disorders, *Am. J. Hematol.* **15:**295–310.
24. Eschbach, J. W., Egrie, J. C., Downing, M. R., Browne, J. K., and Adamson, J. W., 1987, Correction of the anemia of end-stage renal disease with recombinant human erythropoietin: Results of a combined phase I and II clinical trial, *N. Engl. J. Med.* **316:**73–78.
25. Peschle, C., Marmont, A. M., Marone, G., *et al.*, 1975, Pure red cell aplasia: Studies on an IgG serum inhibitor neutralizing erythropoietin, *Br. J. Haematol.* **30:**411–417.
26. Browman, G. P., Freedman, M. H., Blajchman, M. A., and McBridge, J. A., 1976, A complement independent erythropoietic inhibitor acting on the progenitor cell in refractory anemia, *Am. J. Med.* **61:**572–578.
27. Levitt, L. J., Ries, C. A., and Greenberg, P. L., 1983, Pure white cell aplasia. Antibody-mediated autoimmune inhibition of granulopoiesis, *N. Engl. J. Med.* **308:**1146.
28. Boxer, L. A., Greenberg, M. S., Boxer, G. J., and Stossel, T. P., 1975, Autoimmune neutropenia, *N. Engl. J. Med.* **293:**748–753.
29. Hoffman, R., Zaknoen, S., Yang, H. H., Bruno, E., LoBuglio, A. F., Arrowsmith, J. B., and Prchal, J. T., 1985, An antibody cytotoxic to megakaryocyte progenitor cells in a patient with immune thrombocytopenic purpura, *N. Engl. J. Med.* **312:**1170–1174.
30. Karpatkin, S., 1985, Autoimmune thrombocytopenic purpura, *Semin. Hematol.* **22:**260–288.
31. Thomas, E. D., and Storb, R., 1985, Acquired severe aplastic anemia: Progress and perplexity, *Blood* **64:**325–328.
32. Appelbaum, F. R., Fefer, A., Cheever, M. A., Sanders, J. E., Singer, J. W., Adamson, J. W., Mickelson, E. M., Hansen, J. A. Greenberg, P. D., and Thomas, E. D., 1980, Treatment of aplastic anemia by bone marrow transplantation in identical twins, *Blood* **55:**1033–1039.
33. Applebaum E. R., Cheever M. R., Fefer A., Storb R., Thomas E. D., 1985, Recurrence of aplastic anemia following cyclophosphamide and syngeneic bone marrow transplantation: Evidence for two mechanisms of graft failure, *Blood* **65:**553–556.
34. Champlin, R. E., Feig, S. A., Sparkes, R. S., and Gale, R. P., 1984, Bone marrow transplantation from identical twins in the treatment of aplastic anemia: Implications for the pathogenesis of the disease, *Br. J. Haematol.* **56:**455–463.
35. Speck, B., and Gluckman, E., 1977, Treatment of aplastic anaemia by antilymphocyte globulin with and without allogeneic bone-marrow infusions, *Lancet* **2:**1145–1148.
36. Champlin, R., Ho, W., and Gale, R. P., 1983, Antithymocyte globulin treatment in patients with aplastic anemia, *N. Engl. J. Med.* **308:**113–118.
37. Camitta, B., O'Reilly, R. J., Sensenbrenner, L., Rappeport, J., Champlin, R., Doney, K., August, C., Hoffmann, R. G., Kirkpatrick, D., Stuart, R., Santos, G., Parkman, R., Gale, R. P., Storb, R., and Nathan, D., 1983, Antithoracic duct lymphocyte globulin therapy of severe aplastic anemia, *Blood* **62:**883–888.
38. Doney, K., Dahlberg, S. J., Monroe, D., Storb, R., Buckner, C. D., and Thomas, E. D., 1984, Therapy of severe aplastic anemia with anti-human thymocyte globulin and androgens: The effect of HLA-haploidentical marrow infusion, *Blood* **63:**342–348.
39. Bacigalupo, A., Van Lint, M. T., Giordano, C. R., Santini, D., Carella, G., Damasio, M., Rossi, E., Risso, M., Vimercati, R., Podesta, M., Durando, A., Reali, G., Avanzi, G., Barbanti, M., and Marmont, A. M., 1980, Treatment of severe aplastic anemia with bolus 6-methylprednisolone and antilymphocytic globulin, *Blut* **41:**168–171.
40. Speck, B., Gratwohl, A., Nissen, C., Osterwalder, B., Wursch, A., Tichelli, A., Lori, A., Reusser, P., Jeannet, M., and Signer, E., 1986, Treatment of severe aplastic anemia, *Exp. Hematol.* **14:**126–132.
41. Doney, K., Storb, R., Buckner, C. D., McGuffin, R., Witherspoon, R., Deeg, H. J., Appelbaum, F. R., Sullivan, K. M., and Thomas, E. D., 1987, Treatment of aplastic anemia with antithymocyte globulin, high-dose corticosteriods, and androgens, *Exp. Hematol.* **15:**239–242.

42. Zoumbos, N., Ferris, W. O., Hsu, S.-M., Goodman, S., Griffith, P., Sharrow, S. O., Humphries, R. K., Nienhuis, A. W., and Young, N., 1984, Analysis of lymphocyte subsets in patients with aplastic anemia, *Br. J. Haematol.* **58:**95–105.

43. Zoumbos, N. C., Gascon, P., Djeu, J. Y., Trost, S. R., and Young, N. S., 1985, Circulating activated suppressor T lymphocytes in aplastic anemia, *N. Engl. J. Med.* **312:**257–265.

44. Torok-Storb, B., Doney, K., Sale, G., Thomas, E. D., and Storb, R., 1985, Subsets of patients with aplastic anemia identified by flow microfluorometry, *N. Engl. J. Med.* **312:**1015–1022.

45. Ascensao, J., Pahwa, R., Kagan, W., Hansen, J., Moore, M., and Good, R., 1976, Aplastic anemia: Evidence for an immunological mechanism, *Lancet* **1:**669.

46. Kagan, W. A., Ascensao, J. A., Pahwa, R. N., Hansen, J. A., Goldstein, G., Valera, E. B., Incefy, G. S., Moore, M. A. S., and Good, R. A., 1976, Aplastic anemia: Presence in human bone marrow of cells that suppress myelopoiesis, *Proc. Natl. Acad. Sci. USA* **73:**2890–2984.

47. Hoffmann, R., Zanjani, E. D., Lutton, J. D., Zalusky, R., and Wasserman, L. R., 1977, Suppression of erythroid-colony formation by lymphocytes from patients with aplastic anemia, *N. Engl. J. Med.* **296:**10–13.

48. Singer, J. W., Brown, J. E., James, M. C., Doney, K., Warren, R. P., Storb, R., and Thomas, E. D., 1978, Effect of peripheral blood lymphocytes from patients with aplastic anemia on granulocytic colony growth from HLA-matched and -mismatched marrows: Effect of transfusion sensitization, *Blood* **52:**37–46.

49. Torok-Storb, B. J., Sieff, C., Storb, R., Adamson, J., and Thomas, E. D., 1980, *In vitro* tests for distinguishing possible immune-mediated aplastic anemia from transfusion-induced sensitization, *Blood* **55:**211–215.

50. Bacigalupo, A., Podesta, M., Van Lint, M. T., Vimercati, R., Cerri, R., Rossi, E., Risso, M., Carella, A., Santini, G., Damasio, E., Giordano, D., and Marmont, A. M., 1981, Severe aplastic anaemia: Correlation of *in vitro* tests with clinical response to immunosuppression in 20 patients, *Br. J. Haematol.* **47:**423–433.

51. Suda, T., Mizoguchi, H., Miura, Y., Kubota, K., and Takaku F., 1981, Suppression of *in vitro* granulocyte-macrophage colony formation by the peripheral mononuclear phagocytic cells of patients with idiopathic aplastic anemia, *Br. J. Haematol.* **47:**433–442.

52. Ortega, J. A., Ma, A., Shore, N. A., Dukes, P. P., and Merigan, T. C., 1979, Suppressive effect of interferon on erythroid cell proliferation, *Exp. Hematol.* **7:**145–150.

53. Neumann H. A., Fauser A. A., 1982, Effect of interferon on pluripotent hemopoietic progenitors (CFU$_{GEMM}$) derived from human bone marrow, *Exp. Hematol.* **10:**587–590.

54. Broxmeyer, H. E., Lu, L., Platzer, E., Feit, C., Juliano, L., and Rubin, B. Y., 1983, Comparative analysis of the influences of human gamma, alpha and beta interferons on human multipotential (CFU-GEMM), erythroid (BFU-E) and granulocyte-macrophage (CFU-GM) progenitor cells, *J. Immunol.* **131:**1300.

55. Raefsky, F. L., Platanias, L. C., Zoumbos, N. C., and Young, N. S., 1985, Studies of interferon as a regulator of hematopoietic proliferation, *J. Immunol.* **135:**2507–2512.

56. Zoumbos, N. C., Gascon, P., Djeu, J. Y., and Young, N. S., 1985, Interferon is a mediator of hematopoietic suppression in aplastic anemia *in vitro* and possibly *in vivo*, *Proc. Natl. Acad. Sci. USA* **82:**188–192.

57. Torok-Storb, B., Johnson, G. G., Bowden, R., and Storb R., 1987, Gamma-interferon in aplastic anemia: Inability to detect significant levels in sera or demonstrate hematopoietic suppressing activity, *Blood* **69:**629–633.

58. Chang, T. W., McKinney, S., Liu, V., Kung, P. C., Vilceck, J., and Le, J., 1984, Use of monoclonal antibodies as sensitive and specific probes for biologically active human gamma-interferon, *Proc. Natl. Acad. Sci. USA* **81:**5219–5222.

59. Clark, D. A., Dessypris, E. N., and Krantz, S. B., 1984, Studies on pure red cell aplasia. XI. Results of immunosuppressive treatment of 37 patients, *Blood* **63:**277–286.

60. Krantz, S., 1986, New therapies for aplastic anemia, *Am. J. Med. Sci.* **291**:371–379.
61. Dameshek, W., Brown, S. M., and Rubin, A. D., 1967, "Pure" red cell anemia (erythroblastic hypoplasia) and thymoma, *Semin, Hematol.* **4**:222–232.
62. Schmid, J. R., Kiely, J. M. Pease, G. L., and Hargraves, M. M., 1963, Acquired pure red cell agenesis—report of 16 cases and review of the literature, *Acta Haematol.* **30**:255–270.
63. Schmid, J. R., Kiely, J. M. Harrison, E. G. Jr., Bayard, E. D., and Pease, G. L., 1965, Thymoma associated with pure red-cell agenesis, *Cancer* **18**:216–230.
64. Stohlman, G., Queseberry, P. J., Howard, D., Miller, M. E., and Schur, P., 1971, Erythroid aplasia: An autoimmune complication of chronic lymphocytic leukemia, *Clin. Res.* **19**:566.
65. Mangan, K. F., Chikkappa, G., and Farley, P. C., 1982, T gamma (Tγ) cell suppress growth of erythroid colony-forming units *in vitro* in the pure red cell aplasia of B-cell chronic lymphocytic leukemia, *J. Clin. Invest.* **70**:1148–1156.
66. Yoo, D., Pierce, L. E., and Lessin, L. S., 1983, Acquired pure red cell aplasia associated with chronic lymphocytic leukemia, *Cancer* **51**:844–850.
67. Chikkappa, G., Zarrabi, M. H., and Tsan, M. F., 1986, Pure red-cell aplasia in patients with chronic lymphocytic leukemia, *Medicine (Baltimore)* **65**:339–351.
68. Hoffman, R., Kopel, S., Shu, S. D., Dainiak, N., and Zanjani, E. D., 1978, T cell chronic lymphocytic leukemia: Presence in bone marrow and peripheral blood cells that suppress erythropoiesis *in vitro*, *Blood* **52**:255–260.
69. Lipton, J. M., Nadler, L. M., Canellos, G. P., Kudisch, M., Reiss, C. S., and Nathan, D. G., 1983, Evidence for genetic restriction in the suppression of erythropoiesis by a unique subset of T lymphocytes in man, *J. Clin. Invest.* **72**:694–706.
70. Nagasawa, T., Abe, T., and Nakagawa, T., 1981, Pure red cell aplasia and hypogammaglobulinema associated with T-cell chronic lymphocytic leukemia, *Blood* **57**:1025–1031.
71. Hocking, W. G., Singh, R., Schroff, R., and Golde, D. W., 1983, Cell mediated inhibition of erythropoiesis and megaloblastic anemia in T-cell chronic lymphocytic leukemia, *Cancer* **51**:631–636.
72. Hansen, R. M., Lerner, N., Abrams, R. A., Patrick, C. W. Malik, M. I., and Keller, R., 1986, T-cell chronic lymphocytic leukemia with pure red cell aplasia: Laboratory demonstration of persistent leukemia in spite of apparent complete clinicall remission, *Am. J. Hematol.* **22**:79–86.
73. Hunt, F. A., and Lander, C. M., 1975, Successful use of combination chemotherapy in pure red cell aplasia associated with malignant lymphoma, histiocytic type, *Aust. N.Z. J. Med.* **5**:468–471.
74. Morgan, E., Pang, K. M., and Goldwasser, E., 1978, Hodgkin disease and red cell aplasia, *Am. J. Hematol.* **5**:71–75.
75. Krantz, S. B., 1974, Pure red-cell aplasia, *N. Engl. J. Med.* **291**:345–350.
76. Callard, R. E., Smith, C. M., Worman, C., Linch, D., Cawley, J. C., and Beverley, P. C. L., 1981, Unusual phenotype and function of an expanded subpopulation of T cells in patients with hematopoietic disorders, *Clin. Exp. Immunol.* **43**:497–505.
77. Linch, D. C., Cawley, J. C., MacDonald, S. M., Masters, G., Roberts, B. E., Antonis, A. H., Waters, A. K., Sieff, C., and Lydyard, P. M., 1981, Acquired pure red-cell aplasia associated with an increase of T cells bearing receptors for the F_c of IgG, *Acta Haematol.* **65**:270–274.
78. Shionoya, S., Amano, M., Imamura, Y., Nakahara, K., and Okawa, H., 1984, Suppressor T-cell chronic lymphocytic leukemia associated with red cell hypoplasia, *Scand. J. Haematol.* **33**:231–238.
79. Mangan, K. F., Volkin, R., and Winkelstein, A., 1986, Autoreactive erythroid progenitor-T suppressor cells in the pure red cell aplasia associated with thymoma and panhypogammaglobulinemia, *Am. J. Hematol.* **23**:167–173.
80. Abkowitz, J. L., Powell, J. S., Nakamura, J. M., Kadin, M. E., and Adamson, J. W.,

1986, Pure red cell aplasia: Response to therapy with anti-thymocyte globulin, *Am. J. Hematol.* **23:**363–371.

81. Mangan, K. F., and Shadduck, R. K., 1984, Successful treatment of chronic refractory pure red cell aplasia with antithymocyte globulin: Correlation with *in vitro* erythroid culture studies, *Am. J. Hematol.* **17:**417–426.

82. Krantz, S. B., and Kao, V., 1967, Studies on red cell aplasia. I. Demonstration of a plasma inhibitor to heme synthesis and an antibody to erythroblast nuclei, *Proc. Natl. Acad. Sci. USA* **58:**493–500.

83. Krantz, S. B., Moore, W. H., and Zaenta, S. D., 1973, Studies on red cell aplasia. V. Presence of erythroblast cytotoxicity in gamma globulin fraction of plasma, *J. Clin. Invest.* **52:**324–336.

84. Messner, H. A., Fauser, A. A., Curtis, J. E., and Dotten, D., 1981, Control of antibody-mediated pure red-cell aplasia by plasmapheresis, *N. Engl. J. Med.* **304:**1334–1338.

85. Mangan, K. F., Chikkappa, G., Scharfman, W. B., and Desforges, J. F., 1981, Evidence for reduced erythroid burst (BFU-E) promoting function of T lymphocytes in the pure red cell aplasia of chronic lymphocytic leukemia, *Exp. Hematol.* **9:**489–498.

86. Mangan, K. F., and D'Alessandro, L., 1985, Hypoplastic anemia in B cell chronic lymphocytic leukemia: Evolution of T cell mediated suppression of erythropoiesis in early-stage and late-stage disease, *Blood* **66:**533–541.

87. Broun, G. O., Herbig, F. K., and Hamilton, J. R., 1966, Leukopenia in Negroes, *N. Engl. J. Med.* **275:**1410–1413.

88. Starkebaum, G., Singer, J. W., and Arend, W. P., 1980, Humoral and cellular immune mechanisms of neutropenia in patients with Felty's syndrome, *Clin. Exp. Immunol.* **39:**307–314.

89. Reynolds, C. W., and Foon, K. A., 1984, T$_\gamma$-lymphoproliferative disease and related disorders in humans and experimental animals: A review of the clinical, cellular, and functional characteristics, *Blood* **64:**1146–1158.

90. Abdou, N. I., NaPombejara, C., Balentine, L., and Abdou, N. L., 1978, Suppressor cell-mediated neutropenia in Felty's syndrome, *J. Clin. Invest.* **61:**738–743.

91. Bagby, G. C. Jr., and Gabourel, J. D., 1979, Neutropenia in three patients with rheumatic disorders. Suppression of granulopoiesis by cortisol-sensitive thymus-dependent lymphocytes, *J. Clin. Invest.* **64:**72–82.

92. Bagby, G. C., 1981, T-lymphocytes involved in inhibition of granulopoiesis in two neutropenic patients are of the cytotoxic/suppressor (T3$^+$,T8$^+$) subset, *J. Clin. Invest.* **68:**1597–1600.

93. Bagby, G. C. Jr., Lawrence, H. J., and Neerhout, R. C., 1983, T-lymphocyte-mediated granulopoietic failure. *In vitro* identification of prednisone-responsive patients, *N. Engl. J. Med.* **309:**1073–1078.

94. Gewirtz, A. M., Sacchetti, M. K., Bien, R., and Barry, W. E., 1986, Cell-mediated suppression of megakaryocytopoiesis in acquired amegakaryocytic thrombocytopenic purpura, *Blood* **68:**619–626.

95. Brouet, J.-C., Flandrin, G., Sasportes, M., Preud'Homme, J.-L., and Seligmann, M., 1975, Chronic lymphocytic leukaemia of T-cell origin. Immunological and clinical evaluation in eleven patients, *Lancet* **2:**890–893.

96. McKenna, R. W., Arthur, D. C., Gajl-Peczalska, K. J., Flynn, P., and Brunning, R. D., 1985, Granulated T cell lymphocytosis with neutropenia: Malignant or benign chronic lymphoproliferative disorder? *Blood* **66:**259–266.

97. Chan, W. C., Winton, E. F., and Waldmann, T. A., 1986, Lymphocytosis of large granular lymphocytes, *Arch. Intern. Med.* **146:**1201–1203.

98. Newland, A. C., Catovsky, D., Linch, D., Cawley, J. C., Beverley, P., San Miguel, J. F., Gordon-Smith, E. C., Blecher, T. E., Shariari, S., and Varadi, S., 1984, Chronic T cell lymphocytosis: A review of 21 cases, *Br. J. Haematol.* **58:**443–446.

99. Chan, W. C., Winton, E. F., and Check, I. J., 1984, T-cell imbalance in neutropenia of uncertain etiology, *Am. J. Clin. Pathol.* **81:**54–61.

100. Loughran, T. P. Jr., Kadin, M. E., Starkebaum, G., Abkowitz, J. L., Clark, E. A., Dis-

teche, C., Lum, L. G., and Slichter, S. J., 1985, Leukemia of large granular lymphocytes: Association with clonal chromosomal abnormalities and autoimmune neutropenia, thrombocytopenia, and hemolytic anemia, *Ann. Intern. Med.* **102:**169–175.

101. Barton, J. C., Prasthofer, E. F., Egan, M. L., Heck, L. W. Jr., Koopman, W. J., and Grossi, C. E., 1986, Rheumatoid arthritis associated with expanded populations of granular lymphocytes, *Ann. Intern. Med.* **104:**314–323.

102. Wallis, W. J., Loughran, T. P., Kadin, M. E., Clark, E. A., and Starkebaum, G. A., 1985, Polyarthritis and neutropenia associated with circulating large granular lymphocytes, *Ann. Intern. Med.* **103:**357–362.

103. Koizumi, S., Seki, H., Tachinami, T., Taniguchi, M., Matsuda, A., Taga, K., Nakarai, T., Kato, E., Taniguchi, N., and Nakamura, H., 1986, Malignant clonal expansion of large granular lymphocytes with a Leu-11$^+$, Leu-7$^-$ surface phenotype: *In vitro* responsiveness of malignant cells to recombinant human interleukin-2, *Blood* **68:**1065–1073.

104. Chan, W. C., Link, S., Mawle, A., Check, I., Brynes, R. K., and Winton, E. F., 1986, Heterogeneity of large granular lymphocyte proliferations: Delineation of two major subtypes, *Blood* **68:**1142–1153.

105. Tagawa, S., Konishi, I., Kurante, H., Katagiri, S., Taniguchi, N., Tamaki, T., Inoue, R., Kanayama, Y., Tsubakio, T., Machii, T., Yonezawa, T., and Kitani, T., 1983, A case of T-cell chronic lymphocytic leukemia (T-CLL) expressing a peculiar phenotype (E$^+$, OKM1$^+$, Leu 1$^+$, OKT3$^-$, and IgG EA$^-$), *Cancer* **52:**1378–1384.

106. Palutke, M., Eisenberg, L., Kaplan, J., Hussain, M., Kithier, K., Tabaczka, P., Mirchandani, I., and Tenenbaum, D., 1983, Natural killer and suppressor T-cell chronic lymphocytic leukemia, *Blood* **62:**627–634.

107. Winton, E. F., Chan, W. C., Check, I., Colenda, K. W., Bongiovanni, K. F., and Waldmann, T. A., 1986, Spontaneous regression of a monoclonal proliferation of large granular lymphocytes associated with reversal of anemia and neutropenia, *Blood* **67:**1427–1432.

108. Pandolfi, F., Semenzato, G., and De Rossi, G., 1985, Chronic lymphocytosis due to the expansion of granular lymphocytes, *Br. J. Haematol.* **60:**771–772.

109. Waldmann, T. A., Davis, M. M., Bongiovanni, K. F., and Korsmeyer, S. J., 1985, Rearrangements of genes for the antigen receptor on T-cells as markers of lineage and clonality in human lymphoid neoplasms, *N. Engl. J. Med.* **313:**776–783.

110. Grillot-Courvalin, C., Vinci, G., Tsapis, A., Dokhelar, M.-C., Vainchenker, W., and Brouet, J.-C., 1987, The syndrome of T8 hyperlymphocytosis: Variation in phenotype and cytotoxic activities of granular cells and evaluation of their role in associated neutropenia, *Blood* **69:**1204–1210.

111. Rambaldi, A., Pelicci, P.-G., Allavena, P., Knowles, D. M. II, Rossini, S., Bassan R., Barbui, T., Dalla-Favera, R., and Manovani, A., 1985, T-cell receptor beta chain gene rearrangements in lymphoproliferative disorders of large granular lymphocytes/natural killer cells, *J. Exp. Med.* **162:**2156–2162.

112. Flug, F., Pelicci, P.-G., Bonetti, F., Knowles, D. M. II, and Dalla-Favera, R., 1985, T-cell receptor gene rearrangements as markers of lineage and clonality in T-cell neoplasms, *Proc. Natl. Acad. Sci. USA* **82:**3460–3464.

113. Ritz, J., Campen, T. J., Schmidt, R. E., Royer, H. D., Hercend, T., Hussey, R. E., and Reinherz, E. L., 1985, Analysis of T-cell receptor gene rearrangement and expression in human natural killer clones, *Science* **228:**1540–1543.

114. Broder, S., Bunn, P. A. Jr., Jaffe, E. S., Blattner, W., Gallo, R. S., Wong-Staal, F., Waldmann, T. A., and DeVita V. T. Jr., 1984, T-cell lymphoproliferative syndrome associated with human T-cell leukemia/lymphoma virus, *Ann. Intern. Med.* **100:**543–557.

115. Starkebaum, G., Loughran, T. P., Kalyanaraman, V. S., Kadin, M. E., Kidd, P. G., Singer, J. W., and Ruscetti, F. W., 1987, Serum reactivity to human T-cell leukaemia/lymphoma virus type I proteins in patients with large granular lymphocytic leukaemia, *Lancet* **1:**596–599.

116. Chan, W. C., Check, I., Schick, C. C., Brynes, R. K., Kateley, J., and Winton, E. F., 1984, A morphologic and immunologic study of the large granular lymphocyte in neutropenia with T lymphocytosis, *Blood* **63**:1133–1140.
117. Pross, H. F., Pater, I., Giles, A., Gallinger, L. A., Rubin, P., Corbett, W. E. N., Galbraith, P., and Baines, M. G., 1982, Studies of human natural killer cells. III. Neutropenia associated with unusual characteristics of antibody-dependent and natural killer cell–mediated cytotoxicity, *J. Clin. Immunol.* **2**:126–134.
118. Van der Veen, J. P. W., Goldschmeding, R., Miedema, F., Smit, J. W., Melief, C. J. M., and Von dem Borne, A. E. G. K., 1986, K-cell lymphocytosis/neutropenia syndrome: The neutropenia is not caused by autoimmunity, *Br. J. Haematol.* **64**:777–787.
119. Lanier, L. L., Phillips, J. H., Hackett, J., Tutt, M., and Kumar, V., 1986, Natural killer cells: Definition of a cell type rather than a function, *J. Immunol.* **137**:2735–2739.
120. Abkowitz, J. L., Kadin, M. E., Powell, J. S., and Adamson, J. W., 1986, Pure red cell aplasia: Lymphocyte inhibition of erythropoiesis, *Br. J. Haematol.* **63**:59–67.
121. Socinski, M. A., Ershler, W. B., Tosato, G., and Blases, R. M., 1984, Pure red cell aplasia associated with chronic Epstein-Barr virus infection: Evidence for T cell–mediated suppression of erythroid colony forming units, *J. Lab. Clin. Med.* **104**:995–1006.
122. Linch, D. C., Cawley, J. C., Worman, C. P., Galvin, M. C., Robert, B. E., Callard, R. E., and Beverley, P. C. L., 1981, Abnormalities of T-cell subsets in patients with neutropenia and an excess of lymphocytes in the bone, *Br. J. Haematol.* **48**:137–145.
123. Bom–Van Noorloss, A. A., Pegals, H. G., Von Oers, R., Silberbusch, J., Feltkamp-Vroom, T. M., Goudsmit, R., Zeijlemaker, W. P., Von dem Brone, A. K., and Melief, C. J., 1980, Proliferation of T$_\gamma$ cells with killer-cell activity in two patients with neutropenia and recurrent infections, *N. Engl. J. Med.* **302**:933–937.
124. Aisenberg, A. C., Wilkes, B. M., Harris, N. L., Ault, K. A., and Carey, R. W., 1981, Chronic T-cell lymphocytosis with neutropenia: Report of a case studied with monoclonal antibody, *Blood* **58**:818–822.
125. Starkebaum, G., Martin, P. J., Singer, J. W., Lum, L. G., Price, T. H., Kadin, M. E., Raskind, W. H., and Fialkow, P. J., 1983, Chronic lymphocytosis with neutropenia: Evidence for a novel, abnormal T-cell population associated with antibody-mediated neutrophil destruction, *Clin. Immunol. Immunopathol.* **27**:110–123.

18

The Natural Immune System in Autoimmune and Neurological Disease

JEAN E. MERRILL

1. Introduction

Since Erlich's first hypothesis of "horror autotoxicus" and the established paradigm that the immune response does not or should not react to self, immunologists have been made increasingly aware of instances where autoimmunity is neither a rare nor a harmful event. Recognition of modified self seems essential for the removal of virus-infected cells, and idiotype-antiidiotype networks may be regulators of the humoral immune response. Nonetheless, there are numerous autoimmune diseases in which regulation of natural immunity seems to have gone awry. The unanswered question is whether this is in primary association with the disease process or the secondary result of the chronicity of an aberrant immune response. Organ-specific autoimmune diseases include acquired immune hemolytic disorders such as hemolytic anemia, idiopathic thrombocytopenia purpura, or idiopathic neutropenia; diseases of the nervous system or neuromuscular tissue such as multiple sclerosis and myasthenia gravis; diseases of exocrine and endocrine systems such as Hashimoto's thyroiditis, Graves' disease, pernicious anemia, Addison's disease, and diabetes mellitus; and diseases of other organs like pemphigus, bullous pemphigoid, biliary cirrhosis, ulcerative colitis, and uveitis. Systemic diseases that are non-organ-specific include systemic lupus erythematosus, rheumatoid arthritis, Goodpasture's syndrome, and Sjögren's syndrome.

JEAN E. MERRILL • Department of Neurology, UCLA School of Medicine, Los Angeles, California 90024.

411

In all instances except MS, specific autoantibodies have been detected. In the systemic diseases, autoantibodies to multiple self-antigens may be seen. The autoimmune diseases themselves may overlap, with one individual having more than one disease, especially in individuals with autoimmune endocrinopathies. There is no well-established unifying concept to explain the origin and pathogenesis of the various autoimmune disorders. Indeed, autoimmune diseases may originate from genetic abnormalities linked or unlinked to the predisposition to exogenous infections or endogenous hormonal imbalances triggering a normal immune response which becomes abnormal with time.

Thus the role of the natural immune response in autoimmunity remains a mystery. To assign natural killer (NK) cells the task of surveillance and prevention of chronic infections that may trigger an autodestructive cycle, we must identify the viruses causing these diseases; show the relationship between low NK, viral persistence, and autoimmunity; and demonstrate that therapy that induces NK activity prevents disease progression or has some other therapeutic benefit in these patients. This has not yet been done. Nevertheless, this chapter will attempt to discuss and summarize the evidence for defects in NK cell activity and their relationship of these defects to abnormalities in immunoregulatory hormones or other features of autoimmune disease that affect the natural immune response.

2. Abnormalities in the Immune Response Leading to Depressed NK Activity

If one considers the events following antigenic stimulus, alterations of NK activity are a straightforward consequence and not likely to be specific for any antigen triggering a disease process; macrophages are activated by antigen and T lymphokines to produce interleukin 1 (IL1)[1,2] which stimulates IL-2 production,[1,2] which in turn induces IFN$_\gamma$.[3,4] interferon$_\gamma$ induces IL-2 receptors,[3] and the immune response, including NK activity, is then amplified. Interferon,[5,6] especially IFN$_\gamma$[6-9] boosts NK cell activity directly, as does IL2.[10-12] NK cell recruitment, NK cell numbers, and recycling are elevated by IFN.[13] Thus the induction of NK cell activity is IL-1-dependent,[2] IL-2-dependent,[10-12] IFN-dependent,[10,13] and the natural consequence of any immune response. One could propose that a decrease in NK activity may in fact be due to a chronic autoimmune disease process, the disease-associated antigen stimulating antibody production and the resulting immune complexes (1) directly inhibiting NK cells via their Fc$_\gamma$R[14-16] or Fc$_\mu$R[16] or (2) inducing PGE.[17,18] Prostaglandin E can inhibit IL-2 production,[19,20] which may partially explain its suppression of NK cells,[21-23] though PGE must certainly have a direct suppressive effect on NK cells.[24] Immune complexes are the hallmarks of a variety of chronic autoimmune diseases and have been claimed to be responsible for depressed NK activity in several, as shall be discussed.

Thus, induction of NK cell activity is a contiguous event with the rest of the immune response. As many complex reasons for an aberrant NK response could probably be found as there are complex driving mechanisms to stimulate it. Some of these events might even be HLA-linked. In reviewing the autoimmune diseases for explanations about aberrant natural immunity, there are no simple unifying causes for depressed NK cell activity. Indeed, the evidence points to multiple explanations: (1) reduced numbers, lytic activity, or recycling of NK cells; (2) defects in production of or response to IL-1, IL-2, or IFN by NK cells; (3) anti-DR, anti-NK, or anti-IFN antibodies; (4) active suppressive mechanisms like PGE, immune complexes, or other macrophage or T cell suppressors; (5) regulation by neuroendocrine hormones and neurotransmitters derived from the brain and the immune response. These will be discussed in the context of each specific autoimmune disease.

3. Systemic Lupus Erythematosus and Sjögren's Syndrome

If NK cells do in fact produce significant amounts of IFN[25-30] and IFN regulates the pathology of the disease, then NK cells are indirectly important in both virally induced infection and putative virally induced autoimmunity. Viruslike structures or transmissible agents have been isolated from endothelial cells on renal biopsy in systemic lupus erythematosus (SLE) and Sjögren's syndrome (SS).[31] It is suggested that these autoimmunelike diseases are of a viral etiology. Sjögren's syndrome is characterized by lymphoid cells infiltrating into the exocrine gland, reduced endogenous and IFN-inducible NK activity in blood, absence of NK cells in the salivary gland, reduced interferon production, autoantibodies, and benign and malignant lymphomas.[30-35] The immune dysfunction of low IFN production could lead to reduced NK and viral persistence and result in the observed malignancies.[32] Although the numbers of NK cells characterized by their large granular lymphocyte (LGL), phenotype, were the same in SS patients and controls, the SS NK cell activity was only half that of controls. This functional aberration could be corrected *in vitro* by addition of IL2 but not IFN$_\alpha$.[35]

Few if any reports find low NK in SLE to be independent of disease stage[36,37]; the majority find it decreased in active disease.[32,36-43] In SLE, there may be multiple reasons for depressed NK activity. The deficiency is not related to the number of NK cells[44] or their ability to recognize and bind to targets[45,46] but to the inability of active NK cells to kill.[44,45] Recycling of active NK is probably normal.[45] Rather, the defect lies in the release of the natural killer cytotoxic factor (NKCF).[44,47] In addition, it is not clear if an NK cell-bound inhibitor is relevant in SLE, since one report claimed that overnight incubation of effectors increases NK activity[48] whereas other reports have shown that neither incubation[48,46] nor protease treatment[46] reversed low NK function. It is possible that SLE sera do not decrease normal NK activity,[45] though there are reports showing serum-associated inhibitory activity[48,49] and assigning it to anti-NK antibodies,[50,51] prolonged exposure to IFN$_\alpha$[49,52] or immune complexes.[38,50]

Interferon levels in SLE are elevated *in vivo*,[53,57] which may explain the apparent *in vitro* defect in interferon production and interferon-inducible NK in SLE[38,53,54,58,59]; that is, IFN production is already high *in vivo* and cannot be augmented *in vitro*, and NK cells are depressed in activity because of this continuous exposure to IFN.[49,52,54] An unusual interferon system seems to be disproportionately represented in both SLE and acquired immunodeficiency syndrome (AIDS)—that is, an IFN$_\gamma$ deficiency and an acid-labile, IFN$_\alpha$ overproduction.[54,55,57,60,61] A similar pH-sensitive IFN$_\alpha$ is seen in other autoimmune and infectious diseases as well. It is of note that AIDS patients have low NK, though the relationship of low NK to this unusual IFN$_\alpha$ is not clear.[62] In addition, SLE patients have selective defects in IFN production to different inducers.[38,54,58] Interferon-inducible NK may be suppressed by anti-IFN produced by these patients[57] or by insensitivity of their NK cells to IFN resulting from long-term exposure *in vivo*.[59] Sibbett *et al.* have examined the suppressor effects of macrophages in SLE and ruled out direct macrophage suppression or PGE production as a reason for low NK.[46] IL-2 boosts normal NK[10–12] and has been shown to boost low NK in SLE patients.[63] Both IL-1 and IL-2 production[64–66] and IL-2 response[65] are decreased in SLE. The relationship of these defects to NK activity is worthy of further attention.

As with SLE patients, SS patients have elevated serum IFN[67] but low IFN production and low IFN-inducible NK *in vitro*.[32] However, in contrast to SLE, inhibition of NK activity in SS is not associated with suppressive serum factors,[32] antilymphocytic antibodies, or IC.[33]

4. Rheumatoid Arthritis

There appears to be some discrepancy as to whether NK activity is normal[68–73] or reduced[74–77] in rheumatoid arthritis (RA) peripheral blood compared to control blood. Likewise, there are differing findings regarding NK activity in the synovium vs. the blood.[68,69,77–80] Some of the differences cited may be related to (1) subpopulations of cells having NK activity such as LGL,[72] Leu11$^+$, or Leu7$^+$ cells,[78–80] OKT8$^+$ cells,[74] or autologous mixed lymphocyte reaction-induced NK[75]; (2) stage of disease[74,75,77,81]; or (3) NK cell trafficking.[82] NK activity in RA blood is probably depressed in active disease.[74,75,77,81] Nevertheless, even when there are differences in NK activity, there do not seem to be major differences in the proportions of LGL, Leu11$^+$, Leu7$^+$, or OKM1$^+$ cells[72,74,79,80] when comparing RA patients and controls. This suggests function/phenotype discordance in the NK population in arthritis—that is, RA NK cells do not have the surface markers or morphology characteristically seen in normals.

Decrease in NK activity in blood and synovial fluid (SF) of active RA patients has been attributed to depressed production and response to IL-2.[75,77,83] IL-2 can boost NK activity in RA blood and SF cells though to a greater degree in SF cells.[75,77,83] The IL-2 defect, in spite of elevated IL-1 production, may reflect some defect in IL-1 binding or IL-2 inhibitory substances.[83–85] The macrophage plays a negative regulatory role in NK activity of RA blood cells, as *in vitro* treatment with indomethacin or removal of adherent cells augment NK activity.[74–77,81,86] Inexplicably, auranofin, which has a stimulatory effect on RA NK *in vitro*, seems to cause further depression in NK *in vivo*.

Whereas HNK-1$^+$ cells are seen in fibrous synovial and perivascular cells and lymphocyte clusters in synovium,[87] the exact phenotype of SF NK cells is controversial. They are not all Leu11b$^{+(79)}$ nor FcR$^{+(78)}$ and therefore not inhibitable by immune complexes.[70,78] Because SF NK cells kill NK-resistant cell lines,[78,80] respond to IL-2 better than normal or RA blood NK cells,[77] and seem to require macrophage accessory cells (perhaps for IL-1 induction of IL-2 in the synovium),[77,81,86] it is possible that these SF NK cells are the lymphokine-activated killer (LAK) cells described by Grimm *et al.*[88]

5. Hashimoto's Thyroiditis and Graves' Disease

Surgical specimens of thyroid glands from patients with Hashimoto's thyroiditis (HT) have been examined for lymphocyte subpopulations using immunofluorescence and immunoperoxidase techniques. Activated T cells (OKT10$^+$, HLA-DR$^+$) were present in areas of advanced destruction. DR$^+$ thyroid epithelial cells may be an important factor in the progression and self-perpetuation of the disease, which was postulated to be initiated by humor immune components but propagated by cellular immunopathologic mechanisms.[89] Natural killer cells (VEP13$^+$, Leu7$^+$) were noted in the interstitium between thyroid follicles, intruding between thyroid follicular epithelial cells and merging into the thyroid follicular lumen.[89] Indeed, cloning of thyroid infiltrates revealed a majority of T8$^+$ clones with the capacity for lectin-dependent and NK cell cytotoxicity.[90] In spite of this, there was no significant difference in NK activity between HT patients and controls' peripheral blood NK activity[91,92] even when tested against thyrocyte targets.[93] Conclusions about the role of cytotoxic cells in HT center around the ADCC mechanism as a possible contributor to the disease process.[92]

One comment should be made about the identity of populations of NK and K cells. Since NK and K cells can be distinguished functionally by differential sensitivities to immunomodulating molecules,[24,94-96] it cannot be assumed that low NK activity in a given disease automatically means concomitant low ADCC activity. This can be of significance in autoimmunity, where persistent viral infection and low NK are contrasted by an elevated ADCC contributing to the pathology of the disease. ADCC is probably important in the pathology of hepatitis patients[97,98] and also active in RA and HT.[99-102] Interestingly, in spite of low NK activity, K cell ADCC has been observed to be normal or elevated in SLE,[39] MS,[40,103,104] Graves' disease,[105] Crohn's disease,[106] and some forms of immunodeficiency.[107,108]

In contrast to hypothyroid patients with Hashimoto's disease, there is an NK cell defect to eye muscle cell targets in euthyroid and Graves' ophthalmopathy (GD).[105] Enumeration of LGL numbers in patients with autoimmune thyroid disease vs. normals showed that in untreated patients with thyrotoxic GD both the percentages ($11.5 \pm 2.7\%$) and absolute count ($244 \pm 102/mm^3$) of LGL were significantly lower than those in normal women ($17 \pm 3.6\%$; $334 \pm 122/mm^3$). Women have the lowest LGLs of normal subjects. There were no significant differences from normal controls in LGLs in euthyroid GD under treatment, euthyroid, or hypothyroid patients with HT.[109]

As previously mentioned, in spite of low NK activity in Graves' disease, ADCC was increased using serum as a source of antibody and the eye muscle cell as the target in a ^{51}Cr-release assay.[105] These results suggest that this may be one mechanism for the eye muscle cell damage characteristic of this disorder.

6. Relationship of the Natural Immune Response to the Central Nervous System

The nervous system and NK cells share an antigenic determinant associated with the carbohydrates of myelin-associated glycoprotein (MAG)[110] and identified by anti-Leu7 (HNK-1).[111-113] Neuroectodermally derived cells such as oligodendrocytes containing myelin, as well as ependymal and Schwann cells, stain with HNK1 but not another NK marker, VEP13.[111,113] Tumor cells like neuroblastoma, medulloblastoma, retinoblastoma, melanoma, and neural crest-derived adrenal medulla cells are positive, whereas mesenchymally derived adrenal cortex cells are negative.[112] In addition, immunoblots of Leu7 reveal a 75-D-molecular-weight protein different from MAG within the matrices of chromaffin granules, suggesting that secretory granules are targets for the antibody in addition to surface adhesion molecules.[110,114] That this cross-reactive surface antigen may be involved in an autoimmune response directed simultaneously at the nervous and immune systems is suggested by evidence for circulating monoclonal IgM antibodies to MAG in patients with demyelinating neuropathy who have concomitantly decreased circulating HNK-1$^+$ cells.[115]

A second interaction between the nervous system and natural immune response resides in the domain of brain-derived and T lymphocyte–derived neurotransmitters as well as brain cell– or leukocyte-derived lymphokines that affect NK cell activity. Several years ago it became evident that neuroendorine hormones, neurotransmitters, and autocoids were capable of immunoregulation.[116–118] NK cells were shown to be selectively regulated by the sympathetic nervous system *in vitro* and *in vivo* in that low concentrations of adrenaline boosted NK[119–122] whereas high concentrations inhibited NK.[122] Agents that potentiated catecholamines, like cocaine, could also produce an *in vivo* increase in NK cell activity without affecting other lymphocyte functions.[119]

Histamine has been shown to be produced in the CNS and proved to be an effective immunomodulator. Histamine is most likely produced in the CNS by neurons and microvascular endothelium. Histamine's role in the CNS may be to influence cerebrospinal fluid (CSF) production,[121] but it probably acts as a neurotransmitter also, since it is found in some synaptosomal fractions.[121,123] Its role in inflammation is to increase vascular permeability and PGE synthesis.[124] Histamine is probably involved in the expression of experimental autoimmune encephalomyelitis (EAE) in mice.[125] Bordatella pertussis adjuvant increases susceptibility to EAE in these mice by increasing blood vessel sensitivity to histamine and serotonin. Antagonists to these agents inhibit the development of EAE.[125,126] There are histamine receptors on the K cell that mediate ADCC.[127–132] K cells have H_1 receptors (H_1R).[132] T suppressor cells have H_2 receptors (H_2R).[133] Histamine's effect on NK cells is indirect, since NK cells probably do not have H_1R or H_2R.[134–136] Inhibitory effects of NK activity or histamine antagonists suggest that early in the immune response, low doses of histamine may act through macrophage macrophage-mediated induction of IL-2 and IFN$_\gamma$ to increase target-binding capacity and NK killing.[134–137] Later in the immune response, higher concentrations of histamine act through the H_2R on T suppressor cells to induce NK suppressor factors affecting IL-2 and IFN$_\gamma$.[138–140]

The regulatory loop between the immune and neuroendocrine systems has been studied carefully by Blalock and colleagues[141,142] and Hall *et al.*[143] The immune response has the potential to intervene in the hypothalamic-pituitary-adrenal axis and ultimately to affect glucocorticoid production by the adrenal cortex. Blalock and Smith have shown that T lymphocytes produce corticortropin and endorphinlike substances that are similar if not identical to the ACTH and endorphin processed from pro-opiomelanocortin from the pituitary gland.[141,142] T cell-derived ACTH causes a rise in corticosterone levels ultimately leading to decreased inflammation and NK cell activity. Hall *et al.*[143] have shown that thymosin induces a release of endogenous ACTH leading to changes in steroid levels which, depending on concentration, can either increase or decrease NK cell activity.

Beta endorphin and its analogues β-lipotropin and T cell–derived γ endorphin caused a 10–30% increase in NK activity *in vitro* and were three- to fivefold more effective in doing so than Leu-enkenphalin or β-endorphin.[144–148] The effect was specific for LGL NK cells,[148] as there was no effect on ADCC.[145] The increase in NK activity, regulated by the naloxone-reversible effects of β-endorphin, included increases in effector-target conjugates, augmentation of active killing by target-binding cells, increased LGL IFN production, and recycling[147,148] and IFN$_\beta$ boosting of NK.[144] NK stimulation may occur through the interaction of endorphin with calmodulin.[144] Although enkephalins act less well in augmentation of NK, they also may work through the induction of soluble stimulatory factors.[149] *In vivo*, enkephalins have prolonged the survival of mice challenged with L1210 tumor cells.[150]

Finally, NK cells in the circulating blood are also affected by lympho- and monokines IL-1, IL-2, IL-3, and prostaglandin E. These hormones interact and affect each others' synthesis.[151–153] Brain astrocytes produce IL-1, IL-3, and PGE[154]; brain macrophages, the microglia, also produce IL-1.[155]

Thus, NK cells entering the CNS in inflammatory infiltrates will encounter the immunomodulating effects of neurotransmitters and brain-derived cytokines that are identical to or that mimic those derived from the immune response.

7. Multiple Sclerosis and Other Neurological Diseases

In multiple sclerosis (MS), where viral etiology has been suggested but no virus specific to MS patients has ever been isolated, we have only one piece of evidence that the NK-interferon relationship is important in the disease process. This is clouded by the lack of consensus among investigators as to IFN production by MS patients *in vivo* and *in vitro*. IFN was cited in one report as undetectable in sera or CSF[103]; in another report, it was detectably elevated compared with controls.[156] It has been said to be produced *in vitro* by peripheral blood mononuclear cells of MS patients at normal levels in response to a variety of viral and nonviral stimuli,[159–164] though others report abnormally low production to these same stimuli.[63–68] There is also a discrepancy regarding *in vitro* mitogen-induced IFN$_\gamma$ levels.[165–167] At present the majority of data favor a defect in IFN production in MS. It was thus of great interest when Jacobs *et al.*[168,169] treated MS patients with intrathecal injections of purified IFN.

After treatment, the rate of relapse of IFN recipients was less than before treatment and less than in the control group, indicating a positive effect of such an antiviral and immune response–boosting therapy.[168,169]

MS has been the topic of many papers and reviews on NK activity because of the disease association with a possible virus such as measles. Two reports found no defect in NK, ADCC, or IFN-inducible NK or in virus-induced or non-virus-induced IFN production.[157,170] However, there seem to be overwhelming data that MS patients are defective in NK activity to a variety of tumor- and virus-infected targets.[40–42,103,104,158, 161,162,167,171–180] In one study the defect was target cell–related, with lower NK killing to virus-infected targets.[103] There also seems to be general agreement that activity is lowest during active disease[161,104,172–176] and especially more so during an acute relapse than in chronic progressive or remission stages.[104,173–175] Nor can NK activity be detected in cells in cerebrospinal fluid (CSF) of MS patients by a sensitive single-cell assay,[104,172–175] but it is detectable in CSF of patients with acute infections of the CNS.[104,172]

The controversy over HLA-related defects in NK activity and IFN production in MS patients continues. Abb and colleagues have documented that low responsiveness by peripheral blood lymphocytes of any individual to inducers of IFN_α (flu virus, Molt 4) is associated with DR2, though they found no association of IFN_γ production and DR2.[181] It has been shown that MS patients show a higher than normal frequency of DR2 antigen,[103,161,162,171] and thus studies have been conducted to link DR2 antigen as the sole source for defective NK and IFN production in MS patients.

An association of DR2 and low NK and IFN production in the general population would trivialize the findings in MS. However, such an association has not been upheld in the general population, nor has a firm association between DR and defective NK, IFN circuits been established. Initial work correlating low NK and DR2 in MS patients[103,161] is apparently not statistically significant.[162,171] Secondly, the finding that low NK exists in DR2 MS patients but not in normal DR2 controls[103] or in normal (DR+) monozygotic MS twins with normal NK activity[42,158,178,182] makes this HLA association an unlikely explanation for defects in MS patients.

The likelihood that low IFN production is the cause of low NK activity in MS[159,164,171,176] also seems less likely, since IFN_α is not always observed to be low,[42,158,182] and when it is low, it does not correlate with NK activity.[162,176] Although IFN_α corrected low NK activity in MS patients in vitro,[177] the long-term use of IFN_α in vivo caused a decrease in NK activity in vitro and in vivo below pretreatment levels, suggesting a down-regulation of the IFN receptor on NK cells.[183] It would appear that NK and IFN activities are independent phenomena.

Decreased NK activity in MS has been demonstrated using the ⁵¹Crrelease assay,[40,103,104,161, 172-175] but the underlying defect responsible for this result may reside in the defect at the level of target cell binding or killing or effector recycling. At the single-cell level, acute relapse patients during an exacerbation have fewer peripheral blood NK cells; the defect is at both the level of binding to and lysis of the target.[172] There is an apparent defect in recycling as well.[104,172] By a similar assay, there are no binders or killers in CSF i.e., no detectable functional NK cells.[104,172]

It is misleading to quantitate NK cells in a disease state by scoring cells of given phenotypes such as OKM1,[184] LGL,[185,186] or HNK1[187] to make conclusions about activities of such cell populations. This is especially true since NK cells defined by the above phenotypes share markers or functions of helper T cells[185,187] or suppressor cytotoxic T cells.[185,187-189] An example of this discrepancy occurs in MS patients where NK cell activity is depressed and HNK1$^+$ cell numbers are normal[179,190] or elevated[40] or patients have normal numbers of LGL.[104] What is also striking is that LGLs have T cell phenotypes, and T4$^+$ and T8$^+$ cells mediate NK activity in MS patients.[172] Perhaps a better marker is the common brain-immune shared antigen MAG, discussed earlier. MS patients have decreased levels of MAG in the CNS[191] and a decreased percent of circulating MAG-positive cells.[190] Although Leu7$^+$ cells were normal, Leu11$^+$ (MAG$^+$) cells were reduced[190]; since Leu11$^+$ Leu7$^-$ cells are the most potent NK cells, this marker may be more relevant to the hypothesis of an autoimmune event targeting the NK system and oligodendrocytes/myelin. LAK cells may also be reduced in MS, though this is only by indirect inference.

A strong piece of evidence for immunoregulatory defects causing low NK in MS patients is suggested by the work of this laboratory in which MS patients make elevated amounts of PGE *in vitro* (PBL) and *in vivo* (CSF). These elevated PGE levels are associated with decreased NK activity and IFN$_\alpha$ production[173,176] in MS. In addition, NK cells in MS patients are more sensitive to the effects of PGE than controls.[174,175] It has been suggested that circulating IC may activate macrophages to produce PGE. An early observation by Rola-Plesczynski *et al.*[192] regarding the presence of blocking factors in MS sera that prevented NK killing of virus-infected targets may in fact relate to circulating IC or PGE induction.

Not much has been done in studying the natural immune response in other neurological diseases. NK activity is depressed in the female myasthenia gravis patients compared to their female controls.[193] Whereas epileptic patients show a normal proliferative response to mitogens such as PHA, Con A, or PWM, there was a decrease in percent Leu11$^+$ cells and in NK activity. Interestingly, the NK defect was also seen in siblings.[194] Circulating immune complexes[195,196] and autoantibodies[196] in patients with subacute sclerosing panencephalitis (SSPE) and amyotrophic lateral sclerosis (ALS) suggest autoimmune-associated levels in these patients. Examination of the natural immune response will be of interest.

8. Conclusions

Although it is difficult to know in autoimmune diseases whether defective natural immunity has led to a viral infection and loss of self-tolerance or is merely the result of the aberrations in the normal immune response seen in these diseases, it seems clear that the defects are integrally intertwined with anomalous levels of immunomodulatory mediators and in the neurological diseases with altered neurotransmitters. A clearer picture of these interactions will hopefully lead us to the exact mechanism for problems in natural immunity in these chronic diseases.

References

1. Smith, K., 1980, T cell growth factor, *Immunol. Rev.* **51**:337–357.
2. De Vries, J. E., Figdor G. C., and Spits, H., 1983, Regulation of human NK activity against adherent tumor target cells by monocyte subpopulations, IL1 and IFNs, in: *NK Cells and Other Natural Effector Cells* (R. B. Herberman, ed.), Academic Press, New York, pp. 657–668.
3. Kuribayashi, K., Gillis, S., Kern, D. E., and Henney, C. S., 1981, Murine NK cell cultures: Effects of interleukin 2 and interferon on cell growth and cytotoxic reactivity, *J. Immunol.* **126**:2321–2327.
4. Kern, D. E., Gillis, S., Okada, M., and Henney, C. S., 1981, The role of interleukin 2 (IL2) in the differentiation of cytotoxic T cells: The effect of monoclonal anti IL2 antibody and absorption with IL2 dependent T cell lines. *J. Immunol.* **127**:1323–1328.
5. Trinchieri, G., and Santoli, D., 1978, Anti-viral activity induced by culturing lymphocytes with tumor derived or virus transformed cells, *J. Exp. Med.* **147**:1314–1333.
6. Herberman, R. B., Ortalda, J. R., and Bonnard, G. D., 1979, Augmentation by interferon of human natural and antibody-dependent cell mediated cytotoxicity, *Nature* **277**:221–223.
7. Saksela, E., and Timonen, T., 1980. Cellular interactions in the augmentation of human NK activity by interferon, *Ann. N.Y. Acad. Sci.* **350**:102–111.
8. Matheson, D. S., Green, B., and Tan, Y. J., 1981, Human interferons α and β inhibit T cell dependent and stimulate T cell independent mitogenesis and natural cell cytotoxicity: Relationship to chromosome 21, *Cell. Immunol.* **165**:366–372.
9. Herberman, R. B., Ortaldo, J. R., Rubinstein, M., and Pestka S., 1981, Augmentation of natural and antibody-dependent cell-mediated cytotoxicity by pure human leukocyte interferon, *J. Clin. Immunol.* **1**:149–153.
10. Dempsey, R. A., Dinarello, C. A., Mier, J. W., Rosenwasser, L. J., Allegretta, M., Brown, T. E., and Parkinson, D. R., 1982, The differential effects of human leukocytic pyrogen/lymphocyte-activating factor, T cell growth factor, and interferon on human natural killer cell activity, *J. Immunol.* **129**:2504–2510.
11. Donzig, W., Stadler, B. M., and Herberman, R. B., 1983, Interleukin 2 dependence of human natural killer (NK) activity, *J. Immunol.* **130**:1970–1973.
12. Handa, K., Suzuki, R., Matsui, K., Shimizu, Y., and Kumagai, K., Natural killer (NK) cells as a responder to interleukin 2 (IL2). II. IL2 induced interferon production, *J. Immunol.* **130**:988–992.
13. Ullberg, M., Merrill, J., and Jondal, M., 1981, Interferon-induced NK augmentation in human. An analysis of target recognition effector cell recruitment and effector cell recycling. *Scand. J. Immunol.* **14**:285–292.
14. Saksela, E., Timonen, T., Ranki, A., and Hayry, P., 1979, Morphological and func-

tional characterization of isolated effector cells responsible for human natural killer activity to fetal fibroblasts and to cultured cell line targets, *Immunol. Rev.* **44**:71–95.

15. Pape, G. R., Moretta, L., Troye, M., and Perlmann P., 1979, Natural cytotoxicity of human Fc receptor positive T lymphocytes after surface modulation with immune complexes, *Scand. J. Immunol.* **9**:291–299.

16. Merrill, J. E., Ullberg, M., and Jondal. M., 1981, Influence of IgG and IgM receptor triggering on human natural killer cell cytotoxicity measured on the level of the single effector cell, *Eur. J. Immunol.* **11**:536–541.

17. Passwell, J., Rosen, F. S., and Merler, E., 1980, The effect of Fc fragments on IgG on human mononuclear cell responses, *Cell. Immunol.* **52**:395–403.

18. Poleshuck, L. C., and Strausser, H. R., 1980, Immune complex induced prostaglandin production by monocytes of normal human subjects and cancer patients, *Prostaglandins Med.* **4**:3630–375.

19. Baker, P. E., Fahey, J. V., and Munck, A., 1981, Prostaglandin inhibition of T cell proliferation is mediated at two levels, *Cell. Immunol.* **61**:52–58.

20. Rappaport, R. S., and Dodge, G. R., 1982, Prostaglandin E inhibits the production of human interleukin 2, *J. Exp. Med.* **155**:943–948.

21. Brunda, M. J., Herberman, R. B., and Holden, H. T., 1980, Inhibition of murine natural killer cell activity by prostaglandins, *J. Immunol.* **124**:2682–2687.

22. Koren, H. S., Anderson, S. J., Fischer, D. G., Copeland, C. S.., and Jensen, P. J., Regulation of human natural killing. I. The role of monocytes, interferon, and prostaglandins, *J. Immunol.* **127**:2007–2013.

23. Jondal, M., Merrill, J., and Ullberg, M., 1981, Monocyte induced human natural killer cell suppression followed by increased cytotoxic activity during short term *in vitro* culture in autologous serum, *Scand. J. Immunol.* **14**:555–563.

24. Merrill, J. E., 1983, Natural (NK) and other antibody dependent cellular cytotoxicity (ADCC) activities can be differentiated by their different sensitivities to interferon and prostaglandin E_1, *J. Clin. Immunol.* **3**:42–50.

25. Herberman, R. B., and Ortaldo, J. R., 1981, Natural killer cells: Their role in defenses against disease, *Science* **214**:24–25.

26. Welsh, R. M., 1980, Natural killer cells in virus infections, in: *Current Topics in Microbiology and Immunology* (O. Haller, ed.), Springer-Verlag, Berlin, pp. 739–765.

27. Timonen, T., Saksela, E., Virtanen, I., and Cantell, K., 1980, Natural killer cells are responsible for the interferon production in human lymphocytes by tumor cell contact, *Eur. J. Immunol.* **10**:422–427.

28. Peter, H., Dallugge, H., Zavatvsky, R., Euler, S., Leibold, W., and Kirchner, H., 1980, Human peripheral null lymphocytes. II. Producers of type-1 interferon upon stimulation with tumor cells, herpes simplex virus and *Corynebacterium parvum*, *Eur. J. Immunol.* **10**:547–555.

29. Timonen, T, Ortaldo, J. R., and Herberman, R. B., 1981, Characteristics of human large granular lymphocytes and relationship to natural killer and K cells, *J. Exp. Med.* **153**:569–582.

30. Kato, T., and Minagawa, T., 1981, Enahncement of cytotoxicity of human peripheral blood lymphocytes by interferon, *Microbiol. Immunol.* **25**:837–850.

31. Shearn, M. A., Tu, W. H., Stephens, B. G., and Lee, J. C., 1970, Virus-like structures in Sjogren's syndrome, *Lancet* **1**:568–569.

32. Minato, N., Takeda, A., Kano, S., Takaku, F., 1982, Studies of the functions of natural killer-interferon systems in patients with Sjogren's syndrome, *J. Clin. Invest.* **69**:581–588.

33. Goto, M., Tanimoto, K., Shihara, T., and Horiuchi, Y., 1981, Natural cell-mediated cytotoxicity in Sjogren's syndrome and rheumatoid arthritis, *Arth. Rheum.* **24**:1377–1382.

34. Fox, R. I., Hugli, T. E., Lanier, L. L., Morgan, E. L., and Howell, F., 1985, Salivary gland lymphocytes in primary Sjogren's syndrome lack lymphocyte subsets defined by Leu7 and Leu11 antigens, *J. Immunol.* **135**:207–214.

35. Pedersen, B. K., Oxholm, P., Manthorpe, R., and Andersen, V., 1986, Interleukin 2 augmentation of the defective natural killer cell activity in patients with primary Sjögren's syndrome. *Clin. Exp. Immunol.* **63:**1–7.
36. Oshimi, K., Sumiya, M., Gonda, N., Kano, S., and Takaku, F., 1982, Natural killer cell activity in untreated systemic lupus erythematosus, *Ann. Rheum. Dis.* **41:**417–420.
37. Kaufman, D. B., 1982, Natural killer augmentation in systemic lupus erythematosus via a soluble mediator derived from human lymphocytes, *Arth. Rheum.* **25:**562–567.
38. Tsokos, G. C., Rook, A. H., Djeu, J. Y., and Balow, J. E., 1982, Natural killer cells and interferon responses in patients with systemic lupus erythematosus, *Clin. Exp. Immunol.* **50:**239–245.
39. Penschow, J., and MacKay, I. R., 1980, NK and T cell activity of human blood: Differences according to sex, age, and disease. *Ann. Rheum. Dis.* **39:**82–86.
40. McGarry, R. C., Roder, J. C., and Brunet, D., 1982, Mechanisms of natural killer cell depression in multiple sclerosis, in: *NK Cells and Other Natural Effector Cells* (R. B. Herberman, ed.), Academic Press, New York, pp. 1219–1225.
41. Uchida, A., Maida, E. M., Lenzhofer, R., and Micksche, M., 1982, Natural killer cell activity in patients with multiple sclerosis: Interferon and plasmapheresis, *Immunobiology* **160:**392–402.
42. Abb, J., Kaudewitz, P., Zander, H., Ziegler, H.-H. L., Dienhardt, F., and Riethmuller, R., 1982, Interferon (IFN) production and natural killer (NK) cell activity in patients with multiple sclerosis: Influence of genetic factors assessed by studies of monozygotic twins, in: *NK Cells and Other Natural Effector Cells* (R. B. Herberman, ed.), Academic Press, New York, pp. 1233–1240.
43. Hoffman, T., 1980, Natural killer function in systemic lupus erythematosus, *Arth. Rheum.* **23:**30–35.
44. Sibbitt, W. L., Froelich, C. J., and Bankhurst, A. D., 1984, Interferon alpha regulation of lymphocyte function in systemic lupus erythematosus, *Clin. Immunol. Immunopathol.* **32:**70–80.
45. Katz, P., Zaytoun, A. M., Lee, J. H., Panush, R. S., and Longley, S., 1982, Abnormal natural killer cell activity in systemic lupus erythematosus: An intrinsic defect in the lytic event, *J. Immunol.* **129:**1966–1971.
46. Sibbitt, W. L., Mathews, P. M., and Bankhurst, A. D., 1983, Natural killer cell in systemic lupus erythematosus: Defects in effector lytic activity and response to interferon and interferon inducers, *J. Clin. Invest.* **71:**1230–1239.
47. Sibbitt, W. L., Mathews, P. M., and Bankhurst, A. D., 1984, Impaired release of a soluble natural killer cytotoxic factor in systemic lupus erythematosus, *Arth. Rheum.* **27:**1095–1100.
48. Siilverman, S. L., and Cathcart, E. S., 1980, Natural killing in systemic lupus erythematosus inhibiting effects of serum, *Clin. Immunol. Immunopathol.* **17:**219–226.
49. Ytterberg, S. R., and Schnitzer, T. J., 1984, Inhibition of natural killer cell activity by serum from patients with systemic lupus erythematosus: Roles of disease activity and serum interferon, *Ann. Rheum. Dis.* **43:**457–461.
50. Goto, M., Tanimoto, K., and Horiuchi, Y., 1980, Natural cell mediated cytotoxicity in systemic lupus erythematosus, *Arth. Rheum.* **23:**1274–1281.
51. Rook, A. H., Tsokos, G. C., Quinnan, G. V., Balow, J. E., Ramsey, K. M., Stocks, N., Phelan, M. A., And Djeu, J. Y., 1982, Cytotoxic antibodies to natural killer cells in systemic lupus erythematosus, *Clin. Immunol. Immunopathol.* **24:**179–185.
52. Sibbitt, W. L., Gibbs, D. L., Kenny, C., Bankhurst, A. D., Searles, R. P., and Ley, K. D., 1985, Relationship between circulating interferon and anti interferon antibodies and imparied natural killer cell activity in systemic lupus erythematosus, *Arth. Rheum.* **28:**624–629.
53. Strannegard, O., Hermodsson, S., Westberg, G., 1982, Interferon and natural killer cells in systemic lupus erythematosus, *Clin. Exp. Immunol.* **50:**246–252.
54. Preble, O. T., Rothko, K., Klippel, J. H., Friedman, R. M., and Johnston, M. L., 1983, Interferon-induced 2'–5' adenylate synthetase *in vivo* and interferon production *in*

vitro by lymphocytes from systemic lupus erythematosus patients with and without correlating interferon, *J. Exp. Med.* **154:**2140–2146.

55. Preble, O. T., Black, R. J., Friedman, R. M., Klippel, J. H., and Vilcek, S. R., 1982, Systemic lupus erythematosus: Presence in human serum of an unusual acid labile leukocyte interferon, *Science* **216:**429–431.

56. Ytterberg, S. R., and Schnitzer, T. J., 1982, Serum interferon levels in patients with systemic lupus erythematosus, *Arth. Rheum.* **25:**401–406.

57. Panem, S., Check, I. J., Henriksen, D., and Vilcek, J., 1982, Antibodies to interferon in a patient with system lupus erythematosus, *J. Immunol.* **129:**1–3.

58. Neighbour, P. A., and Grayzel, A. I., 1981, Interferon production *in vitro* by leukocytes from patients with systemic lupus erythematosus and rheumatoid arthritis, *Clin. Exp. Immunol.* **45:**576–582.

59. Fitzharris, P., Alcocer, J., Stephens, H. A. F., Knight, R. A., and Snaith, M. L., 1982, Insensitivity to interferon of NK cells from patients with systemic lupus erythematosus, *Clin. Exp. Immunol.* **47:**110–118.

60. Hooks, J. J., Moutsopoulos, H. M., and Geis, S. A., 1979, Immune interferon in the circulation of patient with autoimmune disease, *N. Engl. J. Med.* **301:**5–8.

61. Talal, N., 1983, A clinician and a scientist look at acquired immune-deficiency syndrome (AIDS), *Immunol. Today* **4:**182–183.

62. Rook, A. H., Masur, H., Lane, H. C., Frederick, W., Kasahara, T., Mucher, A. M., Djeu, J. Y., Manischewitz J. F., Jackson, L., Fauci, A. S., and Quinnan, G. V., 1983, Interleukin-2 enhances the depressed natural killer and cytomegalovirus-specific cytotoxic activities of lymphocytes from patients with the acquired immunodeficiency syndrome, *J. Clin. Invest.* **72:**398–403.

63. Tsokos, G. C., Smith, P. L., Christian, C. B., Lipnick, R. N., Balow, J. E., and Djeu, J. Y., 1985, Interleukin-2 restores the depressed allogenic cell-mediated lympholysis and natural killer cell activity in patients with systemic lupus erythematosus, *Clin. Immunol. Immunopathol.* **34:**379–386.

64. Linker-Israel, M., Bakke, A., Kitridou, R. C., Gendler, S., Gillis, S., and Horwitz, D. A., 1983, Defective production of interleukin 1 and interleukin 2 in patients with systemic erythematosus (SLE), *J. Immunol.* **130:**2651–2655.

65. Alcocer-Varela, J, and Alarcon-Segovia, D., 1982, Decreased production of and response to interleukin 2 by cultured lymphocytes from patients with systemic lupus erythematosus, *J. Clin. Invest.* **69:**1388–1392.

66. Whicher, J. T., Gilbert, A. M., Westacott, C., Hutton, C., and Dieppe, P. A., 1986, Defective production of leukocytic endogenous mediator (interleukin-1) by peripheral blood leukocytes of patients with systemic clerosis, systemic lupus erythematosus, rheumatoid arthritis, and mixed connective tissue disease, *Clin. Exp. Immunol.* **65:**80–89.

67. Hooks, J. J., Moutsoupoulos, H. M., and Notkins, A. L., 1982, Circulating interferon in human autoimmune disease, *Texas Rep. Biol. Med.* **41:**164–168.

68. Neighbour, P. A., Reinitz, E., Grayzel, A. I., Miller, A. E., and Bloom, B. K., 1982, Studies of human NK cell functions in chronic diseases, in: *NK Cells and Other Natural Effector Cells* (R. B. Herberman, ed.), Academic Press, New York, pp. 1241–1248.

69. Reinitz, E., Neighbour, P. A., and Grayzel, A. I., 1982, Natural killer cell activity of mononuclear cells from rheumatoid patients measured by a conjugate-binding cytotoxicity assay, *Arth. Rheum.* **25:**1440–1444.

70. Doblong, J. H., Forre, O., Krien, T. K., Egeland, T., and Degre, M., 1982, Natural killer (NK) cell activity of peripheral blood synovial fluid, and synovial tissue lymphocytes from patients with rheumatoid arthritis and juvenile rheumatoid arthritis., *Ann. Rheum. Dis.* **41:**490–494.

71. Pedersen, B. K., Beyer, J. K., Klarlund, J., and Clemmensen, I. H., 1985, Baseline and interferon-enhanced natural killer cell activity in rheumatoid arthritis, *Acta. Pathol. Microbiol. Immunol. Scand.* **93:**79–84.

72. Kibler, R., Poulos, B. T., Stanfield, A. B., and Parsons, J. L., 1986, Cytotoxic activity

of enriched large granular lymphocyte populations in rheumatoid arthritis, *Clin. Exp. Rheumatol.* **4:**17–24.

73. McChesney, M. B., and Bankhurst, A. D., 1986, Cytotoxic mechanisms *in vitro* against Epstein Barr virus infected lymphoblastoid cell lines in rheumatoid arthritis, *Ann. Rheum. Dis.* **45:**546–552.

74. Fantini, F., Valenti, F., Mercuriali, F., Marin, F., and Panajotopoulos, N., 1986, Impaired natural killing activity in patients with rheumatoid arthritis. Clinical characteristics and a study of defective mechanisms, *Boll. Ist. Sieroter Milan* **65:**40–46.

75. Goto, M., and Zvaifler, N. J., 1985, Impaired killer cell generation in the autologous mixed leukocyte reaction by rheumatoid arthritis lymphocytes, *Arth. Rheum.* **28:**731–741.

76. Russell, A. S., and Miller, C., 1984, The activity of natural killer cells in patients with rheumatoid arthritis. I. The effect of drugs used *in vivo, Clin. Exp. Rheumatol.* **2:**227–229.

77. Combe, B., Pope., Darnell, B., and Talal, N., 1984, Modulation of natural killer cell activity in the rheumatoid joint and peripheral blood, *Scand J. Immunol.* **20:**551–558.

78. Silver, R. M., Redelman, D., Zvaifler, N. J., and Naides, S., 1982, Studies of rheumatoid synovial fluid lymphocytes. I. Evidence for activated natural killer (NK) like cells, *J. Immunol.* **128:**1758–1763.

79. Silver, R. M., 1986, Studies of rheumatoid synovial fluid lymphocytes. III. Phenotypic and functional analysis of natural killer cells, *Clin. Immunol. Immunopathol.* **39:**159–167.

80. Goto, M., and Zvaifler, N. J., 1985, Characterization of the natural killer like lymphocytes in rheumatoid synovial fluid, *J. Immunol.* **134:**1483–1486.

81. Combe, B., Pope, R., Darnell, B., Sany, J., and Talal, N., 1985, Regulation of natural killer activity by adherent cells in synovial rheumatoid medium, *Rev. Rheum. Mal. Osteoartic.* **52:**385–390.

82. Fox, R. I., Fong, S., Tsoukas, C., and Vaughan, J. H., 1984, Characterization of recirculating lymphocytes in rheumatoid arthritis patients: Selective deficiency of natural killer cells in thoracic duct lymph, *J. Immunol.* **132:**2883–2887.

83. Combe, B., Pope, R. M., Fishback, M., Darnell, B., Baron, S., and Talal, N., 1985, Interleukin 2 in rheumatoid arthritis: Production of and response to interleukin 2 in rheumatoid synovial fluid, synovial tissue, and peripheral blood, *Clin. Exp. Immunol.* **59:**520–528.

84. Fontana, A., Hengartner, H., Weber, E., Fehr, K., Grob, P. J., and Cohen, G., 1982, Interleukin 1 acitivty in the synovial fluid of patients with rheumatoid arthritis, *Rheumatol. Int.* **2:**49–53.

85. Shore, A., Jaglal, S., and Keystone, E. C., 1986, Enhanced interleukin 1 generation by monocytes *in vitro* is temporally linked to an early event in the onset or exacerbation of rheumatoid arthritis. *Clin. Exp. Immunol.* **65:**293–302.

86. Combe, B., Pope, R., Darnell, B., Kincaid, W., and Talal, N., 1984, Regulation of natural killer cell activity by macrophages in the rheumatoid joint and peripheral blood, *J. Immunol.* **133:**709–713.

87. Koch, B., Locher, P., Burmester, G. R., Mohr, W., and Kalden J. R., 1984, The tissue architecture of synovial membranes in inflammatory and non inflammatory joint diseases. II. The localization of mononuclear cells as detected by monoclonal antibodies directed against T lymphocyte subsets and natural killer cells, *Rheumatol. Int.* **4:**79–85.

88. Grimm, E. A., Mazumder, A., Zhang, H. Z., and Rosenberg, S. A., 1982, Lymphokine activated killer cell phenomenon. Lysis of natural killer resistant fresh solid tumor cells by IL2 activated autologous human peripheral blood lymphocytes, *J. Exp. Med.* **155:**1823–1827.

89. Aichinger, G., Fill, H., and Wick, G., 1985, *In situ* immune complexes, lymphocyte subpopulations, and HLA-DR-positive epithelial cells in Hashimoto's thyroiditis, *Lab. Invest.* **52:**132–140.

90. Del Prete, G. F., Maggi, E., Mariotti, S., Tiri, A., Vercelli, D., Parronchi, P., Macchia,

D., Pinchera, A., Ricci, M., and Romagnani, S., 1986, Cytolytic T lymphocytes with natural killer activity in thyroid infiltrate of patients with Hashimoto's thyroiditis: Analysis at clonal level, *J. Clin. Endocrinol. Metab.* **62**:52–57.

91. Seybold, D., Ryan, E. A., and Wall, J. R., 1981, Natural cytotoxicity of blood mononuclear cells from normal subjects and patients with Hashimoto's thyroiditis against normal thyroid cells, *J. Clin. Lab. Immunol.* **6**:241–244.

92. Chow, A., Baur, R. J., Schleusener, H., and Wall, J. R., 1983, Natural cytotoxicity of peripheral blood leukocytes from normal subjects and patients with Hashimoto's thyroiditis against human adult and fetal thyroid cells, *Life Sci.* **32**:67–75.

93. Sack, J., Baker, J. R., Weetman, A. P., Wartofsky, L., and Burman, K. D., 1986, Killer cell activity and antibody dependent cell mediated cytotoxicity are normal in Hashimoto's disease, *J. Clin. Endocrinol. Metab.* **62**:1059–1064.

94. Santoli, D., and Koprowski, H., 1979, Mechanisms of activation of human natural killer cells against tumor and virus infected cells, *Immunol. Rev.* **44**:125–163.

95. Trinchieri, G., Granato, D., and Perussia, B., 1981, Interferon induced resistance of fibroblasts to cytolysis mediated by natural killer cells: Specificity and mechanism, *J. Immunol.* **126**:335–340.

96. Nair, P. N., and Schwartz, S. A., 1981, Suppression of normal killer activity and antibody dependent cellular cytotoxicity by cultured human lymphocytes, *J. Immunol.* **126**:2221–2229.

97. Hutteroth, T. H., Poralla, T., and Meyerzum Buschenfelde, K.-H., 1981, Spontaneous cell-mediated (SCMC) and antibody-dependent cellular cytotoxicity (ADCC) in patients with acute and chronic active hepatitis, *Klin. Wochensch.* **59**:699–706.

98. Serdengecti, S., Jones, D. B., Holdstock, G., and Wright, R., 1981, Natural killer activity in patients with biopsy-proven liver disease, *Clin. Exp. Immunol.* **45**:361–364.

99. Feldman, J.-L., Becker, M. J., Moutsopoulos, H., Fye, K., Blackman, M., Epstein, W. V., and Talal, N., 1976, Antibody dependent cell mediated cytotoxicity in selected autoimmune disease, *J. Clin. Invest.* **58**:173–179.

100. Michalkiewicz, J., 1978, Immunological characteristics of lymphocytes in snyovial fluid and peripheral blood in patients with rheumatoid arthritis, *Arch. Immunol. Ther. Exp.* **26**:801–805.

101. Calder, E. A., Penhale, W. J., McLeman, D., Barnes, E. W., and Irvine, W. J., 1973, Lymphocyte dependent antibody-mediated cytotoxicity in Hashimoto's thyroiditis, *Clin. Exp. Immunol.* **14**:153–158.

102. Wasserman, T., Von Stedingle, L.-V., Perlmann, P., and Jonsson, J., 1974, Antibody-induced *in vitro* lymphocyte cytotoxicity in Hashimoto's thyroiditis, *Int. Arch. Allergy* **47**:473–482.

103. Hauser, S. L., Ault, K. A., Levin, M. J., Garavoy, M. R., and Weiner, H. L., 1981, Natural killer cell activity in multiple sclerosis, *J. Immunol.* **127**:1114–1117.

104. Merrill, J. E., Scott, A., Myers, L., and Ellison, G., 1982, Cytotoxic activity of peripheral blood and cerebrospinal fluid lymphocytes from patients with multiple sclerosis and other neurological diseases. *J. Neuroimmunol.* **3**:123–138.

105. Wang, P. W., Hiromatsu, Y., Laryea, E., Wosu, L., How, J., and Wall, J. R., 1986, Immunologically mediated cytotoxicity against human eye muscle cells in Graves opthalmopathy, *J. Clin. Endocrinol. Metab.* **63**:316–322.

106. Auer, I. O., and Ziemer, E. 1980, Immune status in Crohn's disease: *In vitro* antibody dependent cell mediated cytotoxicity in peripheral blood, *Klin. Wochenschr.* **58**:779–787.

107. Koren, H. S., Amos, D. B., and Buckley, R. H., 1978, Natural Killing in immunodeficient patients, *J. Immunol.* **120**:796–799.

108. Lipinski, M., Virelizier, H., Turza, T., and Giriscelli, C., 1980, Natural killer and killer cell activities in patients with primary immunodeficiencies or defects in immune interferon production, *Eur. J. Immunol.* **10**:246–249.

109. Iwatani, Y., Amino, N., Kabutomoni, O., Mori H., Tomaki, H., Motoi, S., Izumiguchi, Y., and Miyai, K., 1984, Decrease of peripheral large granular lymphocytes in Graves disease, *Clin. Exp. Immunol.* **55:**239–244.

110. Shy, M. E., Gabel, C. A., Vietorisz, E. C., and Latov, N., 1986, Characterization of oligosaccharides that bind to human anti MAG antibodies and the mouse monoclonal antibody HNK-1, *J. Neuroimmunol.* **12:**291–298.

111. Schuller-Petrovic, S., Gebhart, W., Lassmann, H., Rumpold, H., and Kraft, D., 1983, A shared antigenic determinant between natural killer cells and nervous tissue, *Nature* **306:**179–181.

112. Lipinski, M., Braham, K., Caillaud, J.-M., Carlo, C., and Tursz, T., 1982, HNK-1 antibody detects an antigen expressed on neuroectodermal cells, *J. Exp. Med.* **158:**1775–1780.

113. McGarry, R. C., Helfand, S. L., Quarles, R. H., and Rocter, J. C., 1983, Recognition of myelin associated glycoprotein by the monoclonal antibody HNK-1, *Nature* **306:**376–378.

114. Tischler, A. S., Mobtaker, H., Mann, K., Nunnemachei, G., Jason, W. J. Dayal, Y., Delellis, R. A., Adelman, L., and Woffe H. J., 1986, Anti-lymphocyte antibody Leu7 (HNK-1) recognizes a constituent of neuroendocrine granule matrix, *J. Histochem. Cytochem.* **34:**1213–1216.

115. Murray, N., and Steck, A. J., 1984, Indication of a possible role in a demyelinating neuropathy for an antigen shared between myelin and NK cells, *Lancet* **1:**711–713.

116. Strom, T. B., Lane, M.-A., and Heldeman, J. H., 1978, Immunoregulation by hormones and neurotransmitters, *Proc. VII Int. Cong. Nephrol.* **1978:**585–589.

117. Melmon, K. L., Rocklin, R. E., and Rosenkranz, R. P., 1981, Autocoids as modulators of the inflammatory and immune response, *Am. J. Med.* **71:**100–106.

118. Foris, G., Gyimesi, E., and Komaromi, I., 1985, The mechanism of antibody dependent cellular cytotoxicity stimulation by somatostatin in rat peritoneal macrophages, *Cell. Immunol.* **90:**217–225.

119. Van Dyke, C., Stesin, A., Jones, R., Chuntharapai, A., and Seaman, W., 1986, Cocaine increases natural killer cell activity, *J. Clin. Invest.* **77:**1387–1390.

120. Tonnesen, E., Tonneseu, J., and Christensen, N. J., 1984, Augmentation of cytotoxicity by natural killer (NK) cells after adrenaline administration in man, *Acta Pathol. Microbiol. Immunol. Scand.* **92:**81–83.

121. Crook, R. B., Farber, M. B., and Prusiner, S. B., 1984, Hormones and neurotransmitters control cyclic AMP metabolism in cheroid plexus epithelial cells, *J. Neurochem* **42:**340–350.

122. Hellstrand, K., Hermodsson, S., and Strannegard, O., 1985, Evidence for a beta-adrenoceptor-mediated regulation of human natural killer cells, *J. Immunol.* **134:**4095–4099.

123. Panula, P., Yang, H. Y. T., and Costa, E., 1984, Histamine containing neurons in the rat hypothalamus, *Proc. Natl. Acad. Sci. USA* **81:**1572–1575.

124. Parker, C. W., 1984, Mediators: Release and function, in: *Fundamental Immunology* (W. Paul, ed.), Raven Press, New York, pp. 717–747.

125. Lithicum, D. S., and Frelinger, J. A., 1982, Acute autoimmune encephalomyelitis in mice. II. Susceptibility is controlled by the combination of H_2 and histamine sensitization genes, *J. Exp. Med.* **155:**31–40.

126. Reches, A., Ovadia, H., and Abramsky, O., 1985, Neurotransmitter-depleting agents inhibit the development of experimental allergic encephalomyelitis (EAE), *Neurology* **35:**299–303.

127. Shearer, G. M., Melmon, K. K., Weinstein, Y., and Sela, M., 1982, Regulation of antibody response by cells expressing histamine receptors, *J. Exp. Med.* **136:**1302–1307.

128. Ballet, J. J., and Merter, E., 1976, The separation and reactivity *in vitro* of a subpopulation of human lymphocytes which bind histamine: Correlation of histamine reactivity with cellular maturation, *Cell. Immunol.* **24:**250–269.

129. Schwartz, A., Sutton, S. L., Askenase, P. W., and Gershon, R. K., 1981, Histamine inhibition of conconavalin A–induced suppressor T cell activation, *Cell. Immunol.* **60:**426–439.

130. Krug, U., Krug, F., and Cuatrecasas, P., 1972, Emergence of insulin receptors on human lymphocytes during *in vitro* transformation, *Proc. Natl. Acad. Sci. USA* **69:**2604–2608.

131. Gelfand, E. W., Ipp, M. M., and Riordan, J. R., 1982, Insulin modulation of antibody dependent cytotoxicity and the detection of antireceptor antibodies, *J. Lab. Clin. Med.* **99:**39–45.

132. Lang, I., Torok, K., Gergely, P., Kekam, K., and Petranyl, G., 1981, Effect of histamine receptor blocking on human antibody-dependent cell mediated cytotoxicity, *Scand. J. Immunol.* **12:**361–366.

133. Griswold, D. E., Alessi, S., Badger, A. M., Poste, G., and Hanna, N., 1984, Inhibition of T suppressor cell expression by histamine type 2 (H2) receptor antagonists, *J. Immunol.* **132:**3054–3057.

134. Lang, I., Gergely, P., and Petranyi, G. Y., 1981, Effect of histamine receptor blocking on human spontaneous lymphocyte mediated cytotoxicity, *Scand. J. Immunol.* **14:**573–576.

135. Nair, M. P. N., and Schwartz, S. A., 1983, Effect of histamine and histamine antagonists on natural and antibody dependent cellular cytotoxicity of human lymphocytes *in vitro*, *Cell. Immunol.* **81:**45–60.

136. Hellstrand, K., and Hermodsson, S., 1986, Histamine H_2 receptor mediated regulation of human natural killer cell activity, *J. Immunol.* **137:**656–660.

137. Droge, W., Schmidt, H., Nick, S., and Sonsky, B., 1986, Histamine augments interleukin 2 production and the activation of cytotoxic T lymphocytes, *Immunopharmacology* **11:**1–6.

138. Nair, M. P. N., Cilik, J. M., and Schwartz, S. A., 1986, Histamine induced suppressor factor inhibition of NK cells: Reversal with interferon and interleukin 2, *J. Immunol.* **136:**2456–2462.

139. Carlsson, R., Dohlsten, M., and Sjogren, H. O., 1985, Histamine modulates the production of interferon and interleukin 2 by mitogen activated human mononuclear blood cells, *Cell. Immunol.* **96:**104–112.

140. Flodgren, P., and Sjogren, H. O., 1985, Influence *in vitro* on NK and K cell activities by cimetidine and indomethacine with and without simultaneous exposure to interferon, *Cancer Immunol. Immunotherapy* **19:**28–34.

141. Smith, E. M., and Blalock, J. E., 1981, Human lymphocyte production of corticotropin and endorphin-like substances: Association with leukocyte interferon, *Proc. Natl. Acad. Sci. USA* **78:**7530–7534.

142. Blalock, J. E., and Smith, E. M., 1985, A complete regulatory loop between the immune and neuroendocrine systems, *Fed. Proc.* **44:**108–111.

143. Hall, N. R., McGillis, J. P., Spangels, B. L., Healy, D. L., and Goldstein, A. L., 1985, Immunomodulatory peptides and the central nervous system, *Springer Sem. Immunopathol.* **8:**153–164.

144. Kay, N., Allen, J., and Morley, J. E., 1984, Endorphins stimulate normal human peripheral blood lymphocyte natural killer activity, *Life Sci.* **35:**53–59.

145. Froelich, C. J., and Bankhurst, A. D., 1984, The effect of beta endorphin on natural cytotoxicity and antibody dependent cellular cytotoxicity, *Life Sci.* **36:**261–265.

146. Faith, R. E., Liang, H. J., Murgo, A. J., and Plotnikoff, N. P., 1984, Neuroimmunomodulation with enkephalins: Enhancement of human natural killer (NK) cell activity *in vitro*, *Clin. Immunol. Immunopathol.* **31:**412–418.

147. Wybran, J., 1984, Enkephalins and endorphins as modifiers of the immune system: Present and future, *Fed. Proc.* **44:**92–99.

148. Mandler, R. N., Biddison, W. E., Mandler, R., and Senate, S. A., 1986, Beta endorphin

augments the cytolytic activity and interferon production of natural killer cells, *J. Immunol.* **136:**934–939.

149. Wybran, J., 1985, Enkephalins and endorphins: Activation molecules for the immune system and natural killer activity, *Neuropeptides* **5:**371–374.

150. Plotnikoff, N. P., Murgo, A. J., Miller, G. C., Corder, C. N., and Faith, R. E., 1985, Enkephalins: Immunomodulators, *Fed. Proc.* **44:**118–122.

151. Kalland, T., 1986, Interleukin 3 is a major negative regulator of the generation of natural killer cells from bone marrow precursors, *J. Immunol.* **137:**2268–2271.

152. Birchenall-Sparks, M. C., Farrar, W. L., Rennick, D., Kiliau, P. L., and Roscetti, F. W., 1986, Regulation of expression of the interleuken 2 receptor on hematopoietic cells by interleukin 3, *Science* **233:**455–458.

153. Bernstein, H. A., 1986, Is prostaglandin E_2 involved in the pathogenesis of fever: Effects of interleukin 1 on the release of prostaglandins, *Yale J. Biol. Med.* **59:**151–158.

154. Frei, K., Bodmer, S., Schwerdel, C., and Fontana, A., 1985, Astrocytes of the brain synthesize interleukin 3 like factors, *J. Immunol.* **135:**4044–4047.

155. Giulian, D., Baker, T. J., Shih, L. C. N., and Lachman, L., 1986, Interleukin 1 of the central nervous system is produced by ameboid microglia, *J. Exp. Med.* **164:**594–604.

156. Degre, M., Dahl, H., Vandvik, B., 1981, Interferon in the serum and cerebrospinal fluid of patients with multiple sclerosis and other neurological disorders, *Acta Neurol. Scand.* **53:**152–160.

157. Santoli, D., Hall, W., Kastrukoff, L., Lisak, K. P., Perussia, B., Trinchieri, G., and Koprowski, H., 1981, Cytotoxic activity and interferon production by lymphocytes from patients with multiple sclerosis, *J. Immunol.* **126:**1274–1278.

158. Kaudewitz, P., Sander, H., Abb, J., Siegler-Heitbrock, H. W., and Riethmuller, G., 1983, Genetic influence on natural cytotoxicity and interferon production in multiple sclerosis studies in monozygotic discordant twins, *Hum. Immunol.* **7:**51–58.

159. Neighbour, P. A., and Bloom, B. R., 1979, Absence of virus-induced lymphocyte suppression and interferon production in multiple sclerosis, *Proc. Natl. Acad. Sci. USA* **76:**476–480.

160. Neighbour, P. A., Miller, A. E., and Bloom, B. R., 1981, Interferon responses of leukocytes in multiple sclerosis, *Neurology* **31:**561.

161. Benczur, M., Petranyi, G. G., Palffy, G., Varga, M., Talas, M., Kotsy, B., Foldes, I., and Hollan, S. R., 1980, Dysfunction of natural killer cells in multiple sclerosis: A possible pathogenic mechanism, *Clin. Exp. Immunol.* **39:**657–662.

162. Gyodi, E., Benczur, M., Palffy, G., Talas, M., Petranyi, G., Foldes, I., and Hollan, S. R., 1982, Association between HLA B7, DR2, and dysfunction of natural and antibody-mediated cytotoxicity without connection with the deficient interferon production in multiple sclerosis, *Hum. Immunol.* **4:**209–217.

163. Salonen, R., Ilonen, J., Reuanen, M., and Salmi, A., 1982, Defective production of interferon α associated with HLA DW2 antigen in stable multiple sclerosis, *J. Neurol. Sci.* **55:**197–206.

164. Haehr, S., Moller-Larsen, A., and Pedersen, E., 1983, Immunological parameters in multiple sclerosis patients with special reference to the herpes virus group, *Clin. Exp. Immunol.* **51:**197–206.

165. Vervliet, G., and Schandene, L., 1985, *In vitro* correction of the interleukin-2 and interferon gamma defect in multiple sclerosis, *Clin. Exp. Immunol.* **61:**556–561.

166. Hirsch, R., Panitch, H. S., and Johnson, K. P., 1985, Lymphocytes from multiple sclerosis patients produce elevated levels of gamma interferon *in vitro, J. Clin. Immunol.* **5:**386–389.

167. Neighbour, P. A., 1984, Studies of interferon production and natural killing by lymphocytes from multiple sclerosis patients, *Ann. N.Y. Acad. Sci.* **436:**181–191.

168. Jacobs, L., O'Malley, J., Freeman, A., and Ekes, R., 1981, Intrathecal interferon reduces exacerbations in multiple sclerosis, *Science* **214:**1026–1028.

169. Jacobs, L., O'Malley, J., Freeman, A., Murawski, J., and Ekes, R., 1982, Intrathecal interferon in multiple sclerosis, *Arch. Neurol.* **39**:609–615.

170. Albala, M. M., Davignon, D., Fast, L. D., and Clark, D. D., 1985, Normal T cell subsets and lymphocyte activity in multiple sclerosis, *Clin. Exp. Immunol.* **61**:542–547.

171. Benczur, M., Gyodi, E., Petranyi, G., Hollan, S. R., Palffy, G., Talas, M., Stoger, I., and Foldes, I. 1982, Impaired natural killer cell function in multiple sclerosis and association with the HLA system, in: *NK Cells and Other Natural Effector Cells* (R. B. Herberman, ed.), Academic Press, New York, pp. 1227–1232.

172. Merrill, J. E., Jondal, M., Seeley, J., Ullberg, M., and Siden, A., 1981, Decreased NK killing in patients with multiple sclerosis: An analysis on the level of the single effector cell in peripheral blood and cerebrospinal fluid in relation to disease activity, *Clin. Exp. Immunol.* **47**:419–430.

173. Merrill, J. E., Gerner, R. H., Myers, L. W., and Ellison, G. W., 1983, Regulation of natural killer cell cytotoxicity by prostaglandin E in the peripheral blood and cerebrospinal fluid of patients with multiple sclerosis and other neurological diseases. I. Association between amount of prostaglandin produced, natural killer, and endogenous interferon, *J. Neuroimmunol.* **4**:223–237.

174. Merrill, J. E., Myers, L. W., and Ellison, G. W., 1983, Regulation of natural killer cell cytotoxicity by prostaglandin E in the peripheral blood and cerebrospinal fluid of patients with multiple sclerosis and other neurological diseases. II. Effect of exogenous PGE$_1$ on spontaneous and interferon-induced natural killer, *J. Neuroimmunol.* **4**:239–251.

175. Merrill, J. E., Gerner, R. H., Myers, L. W., and Ellison G. W., 1983, Regulation of NK activity and IFN production by PGE in the peripheral blood and cerebrospinal fluid of patients with multiple sclerosis and other neurological diseases, in: *Intercellular Communication in Leukocyte Function* (J. W. Parker and R. L. O'Brien, eds.), Wiley, New York, pp. 79–84.

176. Neighbour, P. A., Grayzel, A. I., and Miller, A. E., 1982, Endogenous and interferon-augmented natural killer cell activity of human peripheral blood mononuclear cells *in vitro*. Studies of patients with multiple sclerosis, systemic lupus erythematosus, or rheumatoid arthritis, *Clin. Exp. Immunol.* **49**:11–21.

177. Hirsch, R. L., and Johnson, K. P., 1985, The effect of recombinant alpha-2 interferon on defective natural killer cell activity in multiple sclerosis, *Neurology* **35**:597–600.

178. Heltberg, A., Kalland, T., Kallen, B., and Nilsson, O., 1985, A study of some immunological variables in twins discordant for multiple sclerosis, *Eur. Neurol.* **24**:361–373.

179. Oger, J., Kastrukoff, L., O'Gorman, M., and Paty, D. W., 1986, Progressive multiple sclerosis: Abnormal immuno functions *in vitro* and aberrant correlation with enumeration of lymphocyte subpopulations, *J. Neuroimmunol.* **12**:37–48.

180. Merrill, J. E., Ellison, G. W., and Myers, L. W., 1984, Cytotoxic activity of peripheral blood and cerebrospinal fluid lymphocytes from patients with multiple sclerosis and other neurological diseases: Analysis at the single cell level of the relationship of cytotoxic effectors and interferon producing cells, *Clin. Immunol. Immunopathol.* **31**:390–402.

181. Abb, J., Zander, H., Abb, H., Albert, E., and Diehnardt, F., 1983, Associating human leukocyte low responsiveness to inducers of interferon alpha with HLA-DR2, *Immunology* **49**:239–244.

182. Zander, H., Abb, J., Kaudewitz, P., and Riethmuller, G., 1982, Natural killing activity and interferon production in multiple sclerosis, *Lancet* **1**:280–283.

183. Hirsch, R. L., and Johnson, K. P., 1985, Natural killer cell activity in multiple sclerosis patients treated with recombinant interferon 2, *Clin Immunol. Immunopathol.* **37**:236–244.

184. Breard, J., Reinherz, E. L., O'Brien, C., and Schlossman, S. F., 1981, Delineation of an effector population responsible for natural killing and antibody cellular cytotoxicity in man, *Clin. Immunol. Immunopathol.* **18**:145–150.

185. Rumpold, H., Kraft, D., Obexer, G., Radaszkiewicz, T., Majdic, O., Bettelheim, P., Knapp, W., and Bock, G., 1983, Phenotypes of human large granular lymphocytes as defined by monoclonal antibodies. *Immunobiology* **164**:51–62.
186. Gastl, G., Niederwieser, D., Marth, C., Huber, H., Egg, D., Schuler, G., Margreiter, R., Braunsteiner, H., and Huber, C., 1984, Human large granular lymphocytes and their relationship to natural killer cell activity in various disease states, *Blood* **64**:288–295.
187. Abo, T., Cooper, M. D., and Balch, C. M. 1982, Characterization of HNK-1$^+$ (Leu 7) human lymphocytes. I. Two distinct phenotypes of human NK cells with different cytotoxic capability, *J. Immunol.* **129**:1752–1757.
188. Tilden, A. B., Abo, T., and Balch, C. M., 1983, Suppressor cell function of human granular lymphocytes identified by the HNK-1 (Leu 7) monoclonal antibody, *J. Immunol.* **130**:1171–1175.
189. Perussia, B., Fanning, V., Trinchieri, G., 1983, A Human NK and K cell subset shares with cytotoxic T cells expression of the antigen recognized by antibody OKT8, *J. Immunol.* **131**:223–231.
190. Tanaka, M., Sato, S., and Miyatake, T., 1985, Human peripheral lymphocytes defined by anti-myelin associated glycoprotein antiserum in healthy individuals and in patients with multiple sclerosis, *Acta Neurol. Scand.* **71**:278–283.
191. Johnson, D., Sato, S., Quarles, R. H., Inuzuka, T., Brady, R. O., and Tourtellotte, W. W., 1986, Quantitation of the myelin associated glycoprotein in human nervous tissue from controls and multiple sclerosis patients, *J. Neurochem.* **46**:1086–1093.
192. Rola-Plesczynski, M., Abernathy, M., Vincent, M. M., Hansen, S. A., and Bellanti, J. A., 1976, Lymphocyte mediated cytotoxicity to viruses in patients with multiple sclerosis: Presence of blocking factor, *Clin. Immunol. Immunopathol.* **5**:165–172.
193. Rauch, H. C., Montgomery, I. N., and Kaplan, J., 1985, Natural killer cell activity in multiple sclerosis and myasthenia gravis, *Immunol. Invest.* **14**:427–434.
194. Margaretten, N. C., and Warren, R. P., 1986, Reduced natural killer cell activity and OKT4/OKT8 ratio in epileptic patients, *Immunol. Invest.* **15**:159–167.
195. Bartfeld, H., Dharm, C., Donnenfeld, H., Jashnani, L., Carp, R., Kascsak, R., Vilcek, J., Rapport, H., and Wallenstein, S., 1982, Immunological profile of amyotrophic lateral sclerosis responses to viral and CNS antigens, *Clin. Exp. Immunol.* **48**:137–147.
196. Salmi, A., Lynd, G., Ziola, B., and Reunanen, M., 1986, Circulating immune complexes in patients with subacute sclerosing panencephalitis, *Clin. Immunol. Immunopathol.* **41**:16–25.

19

Natural Killer Cells and Diabetes Mellitus

BRUCE A. WODA

1. Introduction

Diabetes mellitus has been recognized since antiquity by the classic syndrome consisting of polyuria, polydipsia, polyphagia, and glycosuria. Diabetes is a group of disorders that are associated with hyperglycemia. It is a common, serious disease of the young with a prevalence rate of 0.3% in individuals under 20 and 0.5% for individuals under 60 years of age.[1,2] Diabetes causes appreciable morbidity secondary to the complications that affect many organ systems.

Diabetes is classified as type I (insulin-dependent (IDDM), or juvenile-onset) or type 2 (non-insulin-dependent, or maturity-onset) diabetes. Type 2 diabetes is associated with the resistance of target tissues to insulin action. This chapter is concerned with our present knowledge of the role of the natural immune system in type I diabetes. Type I diabetes appears to be an autoimmune disease that is characterized by inflammation of the pancreatic islets of Langerhans (insulitis) which results in the destruction of the insulin-producing beta cells.[3,4] The other islet hormone-producing cells, the alpha and delta cells, remain unaffected by this destructive process. Because of the destruction of the insulin-producing beta cells, patients with IDDM depend on insulin for their survival. Insulin prevents death from diabetic ketoacidosis; however, it does not prevent the complications of the chronic disease, which include the generalized vascular disease which affects the coronary and peripheral arterial systems, the

BRUCE A. WODA • Department of Pathology, University of Massachusetts Medical Center, Worcester, Massachusetts 01605.

neuropathy; and the renal and retinal diseases that may result in renal failure and blindness.

The accumulated evidence about IDDM is compatible with the hypothesis that it is an autoimmune disease.[5] However, we understand little about the effector mechanisms that result in beta cell destruction. In this chapter we will examine the role of natural killer (NK) cells in human IDDM and appropriate animal models of the human disease.

2. Type I Diabetes Mellitus

Type I diabetes is a relatively frequent disease with a prevalence rate in Caucasian children and young adults of about 0.3%.[1,6] If a parent has IDDM, the risk to the children for the development of IDDM is between 5% and 10%.[7] The risk for the development of IDDM in siblings is similar. IDDM usually occurs under the age of 25 and has a sharp peak of onset in the early teen years. In patients who succumb during the acute illness, histologic examination of the pancreas reveals insulitis.[3,4] It has been reported that in some of these cases the beta cells express class II histocompatibility proteins[8] and could theoretically act as antigen-presenting cells. The disease is hereditable, and various genetic loci have been implicated in the transmission of the disease. There is an association between IDDM and the HLA complex.[7,9,10] Susceptibility to IDDM is associated with genes on chromosome 6 associated with the HLA-D region. More than 90% of Caucasian subjects with IDDM have HLA-DR3 and/or DR4 alleles.[11,12] The use of HLA-DQ B chain gene probes in a study of restriction fragment length polymorphism has shown differences between controls and IDDM patients.[13]

2.1. Alterations in the Peripheral Immune System

The results of the study of the peripheral immune system have been variable. There does not appear to be a gross alteration in the number of T cells or B cells or of the CD4:CD8 ratio.[14–18] The presence of increased numbers of peripheral blood T cells expressing DR antigens[18–20] has been reported, as has an increased density of DR on both CD4 and CD8 T cells.[18] This increased DR density was correlated with the titer of islet cell autoantibody and the presence of the HLA DR3/4 phenotype.

2.2. Humoral Immunity

Patients with newly diagnosed IDDM have autoantibodies directed against cytoplasmic[21,22] and surface antigens[23–25] of islet cells and against insulin.[26-28] The islet cell cytoplasmic antibody (ICA) is of the IgG type and is present in 0.5% of normal individuals and 70% of newly diagnosed patients with IDDM.[22,29] The presence of autoantibodies may be used as a marker for IDDM. In family studies the presence of ICA may predate the onset of diabetes by many years.[30,31] Relatives of patients who are ICA-positive may show diminished insulin release.[32] Islet cell surface antibody (ICSA) is present in about one-third of newly diagnosed patients with IDDM.[24] Such ICSA may inhibit insulin secretion and mediate beta cell destruction in a complement-dependent mechanism.[23-25]

2.3. Cellular Immunity

As of yet we know little about the pathogenesis of beta cell loss. Beta cell-specific CD4 or CD8 cells have not been found.

It has been shown that islet extracts inhibit lymphocyte migration in patients with IDDM.[33,34] Several studies have shown that peripheral blood lymphocytes from patients with IDDM will inhibit stimulated insulin release from xenogeneic rodent targets[35-37] or show cell-mediated cytotoxicity toward such targets.[38,39] It has been suggested that this activity is mediated by CD3$^+$ T cells and that this activity is enriched in the CD4$^-$ subset[36] or that this cytotoxic activity may be mediated by non-T cells.[38] The mechanism by which human T cells recognize xenogeneic targets has not been explained. Future studies should attempt to block this antiislet activity with anti-CD3 and anti-CD8 to demonstrate that the CD3–T cell receptor complex and the CD8 molecule are involved in this lytic process. These experiments should also more carefully characterize the effector cells responsible for antiislet activity.

2.4. Natural Killer Cells and Antibody-Dependent Cellular Cytotoxicity

The percentage of Leu 7$^+$ cells (presumably NK cells) has been reported to be decreased in newly diagnosed diabetics.[40,41] The patient group in these studies was composed of patients with a duration of IDDM from 2 days to 90 days. The Leu 7$^+$ cells were decreased to 9% from a control of 19% in one study[41] and to 7% from the control level of 17% in a second study.[40] The decrease in Leu 7 cells was correlated with a decrease in NK activity against the K562 target from 56% specific lysis in the control group to 37% in the patient group.[41] Normal NK activity was found in patients with long-term type I IDDM. A second group of investigators has reported increased K cells in 12 of 23 patients with IDDM studied within 3 days of onset.[42] Most patients with IDDM of longer duration had normal levels of K cells. In a second study by this group, the increased K cell levels were accompanied by a somewhat enhanced level of ADCC utilizing red blood cell targets.[43]

It has not been shown whether the alterations measured in the studies that have examined the relative percentage of NK cells and NK and ADCC activity are due to a role for NK cells in the pathogenesis of the disease or are secondary to the stress of the recent onset of diabetes.It should be noted that the differences reported in these studies may be due to the time differential between the onset of disease and the clinical study. Similar studies should be done utilizing the CD16 (Leu 11) and the Leu 19 markers, which are more specific for NK cells than the Leu 7 antibody or the presence of low-affinity sheep erythrocyte receptors. Leu 7 may be found on $CD8^+$ T cells, and the Leu 7^+ $CD16^+$ NK cell is less active than the Leu 7^- $CD16^+$ NK cell.[44,45] Hence, the relationship between the percentage of Leu 7 cells and the percentage of NK cells cannot be determined in the previously published studies.[40,41] It will also be necessary to study the NK system sequentially in a number of individuals at risk for the development of IDDM, such as ICA^+ siblings of patients with IDDM. However, examination of the peripheral activity of the NK system may not accurately reflect the process occurring in the pancreas.

Several studies have shown that ICSA is active in supporting ADCC.[38,46–49] These studies have used a human insulinoma cell line, a human beta cell clone, or rat islets incubated with sera from $ICSA^+$ individuals and lymphocytes from normal controls. In one study,[48] most of the $ICSA^+$ sera supported islet cell ADCC.

3. Animal Models

Animal models are useful for the study of IDDM, because the frequency of diabetes is high, and diabetes may occur at a predictable time. Animal models of IDDM include the administration of multiple doses of streptozotocin,[50] a beta cell cytotoxic nitrosourea, and the infection of certain mouse strains with encephalomyocarditis virus.[51] Two of the best models are the nonobese diabetic mouse (NOD) and the BioBreeding (BB) rat.

3.1. The NOD Mouse

The NOD mouse was derived from the noninbred ICR mouse in Japan.[52] The clinical features of this syndrome are similar to human IDDM. The insulitis in the NOD[52–55] begins at 4–6 weeks of age as a periductal and perivascular accumulation of lymphocytes. At 6–8 weeks, these lymphocytes invade the islets and destroy the beta cells. Diabetes is detectable at 3 months, and by 7 months 80% of the females and 20% of the males develop diabetes. The diabetic syndrome appears to require at least three recessive genes, one MHC linked and one linked to thy-1.[56,57] The NOD appears to have a unique class II type.[58]

Studies of the immune system of the NOD have shown variability. There are reports of decreased[59] and increased[55] T cells. Other studies have shown that the NOD has comparable numbers of T cells to most other inbred mouse strains with an unremarkable Ly 1:Ly 2 ratio and good mitogen responsiveness.[60] NOD mice exhibit antibodies against insulin, ICSA, beta cell proteins, and intracisternal type A virus particle protein.[53,60] ICSA and ICA are detectable prior to the onset of clinical diabetes. Immunopathological studies have shown that the insulitis lesions are composed of both helper and suppressor T cells.[53,55,61] Occasional cells stained positive with the anti-AGM1 antibody and were considered to be NK cells; however, as the distribution of AGM1 is broad, we cannot conclude that the AGM1-positive cells identified in these studies are NK cells.[55] Several groups have reported that the NOD has low peripheral NK activity,[59,62] and little NK activity is seen in lymphocytes isolated from the pancreas of acute diabetic mice.[63] This does not support a role for NK cells in the pathogenesis of diabetes in the NOD mouse but does not rule it out. It has been shown that treatment of young NOD females with the streptococcal cell wall protein OK 432 prevents diabetes in the NOD.[62] OK 432 enhances NK, T cell, and macrophage function.[64]

3.2. The BB Rat

The BB rat is the most widely used animal model of diabetes. It provides a model in which NK cells are the predominant cytotoxic lymphocyte and appear to be important in the mediation of diabetes.

Diabetes in the BB rat was first discovered in 1974 in a commercial colony of outbred Wistar-derived rats at the Bio-Breeding Laboratories of Ottawa, Canada.[65] The BB colony maintained at Worcester, MA (the BB/Wor rat), has two major sublines. The diabetes-prone (DP) rat is lymphopenic and has an incidence of diabetes in its various families that ranges from 60% to 80%.[66,67] The onset of diabetes is abrupt and occurs at about 90 days of age. Less than 0.5% develop diabetes prior to 60 days of age, and less than 15% after 120 days of age. There is an equal sex distribution. The animals are insulin-dependent, nonobese, ketotic, and glycosuric. Like human IDDM, the BB diabetic syndrome is a hereditable disorder. The inheritance of the diabetic syndrome appears to be an autosomal-recessive gene or gene cluster with a penetration of 50%.[66] The diabetes-resistant (DR) subline is nonlymphopenic and has a frequency of diabetes of about 1%.

The pancreatic islets of acute diabetic rats are infiltrated by mononuclear cells.[68] Normoglycemic rats may show insulitis without diabetes,

and as would be expected, insulitis precedes diabetes.[69,70] Not all animals
with insulitis develop overt diabetes. The character of the cellular infiltrate varies with the time of the biopsy. Early lesions are composed predominantly of T helper cells (T_H), macrophages, and OX 8$^+$ NK cells.[71]

3.2.1. Genetics

BB diabetes is a hereditable disease that involves at least one gene
associated with the major histocompatibility complex.[72–74] The class II
region of the RT 1^u (non-RT 1.A) allele appears to be required for diabetes.[74] It had been suggested that the expression of diabetes was linked
to a non-MHC gene responsible for lymphopenia[73]; however, it has recently been shown that spontaneous diabetes can be found at low frequency in the nonlymphopenic DR rat.[75]. DR rats were derived from a
DP subline in the fifth generation of brother-sister mating.[67] The DR rat
has had an incidence of diabetes of about 1% through 17 generations.
The nonlymphopenic DR rats that develop diabetes do so at about 60
days of age, significantly earlier than the lymphopenic DP rat.

3.2.2. Immunologic Abnormalities

The salient immunologic features of the BB/Wor rat are listed in Table I, and a description of the antibodies utilized to phenotype rat lymphocytes is listed in Table II. The DP rat displays severe T lymphopenia[76–79] and has decreased T_H cells and a virtual absence of cytotoxic
T cells (T_C).[80] The deficit in T_C was discovered by performing two-color
flow cytometric analysis of spleen and PBL cells (Fig. 1). Rat NK cells are
W3/13$^+$, OX 19$^-$, OX 8$^+$, and AGM1 bright,[80–85] and T_C are W3/13$^+$,
OX 8$^+$, OX 19$^+$, AGM1 variable.[79,85,86] When BB/Wor lymphocytes were
analyzed after staining with the OX 8 and OX 19 antibodies, the OX 8$^+$
OX 19$^+$ T_C population was virtually absent. In DR spleen, T_C are present
at a frequency of 17%, whereas in DP rats the frequency of phenotypic
T_C is less than 1%. Not only do DP rats have a phenotypic deficiency of

TABLE I
Immunologic Features of BB/Wor Rats

Diabetes-prone	Diabetes-resistant (W line)
Lymphopenic	Nonlymphopenic
Decreased T cells	Normal T cell numbers
Absent T_C cells	Normal T_C cells
Absent RT 6 T cells	Normal RT 6 T cells
Relative increase in NK cells	Low NK cells
Autoantibodies to thyroid colloid smooth muscle, gastric parietal cells. ICSA$^+$, ICA$^-$	Autoantibodies virtually absent

TABLE II
Antigenic Phenotype of Rat Lymphocytes

Cell type	Antigenic phenotype
T helper cells (T_H)	W3/25, W3/13, OX 19, OX 52, OX 34, RT 6±
T cytotoxic cells (T_C)	OX 8, W3/13, OX 19, OX 34, RT 6±
NK cells	OX 8, W3/13, OX 34, AGM1 (high density)

T_C, they also appear to be incapable of generating functional OX 8$^+$ OX 19$^+$ T_C.[87] When DP and DR rats are infected with LCMV, DR rats double their OX 8$^+$ OX 19$^+$ T_C by day 7 after infection. They generate T_C that are RT 1^u-restricted and virus-specific T_C. Such cells cannot be found in DP rats.

A second major cellular immune deficit in the DP rat is the lack of T cells that express the RT 6 alloantigen.[88] RT 6 is a T cell differentiation alloantigen that has two known alleles, the RT 6.1 and the RT 6.2. About half of the T_C and T_H cells in the DR rats express the RT 6.1 antigen.

The responsiveness of DP BB lymphocytes to *in vitro* stimulation by T cell mitogens such as concanavalin A is markedly impaired. Similarly, their ability to recognize alloantigens in a mixed lymphocyte reaction is diminished,[89–94] as is their ability to reject allogeneic transplants. It has been shown that macrophages from DP rats act as strong suppressor elements for the Con A response,[93,94] and the removal of macrophages normalizes the mitogenic response. Such suppressive macrophages are present in DP but not in DR rats. When T lymphocytes are isolated from DP rats by flow sorting of the OX 19$^+$ population, their mitogen response is normal, indicating that DP T cells are not intrinsically abnormal.[94] It is likely that the diminished mitogen responsiveness is due to the T lymphopenia and to the suppressor macrophages.

3.2.3. Autoantibodies

Antibodies to thyroid colloid, gastric parietal cells, and smooth muscle are commonly found in the DP rat.[95,96] ICSA are frequently present and may be detected as early as 40 days of age, and the highest antibody levels are present at the time of onset of diabetes.[97–100] ICA have not been detected. Autoantibodies are virtually absent in the DR rat.

3.2.4. The Natural Killer Cell System of the BB/Wor Rat

Rat NK cells are defined phenotypically by the W3/13$^+$, OX 19$^-$, OX 8$^+$, AGM1 bright phenotype.[80–85] These cells are easily quantitated by measuring the OX 8$^+$ OX 19$^-$ population utilizing two-color flow cytometric analysis. It is noteworthy that the rat NK cell is CD8 (OX 8)–posi-

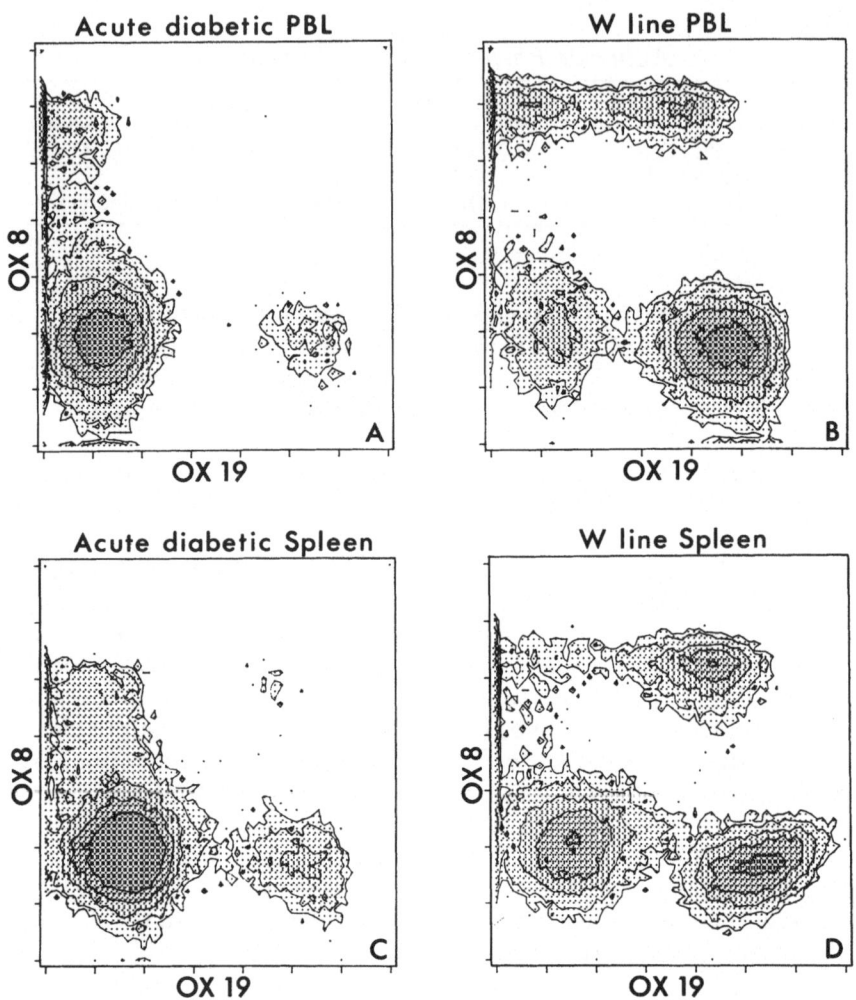

FIGURE 1. Contour plots of spleen and peripheral blood stained with OX 8 and F1 goat anti-mouse IgG and biotin OX 19 and phycoerythrin-avidin. (A) PBL from an acute diabetic shows 13.7% OX 8$^+$, 6% OX 19$^+$, 13.1% OX 8$^+$ OX 19$^-$ and 0.6% OX 8$^+$ OX 19$^+$. (B) PBL from W line (DR) shows 27.6% OX 8$^+$, 68.6% OX 19$^+$, 14.5% OX 8$^+$ OX 19$^-$ and 13.1% OX 8$^+$ OX 19$^+$. (C) Spleen from an acute diabetic shows 10.2% OX 8$^+$, 9.8% OX 19$^+$, 9.2% OX 8$^+$ OX 19$^-$ and 1% OX 8$^+$ OX 19$^+$. (D) Spleen from a W line (DR) shows 24.4% OX 8$^+$, 52.1% OX 19$^+$, 5.7% OX 8$^+$ OX 19$^-$ and 18.7% OX 8$^+$ OX 19$^+$. (Reproduced from Woda et al., J. Immunol. **136:**856, 1986.)

tive. Murine NK cells are CD 8-negative, and only a subset of human NK cells are contained in the CD8 population. When DP PBL are stained with OX 8 and rabbit anti-AGM1, an OX 8, bright AGM1$^+$ island is defined (Fig. 2).[85] The percent of cells in this island is well correlated with the number of OX 8$^+$ OX 19$^-$ phenotypic NK cells, suggesting that the two antibodies are defining the same cell system.

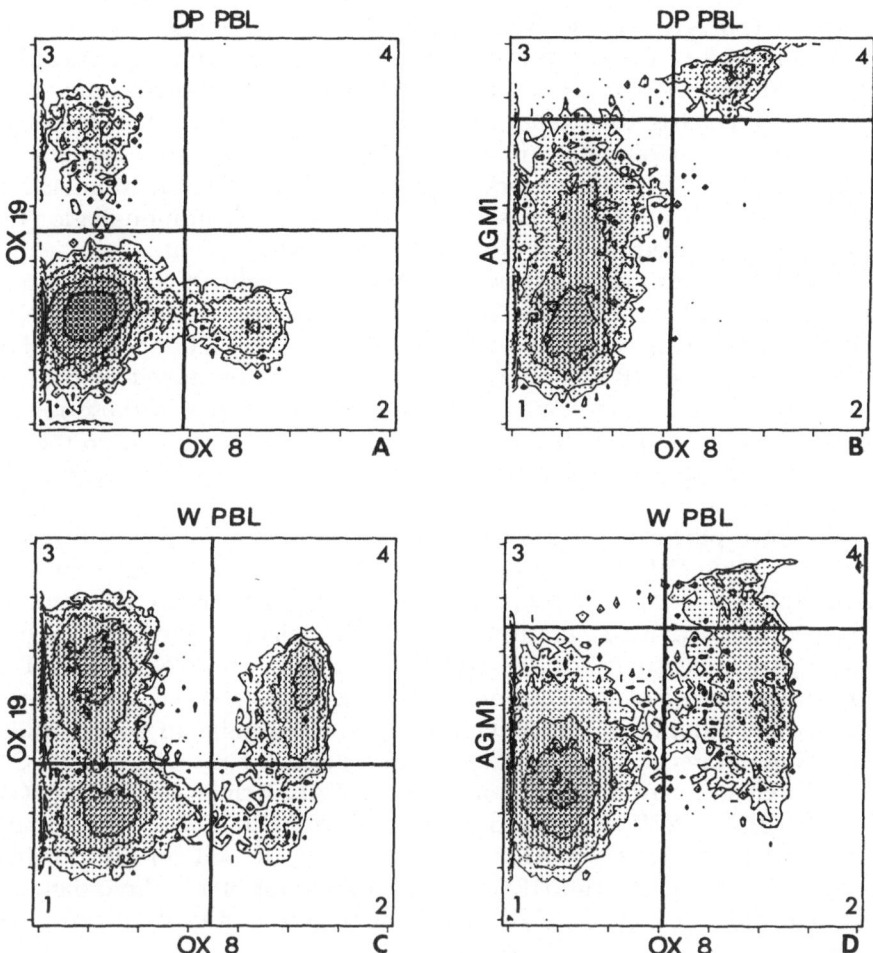

FIGURE 2. Contour plots of PBL stained with OX 19 and AGM1 or OX 19 and OX 8. (A) PBL from a DP rat stained with OX 8 and OX 19 shows (1) 80% OX 8$^-$ OX 19$^-$; (2) 9% OX 8$^+$ OX 19$^-$; (3) 10% OX 8$^-$ OX 19$^+$; (4) % OX 8$^+$ OX 19$^+$. (B) PBL from a DP rat stained with OX 8 and AGM1 shows (1) 85% OX 8$^-$ dull AGM1$^+$ cells; (2) 4% OX 8$^+$ dull AGM1$^+$ cells; (3) 3% OX 8$^-$ bright AGM1$^+$ cells; (4) 8% OX 8$^+$ bright AGM1$^+$ cells. (C) PBL from a W-line (DR) rat stained with OX 8 and OX 19 shows (1) 28% OX 8$^-$ and OX 19$^-$; (2) 4% OX 8$^+$ OX 19$^-$; (3) 46% OX 8$^-$ OX 19$^+$; (4) 22% OX 8$^+$ OX 19$^+$. (D) PBL from a W-line (DR) rat stained with OX 8 and AGM1 shows (1) 72% OX 8$^-$ dull AGM1$^+$ cells; (2) 22% OX 8$^+$ dull AGM1$^+$ cells; (3) 1% OX 8$^-$ bright AGM1$^+$ cells; (4) 5% OX 8$^+$ bright AGM1$^+$ cells. (Reproduced from Woda and Biron, *J. Immunol.* **137**:1860, 1986.)

Cells with NK activity in the BB/Wor rat reside in the OX 8$^+$ OX 19$^-$ compartment. In the DP rat, these cells are present at a frequency of about 8–13% in the peripheral blood and spleen.[80,85] About 55% of these cells are large granular lymphocytes (LGL), and 96% of the OX 8$^+$ cells are contained within this subset. As is true of other rat strains, DP LN contains a few NK cells, with a mean of about 2%. These data indicate that virtually all of the OX 8 cells of the BB/Wor rat are of the NK phenotype and that the DP rat has a virtual absence of phenotypic and functional T_C. We have not examined the unlikely possibility that, although T_C are absent peripherally, they might be expanded within insulitis lesions. The DP rat is a useful model for the study of NK function, as NK cells are relatively abundant, easily identifiable by their immunologic phenotype, and the predominant cytotoxic lymphocyte present.

The DR rat has two- to threefold fewer NK cells in the spleen and PBL than the DP rat. Concomitantly, the lytic activity of DP rats for NK-sensitive targets is greater than that of DR rats.[85,101] The increased NK activity seen in the DP is somewhat more than that predicted by the frequency of NK cells alone. When performing NK studies in these rats, it is advisable to utilize animals less than 90 days of age, because, even though the proportion of NK cells remains relatively constant, their activity diminishes with age.

We have performed several experiments to examine the differences in NK activity in DP and DR rats.[85] The DP rat exhibits low levels of lytic activity for NK-susceptible targets such as K562, whereas the DR rat does not show such activity, suggesting that DP NK cells are activated. To better compare the activity of DP and DR NK cells, the OX 8^+ OX 19^- populations were isolated by flow sorting. DP rats showed greater lytic activity even though the percentage of LGL was similar in both populations. These experiments showed that on a per-cell basis, DP OX 8^+ OX 19^- cells elicited three times more lysis of YAC-1 than DR OX 8^+ OX 19^- cells (DP = 21% ± 3 specific ^{51}Cr release, DR = 7% ± 2 specific ^{51}Cr release; E:T ratio = 5:1). The percentage of LGL in the purified fractions was similar in the DP and DR rat at about 50%. These data are consistent with the view that NK cells must be activated to show good lytic activity.

We then determined whether the increased activity of NK cells in the DP rat was due to *in vivo* blastogenesis. We utilized two assays to determine this. In the first assay, spleen cell populations were size-separated by centrifugal elutriation, and the NK cell activity mediated by the different size classes was measured. If blast cells were present, a shift in lytic activity to large-size cells would be apparent. These experiments showed that the size distributions of NK cells in DP and DR rats were similar, and the NK activity was predominantly in the small and medium-sized fraction, indicating that there were not appreciable blast cells in the population.

In a second assay, we directly determined the cell cycle stage of DP NK cells. In this assay, spleen cells were surface-labeled with the OX 8 antibody, and the DNA was labeled with propidium iodide. Utilizing two-color flow cytometry, we "gated" on OX 8^+ cells and measured the DNA content of these cells. These experiments showed that 98% of the OX 8^+ cells were in the G0–G1 phase of the cell cycle, indicative of a resting population. DP NK cells may be more active than DR spleen cells owing to low level interferon stimulation, which activates NK cells and/or induces low levels of blastogenesis, which we could not detect. These data suggest that while DP cells are activated, there are not appreciable numbers of cells undergoing blast transformation. These studies illustrate that NK cell function is enhanced in DP rats owing to an increase in activity on a per-cell basis and owing to a relative increase in number.

We have examined the NK cell number and activity in a small number of acute diabetic rats. These data showed little difference in NK cell number or activity in rats diabetic for less than 2 days as compared to nondiabetic rats. Similarly, MacKay et al.[101] have shown only a small enhancement in NK function in acute diabetic rats.

NK activity in the DP rat may be abrogated by the in vivo infusion of the OX 8 antibody or with anti-AGM1.[85,102,103] The infusion of OX 8 depletes NK activity and removes OX 8$^+$ cells from the spleen and PBL.[85,102] This treatment is specific for NK cells and does not diminish T cells or B cells. In the DP rat this procedure is selective for NK cells, because these rats lack OX 8$^+$, OX 19$^-$ T$_C$ cells. In the DR rat such treatment removes both NK and T$_C$, and to specifically deplete NK cells, anti-AGM1 must be used.

3.2.5. Transfer and Induction of Diabetes

The BB/Wor rat provides a model in which it is possible to use various modalities to prevent and induce diabetes.

Diabetes can be passively transferred into young DP rats utilizing Con A-stimulated spleen cells from acute diabetic rats.[104] As the recipient rats develop diabetes prior to the age at which the spontaneous disease occurs, this provides good evidence for the transfer or acceleration of diabetes. Diabetes can also be transferred in this manner to DR rats and to BB-F1 hybrids following cyclophosphamide treatment.[105] It has also been shown that the administration of conditioned media produced by the incubation of acute diabetic spleen cells with Con A can induce diabetes prior to 60 days of age in DP and DR rats.[106] As of yet, the cell(s) responsible for the transfer of diabetes or activated by the conditioned media and the active cytokines in the conditioned media are not known. NK cell function, or the role of NK cells in the development of diabetes, has not been studied in the transfer experiments.

A new promising model of diabetes is the ability to induce diabetes in the DR rat by removing T cells bearing the RT 6 alloantigen.[107] As described above, the DP rat does not have T cells that express the RT 6 antigen, whereas DR rats express the RT 6.1 allotype. Biweekly infusion of RT 6.1 antibody into 30-day-old rats induces diabetes in 50–90% of recipients within 4 weeks. This model, in which diabetes can be induced rapidly and in high frequency, should provide an attractive system to determine the immunopathogenesis of diabetes. RT 6, a developmentally regulated protein, is not present on thymocytes, is acquired extrathymically, and is present on about half of peripheral T_H and T_C. Presumably this antigen delineates an important regulatory cell for the control of T cell responses and the control of autoimmune processes.

It has been difficult to propagate T cells *in vitro* with specificity for islet antigens. Prud'homme *et al.* have succeeded in developing T cell hybrids and CD4$^+$ T cell lines with specificity for islet antigens.[108,109]

3.2.6. Prevention of Diabetes

Diabetes can be prevented in the DP rat by several modalities. Diabetes can be prevented by neonatal thymectomy,[110] antilymphocyte serum,[111] total lymphoid irradiation,[112] cyclosporin A,[113,114] the induction of immunogenic tolerance,[115] T cell transfusion,[116] or the removal of OX 19$^+$ T cells or OX 8$^+$ NK cells.[102]

3.2.7. Role of NK Cells in Diabetes

The lack of functional and phenotypic T_C in DP rats suggests that they might adopt a novel effector element, such as an NK cell, for the mediation of diabetes. MacKay, *et al.*[101,117] have shown that DP and acute diabetic splenocytes show lytic activity against syngeneic and allogeneic islet targets and against an allogeneic insulinoma. In general, at a 20:1 E:T ratio they have found about 10% lysis of islet targets and 20% lysis of insulinoma targets in 8- to 14-h assays by DP splenocytes. DR splenocytes showed about 4% lysis of islet and insulinoma targets. Note that the ratio of lysis of islet targets by DP and DR splenocytes is similar to the relative frequency of NK cells in these populations. The effector cells were characterized as NK cells by cold target inhibition with the NK-sensitive target YAC-1, by inhibition of the killing with anti-AGM1 plus complement treatment, and by the presence of the cells with lytic activity in a low-density Percoll fraction.

A lack of MHC restriction in islet cell damage has been reported in islet transplant studies. Islets after *in vitro* culture, under conditions in which they are accepted as allografts, can be transplanted under the subrenal capsule of DP recipients. These islets are rejected[118,119] when the recipient DP rat develops diabetes or shows insulitis.

Experiments by Like *et al.*[102] have shown that the *in vivo* infusion of the OX 8 antibody removes OX 8 cells, depletes NK activity, and reduces diabetes to 12% as compared to 61% in a control group. The incidence of insulitis in the normoglycemic rats was 77%, suggesting that removal of NK cells prevented final beta cell destruction but did not prevent the autoimmune process. The prevention of both diabetes and insulitis by the OX 19 antibody, which does not diminish NK function, indicates that NK cells are not solely responsible for the disease and suggests an important role for T_H cells, which account for virtually all of the OX 19^+ cells. Treatment of DP rats with biweekly intravenous injections of rabbit anti-AGM1 from 30 to 120 days of age also prevents diabetes in BB/Wor rats.[103]

An important question that the studies in the DP raised was, Do DP rats use NK cells in the mediation of diabetes because this is one of the only intact effector elements retained by these immunodeficient animals, and would other animals with an intact immune system use other effector mechanisms? To answer this question, we studied the effect of NK depletion by anti-AGM1 on the incidence of diabetes in the RT 6-depleted DR rat.[103] This study showed that the injection of anti-AGM1 biweekly for 4 weeks in this model had no effect on the development of diabetes. This study indicates that the RT 6.1-depleted DR rat does not use NK cells in mediating diabetes and suggests that there must be different mechanisms of islet destruction in DP- and RT 6-depleted DR rats.

Studies in the BB/Wor rat have shown that islet cells are NK-sensitive targets, that NK cells are the predominant cytolytic lymphocyte in DP rats, and that NK cells are necessary for the development of diabetes in DP rats but not in RT 6.1-depleted DR rats.

4. Mechanisms of Islet Damage

The diversity of the immune system has provided the potential for many mechanisms to damage islet cells. However, to date there is little information about how islet cells may be damaged.

4.1. Natural Killer Cells

Natural killer cells have been shown to kill islet cell targets *in vitro* and are necessary for diabetes in the DP rat. However, the ability of DP NK cells to lyse islet cells does not fully explain the diabetes developed by the BB/Wor rat. Approximately 60–80% of the rats develop diabetes; however, there is no difference in NK cell function in diabetic and non-diabetic rats. Moreover, other rat strains with high NK activity do not develop spontaneous diabetes, and the depletion of T_H cells with the OX 19 antibody prevents diabetes.[102] These data suggest that NK cells are not solely responsible for diabetes and that T_H cells are necessary to stimulate NK cells within the insulitis lesions or to provide help to B lymphocytes for the production of ICSA, which supports ADCC.

The necessity of T_H cells to provide islet cell specificity for an unrestricted lytic mechanism is supported by transplant experiments that show that the islet cell destruction is not MHC-restricted.[118,119] Non-MHC-restricted islet cell damage has also been shown by cytotoxicity in *in vitro* assays.[35–39] These studies suggested that the islet cell destruction was mediated by T cells; however, the non-MHC restriction was not explained. These experiments did not show that the T cells used the CD3-T cell receptor complex or the CD8 structure for target recognition or binding. It is possible that the active cell in these studies was an NK cell. It has been shown that IL-1 is toxic to islet cells[120] and that there is an enhancement of this effect by tumor necrosis factor.[121] It is possible that cytokines secreted by NK cells, including IL-1,[122] may be important in islet damage. In the DP BB/Wor rat, it will be important to show NK cells in the insulitis lesions and to isolate such cells from the pancreas of acute diabetic rats.

Why a differentiated cell, such as an islet cell, is a target for NK cells must be delineated. Islet cells do express class I MHC products, and the expression of class I is enhanced during insulitis, suggesting that these targets might be protected from NK lysis. Moreover, it has been shown that interferon protects islet targets against cellular cytotoxicity, possibly owing to enhanced Class I MHC, which would protect target cells against NK lysis.[123]

4.2. Antibody

ICSA is found prior to the development of diabetes in human IDDM, the NOD mouse, and the BB/Wor rat. The antibodies fix complement and damage islet cells *in vitro*. Some of these sera support ADCC.[38,46–49] The activity of these mechanisms *in vivo* and *in vitro* must be determined.

4.3. T Cells

T_H cells have been shown to be necessary for diabetes induction in the NOD mouse,[58,124] in the DP BB/Wor rat,[102] and in the RT 6-depleted DR rat.[103] In the DR rat, T_H cells appear to be sufficient in the absence of T_C for the development of diabetes. Such a mechanism has also been shown to be active in experimental allergic encephalomyelitis.[125] The mechanisms by which T_H cells damage islet cells in the absence of T_C cells remains to be determined. As rodent islet cells have not been shown to express class II MHC molecules *in vivo*, these cells probably do not become targets for class II restricted T_C. It is possible that T_H damage islets through lymphokines that they secrete, in concert with NK cells and macrophages. To explain diabetes based on such a mechanism, the exquisite specificity for beta cells and the sparing of alpha and delta cells must be explained.

There is no evidence for a role of T_C in the DP BB/Wor rat or the RT 6-depleted DR rat. T_C may play a role in the diabetes of the NOD mouse.[126]

4.4. Macrophages

Macrophages are a prominent cell within insulitis lesions, and these cells may enter the islet early in the course of the insulitis lesions. Macrophages have been shown to be toxic to islet cells,[127] possibly owing to IL-1 secretion.[120]

5. The Future

The immunopathogenesis of diabetes is complex, but progress is being made in our knowledge of the immunologic mechanisms in the BB/Wor rat and the NOD mouse. It appears likely that the cellular mechanisms for islet damage in the rodent systems will be defined in the near future and that the antigenic structures recognized by T_H cells will be determined. Our hope is that such knowledge will open the door to understanding and ameliorating human diabetes.

ACKNOWLEDGMENTS. The work was supported by a grant from the Juvenile Diabetes Foundation.

References

1. LaPorte, R., and Cruickshanks, K., 1985, Incidence and risk factors for insulin dependent diabetes, in: *Diabetes in America*, U.S. Dept. Health and Human Services, NIH publication 85–1468, pp. 1–12.
2. Melton, J., Palumbo, P., and Chu, C., 1983, Incidence of diabetes by clinical type, *Diabetes Care* 6:79–82.
3. Gepts, W., 1965, Pathologic anatomy of the pancreas in juvenile diabetes mellitis, *Diabetes* 14::1619–633.
4. Foulis, A. K., Liddle, C. N., Farquharson, M. A., Richmond, J. A., and Weir, R. S., 1986, The histopathology of the pancreas in type I (insulin dependent) diabetes mellitus: A 25-year review of deaths in patients under 20 years of age in the United Kingdom, *Diabetologia* 29:267–274.
5. Rossini, A. A., Mordes, J. P., and Like, A. A., 1985, Immunology of insulin-dependent diabetes mellitus, *Annu. Rev. Immunol.* 3:289–320.
6. Dorman, J., LaPorte, R., Kuller, L., Cruickshanks, K. J., Orchard, T. J., Wagener, D. K., Becker, D. J., Cavender, D. E., and Drash, A. L., 1984, The Pittsburgh insulin dependent diabetes mellitus (IDDM) morbidity and mortality study. Mortality results, *Diabetes* 33:271–276.
7. Rimoin, D. L., and Rotter, J. I., 1985, The genetics of diabetes mellitus, in: *Immunology in Diabetes* (D. Andreani, U. DiMario, K. F. Federlin, and L. G. Heding, eds.), Krimpton Medical Press, Edinburgh, pp. 45–57.
8. Bottazo, G. F., Dean, B. M., NcNally, J. M., McKay, E. H., Swift, P. G. F., and Gamble, D. R., 1985, *In situ* characterization of autoimmune phenomena and expression of HLA molecules in the pancreas in diabetic insulitis, *N. Engl. J. Med.* 313:356–360.
9. Cudworth, A. G., and Woodrow, J. C., 1975, Evidence of HL-A linked genes in "juvenile" diabetes mellitus, *Br. Med. J.* 3:133–135.

10. Gorsuch, A. N., Spencer, K. M., Lister, J., Wolf, E., Bottazo, G. F., and Cudworth, A. G., 1982, Can future type I diabetes be predicted? A study in families of affected children, *Diabetes* 31::862–866.

11. Platz, P., Jakobsen, B. K., Morling, N., Ryder, L. P., Svejgaard, A., Thomsen, M., Christy, M., Kromann, H., Benn, J., Nerup, J., Green, A., and Hause, M., 1981, HLA-D and Dr antigens in genetic analysis of insulin-dependent diabetes mellitus, *Diabetologia* 21:108–115.

12. Rotter, J. I., Anderson, C. E., Rubin, R., Congleton, J. E., Terasaki, P. I., and Rimoin, D. L., 1983, An HLA genotype study of IDDM: The excess of DR3/DR4 heterozygotes allow rejection of the recessive hypothesis, *Diabetes* 32:169.

13. Owerbach, D., Lernmark, A., Platz, P., Ryder, L. P., Rask, L., Peterson, P. A., and Ludvigsson, J., 1983, HLA-D region β-chain DNA endonuclease fragments differ between HLA-DR identical healthy and insulin dependent diabetic individuals, *Nature* 303:815.

14. Gupta, S., Fikrig, S. M., Khanna, S., and Orti, E., 1982, Deficiency of suppressor T-cells in insulin-dependent diabetes mellitus: An analysis with monoclonal antibodies, *Immunol. Lett.* 4:289–294.

15. Buschard, K., Röpke, C., Madsbad, S., Mehlsen, J., Sorenson, T. B., and Rygaard, J., 1983, Alterations of peripheral T-lymphocyte subpopulation in patients with insulin-dependent (type I) diabetes mellitus, *J. Clin. Lab. Immunol.* 10:127–131.

16. Pozzili, P., Zuccarini, O., Iavicoli, M., Adreani, D., Sensi, M., Spencer, K. M., Bottazo, G. F., Beverly, P. C. L., Kyner, J. L., and Cudworth, A. G., 1983, Monoclonal antibody defined abnormalities of T-lymphocytes in type I (insulin dependent) diabetes, *Diabetes* 32:91–94.

17. Selam, J. L., Clot, J., Andary, M., and Mirouze, J., 1979, Circulating lymphocyte subpopulations in juvenile insulin-dependent diabetes, *Diabetologia* 16:35–40.

18. Hitchcock, C. L., Riley, W. J., Alamo, A., Pyka, R., and Maclaren, N. K., 1986, Lymphocyte subsets and activation in prediabetes, *Diabetes* 35:1416–1422.

19. Jackson, R. A., Morris, M. A., Haynes, B. F., and Eisenbarth, G. S., 1982, Increased circulating Ia-antigen-bearing T cells in type I diabetes mellitus, *N. Engl. J. Med.* 306:785–788.

20. Alviggi, L., Hoskins, P. J., Pyke, D. A., Johnston, C., Tee, D. E. H., Leslies, R. D. G., and Versani, D., 1984, Pathogenesis of insulin-dependent diabetes: A role for activated T-lymphocytes, *Lancet* 2:4–5.

21. MacCuish, A. C., Barnes, E. W., Irvine, W. J., and Duncan, L. J. P., 1974, Antibodies to pancreatic islet cells in insulin-dependent diabetes with coexistent autoimmune disease, *Lancet* 2:1529–1531.

22. Srikanta, S., Rabizadeh, A., Omar, M. A. K., and Eisenbarth, G. S., 1985, Assay for islet cell antibodies: Protein A–monoclonal antibody method, *Diabetes* 34:300–305.

23. Lernmark, A., Freedman, Z. R., Hofmann, C., Rubenstein, A. H., Steiner, D. F., Jackson, R. L., Winter, R. J., and Traisman, H. S., 1978, Islet-cell-surface antibodies in juvenile diabetes mellitus, *N. Engl. J. Med.* 299:375–380.

24. Dobersen, M. J., Scharff, J., E., Ginsberg-Fellner, F., and Notkins, A. L., 1980, Cytotoxic autoantibodies to beta cells in the serum of patients with insulin dependent diabetes mellitus, *N. Engl. J. Med.* 303:1493–1498.

25. Eisenbarth, G. S., Morris, M. A., and Scearce, R. M., 1981, Cytotoxic antibodies to cloned rat islet cells in serum of patients with diabetes mellitus, *J. Clin. Invest.* 67:403–408.

26. Palmer, J. P., Asplin, C. M., Clemons, P., Lyon, K., Talpati, O., Raghu, R., and Paquette, T. L., 1983, Insulin antibodies in insulin-dependent diabetics before insulin treatment, *Science* 222:1337–1339.

26. McEvoy, R. C., Witt, M. E., Ginsberg-Fellner, F., and Rubinstein, P., 1986, Anti-insulin antibodies in children with type I diabetes mellitus. Genetic regulation of production and presence at diagnosis before insulin replacement, *Diabetes* 35:634–641.

28. Dean, B. M., Becker, F., McNally, J., M., Tarn, A. C., Schwartz, G., Gale, E. A. M.,

and Bottazo, G. F., 1986, Insulin autoantibodies in the pre-diabetic period: Correlation with islet cell antibodies and development of diabetes, *Diabetologia* **29**:339–342.

29. Del Prete, G. F., Betterte, C., Padovan, D., Erle, G., Toffolo, A., and Bersahi, G., 1977, Incidence and significance of islet-cell antibodies in different types of diabetes mellitus, *Diabetes* **26**:909–915.

30. Gorsuch, A. N., Spencer, K. M., Lister, J., McHall, J. M., Bottazo, G. F., and Cudworth, A. G., 1981, Evidence for a long prediabetic period in type I (insulin dependent) diabetes mellitus, *Lancet* **2**:1363–1365.

31. Srikanta, S., Ganda, O. P., Eisenbarth, G. S., and Soeldner, J. S., 1983, Islet-cell antibodies and beta-cell function in monozygotic twins initially discordant for type I diabetes mellitus, *N. Engl. J. Med.* **308**:322–325.

32. Srikanta, S., Ganda, O. P., Rabizadeh, A., Soeldner, S. S., and Eisenbarth, G., S., 1985, First-degree relatives of patients with type I diabetes mellitus. Islet cell antibodies and abnormal insulin secretion, *N. Engl. J. Med.* **313**:461–464.

33. MacCuish, A. C., Jordan, J., Campbell, C. J., Duncan, L. J. P., and Irvine, W. J., 1974, Cell-mediated immunity to human pancreas in diabetes mellitus, *Diabetes* **23**:693–697.

34. Nerup, J., Anderson, O. O., Bendixen, G., Egeberg, J., and Poulsen, J. E., 1971, Antipancreatic cellular hypersensitivity in diabetes mellitus, *Diabetes* **20**:424–427.

35. Boitard, C., Debray-Sachs, M., Pouplard, A., Assan, R., and Hamburger, J., 1981, Lymphocytes from diabetics suppress insulin release *in vitro*, *Diabetologia* **21**:41–46.

36. Boitard, C., Chatenoud, L. M., and Debray-Sachs, M., 1982, *In vitro* inhibition of pancreatic B cell function by lymphocytes from diabetics with associated autoimmune diseases: A T cell phenomoenon, *J. Immonol.* **129**:2529–2531.

37. Boitard, C., Saï, P., Debray-Sachs, M., Assan, R., and Hamburger, J., 1983, Anti-pancreatic immunity. *In vitro* studies of cellular and humoral immune reactions directed toward pancreatic islets, *Clin. Exp. Immunol.* **55**:571–580.

38. Charles, M. A., Suzuki, M., Waldeck, N., Dodson, L. E., Slater, L., Ong, K., Kerschnar, A., Buckingham, B., and Golden, M., 1983, Immune islet killing mechanisms associated with insulin-dependent diabetes: *In vitro* expression of cellular and antibody-mediated islet cell cytotoxicity in humans, *J. Immunol.* **130**:1189–1194.

39. Lohmann, D., Krug, J., Lampeter, E. F., Bierwolf, B., and Verlohren, H. J., 1986, Cell-mediated immune reactions against B cells and defect of suppressor cell activity in type I (insulin-dependent) diabetes mellitus, *Diabetologia* **29**:421–425.

40. Chandy, K. G., Charles, M. A., Buckingham, B., Waldeck, N., Kershnar, A., and Gupta, S., 1984, Deficiency of monoclonal antibody (Leu 7) defined NK cells in newly diagnosed insulin-dependent diabetes mellitus, *Immunol. Lett.* **8**:89–91.

41. Negishi, K., Waldeck, N., Chandy, G., Buckingham, B., Kershnar, A., Fisher, L., Gupta, S., and Charles, M. A., 1986, Natural killer cell and islet killer cell activities in type I (insulin-dependent) diabetes, *Diabetologia* **29**:352–357.

42. Pozzilli, P., Sensi, M., Gorsuch, A., Bottazzo, G. F., and Cudworth, A. G., 1979, Evidence for raised K-cell levels in type-I diabetes, *Lancet* **1**:173–175.

43. Sensi, M., Pozzilli,P., Gorsuch, A. N., Bottazzo, G. F., and Cudworth, A. G., 1981, Increased killer cell activity in insulin dependent (type I) diabetes mellitus, *Diabetologia* **20**:106–109.

44. Lanier, L. L., Le, A. M., Phillips, J. H., Warner, N. L., and Babcock, G. F., 1983, Subpopulations of human natural killer cells defined by the expression of the Leu 7 (HNK-1) and Leu 11 (NK-15) antigens, *J. Immunol.* **131**:1789–1796.

45. Lanier, L. L., and Loken, M. R., 1984, Human lymphocyte subpopulations identified by using HNK-1 three color immunoflourescence and flow cytometry analysis: Correlation of Leu 2, Leu 3, Leu 7, Leu 8, and Leu 11 cell surface antigen expression, *J. Immunol.* **132**:151–156.

46. Huang, S. W., and Maclaren, N. K., 1976, Insulin-dependent diabetes: A disease of autoaggression, *Science* **192**:64–66.

47. Sensi, M., Pozzilli, P., and Cudworth, A. G., 1981, Antibody-dependent and natural killer cytoxicity in type I diabetes, *Acta Diabet. Lat.* **18**:339–345.

48. Kruschel, B., Jahr, H., Marx, S., and Michaelis, D., 1984, Cellular immune reactions against rat pancreatic islets mediated by antibodies from type I diabetic patients, *Biomed. Biochem. Acta* **5:**635–639.

49. Maruyama, T., Takei, I., Matsuba, I., Tsuruoka, A., Taniyama, M., Ikeda, Y., Kataoka, K., Abe, M., and Matsuki, S., 1984, Cell-mediated cytotoxic islet cell surface antibodies to human pancreatic beta cells, *Diabetologia* **26:**30–33.

50. Like, A. A., and Rossini, A. A., 1976, Streptozotocin-induced pancreatic insulitis: New model of diabetes mellitus, *Science* **193:**415–417.

51. Craighead, J. E., and McLane, M. F., 1968, Diabetes mellitus: Induction in mice by encephalomyocarditis virus, *Science* **162:**913–914.

52. Makino, S., Kunimoto, K., Muraoka, Y., Mizushima, Y., Katagiri, K., and Tochino, Y., 1980, Breeding of a non-obese, diabetic strain of mice, *Exp. Anim.* **29:**1–13.

53. Kanazawa, Y., Komeda, K., Sato, S., Mori, S., Akanuma, K., and Takaku, F., 1984, Non-obese-diabetic mice: Immune mechanisms of pancreatic β-cell destruction, *Diabetologia* **27:**113–115.

54. Maruyama, T., Takei, I., Taniyama, M., Kataoka, K., and Matsuki, S., 1984, Immunological aspect of non-obese diabetic mice: Immune islet cell-killing mechanism and cell-mediated immunity, Diabetologia **27:**121–123.

55. Miyazaki, A., Hanafusa, T., Yamada, K., Miyagawa, J., Fujino-Kurihara, H., Nakajima, H., Nonaka, K., and Tarui, S., 1985, Predominance of T lymphocytes in pancreatic islets and spleen of pre-diabetic non-obese diabetic (NOD) mice: A longitudinal study, *Clin. Exp. Immunol.* **60:**622–630.

56. Wicker, L. S., Miller, B. J., Coker, L. Z., McNally, S. E., Scott S., Mullen, Y., and Appel, M. C., 1987, Genetic control of diabetes and insulitis in the nonobese diabetic (NOD) mouse, *J. Exp. Med.* **165:**1639–1654.

57. Prochazka, M., Leiter, E. H., Serreze, D. V., and Coleman, D. L., 1987, Three recessive loci required for insulin-dependent diabetes in non-obese diabetic (NOD) mice, *Science* **237:**286–289.

58. Hattori, M., Buse, J. B., Jackson, R. A., Glichmer, L., Dorf, M. E., Minami, M., Makino, S., Moriwaki, K., Kuzuya, H., Imura, H., Strauss, W. M., Seidman, J. G., and Eisenbarth, G. S., 1986, The NOD mouse: Recessive diabetogenic gene in the major histocompatibility complex, *Science* **231:**733–735.

59. Katoaka, S., Satoh, J., Fujiya, H., Toyota, T., Suzuki, R., Itoh, K., and Kumagai, K., 1983, Immunologic aspects of the nonobese diabetic (NOD) mouse: Abnormalities of cellular immunity, *Diabetes* **32:**247–253.

60. Leiter, E. H., Prochazka, M., Coleman, D. L., Serreze, D. V., and Shultz, L. D., 1986, Genetic factors predisposing to diabetes susceptibility in mice, in: *The Immunology of Diabetes Mellitus* (M. A. Jaworski, G. D. Molnar, R. V. Rajotte, and B. Singh, eds.), Elsevier, Amsterdam, pp. 29–36.

61. Koike, T., Itoh, Y., Ishii, T., Ito, I., Takabayashi, K., Maruyama, N., Tomioka, H., and Yoshida, S., 1987, Preventive effect of monoclonal anti-L3T4 antibody on development of diabetes in NOD mice, *Diabetes* **36:**539–541.

62. Toyota, T., Satoh, J., Oya, K., Shintani, S., and Okano, T., 1986, Streptococcal preparation (OK-432) inhibits development of type I diabetes in NOD mice, *Diabetes* **35:**496–499.

63. Appel, M., personal communication.

64. Ishida, N., 1986, Immunopotentiating activities of OK-432, *Excerpta Medica* (monograph).

65. Chappel, C. I., and Chappel, W. R., 1983, The discovery and development of the BB rat colony: An animal model of spontaneous diabetes mellitus, *Metabolism* **32** (Suppl. 1):8–10.

66. Butler, L., Guberski, D. L., and Like, A. A., 1983, The effect of inbreeding on the BB/W diabetic rat, *Metabolism* **32** (Suppl. 1):51–53.

67. Butler, L., Guberski, D. L., and Like, A. A., 1983, Genetic analysis of the BB/W diabetic rat, *Can. J. Genet. Cytol.* **25:**7–15.

68. Nakhooda, A. F., Like, A. A., Chappel, C. I., Murray, F. T., and Marliss, E. B., 1977, The spontaneously diabetic Wistar rat. Metabolic and morphologic studies, *Diabetes* **26:**100–112.
69. Logothetopoulos, J., Valiquett, N., Madura, E., and Cvet, D., 1984, The onset and progression of pancreatic insulitis in the overt, spontaneously diabetic, young adult BB rat studied by pancreatic biopsy, *Diabetes* **33:**33–36.
70. Seemayer, T. A., Tannenbaum, G. S., Goldman, H., and Colle, E., 1982, Dynamic time course studies of the spontaneously diabetic BB Wistar rat. III. Light microscopy and ultrastructural observations of pancreatic islets of Langerhans, *Am. J. Pathol.* **106::**237–249.
71. Like, A. A., Forster, R. M., Woda, B. A., and Rossini, A. A., 1983, T-cell subsets in islets and lymph node of BioBreeding/Worcester(BB/W) rats, *Diabetes* **32** (Suppl. 1): 201A.
72. Colle, E., Guttman, R. D., Seemayer, T. A., and Michel, F., 1983, Spontaneous diabetes mellitus syndrome in the rat. IV. Immunogenetic interactions of MHC and non-MHC components of the syndrome, *Metabolism* **32** (Suppl. 1):56–61.
73. Jackson, R. A., Buse, J. B., Rifai, R., Pelletier, D., Milford, E. L., Carpenter, C. B., Eisenbarth, G. S., and Williams, R. M., 1984, Two genes required for diabetes in BB rats. Evidence from cyclical intercrosses and backcrosses, *J. Exp. Med.* **159:**1629–1636.
74. Colle, E., Guttman, R. D., and Fuks, A., 1986, Insulin-dependent diabetes mellitus is associated with genes that map to the right of the class 1. RT 1 A locus of the major histocompatibility complex of the rat, *Diabetes* **35:**454–458.
75. Like, A. A., Guberski, D. L., and Butler, L., 1986, Diabetic BioBreeding/Worcester (BB/Wor) rats need not be lymphopenic, *J. Immunol.* **136:**3254–3258.
76. Jackson, R., Rassi, N., Crump, T., Haynes, B., and Eisenbarth, G. S., 1981, The BB diabetic rat. Profound T-cell lymphocytopenia, *Diabetes* **30:**887–889.
77. Poussier, P., Nakhooda, A. F., Falk, J. A., Lee, C., and Marliss, E. B., 1982, Lymphopenia and abnormal lymphocyte subsets in the "BB" rat: Relationship to the diabetic syndrome, *Endocrinology* **110:**1825–1827.
78. Naji, A., Silver, W. K., Kimura, H., Bellgrau, D., Markham, J. F., and Barker, F., 1983, Analytical and functional studies on the T cells of untreated and immunologically tolerant diabetes-prone BB rats, *J. Immunol.* **130:**2168–2172.
79. Yale, J. F., and Marliss, E. B., 1984, Altered immunity and diabetes in the BB rat, *Clin. Exp. Immunol.* **57:**1–11.
80. Woda, B. A., Like, A. A., Padden, C., and McFadden, M. L., 1986, Deficiency of T cytotoxic-suppressor cells in the BB/W rat, *J. Immunol.* **136:**856–860.
81. Reynolds, C. W., Sharrow, W. O., Ortaldo, J. R., and Herberman, R. B., 1981, Natural killer activity in the rat. II. Analysis of surface antigens on LGL by flow cytometry, *J. Immunol.* **127:**2204–2208.
82. Reynolds, C. W., Timonen, T., and Herberman, R. B., 1981, Natural killer (NK) cell activity in the rat. I. Isolation and characterization of the effector cells, *J. Immunol.* **127:**282–287.
83. Cantrell, D. A., Robbins, R. A., Brooks, L. G., and Baldwin, R. W., 1982, Phenotype of rat natural killer cells defined by monoclonal antibodies marking rat lymphocyte subsets, *Immunology* **45:**97–103.
84. Woda, B. A., McFadden, M., Welsh, R. M., and Bain, K. M., 1984, Separation and isolation of rat natural killer (NK) cells from T cells with monoclonal antibodies, *J. Immunol.* **132:**2183–2184.
85. Woda, B. A., and Biron, C. A., 1986, Natural killer cell number and function in the spontaneously diabetic BB/W rat, *J. Immunol.* **137:**1860–1866.
86. Gilman, S. C., Rosenberg, J. S., and Feldman, J. D., 1982, Membrane phenotype of the rat cytotoxic T lymphocyte, *J. Immunol.* **129:**1012–1016.
87. Woda, B. A., and Padden, C., 1988, Biobreeding/Worcester (BB/Wor) rats are deficient in the generation of functional cytotoxic T cells, *J. Immunol.* **139:**1514–1517.

88. Greiner, D. L., Handler, E. S., Nakano, K., Mordes, J. P., and Rossini, A. A., 1986, Absence of the RT-6 T cell subset in diabetes prone BB/W rats, *J. Immunol.* **136:**148–151.

89. Elder, M. E., and Maclaren, N. K., 1983, Identification of profound peripheral T lymphocyte immunodeficiencies in the spontaneously diabetic BB rat, *J. Immunol.* **130:**1723–1731.

90. Jackson,R., Kadison, P., Buse, J., Rassi, N., Jegasothy, B., and Eisenbarth, G. S., 1983, Lymphocyte abnormalities in the BB rat, *Metabolism* **32** (Suppl. 1):83–86.

91. Rossini, A. A, Mordes, J. P., Pelletier, A. M., and Like, A. A, 1983, Transfusions of whole blood prevent spontaneous diabetes mellitus in the BB/W rat, *Science* **219:**975–977.

92. Bellgrau, D., Naji, A., Silvers, W. K., Markmann, J. F., and Barker, C. F., 1982, Spontaneous diabetes in BB rats—evidence for the T-cell dependent immune response defect, Diabetologia **23:**359–364.

93. Prud'homme, G. J., Freks, A., Colle, E., Seemayer, T. S., and Guttmann, R. D., 1984, Immune dysfunction in diabetes-prone BB rats: Interleukin-2 production and other mitogen-induced responses are suppressed by activated macrophages, *J. Exp. Med.* **159:**463–478.

94. Woda, B. A., and Padden, C., 1986, Mitogen responsiveness of lymphocytes from the BB/W rat, *Diabetes* **35:**513–516.

95. Elder, M., Maclaren, N., Riley, W., and McConnel, T., 1982, Gastric parietal cell and other autoantibodies in the BB rat, *Diabetes* **31:**313–318.

96. Like, A. A., Appel, M. C., and Rossini, A. A., 1982, Autoantibodies in the BB/W rat, *Diabetes* **31:**816–820.

97. Pollard, D. R., Gupta, K., Mancion, L., and Hynie, I., 1983, An immunofluorescence study of anti-pancreatic islet cell antibodies in the spontaneously diabetic BB Wistar rat, *Diabetologia* **25:**56–59.

98. Martin, D. R., and Logothetopoulos, J., 1984, Complement-fixing islet cell antibodies in the spontaneously diabetic BB rat, *Diabetes* **33:**93–97.

99. Dyrberg, T., Poussier, P., Nakhooda, A. F., Baekkeskov, S., Marliss, E. B., and Lernmark, A., 1984, Islet cell surface and lymphocyte antibodies often precede spontaneous diabetes in the BB rat, *Diabetologia* **26:**159–165.

100. Laborie, C., Sai, P., Feutren, G., Debray-Sachs, M., Quiniou-Debrie, M. C., Poussier, P., Marliss, E. B., and Assan, R., 1985, Time course of islet cell antibodies in diabetic and non-diabetic BB rats, *Diabetes* **34:**904–910.

101. MacKay, P., Jacobson, J., and Rabinovitch, A., 1986, Spontaneous diabetes mellitus in the Bio-Breeding/Worcester rat: Evidence *in vitro* for natural killer cell lysis of islet cells, *J. Clin. Invest.* **77:**916–924.

102. Like, A. A., Biron, C. A., Weringer, E. J., Byman, K., Srocaynski, E., and Guberski, D. L., 1986, Prevention of diabetes in BioBreeding/Worcester rats with monoclonal antibodies that recognize T lymphocytes or natural killer cells, *J. Exp. Med.* **164:**1145–1159.

103. Woda, B. A., Handler, E. S., Padden, C., Greiner, D. L., Reynolds, C., and Rossini, A. A., 1987, Anti–asialo GM₁ (AGM₁) prevents diabetes in diabetes prone (DP) but not RT 6.1 depleted diabetes resistant (DR) Biobreeding/Wor rats, *Diabetes* **36** (Suppl. 1): 66A.

104. Koevary, S., Rossini, A. A., Stoller, W., Chick, W., and Williams, R. M., 1983, Passive transfer of diabetes in the BB/W rat, *Science* **220:**727–728.

105. Like, A. A., Weringer, E. J., Holdash, A., McGill, P., Atkinson, D., and Rossini, A. A., 1984, Adoptive transfer of autoimmune diabetes mellitus in BioBreeding/Worcester (BB/W) inbred and hybrid rats, *J. Immunol.* **134:**1583–1587.

106. Handler, E. S., Mordes, J. P., Seals, J., Koevary, S., Like, A. A., Nakano, K., and Rossini, A. A., 1985, Diabetes in the Bio-Breeding/Worcester (BB/W) rat: Induction and acceleration by spleen cell conditioned media, *J. Clin. Invest.* **76:**1692–1694.

107. Greiner, D. L., Mordes, J. P., Handler, E. S., Angelillo, M., Nakamura, N., and Rossini, A. A., 1987, Depletion of RT 6.1$^+$ T lymphocytes induces diabetes in resistant Bio-Breeding/Worcester (BB/W) rat, *J. Exp. Med.* **166:**461–475.

108. Prud'homme, G. J., Fuks, A., Colle, E., and Guttman, R. D., 1984, Isolation of T-lymphocyte lines with specificity for islet cell antigens from spontaneously diabetic (insulin-dependent) rats, *Diabetes* **33:**801–803.

109. Prud'homme, G. J., Fuks, A., Guttman, R. D., and Colle, E., 1985, T cell hybrids with specificity for islet cell antigens, *J. Immunol.* **136:**1535–1536.

110. Like, A. A., Kislauskis, E., Williams, R. M., and Rossini, A. A., 1982, Neonatal thymectomy prevents spontaneous diabetes mellitus in the BB/W rat, *Science* **216:**644–646.

111. Like, A. A., Rossini, A. A., Appel, M. C., Guberski, D. L., and Williams, R. M., 1979, Spontaneous diabetes mellitus: Reversal and prevention in the BB/W rat with antiserum to rat lymphocytes, *Science* **206:**1421–1423.

112. Rossini, A. A., Slavin, S., Woda, B. A., Geisberg, M., Like, A. A, and Mordes, J. P., 1984, Total lymphoid irradiation prevents diabetes mellitus in the Bio-Breeding/Worcester (BB/W) rat, *Diabetes* **33:**543–547.

113. Laupacis, A., Gardell, C., Dupree, J., Stiller, C. R., Keown, P., and Wallace, A. C., 1983, Cyclosporin prevents diabetes in BB Wistar rat, *Lancet* **1:**10–11.

114. Like, A. A., Dirodi, V., Thomas, S., Guberski, D., and Rossini, A. A., 1984, Prevention of diabetes mellitus in the BB/W rat with cyclosporin-A, *Am. J. Pathol.* **117:**92–97.

115. Naji, A. Silver, W. K., Bellgrau, D., and Barker, C. F., 1981, Destruction of islets is prevented by immunological tolerance, *Science* **213:**1390–1392.

116. Rossini, A. A., Faustman, D., Woda, B. A., Like, A. A., Szymanski, I., and Mordes, J. P., 1984, Lymphocyte transfusions prevent diabetes in the BioBreeding Worcester rat, *J. Clin. Invest.* **74:**39–46.

117. MacKay, P., Boulton, A., and Rabinovitch, A., 1985, Lymphoid cells of BB/W diabetic rats are cytotoxic to islet beta cells *in vitro, Diabetes* **34:**706–709.

118. Weringer, E. J., and Like, A. A., 1985, Immune attack on pancreatic islet transplants in the spontaneously diabetic Bio-breeding/Worcester (BB/W) rat is not MHC-restricted, *J. Immunol.* **134:**2383.

119. Prowse, S. J., Bellgrau, D., and Lafferty, K. J., 1986, Islet allografts are destroyed by disease occurrence in the spontaneously diabetic BB rat, *Diabetes* **35:**110–114.

120. Bendtzen, K., Mandrup-Poulsen, T., Nerup, J., Nielsen, J. H., Dinarello, C. A., and Svenson, M., 1986, Cytotoxicity of human pI interleukin-1 for pancreatic islets of Langerhans, *Science* **232:**1545–1547.

121. Mandrup-Poulsen, T., Bendtzen, K., Dinarello, C. A., and Nerup, J., 1987, Interleukin-1 (IL-1)–mediated beta-cell killing: Potentiating effect of tumor necrosis factor (TNF), *Diabetes* **36** (Suppl. 1):40A.

122. Scala, G., Allavena, P., Djeu, J. Y., Kasahara, T., Ortaldo, J. R., Herberman, R. B., and Oppenheim, J. J., 1984, Human large granular lymphocytes are potent producers of interleukin-1, *Nature* **309:**56–59.

123. Farkus, G., Pusztai, R., Mandi, Y., and Beladi, I., 1983, *In vitro* model for cellular cytotoxicity against rat Langerhans' islets: Protective effect of mouse interferon on rat target cells, *Transplantation* **36:**583–584.

124. Wang, Y., Hao, L., Gill, R. G., and Lafferty, K. J., 1986, Autoimmune diabetes in NOD mouse is L3T4 T-lymphocyte dependent, *Diabetes* **36:**535–538.

125. Sedgwick, J., Brostoff, S., and Mason, D., 1987, Experimental allergic encephalomyelitis in the absence of a classical delayed-type hypersensitivity reaction, *J. Exp. Med.* **165:**1058–1075.

126. Appel, M., personal communication.

127. Schwizer, R. W., Leiter, E. H., and Evans, R., 1984, Macrophage-mediated cytotoxicity against cultured pancreatic islet cells, *Transplantation* **37:**539–544.

20

Role of Natural Effector Cells in Human Gastrointestinal Disease

FERGUS SHANAHAN and STEPHAN TARGAN

1. Introduction

In the gastrointestinal tract, only a single layer of epithelium resting on basement membrane, separates the *internal milieu* from the environment. Through this, the host must absorb nutrients and yet exclude toxic, infectious, and antigenic material. The integrity of mucosal antigenic "barrier" must therefore be protected by a variety of immunological and nonimmunological mechanisms. Despite these protective mechanisms, the antigenic barrier is incomplete, and antigenic material does penetrate and may in some cases cause disease.[1]

Nonimmunological protective mechanisms include the normal intestinal flora, gastric acid, digestive enzymes, mucus secretion, and gut motility. In addition, the humoral and cellular limbs of the local mucosal immune system are uniquely adapted to respond to antigenic stimuli at the mucosal surface in a manner that is largely independent of the systemic immune system. The primary function of the major secretory immunoglobulin (IgA) is the prevention of absorption of dietary antigens (immune exclusion).[2,3] This function is facilitated by its selective transport mechanism through mucosal epithelial cells, where it is packaged with a secretory component to render it resistant to luminal digestive enzymes. IgA may also provide an important backup system for eliminating

FERGUS SHANAHAN and STEPHAN TARGAN • Department of Medicine, UCLA School of Medicine, Los Angeles, California 90024.

circulating gut-derived immune complexes. Since it is selectively trans-
ported across other mucosal surfaces including the biliary tract, this route
may provide a clearance mechanism for IgA-dietary antigen complexes.
In addition to the excretion of immune complexes via the biliary tract,
the liver also has an important backup role in the immunological defense
of the gut. This includes the filtration function of the Kuppfer cells and
the down-regulation of systemic immune responses to dietary antigen ab-
sorbed into the portal circulation.[4]

A comprehensive review of mucosal immunity is beyond the scope of
this chapter; the topic has been the subject of several excellent recent
reviews.[5–10] Our purpose here is to provide a brief overview of recent
advances in our knowledge of the various natural effector cell systems
that protect the integrity of the intestinal mucosa and that may participate
in intestinal inflammatory reactions. Strategically positioned at the interface with the environment, the cellular components of the mucosal immune system are arguably the most important members of the host's army
of natural effector cells. Yet until recently, our understanding of mucosal
natural effectors has lagged considerably behind that of the effectors within
the systemic immune system. Our primary focus here will be on the human gastrointestinal mucosa and the possible role of its natural effector
cells in human disease processes. Animal studies of mucosal natural effector cells are discussed only where they may have particular relevance to
human intestinal diseases. Finally, recent advances in our understanding
of the role of cytotoxic effector cells in various human liver diseases will
be covered.

2. Cytotoxic Lymphocytes in the Human Intestine

Information on human gastrointestinal cytotoxic lymphocytes is limited, and unfortunately, it has become clear that studies of such cells in
animals cannot be extrapolated to the human.[11] Although there is general agreement that lymphocytes isolated from the human intestinal mucosa develop cytotoxic activity when cultured with mitogens or interleukin
2, evidence for the presence of lytically active cells within freshly isolated
mucosal cell preparations has been controversial. The following effector
cells and their possible role in human intestinal disease will be reviewed:
natural killer (NK); killer (K) antibody-dependent cellular cytotoxic (ADCC)
cells; lymphokine-activated killers (LAK); and non-major histocompatibility complex (MHC)-restricted cytotoxic T lymphocytes (CTL).

It is important to appreciate that the cells responsible for each of
these lytic activities differ phenotypically from their counterparts in the
peripheral blood. Therefore, studies of mucosal surface antigen markers
with monoclonal antibodies that have been developed using peripheral

blood lymphocytes should always be correlated with an assessment of functional activity.

2.1. Natural Killer Cells

In contrast to studies in rodents,[12,13] human lymphocytes isolated from the intraepithelial compartment or from the lamina propria of the intestine exhibit little NK activity.[14–23] In addition, cells bearing the Leu 11 or Leu 7 marker are either absent or very sparse in the human intestine.[24–26] Recently, however, a small subset of cells that have lytic activity, have been enriched from freshly isolated mucosal cell preparations.[23] This subset of cells bears the NKH-1 surface marker but is Leu-11$^-$. Interestingly, NKH-1$^+$ Leu-11$^-$ cells represent a similar proportion (2–4%) of the total lymphocyte population in the mucosa, as they do in the peripheral blood.[23] It has been shown that peripheral blood NKH-1$^+$ cells that lack the Leu 11 marker are less lytically efficient than those that have both markers.[27] This may explain why the mucosal NKH-1$^+$ cells were, on a per-cell basis, less lytically efficient than their peripheral blood counterparts.[23] Whether these cells represent a discrete functionally distinct subset of NK cells and why only this subset is found in the mucosa is not clear. It is possible that specialized effectors at the mucosal level, although a small subpopulation, might be uniquely adapted to a specific function at that site.

Peripheral blood NK activity has been reported to be decreased in patients with Crohn's disease[28–30] and ulcerative colitis.[30] These findings are probably the result rather than the cause of the inflammatory disease process and it has been demonstrated that plasma from the mesenteric vein of patients with intestinal inflammation is inhibitory to NK cells.[14]

The possible mechanisms by which mucosal NK cells might influence the pathogenesis of disease states have recently been reviewed,[9,11,31] and it has been emphasized that the "physiologic target" of intestinal NK cells is not known. In addition to direct injury to neoplastic and virally altered epithelial cells, NK cells might influence disease activity indirectly, by their immunoregulatory activities[9,32,33] on the other components of the mucosal immune system. Studies of mucosal NK cells in patients with inflammatory and other intestinal diseases are difficult to interpret because of the uniformly low levels of activity present in unfractionated cell preparations from normal and diseased intestines[14–23] and because the surface antigen phenotypes studied have been limited to the Leu 7 marker.[22] This is now known not to be a specific marker of lytically active NK cells.[34] It has been suggested that direct NK cytolytic effects are not important in intestinal disease largely on the basis of their low numbers in the intestinal mucosa in comparison with the blood.[31] However, it has been argued that such a dismissal of NK involvement in disease pathogenesis may be inappropriate, because the number of cells required for effective tumor and viral infection surveillance and immunoregulation is not known.[35]

2.2. Antibody-Dependent Cellular Cytotoxicity (ADCC)

Whether cells isolated from the intestine can mediate ADCC has been controversial.[17,18,36–38] Differences in results may be technical and related to the influence of the enzymatic digestion procedure on lymphocyte and Fc receptor function.[15,39,40] We have consistently been able to find significant ADCC activity in freshly isolated mononuclear cell preparations from noninflamed human intestinal mucosa (F. Shanahan and S. Targan, unpublished). However, we have not been able to confirm that the effector cell responsible for this activity is either an NK or a Tγ cell. It may be macrophage-mediated.

Questions concerning the presence and nature of intestinal effector cells that are capable of mediating ADCC have considerable relevance to the pathogenesis of inflammatory bowel disease. Serum IgG antibodies capable of participating in ADCC reactions to human carcinoma cells[41] and rat colonic epithelial cells[42] have been demonstrated in ulcerative colitis. Indeed, Das and associates[43] have found a correlation between clinical activity of ulcerative colitis and serum ADCC to a human colon carcinoma line. These investigators also reported a colonic tissue-bound IgG antibody exclusively in patients with ulcerative colitis.[44,45] The antigen recognized by this antibody has been isolated and partially characterized, and it appears to be a colon-specific autoantigen.[46] Whether the tissue-bound antibody and that responsible for the ADCC in the serum are identical is not known.

It has long been known that peripheral blood lymphocytes from patients with Crohn's disease and ulcerative colitis can kill colonic epithelial cells *in vitro*.[47–49] In a series of studies, Shorter and colleagues[50–56] have investigated this phenomenon in detail and provided evidence that the mechanism of lysis is ADCC. More recently, they have found that freshly isolated colonic mononuclear cells from patients with ulcerative colitis and Crohn's disease are cytotoxic to autologous colonic epithelial cells.[56] It was concluded that the mechanism of lysis was a form of ADCC in which intestinal mononuclear cells are "armed" by serum factors. However, the specificity of the "arming factor" and the precise nature of the effector cell were not determined.

2.3. Lymphokine-Activated Killer Cells

Human mucosal lymphocytes when cultured in the presence of interleukin 2 develop significant lytic activity against NK-sensitive and NK-resistant target cells,[23,26,57] including colonic tumor cells.[58] As with peripheral blood lymphocytes, this LAK activity appears to be similar to other forms of culture-activated cytotoxicity that occur in the presence of a variety of mitogens, including fetal calf serum.[23] The precursors of intestinal mucosal LAK cells have the following surface antigen phenotype; Leu-11⁻, Leu 7⁻, CD2⁺, CD3⁻, CD8⁺.[23] In contrast, in the peripheral blood, the majority of the precursors of LAK cells are Leu-11⁺.[59,60]

When cultured in the presence of exogenous IL-2, the generation of LAK activity by mucosal lymphocytes from patients with inflammatory bowel disease has not been found to differ significantly from that of normal mucosal lymphocytes.[26] However, the production of IL-2 by mucosal mononuclear cells from patients with inflammatory bowel disease has been shown to be defective.[61] If LAK activity does occur *in vivo*, then one might speculate that this phenomenon might be defective in patients with inflammatory bowel disease and perhaps explain, in part, the increased frequency of intestinal cancer in these patients.

The therapeutic use of LAK cells for cancer has generated excitement, optimism, and controversy.[62–66] Gastrointestinal cancers including colorectal[66] and hepatic cancer[67] appear to be particularly responsive to this form of therapy.

2.4. T Cell Cytotoxicity

Although cells with a surface antigen phenotype compatible with that of cytotoxic T cells are present within the human intestinal mucosa,[24,25] their lytic function has not been studied. Previous studies using unfractionated intestinal mucosal lymphocytes cultured with mitogens as a measure of cytolytic T cell function[15–17,68] cannot adequately distinguish between T cell-mediated killing and LAK or culture-activated killing.[23]

Evidence for non-major histocompatibility complex (MHC)-restricted cytotoxic T cells is now well established,[69] and a subset of these cells may be conveniently studied *in vitro* by triggering their activity with antibodies to the CD3 component of the T cell antigen receptor complex.[70,71] We have recently used this technique to demonstrate non-MHC-restricted cytotoxic T cell activity in freshly isolated lymphocytes for non-inflamed human intestinal mucosa.[72] The surface antigen phenotype of these mucosal cells is CD2⁺, CD3⁺, CD8⁺, Leu 7⁻ and is therefore different from that of the peripheral blood effectors, which are Leu 7⁺.[70,71] We have also found that in the peripheral blood of patients with Crohn's disease and ulcerative colitis, there is increased activity of these anti-CD3-triggered non-MHC-restricted cytotoxic T cells.[73] Since the anti-CD3-triggered cytotoxicity is thought to reflect a form of antigen mimicry,[70] this phenomenon may be of use in defining *in vivo*-primed cytotoxic T cells, even though the stimulating antigen is not known.[70] This might be of particular importance in inflammatory bowel disease where the antigen provoking the inflammatory response is unknown.[74,75]

The frequency and activity of anti-CD3-triggered cytotoxic T cells in the intestinal mucosa of patients with intestinal diseases has not yet been studied. However, Roche and colleagues[76] have recently shown that intestinal T cells from patients with inflammatory bowel disease but not from control patients could kill red cells coated with epithelial antigens. The significance of these provocative results is uncertain, because the levels of cytotoxicity were low, because lysis occurred only over a restricted range of effector-to-target ratios,[76] and because the relationship between this type of non-MHC-restricted T cell cytotoxicity and that triggered by CD3 monoclonal antibodies is not known.

3. Intestinal Macrophages

Intestinal macrophages and macrophagelike cells and their possible role in intestinal diseases have been the subject of recent detailed reviews.[77–80] Much of our knowledge of intestinal macrophages is still based on conventional morphological or immunohistological studies.[81–84] Intestinal macrophages occur beneath the specialized M cells which overlie Peyer's patches, and although scattered throughout the lamina propria, they tend to localize beneath the columnar epithelium. Within the Peyer's patches and lamina propria, they are in close contact with the other cellular elements of the immune system. Occasionally, they are found within the epithelial layer or may even insert long pseudopodia into the epithelial layer to contact intraepithelial lymphocytes.[85] They are thus ideally located to ingest the antigenic material that crosses the intestinal mucosa and to present it to the mucosal immune system.

Whether the migratory patterns of intestinal macrophages involve a selective "homing" similar to that of intestinal immunoblasts is not known. However, they do migrate to the mesenteric lymph node and are found in the efferent lymph of the mesenteric lymph nodes.[86] Macrophages in animals fed latex particles, as markers, have been found to appear sequentially in the lamina propria, the mesenteric lymph node, and the spleen.[87]

Very little is known of intestinal macrophage function, and few investigators have been successful in developing techniques for their isolation and purification. Human intestinal macrophages have recently been enriched from enzymatically dispersed mucosal cell preparations by Percoll density centrifugation and plastic adherence,[88] by adherence to fibronectin-coated glass,[89] and by elutriation centrifugation.[90] These cells have many of the known characteristics of macrophages at other sites, including complement receptors, expression of Ia antigens, and Fc receptor-mediated phagocytosis. Their cytotoxic function has not been examined.

Macrophages exhibit a wide range of functional and cytochemical heterogeneity, depending on their anatomic location, maturity, and state of activation.[91,92] Within a given anatomic location such as the Peyer's patch or human intestinal lamina propria, considerable histochemical and immunohistological variability has been found.[93,94] Variability of specific macrophage functions may have considerable relevance to certain gastrointestinal diseases and are a subject of considerable current research interest.[95]

Morphological studies have revealed alterations in intestinal macrophages in a variety of diseases.[77–80]. There is impressive circumstantial evidence for a macrophage role in the defense against intestinal cancers,[80] but this has not been directly examined with isolated intestinal macrophages. From animal studies of bacterial, *Giardia*, and other parasitic infections,[78,96–98] it seems reasonable to speculate that the phagocytic capabilities of intestinal macrophages are important in defense against several human intestinal infections. However, the ingestion of infectious agents is not synonymous with killing. In the absence of competent T cell–mediated immunity, macrophages may be unable to eliminate certain bacteria such as the Whipple's disease bacillus[99] and typical and atypical mycobacteria.[78,100] In these disorders, macrophages that are stuffed with viable organisms often become the most prominent cell type within the intestinal mucosa. From there they may transport infectious agents to other sites.

An increase in intestinal macrophages is also seen in inflammatory bowel disease, particularly in Crohn's disease.[78,101,102] The pathological hallmark of this condition is granuloma formation, where macrophages are the most prominent cell type. Indeed, it has even been postulated that a basic defect in macrophage function might explain the clinical differences between ulcerative colitis and Crohn's disease.[103] However, no major monocyte-macrophage defect has been found in either condition, and although several studies indicate macrophage activation in inflammatory bowel disease,[79] it is likely that this is not a primary event but rather is a nonspecific response to the intestinal inflammatory process.[75]

4. Intestinal Mast Cells

The intestinal mucosa is relatively rich in mast cells. However, because of their special staining and fixation requirements,[104,105] intestinal mast cells have in the past been underestimated. In the human intestine, mast cells occur with a frequency of up to $10,000-20,000/mm^3$.[106–108] Although best known for their role in allergic disorders, mast cells have increasingly been implicated in a variety of other immunologic and non-immunologic events.[105,109] They have a complex interrelationship with certain tumor cells, being directly cytotoxic in some situations[110–113] and stimulating tumor cell growth in others.[114] Mast cells enhance eosinophil-[115] and macrophage-mediated cytotoxicity.[116] The possible role of mast cells in delayed hypersensitivity reactions has been debated by several authors.[117,118] In addition, the secretory products of mast cells have important regulatory effects on the immune system[119] and on tissue hemostasis in general.[120] Within the intestine, mast cell secretion may have profound effects on mucus secretion, vascular permeability, and the motility and recruitment of inflammatory cells.[121,122]

In view of the diversity of functions attributed to mast cells, it is not surprising that functionally distinct mast cell subpopulations exist in different tissues.[105,121–126] The best example of mast cell heterogeneity is the distinction between rat intestinal mucosal mast cells and their counterparts in connective tissue at nonmucosal sites.[104,105] The mucosal rat mast cells differ morphologically, cytochemically, and functionally from those within the skin or peritoneal and serosal cavities. For example, unlike peritoneal mast cells, the mucosal mast cell contains chondroitin sulfate rather than heparin proteoglycan.[127] This accounts for differences in the fixation and staining requirements for histological demonstration of the two cell types in tissue sections.[104,105] The mucosal mast cell has a protease content that differs from that of the peritoneal mast cell.[128] In addition, its profile of responsiveness to a variety of secretagogues[129,130] and antiallergic compounds[131] is distinct from that of the connective tissue mast cell.

Heterogeneity might reflect distinct cell lineages, maturational stages, recent secretory activity, or local environmental influences. Using the mast cell-deficient mouse model, Kitamura and colleagues[132–135] have provided compelling evidence for a major role of the local environment in determining the "mucosal" or "connective tissue" phenotype of mast cells.

The evidence for existence of mast cell heterogeneity in humans, although less impressive, is beginning to emerge. Histologic studies of differences in sensitivity to various fixation procedures indicate that two distinct mast cell populations are present in the mucosa, submucosa, and muscle of the human intestine.[106–108] Evidence for the presence of chondroitin sulfate E-containing mast cells (which are thought to be analogous to the rat mucosal mast cell) has recently been reported.[136] In addition, studies of protease content of human mast cells including those in the intestine suggest that distinct subtypes may be present.[137,138] It is not clear, however, whether these subtypes of human mast cells differ functionally, as they do in the rat.[139,140]

Mast cells are important in host defense against a variety of parasitic infections.[98,141] Specific human intestinal diseases that have been associated with an increase in intestinal mast cell numbers include celiac disease,[142,143] Crohn's disease, and ulcerative colitis.[144–147] Their precise pathophysiologic function in these disorders has not been determined, but it is likely they undergo hyperplasia in response to recruitment by other effector cells; mast cell hyperplasia is probably not the primary event in the inflammatory process.[74,75] Intestinal mast cells have also been implicated in the pathogenesis of certain food allergic phenomena.[121,148] Whether one or both types of mast cell subpopulations undergo hyperplasia in these disorders is not known. This question may have important therapeutic implications if human mast cell subsets differ in their responsiveness to antiallergic drugs, as they do in the rat.[131]

5. Role of Cytotoxic Lymphocytes in Liver Disease

Immunological mechanisms have been implicated in the production of hepatocellular damage in an increasing number of liver diseases. Examples that have received considerable research attention recently include the acute and chronic viral hepatitides and chronic active hepatitis.[149–152] However, much of the evidence for the involvement of the immune system in the pathogenetic sequence of these disorders is still somewhat circumstantial, and it is difficult to weave a coherent concept of disease pathogenesis based on the multiple and diverse immunological abnormalities that have been associated with these diseases. Certain advances have been made, however; in particular, the contribution of lymphocyte cytotoxicity to liver cell injury has been clarified recently and is discussed here.

The hepatitis A virus (HAV) has long been thought to be directly cytopathic, destroying the liver cell without any contribution from the host's immune system. However, there is growing evidence that liver injury may be mediated by the immune system.[153,154] Although the effector cells responsible for the cell damage and clearance of the virus are not known, it has been shown that normal human circulating lymphocytes preferentially kill HAV-infected monkey cells.[154] The effector cell has an NK surface antigen phenotype (CD16$^+$ CD3$^-$). Clearance of the virus appears to be an efficient process, and chronic liver disease does not occur.

The hepatitis B virus (HBV) is not directly cytopathic. Liver injury in HBV infections is the result of an interaction between the humoral and cellular immune system with viral antigens on the surface membrane of the infected hepatocyte.[150] HBV core antigen (HBcAg) and HBV surface antigen (HBsAg) are both expressed on the hepatocyte membrane, and recent evidence indicates that HBcAg is the target of the immune assault.[155,156]

In the early 1970s, Dudley et al.[157] postulated that HBV-induced liver cell necrosis was the result of T cell–mediated cytolysis of hepatocytes bearing the appropriate viral antigen. However, initial experimental attempts to demonstrate such lymphocyte cytotoxicity produced conflicting results and have been criticized mainly on the grounds that the effectors and target cells were not histocompatible, NK activity was not distinguished from T cell cytotoxicity, and the antigen specificity of the effector cell was seldom established.[158] Recently, the use of a lymphocyte cytotoxicity assay for autologous hepatocytes has been used to clarify the nature of the effector cells in acute and chronic HBV infections and in other liver diseases in which the immune system may contribute to cell injury.[158] This cytotoxicity assay is a modification of that developed by Takasugi and Klein.[159] It essentially consists of a manual count of freshly isolated human hepatocytes that are adherent to culture plates, before and after exposure to autologous peripheral blood lymphocytes. The detachment of cells is considered an index of cell death. Despite considerable technical limitations, this assay system has the advantage that the expression of the viral antigens on the target cells theoretically simulated the *in vivo* situation, and the effectors and targets are histocompatible.

Although a role for NK cells has not been excluded, the available evidence indicates that in uncomplicated acute HBV infections, T cell–mediated cytolysis of infected hepatocytes bearing HBcAg on their surface occurs with release of virions, which in turn are neutralized and cleared by viral specific antibodies.[150] However, several additional factors may influence the outcome of HBV infection. These include the variable effects of blocking antibodies to HBcAg, serum- and liver-derived immunoregulatory factors, regulatory T cells, and the occurrence in some individuals of virally triggered autoimmune phenomena.[150,160] Whatever the mechanism, unlike HAV infections, HBV in some patients is not effectively cleared, and the disease becomes chronic.

Eddleston and colleagues, using the autologous cytotoxicity assay in a series of studies,[150,156,158,161,162] have examined the interaction between the immune system and the infected hepatocyte in chronic HBV. Their results indicate the presence of several peripheral blood effector cells that are reactive with autologous hepatocytes in these patients. The effect of cytotoxic T cells with specificity for HBcAg on the hepatocyte surface is in some cases offset by the presence of competing antibodies to HBcAG because of either a blocking effect or a modulatory effect on the surface membrane expression of that antigen. Non-T-cell-mediated cytotoxicity for autologous hepatocytes, which appeared to reflect ADCC rather than NK activity, was also found. However, the methods used for T and non-T cell separation in all of these experiments were not rigorous, and contaminant NK cells, which are known to exhibit enhanced activity in chronic HBV infection,[163] might have confounded the results. This and other cautionary notes have been sounded by Dienstag[160] in a critical review of the immunological mechanisms in chronic viral hepatitis.

In contrast to patients with chronic HBV-related liver disease, the effector cell responsible for hepatocyte injury in patients with non-HBV-related "autoimmune or lupoid" chronic active hepatitis appears to be a non-T cell. The mechanism is probably ADCC, and the target antigen to which there is surface-bound antibody is thought to be a lipoprotein autoantigen.[164]

ACKNOWLEDGMENTS. Supported by the UCLA /Harbor Inflammatory Bowel Disease Center (USPHS grant AM 36200), USPHS grant AM 27806. F. S. is a recipient of a research career development award from the National Foundation for Ileitis and Colitis.

References

1. Walker, W. A., 1982, Mechanisms of antigen handling by the gut, *Clin. Immunol. Allergy,* **2:**15–40.
2. Bienenstock, J., and Befus, A. D., 1983, Some thoughts on the biological role of immunoglobulin A, *Gastroenterology* **84:**178–185.
3. Underdown, B. J., and Schiff, J. M., 1986, Immunoglobulin A: Strategic defense initiative at the mucosal surface, *Annu. Rev. Immunol.* **4:**389–417.
4. Triger, D. R., Cyamon, M. H., and Wright, R., 1973, Studies on hepatic uptake of antigen. I. Comparison of inferior vena cava and portal vein routes of immunization, *Immunology* **25:**941–950.
5. Bienenstock, J., and Befus, A. D., 1980, Mucosal immuology, *Immunology* **41:**249–270.
6. Levinsky, R. J., 1983, Natural resistance of the gastrointestinal tract, *Clin. Immunol. Allergy* **3:**441–456.
7. Mayerhofer, G., 1984, Physiology of the intestinal immune system, in: *Local Immune Responses of the Gut* (T. J. Newby and C. R. Stokes, eds.) CRC Press, Boca Raton, FL, pp. 1–96.
8. MacDermott, R. P., 1986, Cell-mediated immunity in gastrointestinal disease, *Hum. Pathol.* **17:**219–233.
9. Elson, C. O., Kagnoff, M. F., Fiocchi, C., Befus, A. D., and Targan, S., 1986, Intestinal immunity and inflammation: Recent progress, *Gastroenterology* **91:**746–768.
10. Kagnoff, M., 1987, Immunology of the digestive system, in: *Physiology of the Gastrointestinal Tract,* 2d Ed. (L. R. Johnson, ed.) Raven Press, New York, pp. 1699–1728.
11. Mowat, A., 1987, Natural killer cells and intestinal immunity, in: *Food Allergy and Intolerance* (J. Brostoff and S. J. Challacombe, eds.), Balliere Tindall, London, pp. 156–166.
12. Nauss, K. M., Pavlina, T. M., Kumar, V., and Newberne, P. M., 1984, Functional characteristics of lymphocytes isolated from the rat large intestine. Response to T-cell mitogens and natural killer cell activity, *Gastroenterology* **86:**468–475.
13. Tagliabue, A., Befus, A. D., Clark, D. A., and Bienenstock, J., 1982, Characteristics of natural killer cells in murine intestinal epithelium and lamina propria, *J. Exp. Med.* **155:**1785–1796.
14. Gibson, P. R., Verhaar, H. J. J., Selby, W. S, and Jewell, D. P., 1984, The mononuclear cells of human mesenteric blood, intestinal mucosa and mesenteric lymph nodes: Compartmentalisation of NK cells, *Clin Exp. Immunol.* **56:**445–452.
15. Chiba, M., Bartnik, W., ReMine, S. G., Thayer, W. R., and Shorter, R. G., 1981, Human colonic intraepithelial and lamina proprial lymphocytes: Cytotoxicity *in vitro* and the potential effects of the isolation method on their functional properties, *Gut* **22:**177–186.

16. Falchuk, Z. M., Barnhard, E., and Machado, I., 1981, Human colonic mononuclear cells: Studies of cytotoxic function, *Gut* **22**:290–294.
17. MacDermott, R. P., Franklin, G. O., Jenkins, K. M., Kodner, I. J., Nash, G. S., and Weinreib, I. J., 1980, Human intestinal mononuclear cells. 1. Investigation of antibody-dependent, lectin-induced, and spontaneous cell-mediated cytotoxic capabilities, *Gastroenterology* **78**:47–56.
18. Bland, P. W., Britton, D. C., Richens, E. R, and Pledger, J. V., 1981, Peripheral, mucosal, and tumour-infiltrating components of cellular immunity in cancer of the large bowel, *Gut* **22**:744–751.
19. Targan, S., Britvan, L., Kendall, R., Vimadalal, S., and Soll, A., 1983, Isolation of spontaneous and interferon inducible natural killer like cells from human colonic mucosa: Lysis of lymphoid and autologous epithelial targert cells, *Clin. Exp. Immunol.* **54**:14–22.
20. Gibson, P. R., Dow, E. L., Selby, W. S., Strickland, R. G., and Jewell, D. P., 1984, Natural killer cells and spontaneous cell-mediated cytotoxicity in the human intestine, *Clin Exp. Immunol.* **56**:438–444.
21. Beeken, W. L., Gundel, M., St.Andre-Ukena, S., and McAuliffe, T., 1984, *In vitro* cellular cytotoxicity for a human colon cancer cell line by mucosal mononuclear cells of patients with colon cancer and other disorders, *Cancer* **55**:1024–1029.
22. Gibson, P. R., and Jewell, D. P., 1986, Local immune mechanisms in inflammatory bowel disease and colorectal carcinoma. Natural killer cells and their activity, *Gastroenterology* **90**:12–19.
23. Shanahan, F., Brogan, M., and Targan, S., 1987, Human mucosal cytotoxic effector cells, *Gastroenterology* **92**:1951–1957.
24. Cerf-Bensussan, N., Schneeberger, E. E., and Bhan, A. K., 1983, Immunohistologic and immunoelectron microscopic characterization of the mucosal lymphocytes of human small intestine by the use of monoclonal antibodies, *J. Immunol.* **130**:2615–2622.
25. Hirata, I.,Berrebi, G., Austin, L. L., Keren, D. F., and Dobbins, W. O., 1986, Immunohistologic characterization of intraepithelial and lamina propria lymphocytes in control ileum and colon and in inflammatory bowel disease, *Dig. Dis. Sci.* **31**:593–603.
26. Fiocchi, C., Tubbs, R. R., and Youngman, K. R., 1985, Human intestinal mucosal mononuclear cells exhibit lymphokine-activated killer cell activity, *Gastroenterology* **88**:625–637.
27. Lanier, L. L., Le, A. M., Civin, C. I., Loken, M. R., and Phillips, J. H., 1986, The relationship of CD16 (Leu-11) and Leu-19 (NKH-1) antigen expression on human peripheral blood NK cells and cytotoxic lymphocytes, *J. Immunol.* **136**:4480–4486.
28. Auer, I. O., Ziemer, E., and Sommer, H., 1980, Immune status in Crohn's disease. V. Decreased *in vitro* natural killer cell activity in peripheral blood, *Clin. Exp. Immunol.* **42**:41–49.
29. Beeken, W. L., MacPherson, B. R., Gundel, R. M., St. Andre-Ukena, S., Wood, S. G., and Sylwester, D. L., 1983, Depressed spontaneous cell-mediated cytotoxicity in Crohn's disease, *Clin. Exp. Immunol.* **51**:351–358.
30. Ginsburg, C. H., Dambrauskas, J. T., Ault, K. A., and Falchuk, Z. M., 1983, Impaired natural killer cell activity in patients with inflammatory bowel disease: Evidence for a qualitative defect, *Gastroenterology* **85**:846–851.
31. James, S. P., and Strober, W., 1986, Cytotoxic lymphoytes and intestinal disease, *Gastroenterology* **90**:235–240.
32. Brieva, J. A., Targan, S. R., and Stevens, R., 1984, NK and T cell subsets regulate antibody production by human *in vivo* antigen-induced lymphoblastoid B cells, *J. Immunol.* **132**:611–615.
33. Kimata, H., Shanahan, F., Brogan, M., Targan, S., and Saxon, A., 1987, Modulation of ongoing human immunoglobulin synthesis by natural killer cells, *Cell. Immunol.* **107**:74–88.
34. Lanier, L. L., Le, A. M., Phillips, J. H., Warner, N. L., and Babcock, G. F., 1983,

Subpopulations of human natural killer cells defined by expression of the Leu-7 (HNK-1) and Leu-11 (NK-15) antigens. *J. Immunol.* **131**:1789–1796.

35. Gibson, P. R., 1986, NK cells, IBD, and cancer, *Gastroenterology* **90**:1314–1315.

36. Clancy., R., and Pucci, A., 1978, Absence of K cells in human gut mucosa, *Gut* **19**:273–276.

37. Chiba, M., Shorter, R. G., Thayer, W. R., Bartnik, W., and ReMine, S., 1979, K-cell activity in lamina proprial lymphocytes from the human colon, *Dig. Dis. Sci.* **24**:817–822.

38. Fiocchi, C., Battisto, J. R., and Farmer, R. G., 1979, Gut mucosal lymphocytes in inflammatory bowel disease. Isolation and preliminary characterization, *Dig. Dis. Sci.* **24**:705–717.

39. Bland, P. W., Richens, E. R., Britton, D. C., and Lloyd, J. V., 1979, Isolation and purification of human large bowel mucosal lymphoid cells: Effect of separation technique on functional characteristics, *Gut* **20**:1037–1046.

40. Gibson, P. R., Hermanowicz, A., Verhaar, H. J. J., Ferguson, D. J. P., Lopez Bernal, A., and Jewell, D. P., 1985, Isolation of intestinal mononuclear cells: Factors released which may affect lymphocyte viability and function, *Gut* **26**:60–68.

41. Nagai, T., and Das, K., 1981, Demonstration of an assay for specific cytolytic antibody in sera from patients with ulcerative colitis, *Gastroenterology* **80**:1507–1512.

42. Hibi, T., Aiso, S., Yoshida, T., Watanabe, M., Asakura, H., Tsuru, S., and Tsuchiya, M., 1982, Anti-colon antibody and lymphocytophilic antibody in ulcerative colitis, *Clin. Exp. Immunol.* **49**:75–80.

43. Das, K. M., Kadono, Y., and Fleischner, G. M., 1984, Antibody-dependent cell-mediated cytotoxicity in serum samples from patients with ulcerative colitis, *Am. J. Med.* **77**:791–796.

44. Das, K. M., Dubin, R., and Nagai, T., 1978, Isolation and characterization of colonic tissue-bound antibodies from patients with idiopathic ulcerative colitis. *Proc. Natl. Acad. Sci. USA* **75**:4528–4532.

45. Nagai, T., and Das, K. M., 1981, Detection of colonic antigen(s) in tissues from ulcerative colitis using purified colitis colon tissue-bound IgG (CCA-IgG), *Gastroenterology* **81**:463–470.

46. Takahashi, F., and Das, K. M., 1985, Isolation and characterization of a colonic autoantigen specifically recognized by colon tissue-bound immunoglobulin G from idiopathic ulcerative colitis, *J. Clin. Ivest.* **76**:311–318.

47. Perlman, P., and Broberger, O., 1963, *In vitro* studies of ulcerative colitis. II. Cytotoxic action of white blood cells from patients on human fetal colon cells, *J. Exp. Med.* **117**:717–733.

48. Watson, D. W., Quigley, A., and Bolt, R. J., 1966, Effect of lymphocytes from patients with ulcerative colitis on human adult colon epithelial cells, *Gastroenterology* **51**:985–993.

49. Kemler, B. J., and Alpert, J. E., 1980, Inflammatory bowel disease: A study of cell-mediated cytotoxicity for isolated human colonic epithelial cells, *Gut* **21**:353–359.

50. Shorter, R. G., Cardoza, M., Huizenga, K. A., ReMine, S. G., and Spencer, R. J., 1969, Further studies of *in vitro* cytotoxicity of lymphocytes for colonic epithelial cells, *Gastroenterology* **57**:30–35.

51. Shorter, R. G., Cardoza, M., Spencer, R. J., and Huizenga, K.A., 1969, Further studies of *in vitro* cytotoxicity of lymphocytes from patients with ulcerative and granulomatous colitis for allogeneic colonic epithelial cells, including the effects of colectomy, *Gastroenterology* **56**:304–309.

52. Shorter, R. G., Cadoza, M. R., ReMine, S. G., Spencer, R. J., and Huizenga, K. A., 1970, Modification of *in vitro* cytotoxicity of lymphocytes from patients with chronic ulcerative colitis or chronic granulomatous colitis for allogenic colonic epithelial cells, *Gastroenterology* **58**:692–698.

53. Shorter, R. G., Huizenga, K. A., Spencer, R. J., Aas, J., and Guy, S., K., 1971, Inflam-

matory bowel disease. Cytophilic antibody and the cytotoxicity of lymphocytes for colonic cells *in vitro, Dig. Dis. Sci.* **16**:673–680.

54. Shorter, R. G., Huizenga, K. A., and Spencer, R. J., 1972, A working hypothesis for
the etiology and pathogenesis of nonspecific inflammatory bowel disease, *Dig. Dis. Sci.*
17:1024–1032.

55. Stobo, J. D., Tomasi, T. B., Huizenga, K. A., Spencer, R. J., and Shorter, R. G., 1976,
In vitro studies of inflammatory bowel disease: Surface receptors of the mononuclear
cells required to lyse allogeneic colonic epithelial cells, *Gastroenterology* **70**:171–176.

56. Shorter, R. G., McGill, D. B., and Bahn, R. C., 1984, Cytotoxicity of mononuclear cells
for autologous colonic epithelial cells in colonic diseases, *Gastroenterology* **86**:13–22.

57. Hogan, P. G, Hapel, A. J., and Doe, W. F., 1985 Lymphokine-activated and natural
killer cell activity in human intestinal mucosa, *J. Immunol.* **135**:1731–1738.

58. Hogan, P. G., Hapel, A. J., and Doe, W. F., 1986, Intestinal lymphokine-activated killer
(LAK) cells—cytotoxicity for colon cancer cells and modulation of their generation,
Gastroenterology, **90**:1462 (abstract).

59. Itoh, K., Tilden, A. B., Kumagai, K., and Balch, C. M., 1985, Leu 11⁺ lymphocytes
with natural killer (NK) activity are precursors of recombinant interleukin 2 (rIL-2)-
induced activated killer (AK) cells, *J. Immunol.* **134**:802–807.

60. Phillips, J. H., and Lanier, L. L., 1986, Dissection of the lymphokine activated killer
phenomenon. Relative contribution of peripheral blood natural killer cells and T lymphocytes to cytolysis, *J. Exp. Med.* **164**:814–825.

61. Fiocchi C., Hilfiker, M. L., Youngman, K. R., Doerder, N. C., and Finke, J. H., 1984,
Interleukin 2 activity of human intestinal mucosal mononuclear cells. Decreased levels
in inflammatory bowel disease, *Gastroenterology* **86**:734–742.

62. Rosenberg, S. A., Lotze, M. T., Muul, L. M., Leitman, S., Chang, A. E., Ettinghausen,
S. E., Matory, Y. L., Skibber, J. M., Shiloni, E., Vetto, J. T., Seipp, C. A., Simpson, C.,
and Reichart, C. M., 1985, Observations on the systemic administration of autologous
lymphokine-activated killer cells and recombinant interleukin-2 to patients with metastatic cancer. *N. Engl. J. Med.* **313**:1485–1492.

63. Bloom, M., 1987, Cancer M. D.'s clash over interleukin therapy, *Science* **235**:154–155.

64. Moertel, C. G., 1986, On lymphokines, cytokines, and breakthroughs, *JAMA* **256**:3141.

65. Durant, J. R., 1987, Immunotherapy of cancer. The end of the beginning? *N. Engl. J.
Med.* **316**:941–943.

66. Rosenberg, S. A., Lotze, M. T., Muul, L. M., Chang, A. E., Avis, F. P., Leitman, S.,
Linehan, W. M., Robertson, C. N., Lee, R. E., Rubin, J. T., Seipp, C. A., Simpson, C.
G., and White, D. E., 1987, A progress report on the treatment of 157 patients with
advanced cancer using lymphokine-activated killer cells and interleukin-2 of high-dose
interleukin-2 alone, *N. Engl. J. Med.* **316**:889–897.

67. Hsieh, K. H., Shu, S., Lee, C. S., Chu, C. T., Yang, C. S., and Chang, K. J., 1987, Lysis
of primary hepatic tumours by lymphokine activated killer cells, *Gut* **28**:117–124.

68. MacDermott, R. P., Bragdon, M. J., Kodner, I. J., and Bertovich, M. J., 1986, Deficient
cell-mediated cytotoxicity and hyporesponsiveness to interferon and mitogenic lectin
activation by inflammatory bowel disease peripheral blood and intestinal mononuclear
cells, *Gastroenterology* **90**:6–11.

69. Lanier, L. L., and Phillips, J. H., 1986, Evidence for three types of cytotoxic lymphocyte, *Immunol. Today* **7**:132–134.

70. Phillis, J. H., and Lanier, L. L., 1986, Lectin-dependent and anti-CD3 induced cytotoxicity are preferentially mediated by peripheral blood cytotoxic T lymphocytes expressing Leu-7 antigen, *J. Immunol.* **136**:1579–1585.

71. Deem, R. L., Shanahan, F., Niederlehner, A., and Targan, S., 1987, Mechanism and
specificity of anti-CD3 triggered T cell killing: Relationship to NK lysis, *Fed. Proc.* **46**:607
(abstract).

72. Shanahan, F., Deem, R., Nayersina, R., Niederlehner, A., Leman, B., and Targan, S.,
1988, Human mucosal T-cell cytotoxicity, *Gastroenterology* **94**:960–967.

73. Leman, B., Shanahan, F., Deem, R., Nayersina, R., Brogan, M., and Targan, S., 1987,

Enhanced peripheral blood non-MHC-restricted cytotoxicity in inflammatory bowel disease, *Gastroenterology* **92:**1499.

74. Shanahan, F., 1987, Inflammatory bowel disease, in: Targan, S. R., moderator. Immunologic mechanisms in intestinal diseases, *Ann. Intern. Med.* **106:**853–870.

75. Strober, W., and James, S., 1986, The immunologic basis of inflammatory bowel disease, *J. Clin. Immunol.* **6:**415–432.

76. Roche, J. K., Fiocchi, C., and Youngman, K., 1985, Sensitization to epithelial antigens in chronic mucosal inflammatory disease. Characterization of human intestinal mucosa-derived mononuclear cells reactive with purified epithelial cell-associated components *in vitro, J. Clin. Invest.* **75:**522–530.

77. Bockman,D. E., 1987, Gut-associated macrophages, in: *Food Allergy and Intolerance* (J. Brostoff and S. J. Challacombe, eds.), Balliere Tindall, London, pp. 67–87.

78. LeFevre, M. E., Hammer, R., and Joel, D. D, 1979, Macrophages of the mammalian small intestine: A review, *J. Reticuloendothelial. Soc.* **26:**553–573.

79. Tanner, A. R., Arthur, M. J. P., and Wright, R., 1984, Macrophage activation, chronic inflammation, and gastrointestinal disease, *Gut* **25:**760–783.

80. Caignard, A., Lagadec, P., Reisser, D., Jeannin, J. F., Martin, M. S., and Martin, F., 1985, Role of macrophage in the defense against intestinal cancers, *Comp. Immunol. Microbiol. Infect. Dis.* **8:**147–157.

81. Donellan, W. L., 1965, The structure of the colonic mucosa. The epithelium and subepithelial reticulohistiocytic complex, *Gastroenterology* **49:**496–514.

82. Sawicki, W., Kucharczyk, K., Szamska, K., and Kujawa, M., 1977, Lamina propria macrophages of intestine of guinea pig, *Gastroenterology* **73:**1340–1344.

83. Takeuchi, A., Sprinz, H., LaBrec, E. H., and Formal, S. B., 1965, Experimental bacillary dysentery. An electron microscopic study of the response of the intestinal mucosa to bacterial invasion, *Am. J. Pathol.* **47:**1011–1044.

84. Deane, H. W., 1964, Some electron microscopic observations of the lamina propria of the gut, with comments on the close association of macrophages, plasma cells, and eosinophils, *Anat. Rec.* **149:**453–474.

85. Collan, Y., 1972, Characteristics of nonepithelial cells in the epithelium of normal rat ileum, *Scand. J. Gastroenterol.* **7** (Suppl. 18):1–66.

86. Pugh, C. W., MacPherson, G. G., and Steer, H. W., 1983, Characterization of nonlymphoid cells derived from rat peripheral lymph, *J. Exp. Med.* **157:**1758–1779.

87. LeFevre, M. E., Olivo, R., Vanderhoff, J. W., and Joel, D. D., 1978, Accumulation of latex in Peyer's patches and its subsequent appearance in villi and mesenteric lymph nodes, *Proc. Soc. Exp. Biol. Med.* **159:**298–302.

88. Golder, J. P., and Doe, W. F., 1983, Isolation and preliminary characterization of human intestinal macrophages, *Gastroenterology* **84:**795–802.

89. Winter, H. S., Cole, S., Huffer, L. M., Davidson, C. B., Katz, A. J., and Edelson, P. J., 1983, Isolation and characterization of resident macrophages from guinea pig and human intestine, *Gastroenterology* **85:**358–363.

90. Beeken, W., Mieremet-Ooms, M., Ginsel, L. A., Leijh, P. C. J., and Verspaget, H., 1984, Enrichment of macrophages in cell suspensions of human intestinal mucosa by elutriation centrifugaton, *J. Immunol. Methods* **73:**189–201.

91. Nelson, D. S., 1981, Macrophages: Progress and problems, *Clin. Exp. Immunol.* **45:**225–233.

92. Hopper, K. E., Wood, P. R., and Nelson, D. S., 1979, Macrophage heterogeneity, *Vox Sang.* **36:**257–274.

93. Selby, W. S., Poulter, L. W., Hobbs, S., Jewell, D. P., and Janossy, G., 1983, Heterogeneity of HLA-DR-positive histocytes in human intestinal lamina propria: A combined histochemical and immunohistological analysis, *J. Clin. Pathol.* **36:**379–394.

94. Spencer, J. MacDonald, T. T., and Isaacson, P. G., 1987, Heterogeneity of non-lymphoid cells expressing HLA-D region antigens in human fetal gut, *Clin. Exp. Immunol.* **67:**415–424.

95. Beeken, W. L., St. Andre-Ukena, S., and Gundel, R. M., 1983, Comparative studies of

mononuclear phagocyte function in patients with Crohn's disease and colon neoplasms, *Gut* **24**:1034–1040.

96. Owen, R. L., 1982, Macrophage function in Peyer's patch epithelium, *Adv. Exp. Med. Biol.* **149**:507–513.
97. Owen, R. L., Allen, C. L., and Stevens, D. P., 1981, Phagocytosis of *Giardia muris* in Peyer's patch epithelium in mice, *Infect. Immun.* **33**:591–601.
98. Befus, D., and Bienenstock, J., 1982, Factors involved in symbiosis and host resistance at the mucosa-parasite interface, *Prog. Allergy* **31**:76–177.
99. Dobbins, W. O., 1982, Current concepts of Whipple's disease, *J. Clin. Gastroenterol.* **4**:205–208.
100. Rodgers, V. D., and Kagnoff, M. F., 1987, Gastrointestinal manifestations of the acquired immunodeficiency syndrome, *West. J. Med.* **146**:57–67.
101. Thyberg, J., Graf, W., and Klingenstrom, P., 1981, Intestinal fine structure in Crohn's disease. Lysosomal inclusions in epithelial cells and macrophages, *Virchows Arch. (Pathol. Anat.)* **39**:141–152.
102. Meurat, G., Bitzi, A., and Hammer, B., 1978, Macrophage turnover in Crohn's disease and ulcerative colitis, *Gastroenterology* **74**:501–503.
103. Ward, M., 1977, The pathogenesis of Crohn's disease, *Lancet* **2**:903–905.
104. Enerback, L., 1981, The gut mucosal mast cell, *Monogr. Allergy* **17**:222–232.
105. Befus, A. D., Bienenstock, J., and Denburg, J. A., 1986, *Mast Cell Differentiation and Heterogeneity*, Raven Press, New York.
106. Strobel, S., Miller, H. R. P., and Ferguson, A., 1981, Human intestinal mucosal mast cells: Evaluation of fixation and staining techniques, *J. Clin. Pathol.* **34**:851–858.
107. Ruitenberg, E. J., Gustowska, L., Elgersma, A., and Ruitenberg, H. M., 1982, Effect of fixation on the light visualization of mast cells in the mucosa and connective tissue of the human duodenum, *Int. Arch. Allergy. Appl. Immunol.* **67**:233–238.
108. Befus, D., Goodacre, R., Dyck, N., and Bienenstock, J., 1985, Mast cell heterogeneity in man. I. Histologic studies of the intestine, *Int. Arch. Allergy. Appl. Immunol.* **76**:232–236.
109. Schwartz, L. B., and Austen, K. F., 1984, Structure and function of the chemical mediators of mast cells, *Prog. Allergy* **34**:271–321.
110. Ferram, E., and Nelson, D. S., 1980, Mouse mast cells as anti-tumor effector cells, *Cell. Immunol.* **55**:294–301.
111. Henderson, W. R., Chi, E. Y., Jong, E. C., and Klebanoff, S. J., 1981, Mast cell–mediated tumor cytotoxicity. Role of the peroxidase system, *J. Exp. Med.* **153**:520–533.
112. Ghiara, P., Boraschi, L., Villa, L., Scapigliati, C., Taddei, C., and Tagliabue, A., 1985, *In vitro* generated mast cells express natural cytotoxicity against tumor cells, *Immunology* **55**:317–324.
113. Tanooka, H., Kitamura, Y., Sado, T., Tanaka, K., Nagase, M., and Kondo, S., 1982, Evidence for involvement of mast cells in tumor suppression in mice. *JNCI* **69**:1305–1309.
114. Roche, W. R., 1985, Mast cells and tumors. The specific enhancement of tumor proliferation *in vitro*, *Am. J. Pathol.* **119**:57–64.
115. Capron, M., Rosseaux, J., Mazingue, C., Bazin, H., and Capron, A., 1978, Rat mast cell–eosinophil interaction in antibody-dependent eosinophil cytotoxicity to *Schistosoma mansoni* schistosomula, *J. Immunol* **121**:2518–2528.
116. Dullens, H. F. J., and Den Otter, W., 1981, A small molecular weight peptide from P815 mastocytoma cells induces macrophage cytotoxicity, *Immunopharmacology* **3**:309–316.
117. Askenase, P. W., and Van Lovern, H., 1983, Delayed-type hypersensitivity: Activation of mast cells by antigen-specific T cell factors initiates the cascade of cellular interactions, *Immunol. Today* **4**:259–264.
118. Galli, S. J., and Dvorak, A. M., 1984, What do mast cells have to do with delayed hypersensitivity? *Lab. Invest.* **50**:365–368.

119. Beer, D. J., and Rocklin, R. E., 1984, Histamine-induced suppressor cell activity, *J. Allergy Clin. Immunol.* **73:**439–452.

120. Lewis, R. A., and Austen, K. F., 1981, Modulation of local homeostasis and inflammation by leukotrienes and other mast cell–dependent compounds, *Nature* **392:**103–108.

121. Barrett, K. E., and Metcalfe, D. D., 1984, The mucosal mast cell and its role in gastrointestinal allergic diseases, *Clin. Rev. Allergy* **2:**39–53.

122. Lemanske, R. F. Jr., Atkins, F. M., and Metcalfe, D. D., 1983, Gastrointestinal mast cells in health and disease. Part II, *J. Pediatr.* **103:**343–351.

123. Bienenstock, J., Befus, A. D., Pearce, F., Denburg, J., and Goodacre, R., 1982, Mast cell heterogeneity: Derivation and function with emphasis on the intestine, *J. Allergy Clin. Immunol.* **70:**407–412.

124. Katz, H. R., Stevens, R. L., and Austen, K. F., 1985, Heterogeneity of mammalian mast cells differentiated *in vivo* and *in vitro*, *J. Allergy. Clin. Immunol.* **76:**250–259.

125. Barrett, K. E., and Metcalfe, D. D., 1984, Mast cell heterogeneity: Evidence and implications, *J. Clin. Immunol.* **4:**253–261.

126. Shanahan, F., Denburg, J. A., Bienenstock, J., and Befus, A. D., 1984, Mast cell heterogeneity, *Can. J. Physiol. Pharmacol.* **62:**734–737.

127. Stevens, R. L., Lee, T. D. G., Seldin, D. C., Austen, K. F., Befus, A. D., and Bienenstock, J., 1986, Intestinal mucosal mast cells from rats infected with *Nippostrongylus brasiliensis* contain protease-resistant chondroitin sulfate di-B proteoglycans, *J. Immunol.* **137:**291–295.

128. Woodbury, R. G., Gruzenski, G. M., and Lagunoff, D., 1978, Immunofluorescent localization of a serine protease in the rat small intestine, *Proc. Natl. Acad. Sci. USA* **75:**2785–2789.

129. Befus, A. D., Pearce, F. L., Gauldie, J., Horsewood, P., and Bienenstock, J., 1982, Mucosal mast cells, I. Isolation and functional characteristics of rat intestinal mast cells, *J. Immunol.* **128:**2475–2480.

130. Shanahan, F., Denburg, J., Fox, J., Bienenstock, J., and Befus, A. D., 1985, Mast cell heterogeneity: Effects of neuroenteric peptides on histamine release, *J. Immunol.* **135:**1331–1337.

131. Pearce, F. L., Befus, A. D., Gauldie, J., and Bienenstock, J., 1982, Mucosal mast cells. II. Effects of anti-allergic compounds on histamine secretion by isolated intestinal mast cells, *J. Immunol.* **128:**2481–2486.

132. Nakano, T., Sonada, T., Hayashi, C., Yamatodani, A., Kanayama, Y., Yamamura, T., Asai, H., Yonezawa, T., Kitamura, Y., and Galli, S. J., 1985, Fate of bone marrow–derived cultured mast cells after intracutaneous, intraperitoneal, and intravenous transfer into genetically mast cell-deficient W/W^v mice, *J. Exp. Med.* **162:**1025–1043.

133. Kobayashi, T., Nakano, T., Nakahata, T., Asai, H., Yagi, Y., Tsuji, K., Komiyama, A., Akabane, T., Kojima, S., and Kitamura, Y., 1986, Formation of mast cell colonies in methylcellulose by mouse peritoneal cells and differentiation of these cloned cells in both the skin and gastric mucosa of W/W^v mice: Evidence that a common precursor can give rise to both "connective tissue–type" and "mucosal" mast cells, *J. Immunol.* **136:**1378–1384.

134. Sonada, S., Sonada, T., Nakano, T., Kanayama, Y., Kanakura, Y., Asai, H., Yonezawa, T., and Kitamura, Y., 1986, Development of mucosal mast cells after injection of a single connective tissue–type mast cell in the stomach mucosa of genetically mast cell-deficient W/W^v mice, *J. Immunol.* **137:**1319–1322.

135. Nakano, T., Kanakura, Y., Asai, H., and Kitamura, Y., 1987, Changing processes from bone marrow–derived cultured mast cells to connective tissue–type mast cells in the peritoneal cavity of mast cell-deficient W/W^v mice: Association of proliferation arrest and differentiation, *J. Immunol.* **138:**544–549.

136. Eliakim, R., Gilead, L., Ligumsky, M., Okon, E., Rachmilewitz, D., and Razin, E., 1986, Histamine and chondroitin sulfate E proteoglycan released by cultured human colonic mucosa: Indication for possible E mast cells, *Proc. Natl. Acad. Sci. USA* **83:**461–464.

137. Huntley, J. F., Newlands, G. F. J., Gibson, S., Ferguson, A., and Miller, H. R. P., 1985, Histochemical demonstration of chymotrypsin like serine esterases in mucosal mast cells in four species including man, *J. Clin. Pathol.* **38**:375–384.
138. Irani, A. A., Schechter, N. M., Craig, S. S., DeBlois, G., and Schwartz, L. B., 1986, Two types of human mast cells that have distinct protease compositions, *Proc. Natl. Acad. Sci. USA* **83**:4464–4468.
139. Fox, C. C., Dvorak, A. M., Peters, S. P., Kagey-Sobotka, A., and Lichtenstein, L. M., 1985, Isolation and characterization of human intestinal mucosal mast cells, *J. Immunol.* **135**:483–491.
140. Befus, A. D., Dyck, N., Goodacre, R., and Bienenstock, J., 1987, Mast cells from the human intestinal lamina propria. Isolation, histochemical subtypes, and functional characterization, *J. Immunol.* **138**:2604–2610.
141. Askenase, P. W., 1980, Immunopathology of parasitic diseases: Involvement of basophils and mast cells, *Springer Semin, Immunopathol.* **2**:417–442.
142. Strobel, S., Busuttil, A., and Ferguson, A., 1983, Human intestinal mucosal mast cells: Expanded population in untreated celiac disease, *Gut* **24**:222–227.
143. Marsh, M. N., and Hinde, J., 1985, Inflammatory component of celiac sprue mucosa. I. Mast cells, basophils, and eosinnophils, *Gastroenterology* **89**:92–101.
144. Zweiman, B., 1983, Mast cells in human disease, *Clin. Rev. Allergy* **1**:417–426.
145. Lloyd, G., Green, F. H. T., Fox, H., Mani, V., and Turnberg, L. A., 1975, Mast cells and immunoglobulin E in inflammatory bowel disease, *Gut* **16**:861–866.
146. Ranlov, P., Nielson, M. H., and Wanstrup, J., 1972, Ultrastructure of the ileum in Crohn's disease: Immune lesions and mastocytosis, *Scand. J. Gastroenterol.* **7**:471–476.
147. Rao, S. N., 1973, Mast cells as a component of the granuloma in Crohn's disease, *J. Pathol.* **109**:79–82.
148. Shanahan, F., and Targan, S., 1986, Food allergy and adverse reactions, in: *Modern Concepts in Gastroenterology*, Volume 1 (A. B. R. Thompson, L. R. DaCosta, and W. C. Watson, eds.), Plenum, New York, pp. 317–333.
149. Eddleston, A. W. L. F., 1980, Immunology of the liver, in: *Clinical Immunology* (C. W. Parker, ed.), W. B. Saunders, Philadelphia, pp. 1009–1050.
150. Mondelli, M., and Eddleston, A. W. L. F., 1984, Mechanisms of liver cell injury in acute and chronic hepatitis B, *Semin. Liver Dis.* **4**:47–58.
151. Paronetto, F., 1986, Cell-mediated immunity in liver disease, *Hum. Pathol.* **17**:168–178.
152. Klingenstein, R. J., and Dienstag, J. L., 1983, Immunologic mechanisms in initiation and the maintenance of chronic active hepatitis, in: *Chronic Active Liver Disease, Contemporary Issues in Gastroenterology*, Volume 2 (S. Cohen and R. D. Soloway, eds.), Churchill Livingstone, New York, pp. 93–115.
153. Lemon, S. M., 1985, Type A viral hepatitis. New developments in an old disease, *N. Eng. J. Med.* **313**:1059–1067.
154. Kurane, I., Binn, L. N., Bancroft, W. H., and Ennis, F. A., 1985, Human lymphocyte responses to hepatitis A virus-infected cells: Interferon production and lysis of infected cells, *J. Immunol.* **135**:2140–2144.
155. Trevisan, A., Realdi,G., Alberti, A., Ongaro, G., Pornaro, E., and Meliconi, R., 1982, Core antigen-specific immunoglobulin G bound to the liver cell membrane in chronic hepatitis B, *Gastroenterology* **82**:218–222.
156. Mondelli, M., Mieli, G., Vergani, M., Alberti, A., Vergani, D., Portman, B., and Eddleston, A. W. L. F., 1982, Specificity of T lymphocyte cytotoxicity to autologous hepatocytes in chronic hepatitis B virus infection: Evidence that T cells are directed against HBV core antigen expressed on hepatocytes, *J. Immunol.* **129**:2773–2778.
157. Dudley, F. J., Giustino, V., and Sherlock, S., 1972, Cell-mediated immunity in patients positive for hepatitis-associated antigen, *Br. Med. J.* **4**:754–756.
158. Mondelli, M., and Eddleston, A. W. L. F., 1984, Lymphocyte cytotoxicity for autologous hepatocytes, *Gut* **25**:109–113.
159. Takasugi, M., and Klein, E., 1970, A microassay for cell-mediated immunity, *Transplantation* **9**:219–227.

160. Dienstag, J. L., 1984, Immunologic mechanisms in chronic viral hepatitis, in: *Viral Hepatitis and Liver Disease* (G. N. Vyas, ed.), Grune and Stratton, Orlando, FL, pp. 135–166.

161. Eddleston, A. W. L. F., Mondelli, M., Mieli-Vergani, G., and Williams, R., 1982, Lymphocyte cytotoxicity to autologous hepatocytes in chronic hepatitis B virus infection, *Hepatology* **2**:122S–127S.

162. Mieli-Vergani, G., Vergani, D., Portman, B., White, Y., Murray-Lyon, I., Marigold, J. H., Woolf, I. Eddleston, A. W. L. F., and Williams, R., 1982, Lymphocyte cytotoxicity to autologous hepatocytes in HBsAg positive chronic active hepatitis, *Gut* **23**:1029–1036.

163. Dienstag, J. L., Savarese, A. M., and Bhan, A. K., 1982, Increased natural killer cell activity in chronic hepatitis B virus infection, *Hepatology* **2**:107S–115S.

164. Mieli Vergani, G., Vergani, D., Jenkins, P. J., Portman, B., Mowat, A. P., Eddleston, A. L. W. F., and Williams, R., 1979, Lymphocyte cytotoxicity to autologous hepatocytes in HBsAg-negative chronic active hepatitis, *Clin. Exp. Immunol.* **38**:16–21.

Summary of Part V

Cells with NK activity have previously been shown to recognize tumor cells, bone marrow cells, and virus-infected cells, with much of the experimental evidence from these studies centered on their cytotoxic activity. In contrast, this section has discussed a variety of diseases in which there is some evidence for the involvement of the NIS in the development or pathogenesis of those diseases.

Within the diverse group of diseases discussed in this section, there are several where NK activity and/or LGL number have been shown to increase (GVHD, neutropenia, pure red cell aplasia, and LGL lymphoproliferative disorders) and several where there is a trend toward a decrease in NK activity and/or LGL number (most autoimmune and some intestinal diseases). Thus in the cases where NK activity is elevated, one could presume that if the LGL are involved in the disease pathology, they will function through an augmenting effect. In contrast, in the situations where NK activity is low, one might presume that disease progression might occur because of the failure of the NIS to perform some normal homeostatic regulatory function. In either case, the involvement of these cells in the pathology of these diseases could be mediated via (1) direct effector cell function or elaboration of toxic cytokines, (2) indirect stimulation of other cell types, or (3) the failure of the NIS to provide a normal regulatory function.

Overall, the data presented in this section for a role of the NIS range from relatively convincing to very circumstantial and/or preliminary. As discussed by Woda (Chapter 19), considerable correlative and direct evidence has been accumulated to indicate that the onset of type I diabetes may be related to an increase in NK activity. Unlike most of the other reviews in this section, the study of the role of LGL and NK activity in diabetes has not been hindered by a lack of appropriate animal models. NK activity is markedly enhanced in BB/Wor rats that spontaneously develop autoimmune diabetes analogous to type I diabetes in humans.

Depression of NK activity by pretreatment of rats with anti-asGM$_1$ serum prevents the onset of disease. However, it appears that some T cells are also required for the development of this disease. The mechanism by which LGL or T lymphocytes may participate in this disease is not clear. However, a recent report demonstrates that increased levels of IL-2 appear to enhance the development of diabetes in BB rats, whereas treatment with anti-IL-2 receptor antibody plus low-dose cyclosporin A cures the rats of this disease.[1] Since both LGL and T lymphocytes make and respond to IL-2, these results are consistent with their role in this disease.

There is also some evidence that LGL with NK activity can kill pancreatic islet cells directly. In contrast, studies with the NOD mouse model of diabetes have not yet indicated a role for LGL in the etiology of the disease. Rather, the production of several potentially pathological antibodies precedes onset of the disease. In summary, the overall data suggest that the etiology of diabetes may be multifactorial and that the disease pathology may develop as a result of a variety of different mechanisms, of which NK activity is only one.

There is also evidence that the NIS participates in graft versus host disease (GVHD) in both rodents and humans. As noted by Clancy (Chapter 16), the suggested relationship between the NIS and GVHD is largely correlative. The major evidence of a relationship is the association of an increase in LGL and NK activity with the onset of GVHD. However, it is possible that the relationship is incidental to, or even a result of, the onset of GVHD and not actually the cause. To better understand the nature of the role played by natural effector cells in GVHD, future studies in this area will need to be performed in situations where the NIS has been selectively depressed prior to the initiation and during the progression of GVHD. Furthermore, it will be important to better characterize the NIS cells that appear early in GVHD reactions to determine their relationship, if any, to classical LGL. Similarly, leukocyte grafts that have been selectively depleted of LGL should be utilized to more clearly address the role of donor cells in the development of GVHD. As with several other diseases, we should also bear in mind that the role of the NIS need not be one of direct cytotoxicity, but rather the effect of the NIS could be indirect via the elaboration of immunoregulatory cytokines.

Studies potentially related to GVHD were presented by Chan and Winton (Chapter 17), who reviewed the evidence for a role of the NIS in hematocytopenias. Interpretation of these results is complicated by the fact that different populations of putative natural effector cells have been studied by various investigators, making historical comparisons between different reports difficult. Generally, there has been a positive correlation between exposure to cells with NK activity and depression of granulocyte or erythroid colony-forming units. These data relate directly to the studies summarized by Trinchieri *et al.* (Chapter 11) elsewhere in this volume. Although the results do not necessarily translate into a pathological effect of the NIS in bone marrow and leukocyte hemostasis, there is a close correlative association between LGL function (or dysfunction) and resultant clinical pathology in LGL lymphoproliferative disorders and neutropenia. There is also some evidence that cells with NK activity contribute to the pathology associated with severe aplastic anemia or pure red cell aplasia.

Overall, there is a need to further address several issues before the relationship between the NIS and hematocytopenias can be adequately evaluated. First, it is apparent that the development and/or utilization of appropriate animal models would be most beneficial, since the cause-and-effect experiments required for relating changes in NIS to the pathology of these diseases cannot be easily evaluated in humans or on human cells *in vitro*. Second, many studies performed with human cells have not adequately characterized the various effector populations. Thus the routine use of the Leu11 and Leu19 (NKH-1) monoclonal antibodies in association with the CD3 antibody would allow for better definition of the subpopulation(s) involved. Third, additional attempts must be made to identify, isolate, and grow the populations of various progenitor cells which may constitute the targets for cells with NK activity. This would allow for studies regarding the recognition and lysis of such cells, as well as the susceptibility of such cells to LGL-produced cytokines.

A role has also been postulated (by Merrill, Chapter 18) for the NIS in the possible prevention of autoimmune disease through the elimination of virus-infected cells (Merrill). It has been suggested that systemic lupus erythematosus (SLE), Sjogren's syndrome (SS), and multiple sclerosis (MS) may all have a viral etiology. Therefore, a case can be made that LGL could mediate surveillance against these diseases through direct lysis of virus-infected cells as well as through the inhibition of virus replication by the production of interferons.

Routinely, NK activity is low in SLE, SS, and MS patients. It is not clear if this depression precedes infection and could, therefore, have contributed to the result, or if it is merely a consequence of the disease pathology. The use of appropriate animal models where immune reactivity in various anatomical compartments can be studied and where selective depletion and reconstitution experiments can be performed would be quite helpful in determining whether a cause-and-effect relationship between the NIS and autoimmunity really exists. In fact, a recent study by Magilavy et al.[2] has utilized the MRL/Rpr and (NZB/NZW)F$_1$ autoimmune mouse models to demonstrate that high levels of NK activity could be found in the liver before the onset of murine SLE. These high levels of NK activity were in contrast to those observed in spleen or peripheral blood, suggesting that increased NK activity may contribute to induction of murine SLE.

The disease pathology of rheumatoid arthritis (RA) correlates with depression of NK activity in blood and synovial fluid (SF). However, NK-like cells isolated from SF have been reported to readily generate LAK activity when incubated in IL-2; however, their role in disease pathology is unknown. NK-like cells have also been isolated from the diseased thyroid in Hashimoto's thyroiditis. There have been suggestions that this pathology may be at least partially mediated via ADCC and that LGL may participate in this function.

Overall, the performance of future clinical trials in these various autoimmune diseases would be strengthened by two considerations. First, the routine use of the CD3, Leu11, and Leu19 (NKH-1) monoclonal antibodies would allow more definitive subset analysis. Second, possible alterations in subset composition should be assessed very early in disease progression. As for many of the other disease situations discussed in this section, it seems important not to lose sight of the potential immunoregulatory role of LGL, since the production of cytokines, which directly cause or ameliorate pathological effects, or trigger other cells to do so, may be more important than direct lysis of target tissue by natural effector cells.

Shanahan and Targan (Chapter 20) have reviewed the evidence for and against a role of the NIS in the pathology of human gastrointestinal disease. Little evidence has been developed to suggest that classical NK cells participate in gastrointestinal disease pathology via direct cytotoxicity on target organs. There is evidence of a role for ADCC in these diseases, but the actual effector cell is unclear and does not appear to be a cell with NK activity. Establishment for a role, or lack thereof, for the NIS in pathology of these diseases will depend on better isolation, characterization, and identification of mucosal effector cells. Similarly, technology to selectively deplete unique mucosal NIS components will be required to determine whether such cells play a cause-and-effect role in the initiation or progression of gastrointestinal disease.

Overall, there is some evidence for a role for the NIS in the pathology associated with several of the diseases discussed in this section. Probably the diseases where there is the strongest evidence for a cause-and-effect role of the NIS include diabetes and GVHD. Largely, the development of such evidence was aided by the effective use of animal models. There is correlative evidence for a role of the NIS in the pathology of several other diseases. However, in all cases it remains unclear how the NIS would actually contribute to disease processes. There is some evidence that NK cells could have a direct effect, such as by lysis of islet cells in diabetes. However, in all of these diseases one must consider the possibility that cells of the NIS might act indirectly by stimulating other cell types or by failing to provide a routine homeostatic regulatory function. For the future, the role of the NIS in these various disease states can be better understood through the development of additional animal models, the use of appropriate monoclonal antibodies, and the examination of NIS function at earlier times in the disease process.

References

1. Hahn, H. J., Lucke, S., Klöting, I., Volk, H. D., Bachr, R., and Diamantstein, T., 1987, Curing BB rats of freshly manifested diabetes by short-term treatment with a combination of a monoclonal anti-interleukin 2 receptor antibody and a subtherapeutic dose of cyclosporin A, *Eur. J. Immunol.* **177**:1075–1078.
2. Magilavy, D. B., Steinberg, A. D., and Latta, S., 1987, High hepatic natural killer cell activity in murine lupus, *J. Exp. Med* **166**:271–276.

Index